Engineering Materials

This series provides topical information on innovative, structural and functional materials and composites with applications in optical, electrical, mechanical, civil, aeronautical, medical, bio- and nano-engineering. The individual volumes are complete, comprehensive monographs covering the structure, properties, manufacturing process and applications of these materials. This multidisciplinary series is devoted to professionals, students and all those interested in the latest developments in the Materials Science field, that look for a carefully selected collection of high quality review articles on their respective field of expertise.

Indexed at Compendex (2021) and Scopus (2022)

Safaa Najah Saud Al-Humairi · Abdollah Bahador ·
Nida Iqbal · Erkan Koç ·
Ameer Najah Saud Al-Huamiri
Editors

Smart Nanomaterials in Biomedical Applications

Properties and Functional Behaviors

Springer

Editors
Safaa Najah Saud Al-Humairi
Faculty of Information Sciences
and Engineering
Management and Science University
Shah Alam, Selangor, Malaysia

Nida Iqbal
Department of Biomedical Engineering
University of Engineering and Technology
Lahore, Pakistan

Ameer Najah Saud Al-Huamiri
Department of Biomedical Engineering
Al-Mustaqabal University
Hillah, Iraq

Abdollah Bahador
Materials Processing Institute
Middlesbrough, UK

Erkan Koç
Department of Biomedical Engineering
Karabuk University
Karabuk, Türkiye

ISSN 1612-1317 ISSN 1868-1212 (electronic)
Engineering Materials
ISBN 978-981-96-8022-1 ISBN 978-981-96-8023-8 (eBook)
https://doi.org/10.1007/978-981-96-8023-8

This Springer imprint is published by the registered company Springer Nature Singapore Pte Ltd.
The registered company address is: 152 Beach Road, #21-01/04 Gateway East, Singapore 189721, Singapore

To the pioneers, researchers, and innovators whose tireless efforts continue to push the boundaries of nanoscience and biomedical engineering.

To the students and future generations of scientists who will build upon this knowledge to create transformative healthcare solutions.

And,

most importantly, to the patients and their families whose lives stand to be improved by the advancements detailed within these pages. May this work contribute to a future of enhanced health, well-being, and hope.

Foreword

We stand at a remarkable juncture in the history of medicine, where the boundaries between physical sciences, engineering, and biology are blurring, giving rise to innovations that were once the realm of science fiction. At the heart of this transformation lies nanotechnology, the ability to design, manipulate, and apply materials at the scale of atoms and molecules. This capability is not merely an incremental advance; it represents a fundamental shift in how we approach the diagnosis, treatment, and prevention of human disease. The unique properties exhibited by materials at the nanoscale—enhanced reactivity, novel mechanical behaviors, distinct optical and electronic characteristics—offer unprecedented tools to interact with biological systems with remarkable precision.

This timely volume provides a compelling overview of the rapidly evolving field of nanomaterials in biomedical science. It navigates the intricate landscape from the fundamental principles governing nanomaterial behavior, exploring their advanced mechanical, thermal, and tribological properties, to their sophisticated applications in real-world healthcare challenges. The chapters within offer critical insights into the synthesis and characterization of these materials, including the crucial development of smart, bioactive, and corrosion-resistant nanomaterials specifically engineered for the demanding physiological environment.

The breadth of applications covered is a testament to the transformative potential of this field. Readers will find detailed explorations of how nanomaterials are revolutionizing drug delivery systems, enabling targeted therapies with reduced side effects; enhancing diagnostic capabilities through highly sensitive biosensors and advanced imaging agents; and pioneering next-generation implants and prostheses in orthopedics, dentistry, and cardiovascular medicine. Furthermore, the crucial role of nanotechnology in tissue engineering—from sophisticated scaffold design and 3D bioprinting to guiding stem cell integration—is thoroughly examined, highlighting pathways to regenerative solutions for complex tissue damage.

However, pioneering the future also requires navigating its complexities. This book commendably addresses the critical challenges accompanying these innovations, including the intricate regulatory pathways necessary for clinical translation, the imperative of ensuring long-term biocompatibility and safety, and the vital ethical considerations that must underpin responsible development. The growing emphasis on sustainable and eco-friendly approaches, explored herein, underscores the commitment to aligning technological progress with environmental stewardship.

Smart Nanomaterials in Biomedical Applications: Properties and Functional Behaviors serves as an invaluable resource, bridging fundamental science with clinical potential. It offers a panoramic view for researchers seeking to push the boundaries of the field, clinicians looking to understand emerging therapeutic options, engineers designing the next generation of medical devices, and students embarking on careers in this exciting interdisciplinary domain. This collection of work not only documents the significant strides already made but also illuminates the path toward a future where nanotechnology profoundly enhances human health and well-being.

Shah Alam, Malaysia

Safaa Najah Saud Al-Humairi

Preface

The dawn of the 21st century has witnessed an unprecedented convergence of disciplines, with nanotechnology emerging as a powerful catalyst reshaping the landscape of biomedical science and healthcare. The ability to engineer materials at the nanoscale—typically between 1 and 100 nm—has unlocked a realm of unique physicochemical properties not observable in bulk materials. This has profound implications for medicine, offering innovative solutions to challenges that have long constrained conventional diagnostic and therapeutic approaches. This book embarks on an exploration of this dynamic frontier, delving into the multifaceted world of nanomaterials and their transformative impact on biomedical applications.

From the fundamental mechanical, thermal, and optical properties that govern their behavior to their intricate interactions with biological systems, nanomaterials present both immense opportunities and complex challenges. This volume brings together diverse perspectives on the design, synthesis, characterization, and application of these advanced materials. We journey through the potential of smart nanomaterials, engineered to respond to specific physiological cues, enhancing targeted drug delivery and enabling sophisticated diagnostic tools. We examine bioactive nanomaterials designed to actively interface with tissues, promoting regeneration and improving the integration of medical implants. Furthermore, the critical aspects of tribology and corrosion resistance are addressed, highlighting how nanotechnology can significantly extend the longevity and performance of biomedical devices operating within the demanding physiological environment.

The scope extends to next-generation implants and prostheses in orthopedics, dentistry, and cardiovascular solutions, where nanostructured alloys and innovative coatings promise enhanced biocompatibility and durability. We explore the burgeoning field of tissue engineering, where nanomaterials are integral to creating sophisticated scaffolds, advancing 3D bioprinting techniques, and guiding stem cell behavior for regenerative therapies. Crucially, the book also navigates the path from laboratory innovation to clinical reality, discussing the essential regulatory pathways, challenges in clinical translation, and the vital ethical considerations that must guide responsible development. The role of sustainable and eco-friendly nanomaterials is also highlighted, reflecting the growing need for green healthcare solutions.

This compilation aims to serve as a comprehensive resource for researchers, clinicians, engineers, students, and industry professionals. It seeks to bridge the gap between fundamental materials science and cutting-edge biomedical applications, offering insights into current advancements and illuminating the path forward. We hope this volume inspires further innovation and collaboration, ultimately harnessing the power of the nanoscale to revolutionize healthcare and improve human well-being.

Shah Alam, Malaysia — Safaa Najah Saud Al-Humairi
Middlesbrough, UK — Abdollah Bahador
Lahore, Pakistan — Nida Iqbal
Karabuk, Türkiye — Erkan Koç
Hillah, Iraq — Ameer Najah Saud Al-Huamiri

Acknowledgments

The successful compilation and completion of this volume, *Smart Nanomaterials in Biomedical Applications: Properties and Functional Behaviors*, represents the culmination of dedicated effort, insightful contributions, and invaluable support from numerous individuals and institutions. We, the editors, Safaa Najah Saud Al-Humairi, Abdollah Bahador, Nida Iqbal, Erkan Koç, and Ameer Najah Saud Al-Huamiri, extend our deepest gratitude to all who have played a part in bringing this comprehensive work to fruition.

First and foremost, our sincere appreciation goes to the esteemed authors who contributed their expertise and research to the individual chapters. Their willingness to share their knowledge and insights forms the very foundation of this book. The depth and breadth of the topics covered, from fundamental properties and synthesis techniques to cutting-edge applications in diagnostics, therapeutics, tissue engineering, and implant technology, reflect the remarkable advancements being made in the field of biomedical nanomaterials. We thank them for their diligence, patience, and commitment throughout the writing and revision process.

We are also indebted to the many unseen experts who served as reviewers. Their critical feedback, constructive suggestions, and meticulous attention to detail were instrumental in refining the content, ensuring scientific rigor, and enhancing the overall quality and clarity of each chapter. Their voluntary contribution of time and expertise is deeply valued.

The support of the institutions affiliated with our contributors—the universities, research centers, and laboratories—cannot be overstated. These environments foster the innovation and discovery showcased within these pages, providing the necessary resources and intellectual climate for such advanced research.

Finally, on a personal note, we wish to thank our colleagues, mentors, friends, and families for their constant encouragement, understanding, and patience. Their support provides the essential foundation upon which such demanding endeavors are built.

It is our hope that this book will serve as a valuable resource for the scientific community, inspiring further research and innovation in the application of smart nanomaterials to address pressing challenges in healthcare and biomedical science.

Contents

Contributors

Manish Baboo Agarwal Sharda School of Basic Sciences and Research, Sharda University Agra, Keetham, Agra, Uttar Pradesh, India

Seema Agarwal Department of Mathematics, SRMIST, Ghaziabad, Uttar Pradesh, India

Safaa Najah Saud Al-Humairi Faculty of Information Sciences and Engineering, Management and Science University, Shah Alam, Malaysia

Elif Esra Altuner Program of Medical Laboratory Techniques, Department of Medical Services and Techniques, Vocational School of European, University of Kocaeli Health and Technology, Kocaeli, Turkey

H. Mohammed Ameen Department of Chemistry (S&H), Bharathidasan Engineering College, Tirupattur District (Affiliated to Anna University, Chennai, TamilNadu, India), India

P. Balakrishnan SRM Institute of Science and Technology, Kattankulathur, Tamil Nadu, India

Manoj Kumar Banjare Department of Chemistry, State Forensic Science Laboratory, Raipur, C.G., India

Benvikram Barman Department of Chemistry, Govt Naveen College, Thelkadih, K.C.G., Chhattisgarh, India

Arvind Kumar Bhardwaj Department of Chemistry, Dr. C. V. Raman University, Kota, Bilaspur, Chhattisgarh, India

Amit Kumar Chaturwedi Department of Chemistry, Dr. C. V. Raman University, Kota, Bilaspur, Chhattisgarh, India

Sarita Diwakar Department of Chemistry, Dr. C. V. Raman University, Kota, Bilaspur, Chhattisgarh, India

G. Ganesh Department of Chemistry (S&H), Bharathidasan Engineering College, Tirupattur District (Affiliated to Anna University, Chennai, TamilNadu, India), India

Milan Hait Department of Chemistry, Dr. C. V. Raman University, Kota, Bilaspur, Chhattisgarh, India

V. H. Haritha National Institute of Plant Science and Technology, Mahatma Gandhi University, Kottayam, Kerala, India

Hareem Hayee Department of Biomedical Engineering, University of Engineering and Technology (UET), Lahore, Punjab, Pakistan

Soufiene Ilahi Pysique, MathématiqueModélisation Quantique, IPEIN, Carthage University, Carthage, Tunisia;
Physics Department, Monastir Faculty of Sciences, Monastir University, Monastir, Tunisia

Nida Iqbal Department of Biomedical Engineering, University of Engineering and Technology (UET), Lahore, Punjab, Pakistan

Saman Iqbal Department of Physics, University of the Punjab, Lahore, Pakistan

Vinita Isad Bombay College of Pharmacy, Mumbai, Maharashtra, India

Jenny Jacob School of Biosciences, Mar Athanasios College for Advanced Studies, Tiruvalla, Kerala, India

Ukasha Jawad Department of Biomedical Engineering, University of Engineering and Technology (UET), Lahore, Punjab, Pakistan

Xiangji Jin Department of Dermatology, School of Medicine, Kyung Hee University, Seoul, Republic of Korea

Chaitanya Kale Industrial Pharmacy, Poona College of Pharmacy, Bhartiya Vidyapeeth (Deemed to Be University), Pune, India

Nand Kumar Kashyap Department of Chemistry, Dr. C. V. Raman University, Kota, Bilaspur, Chhattisgarh, India

Dhekra Loubiri Pysique, MathématiqueModélisation Quantique, IPEIN, Carthage University, Carthage, Tunisia;
Physics Department, Monastir Faculty of Sciences, Monastir University, Monastir, Tunisia

P. Manikandan Department of Chemistry (S&H), Bharathidasan Engineering College, Tirupattur District (Affiliated to Anna University, Chennai, TamilNadu, India), India

Saman Mehmood Department of Biomedical Engineering, University of Engineering and Technology, Lahore, Pakistan

Manu Mehrotra Sharda School of Basic Sciences and Research, Sharda University Agra, Keetham, Agra, Uttar Pradesh, India

Pankaj Mittal Department of Applied Sciences, Anand Engineering College, Keetham, Agra, Uttar Pradesh, India

Arya Muraleedharan School of Biosciences, Mar Athanasios College for Advanced Studies, Tiruvalla, Kerala, India

M. Syed Muzammil Department of Chemistry (S&H), Bharathidasan Engineering College, Tirupattur District (Affiliated to Anna University, Chennai, TamilNadu, India), India

Trang Thi Minh Nguyen Graduate School of Biotechnology, Kyung Hee University, Yongin-Si, Republic of Korea

M. Nishanthi Department of Chemistry (S&H), Bharathidasan Engineering College, Tirupattur District (Affiliated to Anna University, Chennai, TamilNadu, India), India

Muhammad Noman Department of Biomedical Engineering, University of Engineering and Technology, Lahore, Pakistan;
Centre for Healthcare Simulation and Innovation, The University of Lahore, Lahore, Pakistan

Farheen Nusrat Department of Biomedical Engineering, University of Engineering and Technology, Lahore, Pakistan

Manish Kumar Panday Department of Applied Sciences, Anand Engineering College, Keetham, Agra, Uttar Pradesh, India

Deepika Patel Department of Chemistry, Dr. C. V. Raman University, Bilaspur, Chhattisgarh, India

A. Priyadharshini Department of Chemistry (S&H), Bharathidasan Engineering College, Tirupattur District (Affiliated to Anna University, Chennai, TamilNadu, India), India

Neha Raghuwanshi Amity Institute of Pharmacy, Amity University Maharashtra, Mumbai, Maharashtra, India;
School of Pharmacy, RK University, Tramba, Rajkot, Gujarat, India

S. Rahila Department of Chemistry (S&H), Bharathidasan Engineering College, Tirupattur District (Affiliated to Anna University, Chennai, TamilNadu, India), India

M. Safa SRM Institute of Science and Technology, Kattankulathur, Tamil Nadu, India

Pradeep Sahu Department of Chemistry, Dr. C. V. Raman University, Bilaspur, Chhattisgarh, India

G. Saranya SRM Institute of Science and Technology, Kattankulathur, Tamil Nadu, India

R. K. Shanmathi Bombay College of Pharmacy, Mumbai, Maharashtra, India; Poona College of Pharmacy, Bhartiya Vidyapeeth (Deemed to Be University), Pune, India

Kritanjali Singh Subharti Medical College and Research and Development Cell, Swami Vivekanand Subharti University, Meerut, U.P., India

P. Sivasubramaniyan VIMS Hospital, Salem, Tamil Nadu, India

Yasemin Torlak Program of Organic Farming, Department of Agricultural and Livestock Production, Vocational School of Cal, University of Pamukkale, Denizli, Turkey

M. Unnimaya School of Biosciences, Mar Athanasios College for Advanced Studies, Tiruvalla, Kerala, India

Treesa Varghese School of Biosciences, Mar Athanasios College for Advanced Studies, Tiruvalla, Kerala, India

Huma Waheed Department of Biomedical Engineering, University of Engineering and Technology (UET), Lahore, Punjab, Pakistan

Tae-Hoo Yi Graduate School of Biotechnology, Kyung Hee University, Yongin-Si, Republic of Korea

Qiwen Zheng Graduate School of Biotechnology, Kyung Hee University, Yongin-Si, Republic of Korea

Harnessing the Potential of Smart Nanomaterials: Addressing Healthcare Challenges Through Advances Mechanical Properties

Safaa Najah Saud Al-Humairi

Abstract Nanomaterials have emerged as a significant focal point for the scientific and industrial realms, fostering innovative solutions through the precise manipulation of materials at the nanoscale. This chapter reviews recent advancements in the preparation, characterization, and application of nanomaterials, emphasizing their place in medicine, energy, and environmental sustainability. This chapter also discusses various synthesis methods and green and sustainable methods, with the effect of parameters, such as molar ratios, reaction time, and capping agents, on the properties of nanoparticles. Substantial research has been dedicated to nanomaterials' biomedical applications, including targeted drug delivery systems, disease diagnosis, and tissue engineering. The chapter addresses how nanoparticles can go across biological barriers, such as the blood–brain barrier, for treating neurodegenerative diseases and dives into the application of nanomaterials in energy storage and conversion systems, particularly lithium-ion batteries and solar-driven water-splitting technologies. It also addresses applications in environment-related fields, such as pollution control and waste management. In addition, advanced techniques in characterization and simulation in computation have been discussed concerning optimal property behavior and understanding of nanoparticles. Bibliometric analysis indicates the emerging trends and the clusters of research studies in the scientific field, while various case studies demonstrate real applications of nanotechnology in industrial contexts. The chapter highlights the role of the synthesis parameters in achieving desired material properties and looks at grain size performance in an association.

Keywords Nanomaterials · Synthesis parameters · Biomedical applications · Environmental sustainability · Nanoparticles

S. N. S. Al-Humairi (✉)
Faculty of Information Sciences and Engineering, Management and Science University, 40100 Shah Alam, Malaysia
e-mail: safaa_najah@msu.edu.my

S. N. Saud Al-Humairi et al. (eds.), *Smart Nanomaterials in Biomedical Applications*, Engineering Materials, https://doi.org/10.1007/978-981-96-8023-8_1

1 Introduction

Nanomaterials are at the forefront of this transformation, opening new frontiers in nanotechnology for healthcare applications [1]. Particles of these materials, ranging from 1 to 100 nm, exhibit a wide variety of physicochemical properties that distinguish them from their larger bulk counterparts. Enhanced absorption and responsiveness result from this dimensionality reduction [2], unveiling medical applications that were previously unknown, thanks to the satellite's surface area-to-volume ratios and plasmonic programmability. The ability to manipulate matter at the nanoscale can revolutionize our healthcare systems by enabling more precise illness prevention, diagnosis, and treatment in the future. A new concept at the nanoscale suggests that smart nanomaterials are defined by their size-driven responsiveness to a broad range of stimuli derived from the unique characteristics of this building block [3]. Potential adaptations may include changes to morphology, surface area, permeability, and mechanical properties. Temperature, light, and ultrasound are examples of physical stimuli, pH, solvent alterations, and electrochemical processes are chemical stimuli; and enzymes and glucose represent biological stimuli [4]. One driving force behind the creation of these versatile materials is the increasing demand for superior materials to address future technological challenges that society may face. In the complex biological environment, smart nanomaterials could enable precision-targeted and regulated treatments by responding to various stimuli or exhibiting dual or multi-stimuli responsiveness in scientific applications [5]. Polymer systems are particularly suited for these advanced responsive systems due to the flexibility of their structural and chemical components [6].

The enhanced mechanical properties of nanomaterials—strength, flexibility, and durability—are highly significant for applications in the healthcare industry. Due to their exceptional durability and strong resistance to wear, nanoparticles exhibit significantly better mechanical strength, hardness, and flexibility compared to their bulk counterparts [7]. These enhanced mechanical properties are vital in fields such as the development of long-lasting medical implants and stable scaffolding for tissue engineering. Furthermore, the mechanical properties of nanoparticles are crucial in their interaction with biological systems at the nanoscale, influencing processes like cellular uptake in drug delivery and the overall effectiveness of various nano-bio interactions [8]. The explicit mention of "advanced mechanical properties" pertinent to the healthcare challenges discussed underscores the importance of linking these properties with specific advantages throughout this study [9]. This chapter will introduce the intriguing research on smart nanomaterials with extraordinary mechanical properties and their potential to transform healthcare. It will cover foundational definitions and significance regarding these materials, analyze their various forms and synthesis procedures, and provide an overview of their numerous applications in therapeutics and diagnostics. Biocompatibility criteria, safety issues, and the regulatory framework related to these next-generation materials will also be highlighted, paving the way for discussions on future perspectives and challenges in fully harnessing their potential.

2 Understanding Critical Healthcare Challenges Requiring Innovative Nanomaterial Solutions

Modern medicine faces several challenges in achieving early and accurate illness diagnoses and effective treatment plans. There is a significant demand for innovative ideas, as conventional diagnostic and treatment methods often lack specificity and speed and can be inherently intrusive [10]. For instance, despite their benefits, current cancer treatments may negatively affect healthy organs and lead to drug resistance due to their low selectivity for tumor cells and substantial systemic toxicity [11]. Similarly, treating neurological diseases largely relies on efficiently delivering therapeutic medications to specific areas of the body, such as the brain, which is protected by the highly selective blood–brain barrier [12]. These limitations underscore the potential of nanotechnology to offer solutions by facilitating interactions at the cellular and molecular levels with a level of precision typically unattainable through traditional methods [13].

Many illnesses and situations underscore the necessity of creative ideas, such as those provided by nanomaterials, because they reveal the shortcomings of current healthcare methods. In cancer therapy, the challenges of off-target toxicity—where healthy cells are harmed alongside malignant ones—and multidrug resistance—where cancer cells become resistant to previously effective treatments—remain significant. Improving patient outcomes in diseases like cancer and infectious diseases depends on early diagnosis [11]; however, current diagnostic methods may not be sensitive enough to detect biomarkers in the early stages of these conditions. Major challenges also arise in managing neurodegenerative diseases like Alzheimer's disease [14], which require targeted medication delivery across the blood–brain barrier, as well as chronic conditions such as diabetes, which necessitate constant glucose monitoring and precise insulin administration [15]. Moreover, the emergence of infectious diseases, particularly those showing antibiotic resistance, creates a global health emergency and necessitates new therapeutic and diagnostic approaches. The variety of these medical issues across different fields underscores the general relevance of smart nanomaterials with enhanced mechanical properties in addressing unmet therapeutic needs [16].

Transitioning from generalized treatment methodologies, the advancing terrain of healthcare increasingly underscores the necessity for personalized and custom-tailored therapeutic interventions. Nanotechnology essentially facilitates this shift, which enables precise targeting of diseased cells, thereby minimizing adverse effects on healthy tissues and enhancing overall treatment efficacy [17]. For example, nanoparticles can be engineered to deliver medications, genetic material, and other therapeutic agents directly to specific cell types, thereby ensuring a more concentrated and effective therapeutic response [18]. The integration of nanotechnology with personalized medical strategies signifies a substantial paradigm shift and promises a future where therapies are adapted to the unique molecular and genetic profiles of individual patients, ultimately yielding more effective and safer healthcare outcomes [19]. Through the development of nanoparticles that serve as contrast agents in magnetic resonance imaging (MRI) and other imaging modalities, recent advances highlight their importance in enhancing the resolution and specificity of imaging techniques [20]. Multifunctional nanomaterials, for instance, can bridge the gap between diagnosis and therapy by combining imaging and treatment functionalities (theranostics). By facilitating real-time monitoring of disease progression and treatment responses, such advancements enhance diagnostic accuracy and allow for personalized medication. In the realm of therapy, the application of nanomaterials represents a paradigm shift in cancer treatment methodologies [21]. Targeted drug delivery that minimizes systemic toxicity and maximizes therapeutic efficiency through modified nanoparticles as drug carriers help to mitigate systemic toxicity [22]. For photothermal therapy, for example, gold nanoparticles and carbon nanotubes have shown promise in enabling targeted heating to eliminate cancer cells while sparing surrounding healthy tissue [23]. Furthermore, the incorporation of nanomaterials may enhance the absorption and solubility of drugs, thereby addressing the limitations of conventional medications. This personalized drug delivery approach may significantly reduce side effects, improving patient outcomes [24].

Bibliometric data were collected based on the Scopus database within the keyword (KEY (nanomaterials) OR KEY (nanostructure) OR KEY (nanocomposite) OR KEY (nanoparticles) OR KEY (nanotubes)) and subsequently enhanced utilizing VOSviewer software. Figure 1 illustrates the frequency of keyword occurrences derived from the examined works. These occurrences are categorized into nine distinct clusters. Each node, or bubble, corresponds to a particular keyword, with colors assigned to indicate specific clusters. Each node's size reflects the term's frequency, where a larger node signifies a higher frequency. Additionally, the connections between the nodes represent their relationships, with proximity indicating a stronger association.

Wound healing is another important field that has benefited from nanotechnology. Chronic wounds pose significant challenges due to their delayed healing timelines and infection susceptibility. Powerful agents for wound closure and infection prevention now include nanomaterials, such as polysaccharide-based scaffolds and silver nanoparticles [25]. These nanostructures can create a moist wound environment that accelerates healing and possesses antimicrobial properties. The versatility of nanoparticles allows for the development of "smart" dressings that release therapeutic agents in response to specific environmental stimuli, thus optimizing wound treatment methods [26]. Nonetheless, cardiovascular diseases (CVDs) remain a leading global cause of illness and death. Nanotechnology has demonstrated significant potential in cardiovascular therapy by enhancing diagnostic techniques and facilitating targeted treatment approaches. Early detection biosensors for CVD markers may achieve improved sensitivity and specificity with the use of nanomaterials [27]. Furthermore, their role as drug delivery systems may enable new treatments to target vascular lesions while minimizing off-target effects more precisely [28].

Despite the potential advantages, ongoing research has also focused on the underlying toxicity issues associated with long-term human contact with certain materials [29]. Beyond these applications, the biocompatibility and safety of nanoparticles

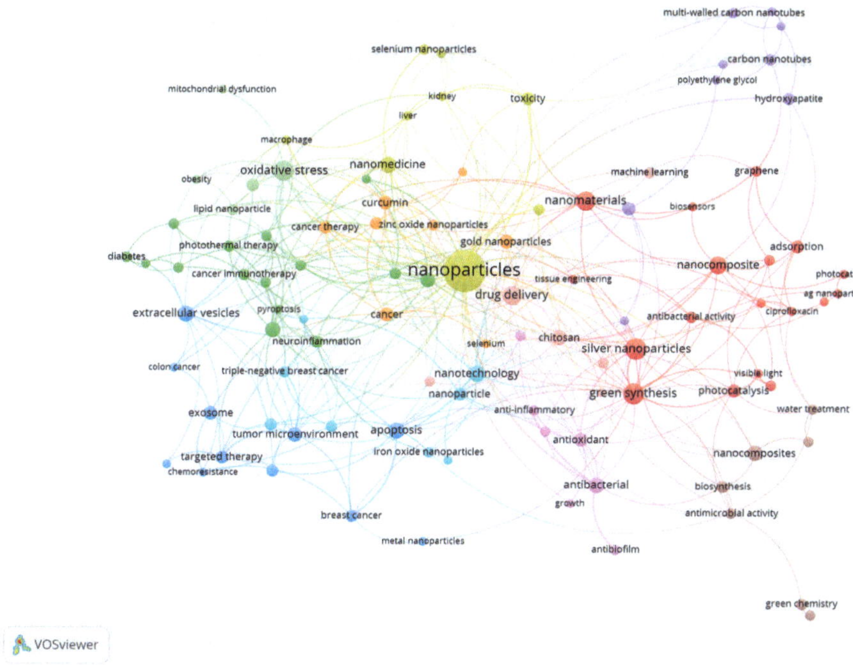

Fig. 1 Chart depicting prolific publishing keywords of publications within the nanomaterials engineering domain

remain paramount. Although the unique characteristics of nanomaterials offer benefits for medical applications, their safety profiles must be rigorously evaluated to mitigate any health risks [30]. The lack of consistent regulations and procedures for assessing their toxicity has hindered the widespread adoption of nanoparticles in clinical practice. Safe nanomedical applications cannot develop without continuous monitoring and risk assessment, as they ensure that innovations do not inadvertently jeopardize patient health [31].

Furthermore, breakthroughs in materials science have produced novel nanomaterials like graphene, specially designed for biomedical applications, and metal–organic frameworks (MOFs). These materials offer significant advantages that expand the potential uses in nanomedicine: large surface areas for drug loading and adjustable pore diameters for targeted delivery. Recent studies emphasize the need to incorporate nanomaterials into existing healthcare systems, thereby addressing logistical challenges and fostering interdisciplinary collaboration between nanotechnology and clinical practice [32].

3 Role of Advanced Mechanical Properties in Smart Nanomaterials for Healthcare Applications

The mechanical characteristics of smart nanomaterials significantly impact their effectiveness in addressing various healthcare challenges, making them highly important for applications ranging from tissue engineering and biosensing to drug delivery. Within the context of drug delivery, the mechanical stability of nanomaterials is essential to ensure the structural integrity of nanocarriers as they navigate the circulatory system and interact with target areas [33]. Additionally, the stiffness and elasticity of nanoparticles directly influence their cellular uptake mechanisms and intracellular transport pathways, consequently affecting drug delivery efficiency. Furthermore, the mechanical properties of drug carriers can be precisely tuned to regulate the release kinetics of the encapsulated therapeutic agents, for instance, through the design of mechanically triggered release mechanisms that respond to specific strains or stresses within the body [17].

Particular mechanical characteristics such as tensile strength and elasticity help solve problems in the medical field. For tissue engineering scaffolds, for instance, high tensile strength is essential as it provides the necessary structural support for cell development, multiplication, and the eventual regeneration of functional tissues [34]. Conversely, flexibility is crucial for creating wearable sensors that must adapt perfectly to the body's dynamic motions while maintaining their functional integrity in monitoring various physiological variables [19]. Concerning medical implants, the long-term performance of the nanomaterials utilized depends on their mechanical stability, which also reduces the likelihood of implant failure due to mechanical stress, corrosion, or wear [35]. The diverse mechanical requirements for different medical applications emphasize the critical need for the customized design of nanomaterials with specific mechanical characteristics to meet their intended usage effectively. The complex interaction between the mechanical characteristics of smart nanomaterials and their response to external stimuli highlights their usefulness [36]. Through targeted triggers such as temperature, light, or pH, these materials can undergo changes in their mechanical properties, like stiffness or shape, thus allowing for controlled drug release or actuation in specified areas [37]. One family of smart nanomaterials, hydrogels, for example, can exhibit significant swelling or shrinking in response to minor temperature or pH changes, thereby altering mechanical stiffness and releasing encapsulated pharmaceuticals in a regulated manner [38]. This sophisticated combination of stimuli-responsiveness and precisely tailored mechanical properties enables more effective and focused healthcare treatments, facilitating an unparalleled degree of control over the behavior of nanomaterials within the complex and dynamic biological environment.

4 Classification and Synthesis of Smart Nanomaterials with Enhanced Mechanical Properties

Smart nanomaterials are understood methodically through their categorization, which can be addressed based on their fundamental composition or response to specific stimuli. The type of stimuli can broadly be categorized into those responsive to physical cues (such as temperature, light, electric fields, and magnetic forces), chemical signals (including pH variations, solvent changes, and electrochemical reactions), and biological agents (such as enzymes and glucose) [39]. Alternatively, categorization based on chemical composition distinguishes between organic nanomaterials (primarily composed of polymeric structures), inorganic nanomaterials (consisting of metallic or non-metallic components), and carbon allotrope-based nanomaterials [40]. Within the realm of physico-responsive nanomaterials, specific examples include materials whose properties change in response to temperature, light, electric field application, or the presence of a magnetic Field. Exploring the diverse landscape of smart nanomaterials and thoroughly investigating their potential applications in healthcare relies on a well-defined categorization system.

A number of advanced nanomaterials (Fig. 2) demonstrate improved mechanical properties within these specific categories, thus rendering them suitable for demanding therapeutic applications. Often demonstrating better mechanical strength and stability than their components, nanocomposites, which combine nanoparticles with a matrix material, show unique cylindrical structures [41]. Carbon nanotubes are well-known for their mechanical, electrical, and thermal characteristics, making them particularly valuable in various biomedical applications [42]. One-dimensional nanostructures with high aspect ratios, such as nanowires, possess distinctive mechanical properties that can benefit both diagnosis and treatment. By incorporating nanoparticle reinforcements, hydrogels—three-dimensional networks capable of storing substantial amounts of water—can be engineered to display improved mechanical strength, thus broadening their applications in tissue engineering and drug delivery [43]. For applications like medical coatings and implants, which critically depend on enhanced mechanical strength and antibacterial properties, metallic nanoparticles—such as those made of gold or silver—can be integrated into composite materials [44]. Ceramic nanoparticles are ideal for orthopedic and dental implants as well as tissue engineering scaffolds, providing a compelling combination of excellent mechanical strength and biocompatibility [45]. Ultimately, with their two-dimensional structure, carbon-based nanomaterials such as graphene exhibit exceptional mechanical, optical, and electrical properties that are highly attractive for numerous biological applications, including tissue engineering and sensing [46].

Utilizing several complex methods, each providing some degree of control over the final material features, numerous smart nanomaterials with customized mechanical properties are synthesized [3]. Generally, these techniques fall into two categories: top-down methods—such as mechanical milling, lithography, and sputtering-that miniaturize bulk materials, and bottom-up approaches—such as sol–gel synthesis,

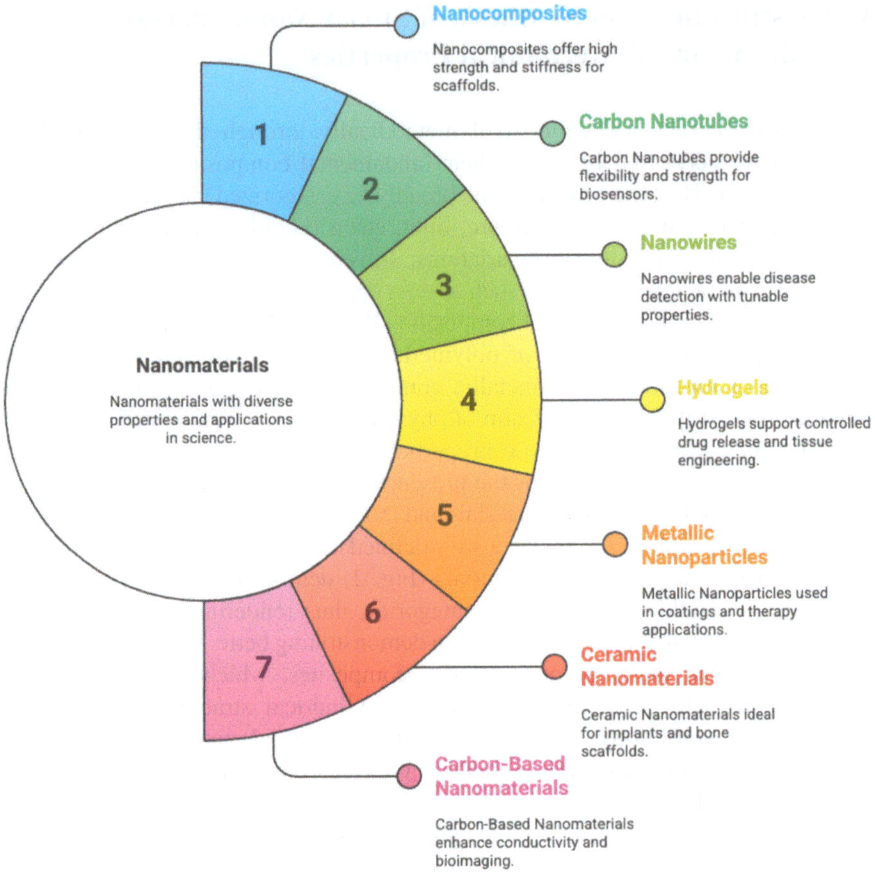

Fig. 2 Versatile applications of nanomaterials

chemical vapor deposition, and self-assembly—that construct nanomaterials from atomic or molecular precursors [47]. The flexible electrospinning method, which frequently incorporates nanoparticles within the fiber structure, is widely used to produce nanofibers with controlled shapes and characteristics. Microwave-assisted synthesis has emerged as a cost-effective and energy-efficient technique for manufacturing high-quality nanomaterials with precise control over size and form [48]. Green synthesis techniques utilizing biological entities such as bacteria, fungi, and plants have gained prominence for producing metallic nanoparticles, presenting an environmentally friendly alternative to conventional chemical processes in response to increasing environmental concerns [49]. Furthermore, mechanical synthesis methods like High-Pressure Tension Extrusion (HPTE) induce extreme plastic deformation, enhancing the mechanical properties of nanomaterials [50]. The selection of an appropriate synthesis technique is crucial, as it directly impacts the mechanical properties of the nanomaterials, underscoring the need to consider the intended application when choosing a production method carefully.

5 Factors Affecting the Mechanical Performance of Nanomaterials

The production process is crucial in determining the performance of nanomaterials, as it affects their structural, mechanical, thermal, electrical, and optical properties. The synthesis method, along with the process parameters and conditions utilized in the fabrication of nanomaterials, can substantially affect their size, shape, composition, and functionality [51]. This section examines the influence of various production processes on the performance of different nanomaterials, emphasizing the essential factors that need to be regulated to attain the desired properties.

5.1 Influence of Process Parameters on Nanomaterial Properties

Nanomaterials' characteristics are sensitive to process variables like temperature, pressure, reaction time, and precursor concentration. For the chemical vapor deposition (CVD) process, for example, temperature and pressure have the most significant bearing on the size and crystallinity of nanoparticles [52]. The results also show that the process parameters influence the nanomaterials' mechanical, thermal, and electrical characteristics. Hence, knowing these factors helps maximize their ability to get customized features for use in catalysis, energy storage, and electronics.

Similarly, in the synthesis of silicon nanoparticles via inductively coupled plasma (ICP), the partial pressure of monosilane and plasma power was found to control the particle size and crystallinity, with higher plasma power leading to more crystalline structures, which was shown that the working pressure primarily affected the production rate of the nanoparticles, with a maximum yield of approximately 76 mg achieved at a working pressure of 1.7 Torr and a SiH4 flow rate of 6 sccm [9]. The work produced very crystalline monodispersed silicon nanoparticles averaging 7.5 nm in size. These find possible uses in wavelength down-converter materials for LEDs and light-absorbing solar cells. In the case of carbon nanotubes synthesized by spray pyrolysis, process variables such as growth temperature and reactor position greatly affected the graphitization degree, diameter, and alignment [53]. An optimum induced power of 80 W was necessary for obtaining a highly crystalline phase. These results emphasize the need for exact control over process parameters to customize the features of nanomaterials for particular uses.

5.2 Role of Synthesis Methods in Shaping Nanomaterial Performance

The field of synthesis of nanomaterials is multifarious and includes many techniques with different benefits and uses. These techniques may be generally distinguished from top-down and bottom-up perspectives. While bottom-up methods, including chemical vapor deposition (CVD), sol–gel processes, and self-assembly, build nanostructures from atomic or molecular precursors [54], top-down techniques, including mechanical milling, lithography, and laser ablation, break down bulk materials into nanoscale structures [19]. Crucially for their applications in sectors like electronics, medicine, and energy, the synthesis technique significantly used affects the characteristics of the resultant nanomaterials, including size, shape, and surface area [54]. A subset of bottom-up techniques, green synthesis uses biological resources such as plant extracts and microorganisms as reducing and capping agents, therefore providing an ecologically acceptable substitute for conventional chemical procedures. This approach is affordable and sustainable as it uses bioactive chemicals for bioreduction metals, lowering environmental impact [10]. Chemical techniques, including co-precipitation, hydrothermal, and sol–gel procedures, remain popular as they may generate nanomaterials with exact control over their physicochemical characteristics [55]. From catalysis and sensing to drug delivery and energy storage, the synthesis of particular nanomaterials, including gold and silver nanoparticles, magnetic nanoparticles, quantum dots, carbon nanotubes, and graphene, is tailored to exploit their unique qualities for applications [56]. Prospects in nanomaterial synthesis concentrate on honing these techniques to increase scalability, biocompatibility, and environmental sustainability, thus guaranteeing the continuing progress of nanotechnology for creative applications [57]. Many elements affect nanomaterials' synthesis, each essential in deciding the nanomaterial's ultimate characteristics, structure, and use.

5.2.1 Temperature

The structural, morphological, and functional characteristics of nanomaterials are influenced by temperature, affecting their synthesis. In the hydrothermal method of synthesis of ZnS nanoparticles, temperature variations between 140 and 160 °C result in changes in crystal structure, morphology, and optical properties, with band gaps ranging from 4.41 to 4.83 eV, suggesting potential applications in optoelectronics has been reported by John et al. [5]. The spatial distribution and size of gold nanoparticles (AuNPs) on ceramic substrates for biosensing applications are similarly influenced by thermal treatment at 450 °C. Shorter treatment times result in smaller, more closely spaced nanoparticles, which in turn enhance the surface-enhanced Raman scattering (SERS) effect [7]. The superparamagnetic characteristics of magnetite nanoparticles in the coprecipitation synthesis are influenced by temperature; samples produced at 40, 50, and 60 °C exhibit consistent magnetization characteristics [58].

Additionally, the synthesis temperature of nickel-copper composite nanoparticles can significantly influence their microstructure; even minor temperature fluctuations can result in substantial structural modifications [59]. The antibacterial properties and morphological qualities of ZnO nanoparticles are improved by biosynthesis at 400 °C using Ocimum sanctum leaf extract, which results in variable crystalline sizes and morphologies due to fluctuating temperatures [60]. Candida macedoniensis synthesizes silver nanoparticles at a 25–30 °C temperature, the optimal range for nanoparticle development. Consequently, the importance of temperature in maintaining the size and morphology of nanoparticles is underscored [61]. Hematite nanoparticles produced through co-precipitation exhibit a larger crystallite size and a reduced coercive field due to increasing temperatures, affecting their magnetic properties [62]. The crystallinity and optical properties of ZnO nanoparticles are enhanced by the development of spherical particles as the temperature increases, resulting from the needle-like structures produced during low-temperature development [63]. The structural properties of Fe–Co nanocomposites are altered by temperature, which influences the production and dispersion of nanoparticles within a carbon matrix during the synthesis process [64]. In conclusion, the synthesis of ZnO–SnO_2 nanocomposites demonstrates that higher calcination temperatures enhance photocatalytic activity due to enhanced phase purity and crystallinity. Consequently, this research addresses a critical need for using these nanocomposites in pollutant degradation [65]. In general, temperature is a crucial factor in developing nanomaterials, as it regulates their physical and chemical properties, which are essential for their application in various industries.

5.2.2 Pressure

Pressure can significantly affect nanomaterials' size, shape, and structural characteristics, influencing their synthesis. High pressure has been shown to control nanoparticles' exsolution from perovskite, influencing their size, population, and composition. This also helps highly scattered, small-sized nanoparticles to develop under typically lower-than-necessary temperatures. In the same line, pressure affects the expansion of ZnO nanoparticles. Pressure first causes particle size to decrease; this minimum is reached at 50 bar and then rises again. This implies a harmony between the nucleation and expansion mechanisms [66]. Pressure affects the diameter of silver nanoparticles in synthesis, larger nanoparticles are generated at more tremendous pressures than those synthesized without pressure [67]. Electronic applications depend on controlling the morphology and degree of order of new nanostructures, including carbon nanorods; high-pressure application helps achieve this process [68].

Furthermore, pressure-induced phase transitions in TiO_2 and other nanomaterials show that pressure can cause permanent amorphization and morphological changes, providing insightful analysis of the evolution of structurally stable nanomaterials [41]. High pressure can create exotic phases in nanomaterials, they are not achievable using regular techniques. Consequently, pressure is a flexible tool in the synthesis of nanomaterials, allowing for the tailoring of physical and chemical properties and the development of materials with unique phases and improved functionalities [8]. The possible applications are thus expanded to include advanced electronics, energy storage systems, and catalysis, where these engineered nanomaterials can enhance performance and efficiency in various technological domains.

5.2.3 PH

pH significantly regulates the synthesis of nanomaterials, therefore impacting properties including particle size, shape, crystallinity, and functional performance. With substantial impacts observed between pH 3 and 5 [69], pH modulates the solubility of gold monomers, nucleation rates, and particle size distribution in the production of gold nanoparticles utilizing plasma-driven solution electrochemistry. The sol–gel synthesis of ZnO nanoparticles shows that higher pH levels produce reduced crystallite and particle sizes, lowering heat conductivity and enhancing thermoelectric performance [39]. A pH of about 13 produces well-crystalline structures with suitable surface characteristics that increase gas sensitivity for ZnO nanoparticles produced for gas sensor uses [70]. Cobalt oxide nanomaterials show pH-dependent morphological changes with differences in dielectric and impedance characteristics at various pH levels [71]. With optimal properties observed at pH 7, the synthesis of α-Fe_2O_3 nanoparticles demonstrates that pH affects size, structure, and antibacterial activity [72]. pH influences the shape and photocatalytic efficiency of $AgVO_3$ nanoparticles, the best degradation of industrial pollutants results from pH 9 [73]. TiO_2 nanorods show that pH affects the crystal structure and photocatalytic activity, altering the energy band gap and improving the photocatalytic degradation efficiency under some pH levels [74].

Moreover, ZnO nanostructured thin films at higher pH show better surface geometry and higher solar cell efficiency [75]. Finally, in anaerobic digestion systems, pH is critical to the performance of nanomaterial-amended processes, with neutral pH conditions supporting rapid methane synthesis relative to higher pH levels, pH is thereby neutral. A key factor in the design of nanomaterials, pH influences their structural, functional, and application-specific properties in many contexts.

5.2.4 Time

Numerous studies on various nanoparticle types indicate that nanomaterials' synthesis duration significantly influences their structural, morphological, and optical properties. An increase in synthesis time was demonstrated to affect the size, shape, and quantity of carbon nanospheres, with modest alterations in their optical characteristics, which are crucial for optoelectronic applications [36]. Similarly, the duration of growth in the production of TiO_2 nanostructures was seen to alter their dimensions and band gap, affecting their potential applications in optoelectronic devices and photocatalysis [76]. The stirring time during electrochemical synthesis influenced silver nanoparticles' size distribution and durability, which are essential for medical devices and antibacterial treatments [34]. The calcination duration in the production of TiO_2 nanoparticles significantly influences particle size reduction and enhances photocatalytic activity [77].

The reaction time in the synthesis of SnO_2 nanospheres and CdTe/CdSe core/shell nanoparticles influenced their crystallinity, particle size, and optical characteristics, which are essential for applications in sensors and bioimaging, respectively [78]. The synthesis duration influenced the formation and degradation of Cu_3N nanocubes, with the observed variations in optical characteristics at different synthesis stages [79]. This research emphasizes synthesis duration's critical role in tailoring nanomaterials' attributes for specific applications. It highlights the necessity for precise regulation of synthesis parameters to achieve desired material properties.

5.2.5 Atmosphere/Environment

The atmospheric and environmental conditions significantly influence the development of nanomaterials. Green synthesis techniques utilize biological resources, such as plant extracts and microorganisms, as an eco-friendly alternative to traditional chemical and physical methods, thereby reducing energy consumption and environmental pollution while serving as effective capping and reducing agents for nanoparticle formation [14]. The stability and properties of nanomaterials are significantly influenced by their air environment. For instance, exposure to air affects the stability of secondary electron emission from materials like magnesium oxide (MgO and alumina (Al_2O_3; MgO exhibits significant degradation over time due to chemical interactions with atmospheric constituents, while Al_2O_3 maintains relative stability [80]. Environmental oxygen comparably influences the creation and agglomeration of nanoparticles, as evidenced by laser ablation techniques, where elevated oxygen levels result in larger aluminum oxide clusters and agglomerates [81]. An alternative method for nanoparticle synthesis involves air-pressure plasma jets, which enable the fast reduction and deposition of metal nanoparticles, such as gold, without the use of toxic chemicals through the interaction of plasma with precursor vapors [82]. This method underscores the importance of the plasma environment in governing nanoparticle size and morphology, which are essential for their applications across several fields.

Furthermore, the environmental implications of nanomaterials, including potential toxicity and bioaccumulation, underscore the necessity for sustainable synthesis methods that mitigate ecological risks while maximizing the benefits of nanotechnology [81]. The synthesis and stability of nanomaterials are typically contingent upon their environmental context; thus, meticulous consideration of air conditions is essential to optimize their production and application [83].

5.3 Impact of Process Engineering on Nanomaterial Structure and Functionality

Large-scale manufacturing of nanomaterials depends much on process engineering. To guarantee consistency and homogeneity in the end product, parameters like mixing, heat transfer, and mass transfer must be precisely tuned [84]. For instance, obtaining the required mechanical and thermal properties in the manufacture of nanocomposite coatings depends critically on the distribution of nanoparticles inside the matrix and their interfacial interactions [85].

Furthermore, intricate nanostructures with customized characteristics are created using sophisticated methods, including laser ablation and microfluidic synthesis [86]. These techniques are practical in electronics, catalysis, and biomedicine since they enable exact control over nanomaterials' size, shape, and composition.

5.4 Differences Across Various Types of Nanomaterials

The production process affects different nanomaterials in distinct ways, depending on their composition and intended application. For example:

5.4.1 Metallic Nanoparticles

Highly sensitive to process parameters, including reaction temperature, precursor concentration, and the application of capping agents [15], Irfan et al. [87] have synthesized metallic nanoparticles, including gold and silver. These elements affect particle size, shape, and surface chemistry, which is crucial for use in catalysis and optics.

5.4.2 Carbon Nanotubes

Careful control of growth conditions is required to achieve the correct chirality, alignment, and graphitization when carbon nanotubes are manufactured via CVD and spray pyrolysis [53]. These characteristics are essential for composite material and electrical applications.

5.4.3 Metal Oxide Nanoparticles

The performance of metal oxide nanoparticles, such as TiO_2 and ZnO, is strongly influenced by their crystal structure, surface area, and composition [88]. These properties can be tailored by precisely controlling synthesis parameters such as reaction temperature, pH, and precursor concentration.

5.4.4 Bimetallic Nanoparticles

The synthesis of bimetallic nanoparticles involves additional complexity due to differences in the constituent metal reduction potentials and atomic sizes [89]. The synthesis method and process conditions must be carefully optimized for uniform alloying and desired catalytic properties, as tabulated in Table 1.

There are still various difficulties, even with significant developments in nanomaterial synthesis. These comprise the development of in situ characterization techniques for real-time process monitoring, the integration of machine learning and artificial intelligence for process optimization, and the necessity of scalable and reasonably priced manufacturing processes [29]. Future studies should concentrate on the creation of sustainable synthesis techniques that minimize environmental effects while preserving high product quality by means of which environmental impact might be minimized. Furthermore, promising for overcoming present constraints in nanomaterial production is the combination of flow chemistry with continuous manufacturing methods [90].

Table 1 Influence of process parameters on nanomaterial properties

Nanomaterial type	Key process parameters	Impact on properties	References
Silicon nanoparticles	Partial pressure, plasma power	Controls particle size and crystallinity	Lee et al. [91]
Carbon nanotubes	Growth temperature, reactor position	Influences graphitization, diameter, and alignment	Schnitzler and Zarbin [53]
Metal oxide nanoparticles	Reaction temperature, pH, precursor concentration	Affects crystal structure, surface area, and composition	Schiopu et al. [92]
Bimetallic nanoparticles	Molar ratio, reaction time, capping agents	Determines particle size, shape, and surface chemistry	Nyabadza et al. [89]
Nanocomposite coatings	Acetylene flow, substrate bias	Controls nanoparticle size, volume fraction, and interfacial interactions	Galvan et al. [10]

5.5 Grain Size

Numerous studies show that the performance of nanomaterials depends critically on their grain size, which also influences other uses, as shown in Table 2. Smaller grain sizes in Ni-rich cathode materials have been proven in lithium-ion batteries, for example, to increase mechanical strength and suppress microcracks, hence increasing capacity retention and cycle stability [93]. In conversion/alloying anodes, smaller particles also show improved reversibility in conversion processes, however, grain size has less of an effect on alloying contributions [94]. The ASTAR method has further shown that single-crystal NCM cathodes outperform poly-crystalline ones, stressing the need for grain size to lower gas evolution and improve lifetime [95]. In catalytic uses, oxygen reduction efficacy is strongly influenced by the size of Co/CoO nanoparticles [96]. In zinc-air batteries, smaller nanoparticles (15.6 nm) show better catalytic activity and, thus, improved power density [97]. This emphasizes how vital grain size control is to maximize catalytic systems. Smaller particle sizes help nanofillers in energy storage devices and improve attributes, including electrical resistivity and catalytic reactivity, thus supporting more effective energy storage solutions [11]. Grain size also affects electrochemical performance in pseudocapacitors, where smaller Co_3O_4 grains improve diffusion capacity and lower activation energy, thereby enhancing the electrochemical performance [98]. When coupled with techniques to avoid aggregation, smaller grain sizes in anode materials, including SnOx nanoparticles, produce better cycle performance and conductivity [99]. Emphasizing the importance of customized grain size distributions, computational studies on polycrystalline nanomaterials have shown an ideal grain size of about 10 nm for mechanical characteristics under deformation. Smaller grain sizes have been proven by molecular simulations on nanocrystalline copper to lower wear; yet, depending on contact size, a transition point exists wherein grain size becomes less important [100]. Grain size significantly affects sensor response in gas sensing applications, smaller grains frequently improve sensitivity, although this is not always true and relies on material and operating circumstances [101]. Larger grain sizes improve field-effect mobility for graphene, while defects and grain boundaries can counteract this advantage [102].

Smaller grain sizes in nanocomposites lower wear rates in tribological coatings, therefore providing the opportunity for application-specific modification [103]. With lower grain sizes (180 nm vs. 700 nm), ceramic materials—such as zirconia noticeably less wear, which is attributed to the development of a nanophase structure under friction [104]. Grain size affects main processes according to deformation mechanisms in nanocrystalline materials; smaller grains encourage grain boundary sliding and dislocation emission [105]. A review of electroformed nanomaterials shows that unique features with specific significant grain size-dependent characteristics result from grain size reduction below 5 nm [106]. Though thermodynamic constraints exist, nanostructuring improves insertion and conversion processes in lithium storage; careful grain size optimization is thus necessary for particular mechanisms [107]. Grain size determines carrier concentration and sensitivity in modeling

Table 2 Grain size effects in various applications

Application	Key finding	Reference
Lithium–ion batteries	Smaller grain sizes improve cycling stability and capacity retention	Ying et al. [99], Zou et al. [93]
Catalysis	Smaller Co/CoO nanoparticles enhance oxygen reduction performance	Li et al. [110]
Gas sensing	Grain size critically influences sensor response and sensitivity	Katoch et al. [109], Korotcenkov et al. [101], Rothschild and Komem [108]
Tribological coatings	Smaller grain sizes reduce wear rates and improve adaptability	Akimov et al. [104], Pan et al. [103]

metal-oxide gas sensors, with a threshold size where surface state density significantly influences performance [108]. At last, in ZnO nanofibers, grain size, and crystallinity interaction control gas sensing capability; ideal results depend on a balance between the two [109]. From energy storage and catalysis to sensing and tribology, grain size is ultimately a crucial factor affecting the performance of nanomaterials across many uses. Although lower grain sizes often improve mechanical strength, catalytic activity, and sensor sensitivity, the ideal size will rely on the particular use and material system. Designing high-performance nanomaterials depends on an awareness of grain size and other parameters, including crystallinity and flaws.

5.6 Grain Boundary Structure

Grain boundary structures significantly impact nanomaterials' performance in several uses, including electrical devices, thermoelectrics, and electrocatalyzes. Grain boundaries in noble metal catalysts, including gold, increase the density of active sites, thereby improving the oxygen reduction and hydrogen evolution reaction performance in electrocatalysis [111]. Grain boundary engineering can decouple thermoelectric parameters in thermoelectric materials, improving the Seebeck coefficient and lowering thermal conductivity, as seen in materials such as $Bi_2Te_{2.7}Se_{0.3}$ and SnTe, producing enhanced figures of merit [112]. Furthermore, the addition of elements such as boron and niobium can change grain boundaries to improve mechanical and thermoelectric characteristics by lowering grain boundary resistivity and thus encouraging grain development [113]. With tailored boundaries enhancing charge mobility and reducing energy consumption in synaptic devices, grain boundaries in polycrystalline materials like MoS_2 influence charge transport and device performance in electronic applications [114]. Moreover, grain boundaries in TiO_2 and metal nanowires improve photocatalytic reactions in photocatalysis by enabling electron transport and active sites for CO_2 reduction [115]. Grain boundary changes in solid electrolytes can enhance ionic conductivity and mechanical strength, which is essential for Li-ion battery applications [116]. Therefore, grain boundary design is vital for maximizing the performance of nanomaterials across various applications, as accurate control and engineering of grain boundaries are fundamental to optimizing material technologies [117].

6 Broad Spectrum of Nanomaterial Applications Across Industries

The adaptability of nanomaterials has resulted in their inclusion into a broad spectrum of industrial uses, influencing almost every industry of the economy. For instance, nanoparticles help create smaller automobile bumpers in the automotive sector, possibly lowering fuel usage. Nanomaterials that give clothes stain-repellent qualities help the textile sector by increasing the lifetime and simplicity of care of clothes [118]. Sunscreen compositions offer a more efficient defense against damaging radiation by the use of nanoparticles. Understanding concrete at the nanoscale has helped to increase its strength and durability even in building with the use of nano-silica [119]. The unique qualities of nanoparticles span many consumer goods. Food packaging may be designed to maximize the shelf life of contents, and cell phone displays can be made smaller. By adding nanomaterials, sports equipment such as balls and tennis rackers become more lightweight and robust [120]. Nano-lubricants enhance machine component performance and production lifetime, lowering maintenance costs. Nanometer-thick coatings are becoming increasingly important in industrial use as they provide better adhesive qualities and increased resistance against heat and UV radiation [121]. By the use of nanoscale additions in textiles for lightweight ballistic energy deflection in body armor, personal safety is also improved.

Beyond these examples, nanoparticles create surfaces with unique qualities such as water and residue repellence, anti-reflection, self-cleaning ability, and UV or infrared light resistance. Additionally, the creation of washable, durable "smart fabrics" is becoming a reality, including flexible nanoscale sensors for energy harvesting and health monitoring [122]. Promising significant fuel savings, the lightweight potential of nanoparticles reaches vehicles, trucks, aircraft, and spacecraft. Products ranging from nanoscale additions in polymer composite materials to lighter, stiffer, more durable ones depend on them. Catalysts—which increase the effectiveness of chemical processes in petroleum refining and vehicle catalytic converters—rely heavily on nanoparticles [123]. Moreover, nano-engineered materials are present in environmental sensors and air cleaners as well as in better-known home goods as stain removers and degreasers. Electronics, construction, packaging, food, energy, healthcare, automotive, and military are just a few of the many industrial fields where cellulosic nanoparticles find possible use [124]. Initially, the wave of nanomaterial applications concentrated on improving the features of current items, showing a strategy of using their unique qualities to value already existing sectors.

6.1 Revolutionizing Medicine with Nanomaterials

Nanotechnology has emerged as a powerful and transformative force in the field of medicine, offering innovative solutions for disease treatment, diagnostics, and drug delivery. The ability to manipulate materials at the nanoscale has opened up new avenues for addressing complex medical challenges and improving patient outcomes.

6.1.1 Targeted and Controlled Drug Delivery Systems

Conventional drug delivery systems can suffer from low drug solubility and non-specific distribution throughout the body, resulting in adverse effects and incapacity to target sick areas [125] efficiently. Through the creation of nanocarriers, nanotechnology presents answers to these problems. These nanocarriers can encapsulate therapeutic chemicals comprising polymeric micelles, liposomes, dendrimers, and nanoparticles and then deliver straight to sick cells, lowering system adverse effects and improving treatment efficiency. These systems work best when they can increase medication solubility, stability, and bio-distribution. Moreover, highly targeted medication delivery is made possible by functionalizing nanocarriers with ligands, especially chemicals designed to attach to target cells' receptors. In cancer treatment, for example, nanoparticles can be made to take chemotherapy medications straight to tumor cells, thereby sparing healthy tissues from the harmful effects of the treatments [126]. The capacity of nanotechnology to produce systems with regulated and continuous release of therapeutic substances is another significant benefit in medication delivery. Commonly utilized in these systems are polymeric nanoparticles and liposomes, which encapsulate the medication and release it progressively when the nanoparticle breaks down or in reaction to outside stimuli such as pH, temperature, or light [127]. This controlled release guarantees that the medication concentration in the bloodstream stays within the therapeutic window for a long time, lowering the need for regular dosage and increasing patient compliance. Nanotechnology also provides fresh approaches to overcoming biological obstacles, including the blood–brain barrier (BBB), which blocks many medications' entrance to the brain [128]. By improving the way medications are delivered across the blood–brain barrier, nanoparticles have shown promise for treating neurological diseases like brain tumors, Parkinson's disease, and Alzheimer's disease [129]. Analogous to this, nanoparticles can pass through the thick tissue environment of solid tumors, therefore enhancing the anticancer medication delivery to these problematic sites. Gene therapy has great promise in treating genetic diseases and cancer when liposomes and polymeric nanoparticles encapsulate DNA or RNA and transport it to cells without inducing an immune response [130]. By precisely targeting sick tissues and regulating the release of therapeutic substances, the development of nanomaterial-based drug delivery systems marks a path towards more customized and less intrusive therapies by finally lowering side effects and raising medicinal effectiveness.

6.1.2 Advanced Diagnostic Tools and Imaging Techniques

Nanotechnology is also revolutionizing medical diagnostics and enabling the early identification of diseases through the use of illness-related biomarkers at extremely low concentrations [131]. Nanosensors and nanoprobes are highly sensitive devices capable of identifying specific proteins, DNA sequences, or other biomolecules indicative of a particular disease, such as cancer, infections, and cardiovascular diseases. For example, gold nanoparticles can be functionalized with antibodies that bind to cancer-related proteins, thereby generating a detectable signal that facilitates the rapid and sensitive detection of cancer markers [132].

Additionally, nanotechnology has enabled significant advancements in medical imaging. Real-time observation of cellular and molecular processes is facilitated by the increased resolution and sensitivity of nanotechnology-enhanced imaging agents, such as quantum dots and superparamagnetic nanoparticles [133]. Quantum dots can function as fluorescent probes to monitor the distribution of medications, imaging cancer cells, and monitoring the response to therapy by utilizing semiconductor nanoparticles that produce light when agitated. In contrast, iron oxide nanoparticles are employed as contrast agents in magnetic resonance imaging (MRI), thereby improving the detection of malignancies, inflammation, and other pathogenic changes in tissues [134]. Nanotechnology-based point-of-care devices are currently in development and offer a rapid and sensitive method of disease diagnosis and treatment monitoring by utilizing nanobiosensors that can identify illness biomarkers in samples such as blood, saliva, or urine within minutes [131]. Nanoparticles can be customized to precisely target specific cells or tissues, thereby increasing the selectivity and sensitivity of detection as imaging agents. Their relevance extends to a variety of disciplines, including transplantation and renal disease. Introducing these advanced diagnostic instruments represents a shift toward earlier diagnosis and more precise molecular-level monitoring of maladies, potentially resulting in improved patient outcomes through prompt therapies [135].

6.1.3 Nanomaterials in Tissue Engineering and Regeneration

Nanomaterials are becoming increasingly critical in tissue engineering and regenerative medicine as they provide creative ways to rebuild or heal damaged organs or tissues. These materials are used to build scaffolds, three-dimensional frameworks that replicate the natural extracellular matrix (ECM) of tissues, thus giving a platform for cells to grow, adhere, proliferate, and differentiate, therefore producing tissue regeneration [136]. Using nanomaterials allows one to precisely regulate the physicochemical properties of these scaffolds, including their porosity, pore size, and mechanical strength, providing an environment fit for the regeneration of particular tissues [137]. Including nanomaterials in tissue engineering scaffolds improves cellular interactions, therefore encouraging adhesion, proliferation, and differentiation of planted cells. Using bioactive compounds like growth factors, peptides, and ECM proteins may also alter the surface of nanomaterials to increase further targeted biological responses and control the regeneration of specific tissues [136]. In bone regeneration, for example, scientists are using bone graft techniques based on nanomaterials such as collagen fibers and biomineralization agents to rebuild muscle 11 and restore bone. First found to improve cell proliferation rates for bone and cardiac tissue regeneration, respectively, are gold nanoparticles (GNPs) and titanium dioxide (TiO_2) [138]. GNPs show better biocompatibility and surface modification ability, making them great candidates for bone tissue regeneration, perhaps replacing bone morphogenetic proteins (BMPs), which have drawbacks like high cost and adverse effects [110]. Moreover, nanoparticles might be flexible carriers for the exact and controlled distribution of pharmaceuticals, including growth factors, to particular sites inside the tissue scaffold, offering advantages like enhanced stability and extended-release kinetics. By incorporating nanotechnology into tissue engineering, damaged tissues and organs may be rebuilt and regenerated, offering possible therapies for many medical disorders [139].

6.2 Nanomaterials Driving Innovation in Electronics

Nanomaterials are at the forefront of innovation in the electronics industry, enabling the development of high-performance devices with enhanced functionality and efficiency. Their unique electrical, optical, and mechanical properties at the nanoscale pave the way for the next generation of electronic components and systems.

6.2.1 Enhancing Semiconductor Performance at the Nanoscale

Nanotechnology is now essential for future advancements, as it is driven by the reduction of semiconductor devices, a fundamental component of the electronics sector. Reducing semiconductor materials to the nanoscale significantly changes their chemical and physical characteristics, resulting in unusual optical and electrical phenomena, most famously the quantum confinement phenomenon [140]. Discrete energy levels and changed electronic characteristics follow from the impact of the semiconductor particle's size approaching that of the electron's wavelength. Transistors and memory devices have been miniaturized thanks to nanomaterials, therefore stretching Moore's Law—which forecasts a doubling of transistors on a microchip every two years [141]. Because of their remarkable electrical conductivity and distinctive two-dimensional structures, carbon nanotubes and graphene are under investigation as possible substitutes for silicon in manufacturing smaller, quicker, more efficient microchips and electronic gadgets. Group II–VI semiconductors include Cadmium sulfide (CdS), Cadmium selenide (CdSe), Zinc oxide (ZnO), and Group III–V semiconductors are among the semiconductor nanomaterials most often utilized in electronics [142]. Faster switching rates, reduced power consumption, and more density—among other benefits—nanoscale electronics provide over their microscale equivalents. Thus, the ongoing progress of electronics depends on the integration of nanomaterials, which helps to enable the creation of ever more potent and effective computer and communication technologies [143].

6.2.2 High-Sensitivity Nanomaterial-Based Sensors

The field of electronic sensors is also being transformed by nanomaterials, which have enabled the development of devices with vastly improved sensitivity, efficacy, and specificity for detecting a variety of analytes. Superior sensing capabilities of nanomaterials, such as metal nanotubes, nanowires, nanofibers, nanocomposites, nanorods, nanoparticles, nanostructured polymers, and carbon allotropes like carbon nanotubes and graphene, are attributed in part to their high surface area to volume ratio and unique electrical properties. Gases, radiation, biomolecules, microbes, and environmental contaminants may be found and measured with these nanomaterial-based sensors [144]. For health monitoring, for example, nanoscale sensors are included in smart textiles to provide non-invasive and continuous tracking of physiological data [1]. Environmental sensors built on nanoparticles can identify minuscule pollution levels in water and air [145]. Nanobiosensors provide the fast and sensitive point-of-care diagnostics needed in the medical sector, facilitating early illness identification and tailored treatment plans [146]. Enhanced sensitivity and selectivity of nanomaterial-based sensors are pushing developments in several fields, including security, industrial safety, environmental monitoring, and healthcare.

6.2.3 Next-Generation Nanomaterial Displays

The development of next-generation electronic displays with enhanced performance and efficiency is also significantly influenced by nanomaterials. By lowering power consumption, weight, and thickness and therefore boosting color quality and resolution, these materials help to enhance display panels [147]. For instance, semiconductor nanocrystals called quantum dots (QDs) show unusual optical characteristics that enable them to generate more vivid and pure colors with more energy economy than conventional materials used in displays. They are being included in other displays, like ultra-high-definition TVs, to improve color gamut and brightness [148]. Operating on a concept similar to cathode ray tubes but at a far smaller scale, carbon nanotubes (CNTs) and silicon nanowires are being investigated for their potential in producing low-energy field emission displays (FEDs), therefore providing the prospect of thin and efficient flat-panel displays [149]. As a transparent conductive electrode in organic light-emitting diodes (OLEDs) and a possibility for flexible touchscreens, graphene's remarkable qualities—including its transparency and electrical conductivity—make perfect sense [150]. Moreover, the manufacture of electronic paper uses organic ink nanoparticles, reducing manufacturing costs, enhancing contrast ratios, and generating excellent brightness. More energy-efficient, high-performance, flexible displays for a variety of electronic devices are being developed by means of nanomaterials' inclusion into display technology [151].

6.3 Powering the Future with Nanomaterials in Energy

The increasing global demand for energy and the urgent need for sustainable energy solutions have positioned nanomaterials as critical components in the development of advanced energy conversion and storage technologies. Their unique properties at the nanoscale offer significant advantages in enhancing the efficiency and performance of solar cells, batteries, and fuel cells.

6.3.1 Boosting Solar Energy Conversion Efficiency

Nanomaterials are very important for making solar cells work better, and they are a key tool for using clean energy. Incorporating nanomaterials like titanium dioxide or nanostructured silicon dramatically increases the surface area of solar cells [152], enabling more effective sunlight absorption. Ideal for increasing energy conversion rates in solar cells, quantum dots—nanoscale semiconductors—can efficiently absorb and emit light. Flexible and lightweight solar cells are also made possible by nanomaterials such as graphene and carbon nanotubes, therefore creating fresh opportunities for their use in other applications [153]. Applied to solar cells, nanocoatings—such as protective layers of nanoparticles—can improve their lifetime and shield them from external elements, including UV radiation and moisture. Because of their excellent efficiency and low-cost manufacturing possibilities, emerging solar cell technologies based on nanomaterials, like perovskite solar cells, have attracted much interest [154]. Thus, the development of solar energy as a more affordable and efficient substitute for conventional energy sources depends on nanotechnology.

6.3.2 Advancements in Battery Technology Through Nanomaterials

Battery technology, which is crucial for operating electric vehicles, portable electronics, and grid-scale energy storage, is also significantly advanced by nanomaterials. Energy density, charging speed, cycle life, and safety may be improved using nanomaterials such as carbon nanotubes, graphene, metal oxides, silicon nanowires, and nanoscale electrolytes in batteries [155]. Faster electron and ion mobility inside the battery, made possible by these nanomaterials' enhanced surface area and excellent conductivity, helps to drive higher power output and quicker charging times [156]. Higher energy–density batteries come from, for instance, nanostructured silicon anodes storing far more lithium ions than typical graphite anodes. Nanoscale electrolytes help increase ion conductivity and stability, enabling solid-state batteries that promise better safety and higher energy density [157]. To provide more reasonably priced and sustainable energy storage options, researchers are also investigating the use of nanomaterials in other battery technologies like sodium-sulfur and zinc-air batteries. Therefore, nanotechnology is essential in overcoming the constraints of present battery technologies and allowing the general acceptance of renewable energy and electric mobility.

6.3.3 Nanomaterials for Efficient Fuel Cells

An additional prospective renewable energy technology that nanomaterials are making substantial contributions to is fuel cells, which directly convert fuel into electricity through an electrochemical reaction [158]. Fuel cells employ metal alloys, metal–organic frameworks, carbon nanotubes, graphene, and platinum nanoparticles among nanomaterials as catalysts to improve the efficiency of the electrochemical interactions [159]. By increasing the catalytic surface area of nanomaterials, their large surface area helps to enable more effective fuel-to-electricity conversion and, hence, lower the demand for costly catalyst materials like platinum. As possible catalysts to reduce the cost of fuel cells, researchers are also vigorously investigating the use of alternative, less expensive nanomaterials such as iron, nickel, and cobalt [160]. Based on the hydrogen fuel cell technology, hydrogen storage is another application for nanomaterials under investigation [161]. Moreover, more effective membranes produced inside fuel cells by means of nanotechnology enhance their efficiency and lifetime. These advancements promise to reduce costs and optimize fuel cell systems' overall performance and sustainability, paving the way for broader adoption in various energy applications.

7 Conclusion

The exploration of smart nanomaterials with advanced mechanical properties reveals a transformative potential for addressing a broad spectrum of healthcare challenges. Their unusual physicochemical properties, response to certain stimuli, and improved mechanical strength provide unheard-of diagnostics, treatments, and regenerative medicine possibilities. Smart nanomaterials are a leading edge in medical innovation, enabling earlier and more exact illness diagnosis, aiding targeted and regulated medication administration, building lasting medical implants, and promoting tissue regeneration. Still, the trip from bench to bedside is difficult. The chapter further represents a movement towards environmental consciousness in the scientific community with the exploration of green synthesis methods. Green strategies for synthesized nanomaterial production mitigate environmental effects and enable industrial-scale realization. The study has been focused on revealing emerging trends through bibliometric analysis and keyword clustering, which identified future directions for the development of nanotechnology.

Significant challenges remain in ensuring these materials' long-term biocompatibility and safety, negotiating the regulatory terrain's complexity, and obtaining scalable and reasonably priced production. Furthermore, much attention has to be paid to the ethical issues related to its use, including privacy, accessibility, and the possibility of unexpected effects. Notwithstanding these difficulties, the new developments and future directions in smart nanomaterials for healthcare show favorable results. The ongoing developments in nanomaterial design, synthesis, and functionalization, as well as the increasing desire for customized and more efficient medical treatments, point to a future in which these small materials will play an ever more critical part in enhancing human health and well-being. Responsibly harnessing the full potential of smart nanomaterials and translating their great promise into concrete benefits for patients globally will depend on cooperative efforts by academics, businesses, regulatory authorities, and the public.

References

1. Narminio, É.: Nanomaterials for Biomedical Engineering Applications (2022). https://doi.org/10.1007/978-981-19-1384-6_5
2. Sarkar, A., Kalita, D., Baruah, S.: Nanomaterials for aquatic contamination sensing and remediation (2021). https://doi.org/10.1016/B978-0-12-821141-0.00007-0
3. Hanlon, J.: Smart Nanomaterials and Their Applications (2023). https://doi.org/10.21741/978 1644902172-5
4. Pogozheva, I.D., Tristram-Nagle, S., Mosberg, H.I., Lomize, A.L.: Structural adaptations of proteins to different biological membranes. Biochim. Biophys. Acta (2013). https://doi.org/10.1016/J.BBAMEM.2013.06.023
5. John, J., Suresh, S., Sivakumar, M., Gopchandran, K., Pillai, V.M.: Ni doping induced property enhancement in laser ablated $BaSnO_3$ films suitable for optoelectronic applications. Heliyon **10**(5) (2024)
6. Vasilev, A., Efimov, M., Muratov, D., Chernavskii, P., Cherednichenko, K., Dzidziguri, E., Karpacheva, G.: Fe–Co alloy nanoparticles dispersed in polymer-derived carbon support: effect of initial polymer nature on the size, structure and magnetic properties. Materials **16**(20), 6694 (2023)
7. Hernández-Rodríguez, Y.M., Baños-López, E., Damián-Matsumura, P., González de la Rosa, C.H., Cigarroa-Mayorga, O.E.: The influence of temperature on the spatial distribution of AuNPs on a ceramic substrate for biosensing applications. Chemosensors **12**(10), 212 (2024)
8. Meng, L., Vu, T.V., Criscenti, L.J., Ho, T.A., Qin, Y., Fan, H.: Theoretical and experimental advances in high-pressure behaviors of nanoparticles. Chem. Rev. (2023). https://doi.org/10.1021/acs.chemrev.3c00169
9. Lee, H.K., Jang, B.Y., Ko, C.H., Lee, J.C., Park, Y.K.: Effects of process conditions on the properties of silicon nanoparticles synthesized by gas phase reactions using inductive coupled plasma. J. Nanosci. Nanotechnol. **12**(2), 1589–1593 (2012)
10. Galvan, D., Pei, Y., De Hosson, J.T.M.: Influence of deposition parameters on the structure and mechanical properties of nanocomposite coatings. Surf. Coat. Technol. **201**(3–4), 590–598 (2006)
11. Riascos, P., Llamosa, D., Amaya, J., Murcia, H.: Preliminary study on pressure and pH control of the size of silver nanoparticles: a rapid and cost-effective ultrasound-assisted synthesis for further applications (2025)
12. Yu, T., Huang, X., Zhang, X.F., Li, K., Liu, S.P., Dai, N., Zhang, K., Zhang, Y.X., Li, H.: A review of nanomaterials with excellent purification potential for the removal of micro-and nanoplastics from liquid. DeCarbon 100064 (2024)
13. Jafarov, M., Ganbarov, K., Kariptaş, E., Huseynova, S., Guliyeva, S.: The impact of temperature on the synthesis of silver nanoparticles by candida macedoniensis. Black Sea J. Health Sci. **6**(1), 163–166 (2023)

14. Chauhan, S.B.: Current status of smart nanoparticles in drug delivery. In: Biogenic Wastes-Enabled Nanomaterial Synthesis: Applications in Environmental Sustainability, pp. 351–372 (2024)
15. Xiong, J., He, S., Zhang, S., Qin, L., Yang, L., Wang, Z., Zhang, L., Shan, W., Jiang, H.: A label-free aptasensor for dual-mode detection of aflatoxin B1 based on inner filter effect using silver nanoparticles and arginine-modified gold nanoclusters. Food Control **144**, 109397 (2023)
16. Guo, P.: Application of smart nanomaterials in rheumatism drug delivery systems. Adv. Mater. Res. (2014). https://doi.org/10.4028/WWW.SCIENTIFIC.NET/AMR.893.141
17. Yu, F., Li, X., Sheng, C., Li, L.: DNA nanotechnology targeting mitochondria: from subcellular molecular imaging to tailor-made therapeutics. Angew. Chem. **136**(36), e202409351 (2024)
18. Nishimura, Y., Ishii, J., Ogino, C., Kondo, A.: Genetic engineering of bio-nanoparticles for drug delivery: a review. J. Biomed. Nanotechnol. (2014). https://doi.org/10.1166/JBN.2014.1951
19. Kamarajan, D., Anburaj, B., Porkalai, V., Muthuvel, A., Nedunchezhian, G., Mahendran, N.: Green synthesis of ZnO nanoparticles and their photocatalyst degradation and antibacterial activity. J. Water Environ. Nanotech. **7**(2), 180–193 (2022)
20. Soufi, G., Hekmatnia, A., Iravani, S., Varma, R.: Nanoscale contrast agents for magnetic resonance imaging: a review. ACS Appl. Nano Mater. (2022). https://doi.org/10.1021/acsanm.2c03297
21. Paramasivam, G., Palem, V.V., Sundaram, T., Sundaram, V., Kishore, S.C., Bellucci, S.: Nanomaterials: synthesis and applications in theranostics. Nanomaterials (2021). https://doi.org/10.3390/NANO11123228
22. Barui, A.K., Jana, B., Ryu, J.-H.: An insight into characterizations and applications of nanoparticulate targeted drug delivery systems (2019).https://doi.org/10.1007/978-3-662-59596-1_11
23. Mohammed, S., Mohammed, A.M., Al-Rawi, K.F.: Novel combination of multi-walled carbon nanotubes and gold nanocomposite for photothermal therapy in human breast cancer model. Steroids (2022). https://doi.org/10.2139/ssrn.4031273
24. Bhat, S., Guddadar, V.: Targeted therapy in chronic diseases treatment by nanomaterial based drug delivery (2021). https://doi.org/10.48175/IJARSCT-799
25. Girija, A.R., Balasubramanian, S., Cowin, A.J.: Nanomaterials-based drug delivery approaches for wound healing. Curr. Pharm. Des. (2022). https://doi.org/10.2174/1381612828666220328121211
26. Ruggeri, M., Vigani, B., Boselli, C., Icaro Cornaglia, A., Colombo, D., Sánchez-Espejo, R., Del Favero, E., Mandras, N., Roana, J., Cavallo, L., Cantù, L., Viseras, C., Rossi, S., Sandri, G.: Smart nano-in-microparticles to tackle bacterial infections in skin tissue engineering. Mater. Today Bio. (2022). https://doi.org/10.1016/j.mtbio.2022.100418
27. Lu, S., Yu, T., Wang, Y., Liang, L.-G., Chen, Y., Xu, F., Wang, S.: Nanomaterial-based biosensors for measurement of lipids and lipoproteins towards point-of-care of cardiovascular disease. Analyst (2017). https://doi.org/10.1039/C7AN00847C
28. Marlevi, D., Edelman, E.R.: Vascular lesion-specific drug delivery systems: JACC state-of-the-art review. J. Am. Coll. Cardiol. (2021). https://doi.org/10.1016/J.JACC.2021.03.307
29. Vasudevan, A., Shvalya, V., Košiček, M., Zavašnik, J., Jurov, A., Santhosh, N.M., Zidanšek, A., Cvelbar, U.: From faceted nanoparticles to nanostructured thin film by plasma-jet redox reaction of ionic gold. J. Alloy. Compd. **928**, 167155 (2022)
30. Shafique, B., Rehman, A., Mehmood, A., Ali, A.H., Zahra, S.M., Roobab, U., Singh, A.K., Ibrahim, S.A., Siddiqui, S.A.: Biocompatible nanomaterials in food science, technology, and nutrient drug delivery: recent developments and applications. Front. Nutr. (2022). https://doi.org/10.3389/fnut.2021.778155
31. Peixe, T.S., de Souza Nascimento, E., Bulcão, R.P., dos Santos, C.E.M., Novo Pais, M.C.: Regulations for safety assessment of nanomaterial (2018). https://doi.org/10.1201/B21545-23
32. Siddique, S., Alexander, A., Yadav, P., Agrawal, M., Shehata, A.M., Shaker, M.A., Rahman, S.A.U., Mohammad Abdul, M.I., Shaker, M.A.: Nanomedicines: challenges and perspectives

for future nanotechnology in the healthcare system. Sci. Res. Essays (2019). https://doi.org/10.5897/SRE2019.6604

33. Karimi, M.: Electrical and mechanical-responsive nanocarriers (2015). https://doi.org/10.1088/978-1-6817-4202-1CH6
34. Albashir, A.I.M., Lu, X., Dai, X., Qi, W.: Effects of porous structure and oxygen functionalities on electrochemical synthesis of hydrogen peroxide on ordered mesoporous carbon. Commun. Chem. **7**(1), 111 (2024)
35. Khajuee, S., Memarian, N.: Hydrothermal synthesis of ultrafine SnO_2 nanospheres: effect of reaction time on physical properties. Eur. Phys. J. Plus **136**(5), 1–12 (2021)
36. Ordóñez-Casanova, E.G., Trejo-Mandujano, H.A., Saucedo-Acuña, R.A., Villanueva-Montellano, A.: Morphological analysis and its effect on the optical properties of carbon nanospheres as a function of synthesis time. J. Syst. Innov. **8**(22), 1–7 (2024)
37. Petlin, D.: Controlled Drug Release from Biomaterials and Its Mathematical Modeling (2017)
38. Wang, Y., Li, Z., Ouyang, J., Karniadakis, G.E.: Controlled release of entrapped nanoparticles from thermoresponsive hydrogels with tunable network characteristics. Soft Condens. Matter (2020). https://doi.org/10.1039/D0SM00207K
39. Suleiman, M.H., El-Sheikh, S.M., Mohamed, E.T., El Raey, M.A., El Sherbiny, S., Morsy, F.A., El-Hout, S.I., Sheta, S.M.: Green synthesis of ZnO-NPs using sugarcane bagasse waste: phytochemical assessment of extract and biological study of nanoparticles. Dalton Trans. **53**(46), 18494–18505 (2024)
40. Rizwan, M., Shoukat, A., Ayub, A., Razzaq, B., Tahir, M.B.: Types and classification of nanomaterials (2021). https://doi.org/10.1016/B978-0-12-823823-3.00001-X
41. Meng, F., Ling, Y., Li, Y., Liu, D., Wei, K., Sun, L., Sang, Z.: Synthesis of visible-light-driven photocatalyst of TiO_2 modified waste red mud and its application in tetracycline hydrochloride removal. Surf. Interfaces **35**, 102482 (2022)
42. Bekyarova, E., Ni, Y., Malarkey, E.B., Montana, V., Mcwilliams, J.L., Haddon, R.C., Parpura, V.: Applications of carbon nanotubes in biotechnology and biomedicine. J. Biomed. Nanotechnol. (2005). https://doi.org/10.1166/JBN.2005.004
43. Cordero, M., Ruiz, C., Palacio, D.A., Turunen, P., Rowan, A.E., Urbano, B.F.: Effect of low aspect ratio one-dimensional nanoparticles on properties of photocrosslinked alginate nanocomposite hydrogels. Int. J. Biol. Macromol. (2022). https://doi.org/10.1016/j.ijbiomac.2022.02.059
44. Mobed, A., Hasanzadeh, M., Seidi, F.: Anti-bacterial activity of gold nanocomposites as a new nanomaterial weapon to combat photogenic agents: recent advances and challenges. RSC Adv. (2021). https://doi.org/10.1039/D1RA06030A
45. Wang, N., Fuh, J.Y.H., Dheen, S.T., Kumar, A.S.: Functions and applications of metallic and metallic oxide nanoparticles in orthopedic implants and scaffolds. J. Biomed. Mater. Res. Part B (2021). https://doi.org/10.1002/JBM.B.34688
46. Ghosal, K., Mondal, P., Bera, S., Ghosh, S.: Graphene family nanomaterials- opportunities and challenges in tissue engineering applications. FlatChem (2021). https://doi.org/10.1016/J.FLATC.2021.100315
47. Khan, F.A.: Synthesis of Nanomaterials: Methods & Technology (2020). https://doi.org/10.1007/978-981-15-4802-4_2
48. Le, Q., Cheng, Z.: Microwave-assisted rapid growth of corncob-like nano-Ag2O/ZIF-8 on PAN electrospinning nanofibers enabled highly efficient selective adsorption desulfurization. Appl. Surf. Sci. (2023). https://doi.org/10.1016/j.apsusc.2023.157109
49. Jeevanandam, J., Kiew, S.F., Boakye-Ansah, S., Lau, S.Y., Barhoum, A., Danquah, M.K., Rodrigues, J.: Green approaches for the synthesis of metal and metal oxide nanoparticles using microbial and plant extracts. Nanoscale (2022).https://doi.org/10.1039/d1nr08144f
50. Ivanisenko, Y.: Perspectives of scaling up of severe plastic deformation: a case of high pressure torsion extrusion. Mater. Trans. (2023). https://doi.org/10.2320/matertrans.mt-mf2022057
51. Popov, G., Glukhoy, Y.: Nanoparticle production line with in situ control of the chemical composition, size, and morphology of nanoscale particles (2005).https://doi.org/10.1117/12.579234

52. Smith, J., Wang, S., Eldred, T.B., DellaRova, C., Gao, W.: Characterization of nanomaterials dynamics with transmission electron microscope (2023). https://doi.org/10.1016/b978-0-12-822425-0.00049-x
53. Schnitzler, M.C., Zarbin, A.J.G.: The effect of process variables on the characteristics of carbon nanotubes obtained by spray pyrolysis. J. Nanopart. Res. **10**, 585–597 (2008). https://doi.org/10.1007/S11051-007-9289-4
54. Auerbach, S.M.: An overview of the synthesis of nanomaterials. Mater. Res. Found. (2023). https://doi.org/10.21741/9781644902370-2
55. Singh, B.P., Perala, R.S., Srivastava, M., Ningthoujam, R.S.: Synthesis of nanostructured materials by thermolysis (2021). https://doi.org/10.1007/978-981-16-1807-9_12
56. Ouyang, Y., O'Hagan, M.P., Willner, I.: Functional catalytic nanoparticles (nanozymes) for sensing. Biosens. Bioelectr. (2022). https://doi.org/10.1016/j.bios.2022.114768
57. Gupta, D.: Green and sustainable synthesis of nanomaterials: recent advancements and limitations. Environ. Res. (2023). https://doi.org/10.1016/j.envres.2023.116316
58. Nkurikiyimfura, I., Wang, Y., Safari, B., Nshingabigwi, E.: Temperature-dependent magnetic properties of magnetite nanoparticles synthesized via coprecipitation method. J. Alloys Compd. (2020). https://doi.org/10.1016/J.JALLCOM.2020.156344
59. Isaeva, I.Y., Ostaeva, G.Y., Eliseeva, E.A., Golovin, A.L., Vasiliev, A.: Effect of synthesis temperature on the formation of nickel-copper composite nanoparticles. Lett. Mater. (2023). https://doi.org/10.22226/2410-3535-2023-2-149-152
60. Kamarajan, G., Anburaj, D.B., Porkalai, V., Muthuvel, A., Nedunchezhian, G.: Effect of temperature on optical, structural, morphological and antibacterial properties of biosynthesized ZnO nanoparticles. J. Niger. Soc. Phys. Sci. (2022). https://doi.org/10.46481/jnsps.2022.892
61. Kariptaş, E.: The impact of temperature on the synthesis of silver nanoparticles by candida macedoniensis. Black Sea J. Health Sci. (2022). https://doi.org/10.19127/bshealthscience.1189775
62. Utari, R.A., Suharno, H.W., Budi, P.: Effect of synthesis temperature on structural and magnetic properties in hematite (α-Fe_2O_3) nanoparticles produced by Co-precipitation method. Defect Diffus. Forum Ser. (2022). https://doi.org/10.4028/p-e8v51k
63. Dejene, F.: Characterization of low-temperature-grown ZnO nanoparticles: the effect of temperature on growth. J. Phys. Commun. (2022). https://doi.org/10.1088/2399-6528/ac8049
64. Vasilev, A.A., Dzidziguri, E.L., Efimov, M.N., Muratov, D.G., Karpacheva, G.P.: Effect of the synthesis temperature and metal ratio on structural characteristics of nanocomposites based on pyrolyzed chitosan and bimetallic Fe–Co nanoparticles. Russ. J. Phys. Chem. B (2021).https://doi.org/10.1134/S1990793121030313
65. Jansanthea, P., Kanthabangharn, J., Chomkitichai, W., Ketwaraporn, J., Saovakon, C., Wansao, C., Wanaek, A., Kraivuttinun, P., Pookmanee, P., Phanichphant, S.: Temperature-controlled synthesis and photocatalytic properties of ZnO–SnO_2 nanocomposites. J. Aust. Ceram. Soc. (2021).https://doi.org/10.1007/S41779-021-00567-4
66. Deuri, J.K., Sahu, P., Manju, U.: In situ pressure controlled growth of ZnO nanoparticles: tailoring sizes, defects, and optical properties. Inorg. Chem. (2023). https://doi.org/10.1021/acs.inorgchem.3c00586
67. Wang, L.: High-pressure synthesis of nanomaterials with exotic phases (2023). https://doi.org/10.1016/b978-0-12-822425-0.00066-x
68. Liang, H., Dong, Y., Ding, Q., Li, X., Yu, M., Li, P., Duan, L., Wang, Y.: Dimetallic praseodymium-cobalt carbon nanotubes as highly efficient electrocatalyst for oxygen reduction reaction. Carbon Trends **15**, 100350 (2024)
69. Nam, J.H., Bruggeman, P.: Effect of the pH on the formation of gold nanoparticles enabled by plasma-driven solution electrochemistry. Plasma Process. Polym. **22**(1), 2400140 (2025)
70. Mechai, F., Al Shboul, A., Bensidhoum, M.O., Anabestani, H., Ketabi, M., Izquierdo, R.: Influence of pH on room-temperature synthesis of zinc oxide nanoparticles for flexible gas sensor applications. Chemosensors **12**(5), 83 (2024)

71. Chrisma, R.B., Avani, A., Anila, E.I.: Effect of pH on the structural and optical properties of cobalt oxide nanoparticles synthesized by hydrothermal method. Mater. Today Proc. (2022). https://doi.org/10.1016/j.matpr.2022.11.332
72. Pavithra, S., Keerthana, S., Yuvakumar, R., Senthil Kumar, P., Rajesh, S., Vidhya, B., Sakunthala, A.: Preparation of eta-FeOOH by ultrasound assisted precipitation route for aqueous supercapacitor applications. Inorg. Nano-Metal Chem. **53**(2), 152–158 (2023)
73. Mansha, M.S., Iqbal, T., Farooq, M., Riaz, K., Afsheen, S., Sajjad, M., Al-Zaqri, N., Warad, I.: PH mediated simple synthesis of $AgVO_3$ nanomaterials for degradation of industrial waste. Optik **290**, 171285 (2023)
74. Baqaei, A., Sabbagh Alvani, A.A., Sameie, H., Rosei, F.: Role of pH in the hydrothermal synthesis of TiO_2 nanorod photocatalysts. ChemistrySelect. https://doi.org/10.1002/slct.202300792
75. Garcia-Barrientos, A., Ambrosio-Lazaro, R.C., García-Ramírez, M.A., Pérez-Cortés, O., Plaza-Castillo, J.: pH-effect in the fabrication of ZnO nanostructured thin films by chemical bath deposition for increasing the efficiency of solar cells. Materials (2023). https://doi.org/10.3390/ma16083275
76. Paul, A., Joseph, N., Sebastian, T., Sreekala, C.O., Bindu, R.G., Augustine, S.: The impact of quantum-sized nickel nanoparticles on TiO_2 in photovoltaic and photocatalytic systems. J. Appl. Phys. **136**(17) (2024)
77. Alhalili, Z., Smiri, M.: The influence of the calcination time on synthesis of nanomaterials with small size, high crystalline nature and photocatalytic activity in the TiO_2 nanoparticles calcined at 500 °C crystals (2022). https://doi.org/10.3390/cryst12111629
78. Kiprotich, S., Dejene, F., Onani, M.O.: Effects of growth time on the material properties of CdTe/CdSe core/shell nanoparticles prepared by a facile wet chemical route. Mater. Res. Expr. (2022). https://doi.org/10.1088/2053-1591/ac5073
79. Kadzutu-Sithole, R., Machogo-Phao, L.F.E., Kolokoto, T., Zimuwandeyi, M., Gqoba, S.S., Mubiayi, K.P., Moloto, M.J., van Wyk, J.L., Moloto, N.: Elucidating the effect of precursor decomposition time on the structural and optical properties of copper(I) nitride nanocubes. RSC Adv. (2020). https://doi.org/10.1039/C9RA09546B
80. Lian, Z., Zhu, X., Wang, D., Meng, X., He, Y.: Effect of atmospheric environment on the stability of secondary electron emission from magnesium oxide and alumina surfaces. J. Phys. D Appl. Phys. **57**(12), 125302 (2023)
81. Kautz, E.J., Zelenyuk, A., Gwalani, B., Olszta, M.J., Phillips, M.C., Manard, M.J., Kimblin, C.W., Harilal, S.S.: Impact of environmental oxygen on nanoparticle formation and agglomeration in aluminum laser ablation plumes. J. Chem. Phys. **159**(17) (2023)
82. Vasudevan, A.: Single step metal nanoparticle fabrication using atmospheric pressure plasma jets. Meet. Abstracts (2022).https://doi.org/10.1149/ma2022-0219898mtgabs
83. Yagyaeva, E.B.: Advances of nanomaterials for air pollution remediation and their impacts on the environment. Chemosphere (2022).https://doi.org/10.1016/j.chemosphere.2021.132083
84. Li, C.: Structure controlling and process scale-up in the fabrication of nanomaterials. Front. Chem. Eng. China (2010).https://doi.org/10.1007/S11705-009-0305-3
85. Zheng, G., Jakumeit, J., Pabel, T., Kneissl, C., Magagnin, L.: Computational Simulation of Nanoparticle Distributions in Metal Matrix Composite Casting Processes (2021). https://doi.org/10.1007/978-3-030-65396-5_25
86. Parthasarathy, S., Behera, S., Das, D.K., Patnaik, P., Das, N., Mallick, P., Nayak, S.K.: Nanoparticle synthesis methods. In: Discovery, Disruption, and Future Implications of Nanomaterials, pp. 33–68. IGI Global (2024)
87. Irfan, M., Nabi, R.A.U., Hussain, H.K., Naz, M.Y., Shukrullah, S., Khawaja, H.A., Rahman, S., Althobiani, F.: Numerical sensitivity analysis of temperature-dependent reaction rate constants for optimized thermal conversion of high-density plastic waste into combustible fuels. Can. J. Chem. Eng. (2023). https://doi.org/10.1002/cjce.24883
88. Hernández, S., Hernández, S., Hidalgo, D., Hidalgo, D., Sacco, A., Chiodoni, A., Lamberti, A., Lamberti, A., Cauda, V.A., Tresso, E.M., Tresso, E.M., Saracco, G.: Comparison of photocatalytic and transport properties of TiO_2 and ZnO nanostructures for solar-driven water splitting. Phys. Chem. Chem. Phys. (2015). https://doi.org/10.1039/C4CP05857G

89. Nyabadza, A., Vazquez, M., Brabazon, D.: A review of bimetallic and monometallic nanoparticle synthesis via laser ablation in liquid. Crystals **13**(2), 253 (2023)
90. Długosz, O., Banach, M.: Inorganic nanoparticle synthesis in flow reactors—applications and future directions. React. Chem. Eng. (2020). https://doi.org/10.1039/D0RE00188K
91. Lee, H.K., Jang, B.Y., Ko, C.H., Lee, J.C., Park, Y.K.: Effects of process conditions on the properties of silicon nanoparticles synthesized by gas phase reactions using inductive coupled plasma. J. Nanosci. Nanotechnol. (2012).https://doi.org/10.1166/JNN.2012.4594
92. Schiopu, A.-G., Iordache, D.M., Oproescu, M., Cursaru, L.M., Ioţa, A.-M.: Tailoring the synthesis method of metal oxide nanoparticles for desired properties. Crystals **14**(10), 899 (2024)
93. Zou, B., Zhang, L., Xue, X., Tan, R., Jiang, P., Ma, B., Song, Z., Hua, W.: A review on the fault and defect diagnosis of lithium-ion battery for electric vehicles. Energies **16**(14), 5507 (2023)
94. Umirov, N., Seo, D.-H., Kim, H.-Y., Kim, S.-S.: Grain size effect of nanocrystalline-Si embedded in buffering alloy-matrix as anode for Li-ion batteries. J. Alloys Compd. (2021). https://doi.org/10.1016/J.JALLCOM.2021.160558
95. Hu, J., Li, L., Bi, Y., Tao, J., Lochala, J., Liu, D., Wu, B., Cao, X., Chae, S., Wang, C., Xiao, J.: Locking oxygen in lattice: a quantifiable comparison of gas generation in polycrystalline and single crystal Ni-rich cathodes. Energy Storage Mater. (2022). https://doi.org/10.1016/j.ensm.2022.02.025
96. Du, H., Ding, F., Gu, L., Zhao, J., Zhang, X., Qu, K., Li, Y.-W., Lan, T., Li, J., Zhang, Y., Qin, Y., Lu, W.: CoO/Co/N-C nanoparticles embedded in carbon as mediate for oxygen reduction electrocatalysts. J. Alloys Compd. (2021). https://doi.org/10.1016/J.JALLCOM.2021.161174
97. Zinzani, P.L.: Chemically embedding Co nanospheres in N-doped carbon nanosheets for enhanced zinc-air batteries. J. Power Sour. (2023). https://doi.org/10.1016/j.jpowsour.2022.232381
98. Hao, J., Wu, W., Wang, Q., Yan, D., Liu, G., Peng, S., Peng, S.: Effect of grain size on electrochemical performance and kinetics of Co_3O_4 electrode materials. J. Mater. Chem. (2020). https://doi.org/10.1039/D0TA02032J
99. Ying, H., Yang, T., Zhang, S., Guo, R., Wang, J., Han, W.-Q.: Dual immobilization of SnOx nanoparticles by N-doped carbon and TiO_2 for high-performance lithium-ion battery anodes. ACS Appl. Mater. Interfaces (2020). https://doi.org/10.1021/ACSAMI.0C15670
100. Li, A., Szlufarska, I.: How grain size controls friction and wear in nanocrystalline metals. Phys. Rev. B **92**(7), 075418 (2015)
101. Korotcenkov, G., Cho, B.: The role of grain size on the thermal instability of nanostructured metal oxides used in gas sensor applications and approaches for grain-size stabilization. Progr. Cryst. Growth Charact. Mater. (2012). https://doi.org/10.1016/J.PCRYSGROW.2012.07.001
102. Yagi, K., Yamada, A., Hayashi, K., Harada, N., Sato, S., Yokoyama, N.: Dependence of field-effect mobility of graphene grown by thermal chemical vapor deposition on its grain size. Japan. J. Appl. Phys. (2013). https://doi.org/10.7567/JJAP.52.110106
103. Pan, L., Dawson, B., Krim, J., Baker, C.C., Pearson, J., Zikry, M.A., Vovoedin, A.: Nanoscale Design of Adaptive Tribological Coatings (2008). https://doi.org/10.1115/IJTC2008-71196
104. Akimov, G.Y., Marinin, G.A., Chaika, E.V., Varyukhin, V.: Effect of the grain size on the formation of a nanophase structure and tribological properties of the friction surface of a ceramic made of partly stabilized zirconia. Tech. Phys. (2007). https://doi.org/10.1134/S1063784207100180
105. Zhu, Y., Langdon, T.G.: Influence of grain size on deformation mechanisms: an extension to nanocrystalline materials. Mater. Sci. Eng. A-Struct. Mater. Prop. Microstruct. Process. (2005). https://doi.org/10.1016/J.MSEA.2005.05.111
106. Erb, U.: Size effects in electroformed nanomaterials. Key Eng. Mater. (2010). https://doi.org/10.4028/WWW.SCIENTIFIC.NET/KEM.444.163
107. Hariharan, S., Ramar, V., Balaya, P.: Influence of nanosize and thermodynamics on lithium storage in insertion and conversion reactions (2012). https://doi.org/10.1117/12.921157

108. Rothschild, A., Komem, Y.: On the relationship between the grain size and gas-sensitivity of chemo-resistive metal-oxide gas sensors with nanosized grains. J. Electroceram. (2004). https://doi.org/10.1007/S10832-004-5178-8
109. Katoch, A., Sun, G.-J., Choi, S.-W., Byun, J.-H., Kim, S.S.: Competitive influence of grain size and crystallinity on gas sensing performances of ZnO nanofibers. Sens. Actuators B Chem. (2013). https://doi.org/10.1016/J.SNB.2013.05.030
110. Li, K., Liu, X., Mou, X., Guo, Y., Yang, L., Chen, Y., Zhou, Y., Yang, Z., Chen, Z.: Advances in the application of bone morphogenetic proteins and their derived peptides in bone defect repair. Compos. Part B Eng. (2023).https://doi.org/10.1016/j.compositesb.2023.110805
111. Dong, C.-K.: Grain boundary—a route to enhance electrocatalytic activity for hydrogen evolution reaction. Appl. Sci. (2022). https://doi.org/10.3390/app12094290
112. Li, S., Huang, Z., Wang, R., Zhao, W., Luo, J., Xiao, Y., Pan, F.: Highly distorted grain boundary with an enhanced carrier/phonon segregation effect facilitates high-performance thermoelectric materials. ACS Appl. Mater. Interfaces (2021). https://doi.org/10.1021/ACSAMI.1C15422
113. Zhang, C., Geng, X., Chen, B., Li, J., Meledin, A., Hu, L., Liu, F., Shi, J., Mayer, J., Mayer, J., Wuttig, M., Wuttig, M., Cojocaru-Mirédin, O., Yu, Y.: Boron-mediated grain boundary engineering enables simultaneous improvement of thermoelectric and mechanical properties in N-type Bi_2Te_3. Small (2021).https://doi.org/10.1002/SMLL.202104067
114. Wang, X., Wang, B., Zhang, Q., Sun, Y., Wang, E., Luo, H., Wu, Y., Gu, L., Li, H., Liu, K.: Grain-boundary engineering of monolayer MoS_2 for energy-efficient lateral synaptic devices. Adv. Mater. (2021). https://doi.org/10.1002/ADMA.202102435
115. Zhu, Y., Xu, Z., Lang, Q., Jiang, W., Yin, Q., Zhong, S., Bai, S., Bai, S.: Grain boundary engineered metal nanowire cocatalysts for enhanced photocatalytic reduction of carbon dioxide. Appl. Catal. B Environ. (2017). https://doi.org/10.1016/J.APCATB.2017.01.035
116. Zheng, C., Lu, Y., Su, J., Song, Z., Xiu, T.P., Jin, J., Badding, M.E., Wen, Z.: Grain boundary engineering enabled high-performance garnet-type electrolyte for lithium dendrite free lithium metal batteries. Small Methods (2022). https://doi.org/10.1002/smtd.202200667
117. Kobayashi, S., Tsurekawa, S., Watanabe, T.: A new approach to grain boundary engineering for nanocrystalline materials. Beilstein J. Nanotechnol. (2016). https://doi.org/10.3762/BJNANO.7.176
118. Dhinakaran, V., Varsha Shree, M.: The Role and Applications of Nanomaterials in the Automotive Industry (2021). https://doi.org/10.1201/9781003160946-5
119. Santos, A.C., Martõ, J., Chá-Chá, R., Martins, A.M., Pereira-Silva, M., Ribeiro, H.M., Veiga, F.: Nanotechnology-based sunscreens—a review. Mater. Today Chem. (2022). https://doi.org/10.1016/j.mtchem.2021.100709
120. Application of Nanotechnology in Food Packaging and Food Quality (2022). https://doi.org/10.1201/9781003207641-2
121. Ye, X.: Nanoparticle additives for improving lubricant performance. In: Highlights in Science, Engineering and Technology (2022). https://doi.org/10.54097/hset.v17i.2580
122. Fouda, A.H., Hamza, M.F., Shaheen, T.I., Wei, Y.: Editorial: nanotechnology and smart textiles: sustainable developments of applications. Front. Bioeng. Biotechnol. (2022). https://doi.org/10.3389/fbioe.2022.1002887
123. Satria, D.: Nanomaterials in the automobile sector. Mater. Res.. Found. (2023). https://doi.org/10.21741/9781644902288-6
124. Selvaraj, T., Perumal, V., Khor, S.F.: Non-metallic nanomaterial productions from natural resources (2021). https://doi.org/10.1016/B978-0-12-821163-2.00013-3
125. Lakavath, S.K.: Novel delivery system used for oral bioavailability enhancement of poorly water soluble drugs. J. Drug Deliv. Ther. (2020). https://doi.org/10.22270/JDDT.V10I6-S.4613
126. Shinde, A.S., Shinde, P., Santra, T.S.: Nanomaterials: versatile drug carriers for nanomedicine (2021). https://doi.org/10.1007/978-981-33-6252-9_10
127. Saokar, S., Saudagar, R.B.: Nanoparticle in pharmaceutical drug delivery system: a review. J. Drug Deliv. Ther. (2019). https://doi.org/10.22270/JDDT.V9I3.2772

128. Tan, Y.F., Lao, L.L., Xiong, G.M., Venkatraman, S.S.: Controlled-release nanotherapeutics: state of translation. J. Controlled Release (2018). https://doi.org/10.1016/J.JCONREL.2018.06.014
129. Hasannejad-Asl, B., Pooresmaeil, F., Choupani, E., Dabiri, M., Behmardi, A., Fadaie, M., Fathi, M., Moosavi, S.A., Takamoli, S., Hemati, E., Yaghoubi Naei, V., Kazemi Lomedasht, F.: Nanoparticles as powerful tools for crossing the blood-brain barrier. CNS Neurol. Disord. Drug Targets (2022).https://doi.org/10.2174/1871527321666220222092655
130. Luiz, M.T., Dutra, J.A.P., Tofani, L.B., de Araújo, J.T.C., Di Filippo, L.D., Marchetti, J.M., Chorilli, M.: Targeted liposomes: a nonviral gene delivery system for cancer therapy. Pharmaceutics (2022). https://doi.org/10.3390/pharmaceutics14040821
131. McCans, K.: Nanobiosensor: Advancement in Disease Diagnostic (2022). https://doi.org/10.1007/978-981-19-5141-1_12
132. Adam, T., Gopinath, S.C.B.: Nanosensors: recent perspectives on attainments and future promise of downstream applications. Process Biochem. (2022). https://doi.org/10.1016/j.procbio.2022.03.024
133. Moyosore, A.A.: Application of Nanoparticles in Biomedical Imaging (2022). https://doi.org/10.1201/9781003277255-3
134. Ilaiyaraja, N., Fathima, S.J., Khanum, F.: Quantum dots: a novel fluorescent probe for bioimaging and drug delivery applications (2018). https://doi.org/10.1016/B978-0-12-813661-4.00012-2
135. Dizaj, S., Eftekhari, A., Mammadova, S., Ahmadian, E., Ardalan, M., Davaran, S., Nasibova, A., Khalilov, R., Valiyeva, M., Mehraliyeva, S., Mostafavi, E.: Nanomaterials for chronic kidney disease detection. Appl. Sci. (2021). https://doi.org/10.3390/APP11209656
136. Aziz, R.: Applications of Nanomaterials in Tissue Engineering and Regenerative Medicine (2021).https://doi.org/10.1007/978-3-030-67223-2_9
137. Achachelouei, M.F., Keskin, D., Tezcaner, A.: Use of nanoscale-delivery systems in tissue/organ regeneration (2020). https://doi.org/10.1016/B978-0-08-102906-0.00026-X
138. Masne, N., Ambade, R., Bhugaonkar, K.: Use of nanocomposites in bone regeneration. Cureus (2022). https://doi.org/10.7759/cureus.31346
139. Sher, E.K., Farhat, E.K., Sher, F.: Nanostructured materials for drug delivery and tissue engineering applications. Mol. Biotechnol. (2023). https://doi.org/10.1007/s12033-023-00784-1
140. Feliczak-Guzik, A., Szczyglewska, P.: Nanotechnology–general aspects: a chemical reduction approach to the synthesis of nanoparticles. Molecules (2023). https://doi.org/10.3390/molecules28134932
141. Beausoleil, A.: Moore's law and social theory: deconstructing and redefining technology industry's innovation edict. Int. J. Actor Netw. Theory Technol. Innov. (2014).https://doi.org/10.4018/IJANTTI.2014100101
142. Kolahdouz, M., Xu, B., Faghih Nasiri, A., Fathollahzadeh, M., Manian, M., Aghababa, H., Wu, Y., Radamson, H.H.: Carbon-related materials: graphene and carbon nanotubes in semiconductor applications and design. Micromachines (2022). https://doi.org/10.3390/mi13081257
143. Barik, T.K., Maity, G.C., Gupta, P., Mohan, L., Mohan, L., Santra, T.S.: Nanomaterials: An Introduction (2021). https://doi.org/10.1007/978-981-33-6252-9_1
144. Panigrahi, B., Suryakanta, U., Mishra, S., Singh, R.K., Mandal, D.: Role of Nanotechnology in Tissue Engineering and Regenerative Medicine (2022).https://doi.org/10.1201/9781003161158-24
145. Wang, L., Ma, W., Ma, W., Xu, L., Xu, L., Chen, W., Chen, W., Zhu, Y., Zhu, Y., Xu, C., Kotov, N.A.: Nanoparticle-based environmental sensors (2010). https://doi.org/10.1016/J.MSER.2010.06.012
146. Akrema, R., Khan, S.: Nanosensors Based on DNA as an Emerging Technology for the Detection of Disease (2020). https://doi.org/10.1007/978-981-15-8346-9_5
147. Vidales-Herrera, J., López, I.: Nanomaterials in coatings: an industrial point of view (2020).https://doi.org/10.1016/B978-0-12-821381-0.00003-X

148. Ahamed, M.I., Ahamed, S., Prathap, N., Srinivasan, M.N., Mathuvanesan, C.: Quantum dots and their applications in television display technologies. World J. Adv. Res. Rev. (2022). https://doi.org/10.30574/wjarr.2022.16.3.1455
149. Liang, X., Xia, J., Dong, G., Tian, B., Peng, L.-M.: Carbon nanotube thin film transistors for flat panel display application. Top. Curr. Chem. (2016).https://doi.org/10.1007/S41061-016-0083-6
150. Miao, J., Fan, T.: Flexible and stretchable transparent conductive graphene-based electrodes for emerging wearable electronics. Carbon (2023).https://doi.org/10.1016/j.carbon.2022.11.018
151. Sharifi, A.R., Naghdi, T., Yousefi, H., Kiani, M.A., Golmohammadi, H.: Nanopapers Toward Green Photonic and Optical Applications (2022).https://doi.org/10.1021/acssuschemeng.2c04224
152. Mhetre, H., Kanse, Y.K.: Contribution of nanomaterials to the development of solar cell technology: a review. J. Nano Electr. Phys. (2022). https://doi.org/10.21272/jnep.14(3).03017
153. Riaz, S., Park, S.-J.: Carbon quantum dots-embedded graphitic carbon nitride nanotubes for enhancing the power conversion efficiency of sensitized solar cells. Mater. Today Chem. (2022).https://doi.org/10.1016/j.mtchem.2021.100763
154. Cherches, T.: Nanomaterials' Architectural Study in Perovskite Solar Cells (2022).https://doi.org/10.1007/978-981-19-1384-6_4
155. Zhao, Z.: Applications of nanotechnology: lithium-ion based batteries in electric vehicles. In: Highlights in Science, Engineering and Technology (2023). https://doi.org/10.54097/hset.v32i.4946
156. Varman, R.: Application of Nanomaterials for Electrochemical Devices (2022).https://doi.org/10.1201/9781003216308-9
157. Shearing, P.R., Johnson, L.: Toward Practical Demonstration of High-Energy-Density Batteries. Joule (2020).https://doi.org/10.1016/J.JOULE.2020.06.019
158. Hassan, A.I., Saleh, H.M.: Nanomaterials for the conversion of carbon dioxide into renewable fuels (2021). https://doi.org/10.1016/B978-0-323-90150-5.00005-4
159. Stephen, A.J., Rees, N.V., Mikheenko, I.P., Macaskie, L.E.: Platinum and palladium biosynthesized nanoparticles as sustainable fuel cell catalysts. Front. Energy Res. (2019). https://doi.org/10.3389/FENRG.2019.00066
160. Navinprasad, A.T., Prakash, R., Yoganandh, J., Meenakshipriya, B., Ramakrishnan, K.: Experimental Investigation of Nickel Oxide Nanomaterials with n-Pentane Diesel Blends in Compression Ignition Engine (2021). https://doi.org/10.1088/1757-899X/1070/1/012126
161. Epelle, E.I., Desongu, K.S., Obande, W., Adeleke, A.A., Ikubanni, P.P., Okolie, J.A., Gunes, B.: A comprehensive review of hydrogen production and storage: a focus on the role of nanomaterials. Int. J. Hydrogen Energy (2022). https://doi.org/10.1016/j.ijhydene.2022.04.227

Exploring the Mechanical Behavior of Nanomaterials: Elasticity, Strength, and Deformation for Biomedical Applications

Safaa Najah Saud Al-Humairi

Abstract This chapter comprehensively addressed nanomaterials and their mechanical characteristics, deformation processes, and uses in many different disciplines. With a particular focus on grain boundaries, dislocation dynamics, and grain size effects, the chapter examines new studies about how nanoscale features influence material behavior. Modern experimental characterization of nanomaterials and molecular dynamics simulations—has made it feasible to understand atomic-level mechanisms controlling nanomaterials' performance. With nanocrystalline microstructures, the chapter emphasizes significant developments in knowledge of fatigue resistance, fracture toughness, and strengthening processes in metals, ceramics, and composites. One of the main goals is to improve mechanical properties by using interfaces, twin barriers, and networks of nanoparticles. It also addresses the relevance of the acquired results in practical applications spanning structural materials to biomedical devices. We address how multiscale modeling approaches and computational methodologies forecast and maximize the behavior of nanomaterials. Integration of experimental data with several theoretical models generated discoveries in the creation of materials with tailored-made qualities for specific purposes. This thorough analysis confirms that, compared to traditional materials, nanoscale engineering creates opportunities for the creation of next-generation materials with improved strength, ductility, and durability.

Keywords Nanomaterials · Mechanical properties · Grain boundaries · Deformation mechanisms · Computational modeling

S. N. S. Al-Humairi (✉)
Faculty of Information Sciences and Engineering, Management and Science University, 40100 Shah Alam, Malaysia
e-mail: safaa_najah@msu.edu.my

S. N. Saud Al-Humairi et al. (eds.), *Smart Nanomaterials in Biomedical Applications*, Engineering Materials, https://doi.org/10.1007/978-981-96-8023-8_2

1 Introduction

In recent years, nanomaterials have played a significant role in developing established and emerging industries. Nevertheless, a number of researchers operating within applied and industrial contexts and the general public possess a comprehensive understanding of the history and potential—particularly that which remains unfulfilled—of nanomaterials and the associated technology [1]. The history of nanomaterials consists of microscopic observations and measurements. Following the invention of the electron microscope in 1931, it became possible to visualize structures just a few nanometers in size, which kickstarted the birth of the field of nanotechnology [2]. From a technological perspective, it is worth noting that microchips were independently developed after a landmark lecture in 1959, just three years later. Suppose nanotechnology is characterized as the technology and science concerned with controlling and manipulating structures on the atomic and molecular scale [3]. In that case, the foundation of this field can perhaps be traced back to the mid-1950s, following the discovery of diffraction by evanescent electromagnetic waves. Experiments on pyroelectric crystals and catalysis also underscore the foundations of the nanomaterials era. Nonetheless, it would be somewhat awkward to try to romanticize the idea of "early nanotechnology." [4]. Nanomaterials have revolutionized various scientific and technological domains with particularly profound implications for biomedical engineering. These materials exhibit a suite of physical and chemical properties markedly distinct from their macroscopic counterparts, primarily arising from their exceptionally high surface area-to-volume ratio and the manifestation of quantum size effects [5]. The expansive surface area of nanomaterials enhances their interaction with the surrounding environment, leading to improved catalytic activity, increased reactivity, and enhanced adsorption capabilities [6]. Furthermore, when materials are reduced to the nanoscale, quantum mechanical phenomena become dominant, resulting in alterations in their electronic, optical, and magnetic characteristics [7].

Nanomaterials are increasingly significant in various domains of contemporary biomedical sciences. They perform a vital function in facilitating numerous applications, including drug delivery systems, biomarkers for high-resolution bioimaging, sensitive diagnostic tests, and the fabrication of innovative implants and tissue engineering solutions constructs [8]. Nanomedicine, a direct derivative of the advances in nanotechnology in making, developing, and applying nanoscale tools and materials, is mainly meant for advancing disease diagnosis, treatment, and prevention at the basic cellular level of the human body [9]. The potential of understanding the structural and physicochemical property manipulations at such a minute scale makes it extremely valuable in addressing the entire field of biomedicine. The combination of nanotechnology and biomedical engineering, therefore, holds the promise of transforming approaches to treat medical conditions that were previously unable to be addressed by using nanomaterials' remarkably unique, scale-dependent properties [10]. It would enable designing medical solutions directly from the bottom up using nanoscale building blocks to interact with biological systems at their natural length scales. In addition to their well-known optical, electronic, and magnetic characteristics, nanomaterials possess unusual mechanical behaviors relevant to their role in biomedical applications. Nanomaterials may be stronger mechanically than their bulk forms, they may, for example, exhibit carbon nanotubes that display exceptional tensile strength. Moreover, the elasticity of nanomaterials might significantly differ from that of bulk materials and could often demonstrate a size-dependent behavior [11]. The nanomaterials can also show different deformation mechanisms under stress compared to those seen in macroscale materials, which are influenced by the increased surface stresses and the modified defect dynamics operating at a nanoscale. These atypical mechanical properties do not only present scholarly problems [12]. Still, they are also critical determinants of the performance and usefulness of nanomaterials in a wide variety of biomedical devices and therapies. For example, the transition from bulk to nanoscale will deviate from classical material behavior, allowing the engineering of materials that may possess completely new mechanical properties tailored for biological environments and functions [7]. This chapter focuses on the mechanical properties of nanomaterials, with special emphasis on elasticity, strength, and deformation, which are extremely important in biomedical applications. Applications that need flexibility and resilience, for example, wearing biosensors and nanoparticles intended for effective drug delivery, rely upon elasticity, the ability of a nanomaterial to undergo reversible deformation under stress. Strength, which operates on the principle that a material resists applied stresses without breaking, is therefore critical for load-bearing applications like orthopedic implants and tissue engineering scaffolds. Finally, understanding the deformation mechanisms in nanomaterials confronted with different load types is essential for biomedical devices' long-term reliability and performance. By exploring three essential mechanical characteristics, this chapter will highlight their relevance to developing this area of the biomedical engineering discipline.

2 Mechanical Properties of Nanomaterials

In recent years, characterizing the mechanical properties of nanomaterials has emerged as a key objective, as their distinct behaviors present opportunities unattainable with bulk materials. This chapter examines the mechanical properties of nanomaterials, emphasizing elasticity, strength, and deformation. First, the challenges and limitations of standard techniques for measuring the elastic moduli of nanomaterials are introduced. Then, an overview of recent advances in measuring the strength and deformation of nanomaterials is provided, concluding with an outlook on future developments. Nanomaterials have become an extraordinary area of research with the advent of nanotechnology. Numerous methods exist for creating these materials, but most involve downscaling bulk materials into nano-sized entities. Such materials are of great interest due to their unique chemical, physical, electronic, and mechanical properties [13]. Young's modulus, strength, and ductility are essential characteristics of nanomaterials for their integration into applications, particularly in biomedical and structural uses. The elastic properties play a significant role in the mechanobiology of cells [14]. Therefore, understanding Young's modulus, hardness, and the H/E* ratio of nanomaterials, which govern their deformation behavior, is crucial for developing advanced devices. However, measuring mechanical properties at the nanometer level is quite challenging because these properties can vary with changes in the local environment [15].

Significant efforts have been made to calculate the mechanical properties of nanomaterials. Computational models have been developed to determine the strength, hardness, stress–strain behavior, and Young's modulus of nanostructures [16]. These models provide valuable information, given the challenges and constraints of experimentation. Recent work features a concise introduction to computational methods that precedes the main results, which helps assess the accuracy of the calculations where applicable. With the emergence and success of computational simulation methods, a wide range of nanomaterials can be characterized. The lower size limit that can be examined experimentally is approximately 1 μm. Beyond this limit, traditional mechanical testing becomes increasingly challenging [17].

2.1 Elasticity Characteristic of Nanomaterials

Elasticity, in its fundamental sense, refers to the inherent ability of a material to deform under the influence of applied stress and subsequently return to its original shape once that stress is removed [18]. This reversible deformation directly results from the stretching or compression of the interatomic bonds within the material's structure. Young's modulus, a key quantitative measure of elasticity, describes explicitly a material's stiffness or resistance to deformation under uniaxial tension or compression. Young's higher modulus indicates a greater stiffness [19]. Nanomaterials hold a unique position in elasticity, making them an innovative design parameter for various biomedical applications. For instance, nanomaterials must exhibit sufficient elasticity in the development of flexible electronics, such as implantable biosensors [20], where mechanical effects like bending and stretching occur without breaking or losing functional integrity. The elastic properties of nanoparticles play significant roles in drug delivery systems, as they greatly influence propulsion through the circulatory system's complex and often tortuous routes and their effective interactions with biological cells [21]. The elastic response of nanomaterials determines their compatibility with the mechanical environment in dynamic biology, ensuring their functionality under physiological stress and strain conditions [22].

Elastic properties of nanomaterials often differ significantly from those of bulk materials due to the more significant influence of nanoscale phenomena. One such effect is that Young's modulus strongly depends on size. While this property is typically considered size-independent for bulk materials [23], nanomaterials—particularly those less than 50 nm in dimension—can enhance their elastic modulus by 1–50% due to various nanoscale effects. Notable among these effects are the increasing significance of bond length and energy changes, surface tension stress, and lattice contraction common in nanocrystals. Conversely, studies on one-dimensional nanostructures, such as nanowires and nanotubes, indicate that their elastic modulus decreases as their radial diameter increases [24]. This relationship highlights how the type and size of the nanomaterial affect its elastic properties. Another critical factor affecting the elastic properties of nanomaterials is the significant increase in the surface area-to-volume ratio at the nanoscale [25]. With this high ratio, surface energy and stress become the most relevant quantities that can substantially alter mechanical behavior, including elasticity. A new approach based on surface-energy density suggests that the elastic behavior of nanomaterials can be described using surface traction-dependent forces that do not consider any volumetric effects and rely solely on Eulerian surface-energy density [26]. This contrasts with previous concepts that claimed the surface layer behaved in a nearly linear elastic manner. Additionally, when surface effects are nonlinear at the nanoscale, some nanomaterials may exhibit nonlinearly elastic behavior, even if the corresponding bulk material is linearly elastic [27]. While quantum-size confinement is primarily known to affect optical and electronic properties, as the size of a material approaches the de Broglie wavelength of electrons, these quantum effects also indirectly influence the elastic behavior of nanomaterials [28]. The confinement of electrons in nanoscale dimensions alters the electronic band structure, which, in turn, modifies the strength and stiffness of interatomic bonds, ultimately affecting the material's elastic modulus. These size, surface, and quantum effects combine to produce complex and often nonintuitive elastic behavior in nanomaterials [29], necessitating careful consideration in the design and application of nanomaterials in biomedicine. Elasticity is one of the most critical features of nanomaterials for many vital biomedical applications [30]. In flexible biosensors, these devices need to adapt to biological surfaces and withstand the dynamic movements of the human body without failure. Flexible devices based on proteins, for instance, possess excellent electrical and mechanical properties, making them suitable candidates for biosensor production due to their inherent pressure sensitivity and insulating characteristics [31]. The use of organocomposites such as carbon nanotubes, graphene, and metallic nanoparticles in nanosensors demonstrates unmatched sensitivity in detecting minute amounts of biological and chemical analytes [32]. These nanomaterials' natural flexibility lets the sensors bend and flex without sacrificing their detecting power. Printing conductive polymers and nanoparticles onto biocompatible substrates results in flexible wireless bioelectronics that have been shown to detect physiological signals on the skin much more effectively than others, stressing the need for mechanical flexibility in skin-contact uses [33]. Although graphene's remarkable strength and flexibility have been well investigated, it is also under investigation for application in wearable piezoresistive sensors capable of detecting minute physiological changes like pulse waveforms [34]. Because of their low cost, simplicity of production, and adjustable electrical characteristics, metal-oxide nanoparticles are also attracting more and more attention as a means of developing flexible and wearable sensors [35]. In drug delivery uses, the capacity of nanoparticles to negotiate the biological environment and interact with cells depends critically on their flexibility. Research on blood circulation times, biodistribution patterns, efficacy of antibody-mediated targeting, and processes of cellular absorption—including endocytosis and phagocytosis—has revealed that the elasticity of nanoparticles is a fundamental determinant of these factors [20]. With more elastic nanoparticles (MONs100-HA and MONs200-HA) demonstrating better performance in these areas than semi-elastic and rigid variants [30], it was found that the elasticity of mesoporous organosilica nanoparticles (MONs-HA) considerably influences their cellular absorption, blood circulation, and tumor accumulation. The results imply that adjusting the elasticity of nanoparticles makes it feasible to increase their transport efficiency to tumors, suggesting that elasticity should be a significant factor in designing nanoplatforms for better therapeutic results.

In in vivo studies, softer nanoparticles with reduced elastic moduli typically show longer circulation durations and more efficacy in targeting than their harder counterparts [36]. Similar in vitro experiments indicate that softer nanoparticles are less likely to be phagocytosed by immune cells and other cell types, which might be advantageous in extending circulation in the bloodstream while lowering undesired off-target effects [37]. Moreover, elastic deformation is crucial as these nanomaterials must negotiate the small capillary limits and interact with internalizing cellular membranes [20]. Many examples show how the specifically designed elasticity of nanoparticles is used in biological applications. Synthesized to examine how elasticity affects drug delivery efficiency, polyethylene glycol (PEG)-based hydrogel nanoparticles with precisely regulated diameters and adjustable elastic moduli spanning from 0.255 to 3000 kPa have been developed [38]. Flexible electronic devices and biosensors are made from carbon nanotubes, which are well-known for their high tensile strength and natural flexibility. Promising for flexible and stretchy sensor uses is graphene, known for its extraordinary flexibility. Moreover, protein-based nanomaterials are under investigation for their natural biocompatibility and adaptability in several biomedical uses, including flexible sensor creation [30].

2.2 *Strength Characteristics of Nanomaterials*

Nanomaterials' strength and damage-free behavior under applied forces is critical in their suitability for various biomedical applications. The strength characteristics of nanomaterials are linked to their size, shape, material composition, and specific crystalline structure [39]. Notably, nanoparticles, particularly those with a nanocrystalline structure characterized by minimal dimensions of the grains, are often hard and wear-resistant compared to their bulk forms. Grain boundaries' high density in nanocrystalline materials usually produces obstacles to dislocation movement, which is the key carrier of plastic deformation, thereby heightening strength [40]. Additionally, incorporating nanoparticles into bulk nanostructured materials may refine grain structure and improve mechanical properties, including strength and toughness. The efficiency of this approach may depend on the uniform dispersion of nanoparticles within the matrix and the strength of the interfacial bonding between the nanoparticles and the neighboring material [41]. Carbon nanotubes are nanomaterials that are different from all other nanomaterials in terms of mechanical properties [42]. Their tensile strengths and moduli are notably remarkably higher than other nanomaterials production. These features enable carbon nanotubes to have an excellent strength-volume ratio across the entire density range, rendering them quite beneficial when weight-minimized applications are a significant concern. Their Young's modulus may be much greater than steel's, thus showing high stiffness and excellent resistance to deformation under mechanical load. The relatively high strength of carbon nanotubes is a consequence of the covalent bonds that structurally exist between the carbon atoms within their unique tubular structure [43]. This can also be found in some nanowires, which are considered one-dimensional nanomaterials, because they could also show size-dependence strength properties, with smaller diameter steel nanowires having higher strength than thicker ones due to their lower density of structural defects [44]. In addition to such electrical conductivity, metal nanowires could possess significant mechanical strength along their longitudinal axis.

Silicon nanowires are also high on research agendas for many applications in biomedicine, with their mechanical properties being critical for effective integration into devices and scaffolds. Graphene, the two-dimensional nanomaterial comprised of a single layer of carbon atoms arranged in a hexagonal lattice, is now well-known for its extremely high mechanical strength and Young's modulus, which make it one of the strongest materials known. Graphene's strength is the strong sp^2 hybridized carbon–carbon bonds that form its unique two-dimensional structure [45]. The diverse strength characteristics of nanomaterials make them suitable for a wide range of load-bearing biomedical applications, particularly in the realm of implants and tissue engineering scaffolds [46]. Nanomaterials are increasingly incorporated in orthopedic and dental implants to enhance biocompatibility and overall functionality. Surface coatings of titanium implants with nanoparticles such as titania and zirconia can improve the integration of the implant with the surrounding bone tissue (osseointegration) and enhance mechanical properties like hardness and resistance to wear [47]. Due to their increased hardness and wear resistance, nanocrystalline materials promise to improve the longevity of joint replacements. Carbon nanotubes are being explored as reinforcing agents within implant materials to enhance their strength and toughness without compromising their biocompatibility. Hydroxyapatite nanoparticles, which closely mimic the mineral composition of bone, are utilized in coatings and composite materials for implants to promote better bonding and integration with bone tissue.

Furthermore, graphene nanoparticles are being investigated for their potential to reinforce biodegradable magnesium alloys used in temporary implants like cardiovascular stents to improve their initial strength and control their degradation rate [48]. Tissue engineering scaffolds, which serve as three-dimensional substrates for cell growth and tissue regeneration, require mechanical properties that closely resemble those of the specific target tissue to provide adequate structural support and guide the formation of new tissue [49]. Nanofiber scaffolds, often fabricated from polymers, can mimic the structure of the natural extracellular matrix, and their mechanical properties, including strength, can be tailored by incorporating various nanomaterials. Carbon-based nanostructures, such as graphene oxide and carbon nanotubes, have demonstrated significant potential for enhancing the mechanical properties of biocomposite scaffolds due to their inherent high strength and the ease with which they can be functionalized [50]. Nanocellulose, a naturally derived nanomaterial, can reinforce scaffolds, providing strength and stiffness comparable to specific native tissues. The enhanced strength observed in many nanomaterials can be attributed to several factors related to their nanoscale structure. The probability of finding structural defects, such as dislocations and cracks [51], is significantly lower in the small volumes characteristic of nanomaterials compared to bulk materials. This reduced defect density often allows nanomaterials to approach their theoretical strength limits. In nanocomposite materials, incorporating nanoparticles can lead to a refinement of the grain structure within the matrix [52]. According to the Hall–Petch relationship in materials science, a smaller grain size typically increases strength and hardness by impeding the movement of dislocations across grain boundaries [53]. Additionally, materials like carbon nanotubes and graphene possess powerful covalent bonds within their atomic structures, directly contributing to their remarkable tensile strength and stiffness [54]. The high surface area of nanoparticles in composite materials also allows for a larger interfacial area, facilitating a more efficient transfer of stress between the nanoparticles and the surrounding matrix, ultimately leading to an enhancement of the overall strength of the composite. Several case studies highlight the practical application of high-strength nanomaterials in biomedical implants. Carbon nanotubes are being actively investigated to reinforce polymers used in orthopedic implants and create more potent and more durable devices capable of withstanding the substantial loads encountered within the body [55]. Nanocrystalline diamond and diamond-like carbon coatings are being explored for their extreme hardness and biocompatibility, offering a potential solution for creating highly wear-resistant coatings on metallic implants [56]. Graphene and its derivatives are being researched for their ability to reinforce biodegradable magnesium alloys used in temporary cardiovascular stents. The aim is to provide sufficient initial mechanical strength to support the vessel while allowing for controlled stent degradation over time.

2.3 Deformation Mechanisms in Nanomaterials

Nanomaterials, due to their smaller size and larger surface area, can experience deformation—changes in shape and size from applied forces—through a variety of processes that often differ significantly from those observed in bulk materials [57]. Although such traditional deformation mechanisms as dislocation, the movement of defects in the crystal lattice under applied stress, are present, their voice is muted in nanomaterials due to their tiny dimensions and/or the vicinity of surfaces and grain boundaries [58]. In nanocrystalline materials, the high-density grain borders can also move towards grain boundary sliding as a primary source of deformation with the sliding of individual grains past one another and the support of diffusion or dislocation activity along these boundaries [59]. For low-stacking fault energy materials, particularly smiling twins, which develop when there is a mirror reflection of the crystal structure across a given crystallographic plane, it can also be a significant deformation process initiated by stress [60]. Diffusion-related creep processes become important under a microscope because they involve atoms moving under stress and producing microscopic microdeformation that grows with time, especially at high temperatures [61]. Mechanical stress can cause some kinds of nanomaterials to phase shift, therefore altering their shape and material properties—usually depending on their chemical compositions or crystal structure. Hence, in the biomedical use's framework, nanoparticles' deformation behavior might be relevant to the device and treatments' lifetime performance and dependability [62]. Thus, foretelling lifespan and structural integrity, for example, in load-bearing implants or tissue engineering scaffolds, depends on a solid knowledge of how nanomaterial components deform under physiological stresses [22]. While variations in drug release kinetics for drug-eluting devices may arise, the most significant and uncontrolled deformation might cause malfunctioning or mechanical support loss. Conversely, developing nanoparticles with particular, predictable deformation behaviors might enhance performance in applications such as self-tying surgical sutures that deform confidently in response to temperature-triggered deformation of shape memory polymers.

These include several deformation modes rarely seen or non-existent in their bulk counterparts. Such is the case of rotational deformation, reported, for example, in the case of nanocrystalline materials, wherein the application of mechanical load can lead to or create disclination dipoles such as and coupled, causing individual crystallites to fragment and to rotate at the nanometer scale [63]. Usually, the onset and course of plastic deformation in a nanowire are notably size-dependent together with surface stresses and limited availability of defects, thus making them strong determinants of nanomaterials' mechanical response. Furthermore, the emergence of smart nanomaterials has introduced the concept of intelligent deformation, where these materials can undergo reversible changes in shape, size, or other properties in response to external stimuli such as temperature, light, or pH, thus opening novel possibilities for biomedical applications like targeted drug release and self-actuating medical devices [64]. Additionally, interactions with the biological environment, such as the adsorption of proteins onto the surface of nanomaterials, can induce subtle deformations that may affect their overall behavior [65]. Examples of nanomaterials specifically designed for deformation behavior are shape memory polymers of polyurethane for self-knotting surgical sutures, which can be tightened automatically upon exposure to the body temperature and thus facilitate surgical operations [66]. Nanoparticles with these precisely designed elastic properties are made to deform such that they optimally traverse biological barriers toward enhancing uptake by target cells during drug delivery. Wafer-thin nanoparticles may also be made to actuate in response to a stimulus, such as loading and unloading drug molecules for controlled release.

2.3.1 Elastic Deformation

Materials deform under the application of forces or moments. Typically, materials return to their original shape after being deformed by an external force moving away from them. The deformation is generally elastic, and the elastic state is recoverable and reversible up to a primal condition known as the before-start condition [67]. Understanding elastic deformation and structural stability at the microscale is crucial in many fields, including materials science, engineering, physics, and biology [68]. However, due to their minuscule size, some nanomaterials exhibit unique mechanical deformations beyond elastic [39]. Recent advances in nanotechnology allow us to utilize elastic strain engineering to control the physicochemical properties of hard materials by applying externally imposed elastic strains. It is essential to suppress the occurrence of inelastic deformations and failures, which requires fundamental knowledge of fracture behavior at the nanoscale [69]. Any significant impairment of structural integrity could lead to equipment failures, traffic accidents, or severe health issues, which can be catastrophic. Hard nanomaterials are used as antimicrobial agents and applied to the surfaces of medical instruments and household items. However, it is also noted that there may be potential cytotoxic effects, adverse tissue responses, and cellular dysfunction resulting from using hard nanomaterials in various applications or accidental exposure exposures [70]. The mechanistic understanding of the toxicity of nanoparticles and the mechanisms through which they might injure biological targets is still in its infancy. Understanding the fundamental concepts of nanocomposites' structural deformation and stability responses on these vital life science issues is of paramount interest. These necessitate the study of the elasticity, strength, and deformation behavior of nanomaterials, including biofilms, bacteriophages, and soft nanocomposites [71]. No prominent commercial device or product uses for hard materials with sizes below 200 nm were identified. Analyzing the side effects of these nanoscale hard materials and identifying key components that influence their properties provides knowledge that could open new markets or extend current markets [72]. Bian et al. [73] study investigates the behavior of single-crystalline copper nanoparticles when compressed, as shown in Fig. 1. Their work explored how these tiny particles deform, highlighting that the deformation characteristics vary with the crystallographic orientation of the applied load. It examined three main compression directions: [111], [110], and [112]. They discovered the elastic response in each direction—the way the nanoparticles react initially before any permanent change [73].

2.3.2 Plastic Deformation

Plastic deformation (PD) of a material is an irreversible change in the shape and dimensions of a body due to the effect of external forces. In most typical cases, material deformation, that is, its transition into a state beyond the elastic limit, occurs as the body deforms. In the most straightforward cases, the deformation is modeled as proportional to the intensity of the internal forces, which ultimately create tensional or compressive stress. Over time, through various processes and under specific conditions, the material is slowly loaded, possibly persistently deformed only elastically, for an extended period. When the subsequently applied tensile force exceeds the value of elastic strength, the body experiences both elastic and plastic deformation.

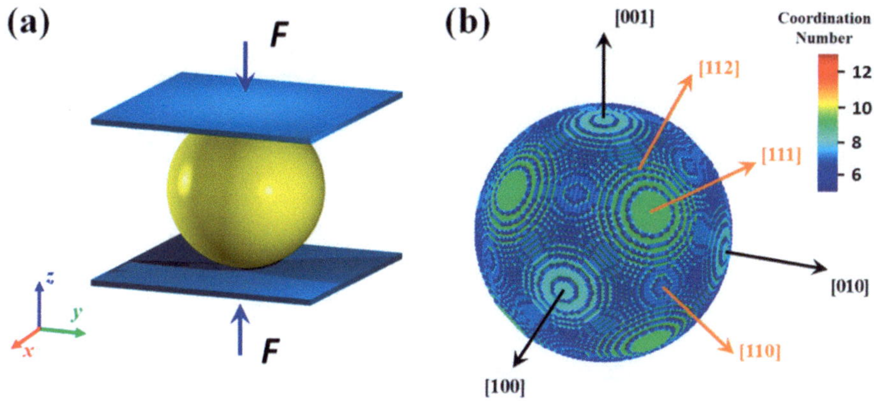

Fig. 1 **a** Schematic of the uniaxial compression of a nanoparticle, and **b** initial configuration of a copper nanoparticle (surface atoms are colored by their coordination number) [73]

The plastic deformation proceeds, but only a small part of the accumulated plastic strain remains irreversibly stored in the neck of the sample, in the so-called work hardening zone that progresses according to Wick's principle [74].

Within the framework of theoretical mechanics related to deformable solids, the study of plastic deformation also considers the material's internal structure. The processes involved in the formation of dislocations under stress are proposed. The impact of dislocations and other defects within the crystal lattice on the mechanical properties of the metal is demonstrated. The relationship between the positioning of atomic defects and the emergence of nascent (effective or atomic) microlevels of plasticity (macrohardness) in a material, influenced by various factors, is established, allowing the mechanical characteristics of the material to be determined by the spectrum of dislocation sections. The elastic nature of structural defect accumulation processes is examined, and the Belly-John mechanism for the transition of plastic deformation in the crystal lattice is introduced. Analytical dependencies are derived for the parameters governing the power law of the accumulation of microgroups of dislocation density in irradiated metal [75]. At the nanoscale, the plastic deformation of nanomaterials is governed by numerous physical processes and mechanisms that may differ significantly from those in macroscopic materials. Consequently, theoretical frameworks have been developed to model these processes and predict material behavior under applied stress [12]. Continuum mechanics has been crucial in establishing a macroscopic understanding of material deformation. It has been shown that various complex deformation behaviors can be described using simple mathematical models, enabling predictions of stress–strain relationships and macroscopic deformation for multiple materials and loading conditions [76]. Continuum models rely on averaging procedures over many atomic or microstructural interactions, resulting in constitutive relationships between stress and strain tensors [77]. However, these models often involve ad hoc assumptions that may not hold at the nanoscale, limiting their direct applicability in predicting behavior at this level. While continuum models

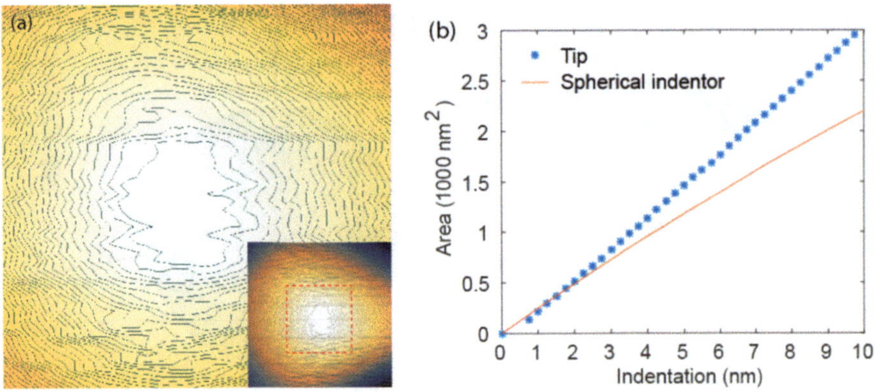

Fig. 2 **a** Atomic force microscopy (AFM) height image of the apex of the sharp tip with contour lines, height step 0.25 nm, obtained by scanning on TGT1 and **b** Area of the tip contours at 0–10 nm tip height determined from AFM image, compared with expected contour area from a spherical indenter with the effective tip radius calibrated on FS (40 nm) [79]

are not typically used for making detailed predictions of nanoscale deformation and stress, they are a helpful starting point for developing and validating models with fewer ad hoc assumptions [78]. High-frequency mechanical mapping benefits speed and low noise levels compared to classical force volume mapping, Zeng et al. [79] show the relevance of acquisition frequency in atomic force microscopy (AFM) for accurate elasticity mapping of rigid nanomaterials, as shown in Fig. 2. The measured elasticity was much influenced by loading force; moreover, the geometry of the AFM tip is quite vital since blunt tips (200 nm radius) showed a deviation of up to 13% from nominal values. By contrast, sharp points (40 nm radius) showed more variation of 121%, primarily from plastic deformation [79].

Generally, most deformation in metal forming occurs at the macroscopic scale, several orders of magnitude larger than the typical crystallographic size. Continuum mechanics provides a framework for treating materials as continuous media. In classical continuum mechanics, materials are described by macroscopic stresses and strains, while microscopic structures are overlooked [80]. This approach implies that quantities such as strains and stresses are average values, neglecting atomic-scale variations of these quantities. Dislocations play a critical role in the plastic properties of metals. The stress increments (resolved stress) required to move these line defects arise from the applied stress, the current macroscopic stress state, and the underlying material [81]. In simple material behavior, stress increments are a linear function of the applied stress state, with the proportionality represented by a material constant known as the Schmid factor. For example, dislocations in a cubic material move under resolved shear stress on one of their family planes. In conventional microstructured materials (with grains or small inclusions, a mean representation for the microstructure is typical, allowing the macroscopic stress–strain response to be predicted. However, at smaller length scales, it becomes essential to predict mechanical behavior without ignoring the physical structure. There is some ambiguity regarding the term "continuum mechanics" in contexts where scales are separated from the microscopic by several orders of magnitude. In this context, classical continuum mechanics provides a statistical description of the material, that is, macroscopic stress and strain are treated as mean values [82]. It is well recognized that such a treatment is justified for describing and predicting mechanical behavior at length scales substantially more significant than those of the microstructure [83]. Classical continuum mechanics can be applied efficiently and accurately at these larger scales [84]. Nevertheless, the macroscopic approach yields a problematic description of materials at smaller length scales, where mechanical behavior is highly sensitive to the microstructure and cannot be captured by conventional continuum models. Consequently, an important and active area of current research focuses on developing models that account for scale effects and microstructural complexities [85].

Molecular dynamics (MD) simulations involve the motion of thousands to over a million atoms simultaneously, with particles interacting through a given interatomic potential while solving the equations of motion for a relatively short time interval [86]. Observables can then be calculated from the atoms, such as the stress tensor, elastic modulus, dislocation structures, and atomic trajectories. Although this is a powerful tool for investigating the fundamental atomic-level interactions and mechanisms behind the plastic deformation of materials, limitations currently restrict simulations to length scales of only a few nanometers. Consequently, there has been considerable interest in developing models for the bulk mechanical properties of nanomaterials that interface with data sets, enabling experimental macroscopic data on deformation to enhance understanding and validate comparisons with other models. According to Li et al. [87], grain size effects depend on strain rates in the deformation behavior of nano-polycrystalline Al/Mg layered composites. Larger grains produce lower yield stress at low strain rates, but they give rise to high tensile strength at very high rates. An optimal thickness of the intermediate intermetallic compounds (IMCs) layer (Fig. 3) improves the composite strength. Yet, excessive thickness results in lower tensile strength as fewer boundaries are available for dislocations, particularly at high strain rates when grains cannot adapt quickly to large strain conditions [87].

Atomistic simulations play a key role in studying plastic deformation and the structural evolution of deformed materials at the atomic level [88]. Molecular dynamics simulations have been the most intensive atomistic method for investigating such problems. With a sufficiently detailed and accurate many-body potential, molecular dynamics enables the time evolution of the atoms in a system to be tracked under the influence of externally applied forces. Outputs from molecular dynamics simulations include time-dependent trajectories of trapped atoms and various local and global quantities that characterize the system's structure and mechanics [89]. Mechanical properties are then analyzed concerning multiple parameters of interest, such as temperature, strain state, strain rate, and the influence of grain boundaries or other defects, allowing optimal mechanical properties for specific applications or processing needs [90]. Milestones, along with sound and sensible projects, ensure the continued success of this enterprise. Simulations revealed that graphene sheets exhibited the maximum encapsulation effect in the flower-like shape. As a result,

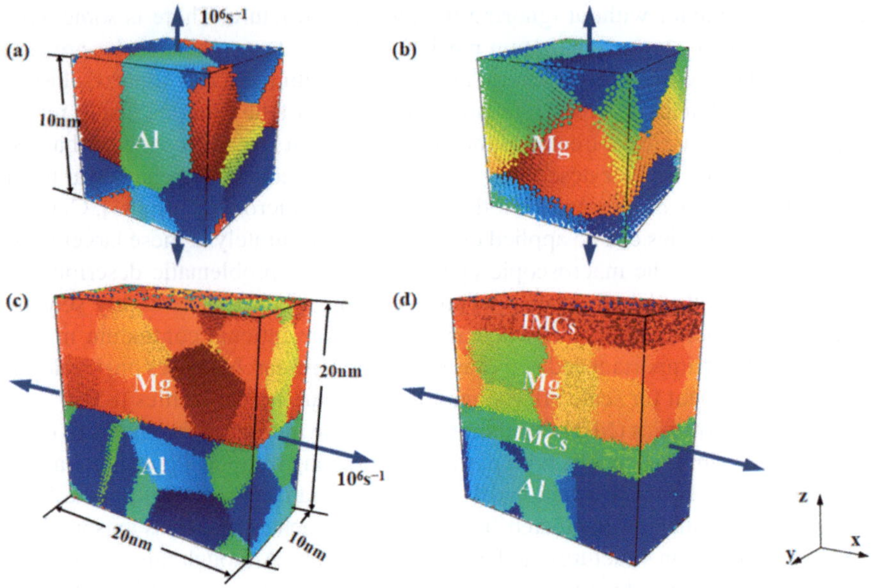

Fig. 3 Simulation models and the loading method for **a** polycrystalline Al, **b** polycrystalline Mg, **c** Al/Mg layered composites with no IMCs interlayer, and **d** Al/Mg layered composites with IMCs interlayer [87]

the compact secondary structure of the dimeric protein was disrupted. The cofactors and the active residues moved away from the biologically relevant orientations, and the catalytic activity of the protein was damaged. Results highlighted that even if immobilized correctly in the presence of the graphene sheets, the protein will be perturbed, and its capability of indirectly measuring glucose from a blood sample would be damaged. The interaction energies between the graphene sheet and the charged amino acids were calculated using the DFT method. We envisage that the insights provided by this study would provide additional knowledge when developing novel graphene–enzyme biosensors [91].

The strength and damage of nanomaterials can only be understood by assessing how complex interactions among the atoms determine the time evolution of a deforming material. Such processes are best studied with time-dependent many-body simulations, with molecular dynamics being the most widely used. Among the time-dependent material behaviors accessible through molecular dynamics, attention is focused on those involving the material's microscopic temperature, velocity, and stress fields. The time-dependent field variables employed here provide more detail on the spall process and allow for the analysis of permanent deformation and damage formation.

2.3.3 Mechanisms of Deformation in Nanomaterials

i. **Dislocation Movement**

Plastic deformation deals with the deformation process that permanently alters a material's shape. Energy is needed to overcome the energy required to cause this change, which is extensive in the case of low melting point materials. The dislocations grow, multiply, react, and form new grains at the nanoscale. However, most physical properties of bulk and nanomaterials are different due to the size effect. It is well known that many nano properties are superior to those of bulk materials. Understanding the movement of dislocations is fundamental to grasping plastic deformation; dislocations were initially identified through the careless etching of giant pairs in early microscopy [92]. Thus, plastic deformation primarily involves the movement of dislocations, and numerous reports have been published on this topic. Many models of dislocation movement have been proposed to date. The movement of dislocations is influenced by applied stress, temperature, crystal structure, solute, alloy composition, and grain size [93]. Li et al. [94] propose that grain boundaries (GBs) (shown in Fig. 4) act as nanomaterials' primary source of dislocations, replacing the more conventionally recognized FR sources. They observed that in nanocrystalline (NC) metals, especially those with grain sizes below 30 nm, dislocation emissions occur predominantly from GB ledges rather than from dislocation pile-ups observed in larger grains [94].

Furthermore, dislocations increase and interact in various ways when a material undergoes plastic deformation. Nanomaterials have several unique characteristics, including size and surface effects; the surface effect is particularly dominant at the nanoscale, making the PD model applicable for defining nano deformation. Although nanomaterials exhibit excellent mechanical properties, the contradiction observed in experiments—namely, the bulk-like strength of nanopillars—has not yet

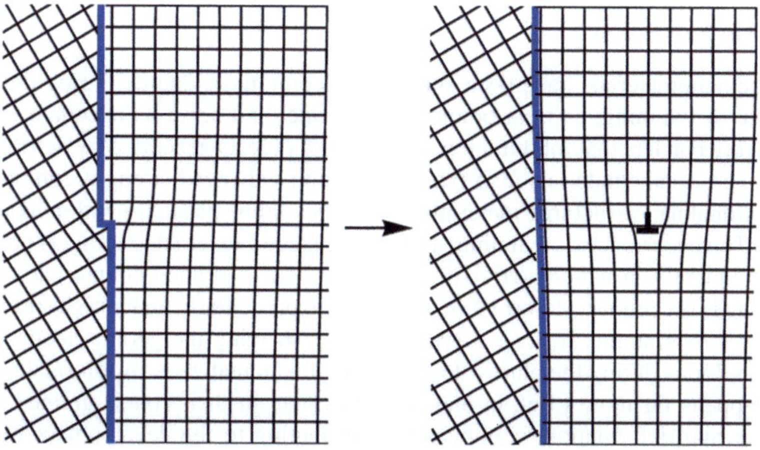

Fig. 4 Grain boundary ledge model [94]

been explained in plastic terms, leaving room for further discussion to explore nano deformation [95].

ii. **Grain Boundary Sliding**

During plastic deformation, materials change shape due to the relative motion of microstructural elements. This process can lead to various distinct phenomena in solids across different length scales [96], and it is essential to identify the fundamental mechanisms critical to nanomaterials, as shown in Table 1. The ways in which individual atoms or molecules interact and move on a nanometer scale open up an entirely new dimension for material properties. Grain boundary sliding has been proposed as a mechanism contributing to nanomaterials' plastic deformation [97]. It refers to the relative motion of grains at their boundaries under externally applied stress. Numerous controversial issues persist in our understanding of this phenomenon in crystalline materials due to variability in temperature, stress level, purity, dislocation density, and ambiguity in identification techniques [98]. The significance of grain size, boundary characteristics, and temperature on this mechanism is well established. A crucial aspect in this regard is the discovery that physical properties can differ dramatically for samples in nanoparticle form, especially when the average particle size is sufficiently small to fall below a critical threshold [99]. Observing enhanced sliding phenomena due to small grain size suggests that this mechanism may play an essential role in the deformation of nanostructured materials. It is generally accepted that microstructural characteristics influence various deformation modes in polycrystalline materials. It has been suggested that grain boundary sliding and other deformation modes coexist in nanostructured materials, with experimental evidence indicating that this mechanism is one of the possible contributors to the plastic behavior of nanomaterials [100]. Experimentally observed features of this phenomenon, including temperature-dependent flow stress and strain rate sensitivity, have been reported and share similarities with observations of this mechanism in polymeric materials.

iii. **Twin Formation**

Metallic nanomaterials have been shown to become more ductile during deformation due to twinning, as twin boundaries migrate, and serve as adequate energy-dissipating barriers that inhibit the propagation of defects generated by deformation [111]. These nanomaterials also typically exhibit enhanced strength compared to conventional nanostructured materials since they confine plasticity within nanoscale grains. The occurrence of twinning, as a direct result of external loading, is observed more frequently in nanomaterials than in their coarse-grained counterparts [112]. Twinning is the primary deformation mechanism, particularly under high strain rates and low temperatures. It plays a crucial role in meeting the geometrically necessary conditions during the plastic deformation of nanomaterials. Additionally, twinning happens far more often in nanomaterials than in their coarse-grained equivalents during deformation, significantly contributing to the development of more favorable nanoscale microstructures within these materials [113]. Gaining a deeper

Table 1 Advancements and applications of grain boundary sliding in nanomaterials

References	Findings	Challenges	Contributions	Methods Used
Atul [101]	Grain boundaries strengthen materials at low temps, weaken them at high temps via sliding/creep	Maintaining material strength across temps; limited superplasticity studies in fine-grain alloys	Shows how grain boundaries affect strength and strain (40–80% from sliding in fine grains)	Not specified in detail
Qishan et al. [102]	Grain boundaries adjust via twinning, boosting mobility in nanocrystals under load	Understanding dynamic grain boundary changes and twinning under load	Reveals twinning enhances grain boundary mobility and plasticity in nanocrystals	In situ nanomechanical testing, molecular dynamics simulations
Singh et al. [103]	Grain boundary sliding boosts ductility in Mg–0.6Ca alloy, more at low strain rates	Not specified beyond strain rate sensitivity and sliding relation	Shows sliding increases ductility, strain rises as rate drops in Mg–0.6Ca alloy	In situ analysis of strain rate effects
Shufen et al. [104]	Dislocation climb in gold occurs via two-column reconstruction, not single-column change	Understanding atomic-scale climb; revising old climb theories	New climb mechanism found, suggests ways to control nanostructured metal properties	In situ atomic-scale observation, Monte Carlo simulations
Ian and Yuri [105]	Grain boundary diffusion is intermittent, tied to system size and defect avalanches	Explaining diffusion at mid-temps; measuring full avalanche scale	Links diffusion behavior to defect avalanches, improves atomic transport understanding	Molecular dynamics simulations (LAMMPS), OVITO visualization
Zheng et al. [106]	Grain boundaries in gold transform reversibly under shear, tied to steps/dislocations	Unclear atomic-scale changes during stress-driven migration	Observes reversible boundary changes and dissociation under load in gold nanocrystals	In situ high-res TEM, molecular dynamics simulations
Qi et al. [107]	Low-angle grain boundaries in bicrystals enable reversible plasticity via dislocations	Preventing irreversible damage in nanodevices; complex boundary engineering	Introduces boundary design for reversible plasticity, reduces damage in bicrystals	In situ nanomechanical testing, atomistic simulations

(continued)

Table 1 (continued)

References	Findings	Challenges	Contributions	Methods Used
Hyeon et al. [108]	Grain boundary sliding in Fe-6.6Mn–2.3Al steel starts at 0.69 strain, uses dislocation process	Understanding transition to sliding at critical strain; complex accommodation	Identifies sliding onset and process, preserves grain shape in steel	High-temp tensile tests, microstructural observation
Anping et al. [109]	Grain boundary migration reversibility in copper depends on grain size, shifts at threshold	Explaining size effects; modeling irreversible grooving with size increase	Shows size-driven migration shift, refines model for cyclic deformability	Molecular dynamics simulations, refined disconnection model
Andrew et al. [110]	Imperfect attachment in gold nanoparticles aligns via dislocation glide, forms single crystal	Unclear attachment pathways; revising energy models for nanoscale boundaries	Explains imperfect attachment via dislocation quantization, aids alignment in nanoparticles	Molecular dynamics simulations, in situ TEM observation

understanding of the link between twin formation and the overall mechanical performance of nanostructured materials is essential for their continued advancement and application [114].

Under different loading conditions, there is clear experimental evidence of twin formation in various nanomaterials, including ultrafine-grained metals, nanograined alloys, and nanoscale granular films [115]. Research indicates that twin boundaries in nanomaterials can significantly influence their mechanical responses and future designs. Consequently, extensive studies have focused on the formation and behavior of twin boundaries, especially in FCC metals and alloys with controlled stacking fault energy, under various deformation conditions [116]. With recent advancements in experimental techniques and methodologies, exploring twin formation and its effects within metallic nanoparticles and nano-composite materials is particularly compelling. Among all plastic deformation methods, in situ, high-resolution transmission electron microscopy observations of twin boundary migration in nanoparticles are highly sought after. In FCC materials with high stacking fault energy, such as copper and nickel single crystals, complete dislocations are often reported as precursors to the formation of multiple twins [117]. Dislocations nucleate near the center of a free surface or surface-triple junction, gliding to the surface–triple junction where their intersections annihilate, creating an immobile Lomer-Cottrell sessile dislocation [118].

2.4 Fracture Mechanics of Nanomaterials

Fracture constitutes the loss of load-bearing capacity resulting from the propagation of atomic-scale cracks within nanostructured materials. Upon initial examination, the fracture may present one of the most straightforward mechanical challenges, as it is inherently binary; a material either forfeits its load-bearing capability in one state or regains it in another [119]. Indeed, fracture mechanisms, including cavitation and decohesion in well-bonded materials, can exhibit greater complexity and incur higher computational costs. Nevertheless, drawing from the foundational concepts of continuum fracture mechanics, the characterization of material failure in fundamental terms of' what breaks where' remains steadfast [120]. Numerous transitions from macro to nano can be identified within linear elasticity, plasticity, damage mechanics, and fatigue across various material combinations [121]. In these contexts, while the complexities of constitutive modeling and data acquisition for specific material systems are circumvented, the correlation of applied stress or strain scenarios to a safety measure is critical. Fracture signifies the loss of load-carrying capacity due to the propagation of a crack that achieves a geometrically sharp configuration over a characteristic length [122]. For the characterization of such structures under deformation, infinitely differentiable stress and strain fields are conventionally defined. Within the classical framework of linear elasticity, the continuity of second-order stresses across the region of interest is similarly assumed [123]. The concept of a crack was implicitly omitted from micromechanical descriptions, and corresponding material parameters, including Young's modulus, Poisson's ratio, and stress concentration factors, were derived uniformly from far-field loading conditions [124]. This methodology proves effective in numerous practical mechanical applications where the dimensions of cracks and notches are smaller than those of the microstructural size.

Nanotwinned and nanolayered materials have improved fracture resistance owing to their hierarchical interfaces, which may be tuned by adjusting microstructural parameters such as twin bundle size and loading orientation for twin borders [125]. Using nanomaterials, such as nano-silica particles and carbon nanotubes, markedly enhances the fracture toughness of cementitious materials, particularly in severe situations like early-age frost attacks. On the other hand, Jiang and Li highlighted that these nanoparticles improve micro and macro fracture characteristics by fortifying the interfacial transition zones and altering the microstructure [126]. Nanostructured metals and alloys, defined by nanoscale grain dimensions or twin thicknesses, exhibit remarkable strength and distinctive deformation processes. Challenges such as inhibited dislocation slip and grain-boundary instability can influence fatigue and fracture characteristics, requiring microstructural engineering approaches such as gradient grain sizes and regulated border mobility to improve toughness [127]. Advanced imaging techniques, including in situ scanning transmission electron microscopy, have yielded atomic-scale insights into fracture mechanisms, elucidating brittle fracture patterns and crack healing in materials such as 2D rhenium disulfide [119]. The creation of lightweight nano-Bouligand materials with regulated nanostructures demonstrates that nanoscale heterogeneity and size-induced ductility may substantially boost toughness [128]. Carbon nanotube bundles, albeit robust individually, demonstrate diminished strength when aggregated, with fracture patterns affected by tube diameter and chirality, underscoring the intricacies of scaling these materials [129]. The combined impacts of carbon nanofibers and short carbon fibers in composites boost mechanical and fracture characteristics, resulting in notable improvements in tensile, flexural, and impact strength [130].

2.4.1 Fracture Toughness

The smallest viable samples that can be produced and subjected to mechanical testing considerably augment the understanding of how imperfections influence the properties of a material's strength. This influence involves the likelihood of observing a pre-existing flaw at a given stress and the sensitivity of strength to the presence of such a flaw. Simple scaling arguments based on the Griffith criterion suggest that smaller samples should exhibit higher strength [131], however, too small samples will not display size-dependent strength because intrinsic material length scales will completely dominate them. This size-dependent behavior is predicted only by considering extreme-value statistics [85]. A focused ion beam fabricates large, medium, and small single-crystal aluminum tensile specimens, with small samples as tiny as 3 μm in cross-section.

The tensile strength of perfect aluminum increases with decreasing sample size, aligning with previous single-crystal measurements and simulations. It is evident that the probability of accurately representing the material with small samples is low due to an upper bound determined by a sample characteristic length scale related to the machined notches; increasing this length scale through a different geometry or preparation method would negate the advantages of utilizing smaller samples [132]. In a regime where a sample does not represent the material, the size effect on sample strength is absent. These new results provide a broader context for understanding other seemingly contradictory findings and offer significant guidelines for future investigations and applications in submicron or small-scale material research and development efforts, both in experiments and simulations. Bortz et al. [133] synthesized graphene by opening helical carbon nanofibers, forming graphene nanoribbons, which were subsequently oxidized to produce graphene oxide (GO). The tensile strength and modulus were reported to be 73.57 MPa and 3.32 GPa at 0.5 wt% and 0.1 wt% GO, respectively. The fracture toughness (KIC) and the critical stress intensity factor (GIC) exhibited 62% and 110% improvements, respectively, for 1 wt% GO/ epoxy composites compared to neat epoxy [133]. Figure 5 represents toughening mechanisms in epoxy reinforced with 2D fillers. The predominant toughening mechanisms vary depending on the filler added. To date, a number of fillers have been explored for reinforcing epoxy, with a focus on enhancing its fracture properties.

Murugan et al. examined the rGO incorporation in OPC cement paste and other nanomaterials and discovered that rGOs resulted in pore refinement, which helped improve the mix's durability. However, rGO and GO were not compared. Due to better

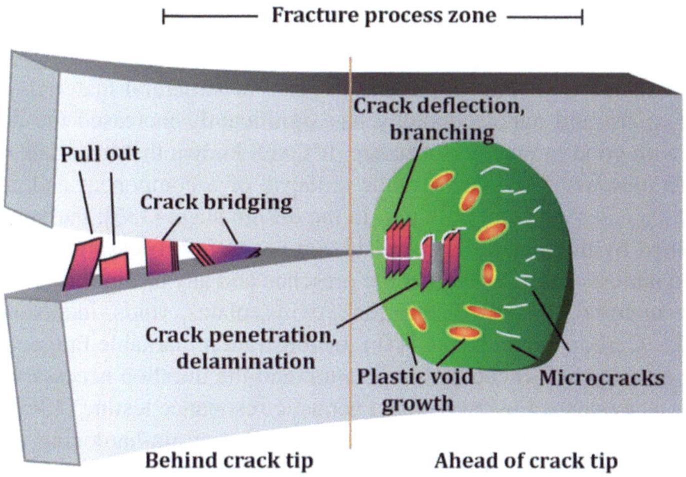

Fig. 5 Toughness mechanism in the composite of epoxy that 2D nanomaterials have reinforced [133]

and greater C–S–H crystal growth in both GO-cement (due to increased water adsorption sites) and rGO-cement (increased surface area for C–S–H growth), cracks can be deflected very early on and do not propagate, also making the matrix much more tortuous and harder for fluid penetration, hence improving durability, as simplified in Fig. 6 [134].

Fig. 6 Illustrating how nano-reinforcement encourages crack deflection and stops crack propagation through the structure

2.4.2 Fatigue Resistance

Fatigue resistance is another important property of structural materials. The rapid growth of micro and nanotechnology has significantly increased the demand for materials with good fatigue performance. It's well known that materials with better fatigue life may avoid the catastrophic collapse of a component and can significantly reduce environmental pollution in the design stage [135]. Fatigue resistance is determined by the material's chemical composition, processing method, and its micro- and nanostructure, as well as the presence and any change of various defects within the material, such as dislocations, twins, plates, voids, inclusions, cracks, and the like. Carbon nanotubes (CNTs) demonstrate remarkable fatigue resistance, with their efficacy influenced by temperature and the duration necessary for defect formation, as evidenced by noncontact acoustic resonance testing [136]. Likewise, metallic glass-based nanomaterials exhibit superior strain-hardening capabilities, which enhance their fatigue resistance by establishing localized shear transformation zones that increase material density under cyclic loading [137]. Reinforced with elastomeric nanoparticles, epoxy-based nanocomposites also reveal heightened fatigue resistance, significantly improving crack propagation resistance compared to pure epoxy resin [138]. Nanostructured metals and alloys, including nanotwinned high-entropy alloy films [139], benefit from distinctive deformation mechanisms that stabilize twin boundaries and maintain slip systems, culminating in exceptional fatigue resistance under cyclic loading on the coarse-grained HEA films with different impact energy, as shown in Fig. 7.

Furthermore, surface nanolaminates in gradient pure titanium markedly delay crack nucleation and propagation, providing a scalable approach for developing fatigue-resistant alloys [140]. Combining mechanical training and nanocrystallization in NiTi shape memory alloys, [141] found, enhances fatigue life by diminishing residual strain and modifying crack growth behaviors [141]. Wan et al. [142] found that the bioinspired graphene-based nanocomposites exploit synergistic interfacial

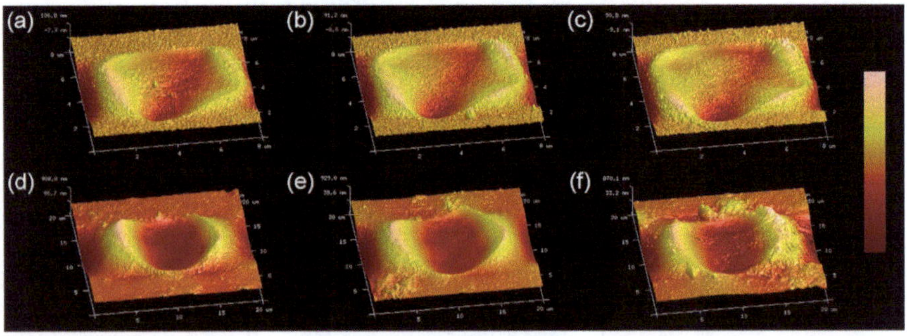

Fig. 7 AFM micrographs obtained after 103 indentation cycles on the nanotwinned HEA film with an impact energy of **a** 150 nJ, **b** 187.5 nJ, and **c** 225 nJ (size:8 μm × 8 μm), and on coarse-grained HEA films with an impact energy of **d** 150 nJ, **e** 187.5 nJ, and **f** 225 nJ (size: 20 μm × 20 μm) [139]

interactions to mitigate crack propagation, achieving remarkably high fatigue life while preserving electrical conductivity [142]. In summary, the fatigue resistance of nanomaterials is augmented through various methodologies, including microstructural engineering, mechanical training, and the integration of reinforcing phases, collectively improving their performance under cyclic stresses [143]. These developments underscore the potential of nanomaterials in applications necessitating high durability and resilience under repeated mechanical loading [144].

3 Factors Influencing the Mechanical Properties of Nanomaterials

3.1 Size and Dimensionality

Nanomaterials are of great interest to scientists and engineers due to the increasing demand for materials with superior characteristics and performance capabilities in emerging fields. Compared to their bulk counterparts, novel properties can be observed in nanostructured materials, such as enhanced mechanical properties, high strength-to-weight ratios, and exceptional thermal stability. The ability to fine-tune the microstructures of these materials allows for optimization of their mechanical behavior for specific applications [145]. Despite numerous investigations conducted over the past two decades, a complete understanding of the size-dependent mechanical behavior of nanomaterials remains elusive, motivating further research in this area.

Numerous factors can affect the mechanical deformation of a nanomaterial. The most influential ones can be categorized into size and dimensionality effects, surface effects, and internal/external boundary effects [39]. These effects are not independent, in many cases, they depend on and are interconnected with each other to some degree. In thin-film nanomaterials, grain size significantly influences mechanical deformation. As grain size decreases, the Hall–Petch equation predicts an increase in material strength. It is widely accepted that the mechanical behavior of nanostructured thin films is governed by a grain-boundary-mediated process [146].

Furthermore, the microstructure of thin-film nanomaterials differs from that of their bulk forms. Mismatches in the elastic modulus and thermal expansion coefficient often occur at the grain boundaries of thin films. In nanostructured thin films, both dislocation nucleation and subsequent expansion are inhibited by the presence of grain boundaries [147]. Simulations show that the microstructures between amorphous zones and single-crystal-like nanowires hinder efficient S-PSB nucleation when partial dislocations impact the interface [148]. While theoretical predictions remain debated, they cannot fully explain the mechanical experimental data of nanofilms. The mechanical deformation of nanostructured thin-film systems likely depends on a complex and still unpredictable interplay of microstructural elements and specific material properties.

The Hall–Petch relationship states that the yield strength of a material is directly proportional to the square root of its average grain size [149]. It takes the analytical form $\sigma = \sigma 0 + kd\text{-}1/2$, where σ is the yield/tensile strength, $\sigma 0$, k, and α are all material properties, and k is a constant; and d is the mean diameter of each grain in the material. The Hall–Petch relationship describes the hardening of polycrystalline materials as a function of their grain size. As a rule of thumb, and all else being equal, smaller grains will lead to higher strength [146].

There are, however, certain limitations to this concept. Initially, at substantial grain sizes, the grain boundary is considered to have a weak or negligible effect on strength, and therefore, the Hall–Petch relationship will not apply. This is why small grains are strong: grain boundaries and triple junctions will pin or push dislocations in large grain materials to create larger offsets [150]. As with all mathematical models and theories that are not essentially universal laws, there will inevitably be exceptions and areas where they do not hold. The Hall–Petch relationship was first discovered as a result of a series of tests on bulk polycrystalline engineered materials, as such, their grain sizes lie predominantly in the range of 1 mm to hundreds of microns. With the development of newer processing techniques such as mechanical alloying, electrodeposition, and acceleration-induced grain refinements, as well as new materials with nanostructures, it has since been found that the Hall–Petch relationship no longer applies or holds for the minimal grain size materials found on the nanoscale [150].

Numerous efforts and studies have aimed to quantify the Hall–Petch effect and understand its mechanistic background. This has led to the Hall–Petch coefficient, simplistically expressed as k in the equation above [151]. Recent analyses indicate that the values of the Hall–Petch coefficient are not fixed properties of a material, as is commonly assumed, but are heavily influenced by the average grain size in experiments and the experimental techniques and analysis methods employed. Moreover, various experimental results do not always produce the quadratic dependence predicted by the Hall–Petch relationship [152]. However, this does not disprove the Hall–Petch effect or its overall validity. The Hall–Petch equation remains useful for predicting trends and facilitating the design of new materials with engineered grain sizes, in many cases, material properties and mechanical behaviors still strongly correlate.

3.2 Strengthening Mechanisms

The mechanical properties of nanostructured materials are critical for their wide application in structural materials, sensors, and microelectromechanical systems. Strengthening is a vital aspect of the mechanical properties of nanomaterials [153]. The mechanisms involved in strengthening nanomaterials include grain refinement, solid solution strengthening, second-phase precipitation strengthening, and more. The interaction of various crystal defects is a significant method for enhancing material strength. The interaction mechanism of nanocrystalline materials comprises three elements: dislocations, atoms, and grain boundaries [94]. The interaction among these three defects becomes quite complex when considering the factors related to crystal defects.

Furthermore, the influence of temperature and strain rate must also be taken into account. Theoretical analyses highlight the crucial role of the interaction among dislocations, disinclination, and twin boundaries in impacting deformation mechanisms [154]. The mediating effect of disinclination is quantitatively assessed through the angle metric and corresponding deformation fields. Numerical results from a novel finite element method support the validity of these analyses. This research broadens the application fields of commonly used finite element methods and discrete models while also enhancing the comprehensive understanding of the interactions among different crystal defects in relation to material strength [155].

Nanomaterials' deformation and fracture behavior is a significant subject in materials science and engineering because of their potential for high strength and toughness. The microstructure of materials directly impacts their properties; therefore, a comprehensive study of the relationship between microstructure and properties is essential [156]. The deformation and fracture mechanisms of nanocrystalline materials have attracted considerable attention since the report of superstrength in nanocrystalline nickel. A number of successful studies on the microstructure and mechanical properties of nanocrystalline materials have been carried out since then [157]. However, the nanomaterials field still faces several key scientific challenges in combining advanced large-scale characterization techniques. The intriguing properties of nanocrystalline materials are directly size-dependent since grain size in a nanocrystalline material is an intrinsic microstructural parameter [158]. However, there is a fundamental misunderstanding regarding the relationship between microstructure and the properties of nanocrystalline materials. To better understand the mechanical behavior of nanomaterials, in-situ investigations were conducted on the deformation and fracture behavior of nanocrystalline Ni in a low cycle fatigue (LCF) test with a longitudinal orientation to the tensile shaft axis [159]. These experiments demonstrate the importance of understanding the deformation and fracture mechanisms in order to design nanocrystalline materials with desired mechanical properties and reliability, including ductility and strain rate sensitivity. The technology and knowledge employed aid in better understanding the deformation and fracture behavior of nanocrystalline materials.

3.3 Surface-To-Volume Ratio

With the advent of nanotechnology, accurately predicting and designing the mechanical behavior of nanomaterials has become essential. A research gap exists regarding the influence of the surface-to-volume ratio on these materials' mechanical properties. The effects of the surface-to-volume ratio on the mechanical behavior of nanomaterials are examined using alternating molecular dynamics, continuum mechanics, and analytical solutions [160]. By analyzing the stress and strain distributions alongside the yield strengths obtained through both methods, it becomes clear that the surface-to-volume ratio significantly affects the yield strengths of nanomaterials [161]. Moreover, non-matching yield strengths may be observed for the same materials when employing the energy and elemental energy methods [162]. To design nanomaterials with the desired mechanical properties, the results indicate that maintaining a consistent surface-to-volume ratio during numerical simulations and experiments effectively investigates their mechanical responses [163].

3.4 Defects and Imperfections

Comprehending the impact of defects on the mechanical properties of nanomaterials is essential for the advancement of high-quality materials nanostructures. Significant research efforts have focused on studying the mechanical properties of defect-free nanostructures, which provide insights into the limits of material performance and may be prerequisites for high-volume materials manufacturing. However, defects and imperfections inevitably exist in newly synthesized nanomaterials and those that have undergone long-term use or service. The presence of defects and imperfections plays a decisive role in the mechanical behavior of nanomaterials, as they govern properties such as strength and ductility [164]. Nanocrystalline and ultrafine-grained materials have attracted much attention recently due to their superior mechanical properties, such as high strength and good ductility [165]. However, because of their small grain sizes and high-energy grain boundaries, these materials are generally more prone to deformation-induced grain growth during plastic deformation and subsequent hardness, strength, and ductility changes [166]. Various approaches have been proposed to stabilize nanograin sizes in metallic materials, including solute addition and severe plastic deformation combined with q-phase insertion [167]. Many studies have revealed that metallic materials strengthened by nanoscale grain sizes and high-density dislocations tend to have high hardness but low ductility [41]. In addition to the significant adverse effects on ductility, nanoprecipitates are commonly prone to aggregation and/or coarse growth during heat treatment. Iterative processing will harm the stability of nano-structural features, resulting in a drop in hardness and strength [168].

3.5 Material Composition

In the realm of nanomaterials, the volume fraction of nanoparticles (NPs) is a crucial factor in determining the mechanical properties of metal matrix nanocomposites. NP spacing is an additional, equally important factor often obscured by the emphasis on NP type or shape [169]. At sufficiently high-volume fractions, the NPs form a lattice-like structure that provides remarkable improvements in strength within the metal matrix [170]. Variations in NP volume fraction and spacing alter the mechanisms of plastic deformation. With low volume fractions and large interparticle spacing, the NPs merely limit dislocation glide, increasing the strength of the composite [171]. With large volume fractions or small NPs, or more commonly both, the NPs begin to control dislocation motion by pinning them, leading to significant increases in yield stress [172]. They also cause a transition in the spacing of dislocation slip bands from larger to smaller than the NP diameter. With more NPs, solutes segregate into the interfacial regions between NPs and the matrix [173], acting as "superpins". Collisions of superpins and NPs increase the average mechanical equilibrium spacing of NPs, changing the potential slip system, and the greater the RMES, the more random that system becomes [174].

While volume fraction and spacing of NPs largely govern the mechanical properties of a nanoparticle-reinforced material, the mechanical properties of configurations can be altered by changing the geometry of the meta-NP composites [175]. Moreover, fine-scale topological features that preclude optimal packing also drastically reduce strength [176]. For a metal matrix, stacking fault energy dictates which excessive mechanism(s) will dominate and cause strain softening. The beneficial impact of the Latin-like NP structures likely arises from significantly increased hardness and more uniform stress distribution within the particle chains [177].

4 Conclusion

The wide range of studies of nanomaterials in this chapter reveals a rich terrain of interactions from the atomic and nanoscale worlds, which will significantly define the properties and performance of materials. This research indicates that modifications of microstructures at the nanoscale are giving much inspiration for improvements in mechanical characteristics, particularly for strength, toughness, buoyancy, and fatigue. Grain boundary engineering held the most crucial place in determining the behavior of materials, with experimental evidence showing how careful control of such interfaces leads to unprecedented levels of plastic reversibility and superior mechanical performance. Dislocation dynamics examination and twin boundary interaction analysis highlight deformation mechanisms, which differ entirely from bulk behavior under deformation.

Modern techniques for advanced characterization, especially in situ observations coupled with computational modeling, have changed the paradigm of how we interpret phenomena at the nanoscale. They now allow the observation of atomic-scale events in real-time and speculation into novel deformation and fracture mechanisms. Highly refined simulation tools have supplemented experimental observations, proving that the combination works for predicting materials' behavior under diverse conditions. They are spearheading the development of sophisticated constitutive models. These results will be applicable in almost every industry, from aerospace and automotive to biomedical. Their unique features will help them develop lightweight, high-strength components and further functional devices. Scaling up production processes without compromising quality and performance will be another challenge. Additional work will investigate these materials' long-term stability and environmental assessment.

Future research efforts should be aimed at closing the gap between pure science and industry with regard to cost-effective methods of producing nanomaterials in order to create synergy from different nanoscale features into hybrid materials. The prospect of "hybrid" materials with desirable properties continues to excite. At the same time, including machine learning algorithms in more traditional materials science would hasten the discovery and optimization of new nanomaterial systems. Typically, the more integrated our understanding of nanoscale phenomena at work becomes, the more likely it is that new solutions for our constraints on materials will emerge and new routes to possible future technology influences.

References

1. Subhan, A., Choudhury, K.P., Neogi, N.: Advances with Molecular Nanomaterials in Industrial Manufacturing Applications (2021). https://doi.org/10.3390/NANOMANUFACTURING1020008
2. Kreuter, J.: Nanoparticles—a historical perspective. Int. J. Pharm. (2007). https://doi.org/10.1016/J.IJPHARM.2006.10.021
3. Momose, S.: Microchip and method of manufacturing microchip (2008)
4. Schaming, D., Remita, H.: Nanotechnology: from the ancient time to nowadays. Found. Chem. (2015). https://doi.org/10.1007/S10698-015-9235-Y
5. Narminio, É.: Nanomaterials for Biomedical Engineering Applications (2022). https://doi.org/10.1007/978-981-19-1384-6_5
6. Manyangadze, M., Chikuruwo, N.H.M., Narsaiah, T.B., Chakra, Ch.S., Radhakumari, M., Danha, G.: Enhancing adsorption capacity of nano-adsorbents via surface modification: a review. S. Afr. J. Chem. Eng. (2020). https://doi.org/10.1016/J.SAJCE.2019.11.003
7. Chava, A.I.: Electronic and Mechanical Properties of Nanoparticles (2022). https://doi.org/10.1201/9781003216308-6
8. Chen, J., Li, L., Song, K.: Application of different nanomaterials in biomedical field. In: Highlights in Science, Engineering and Technology (2022). https://doi.org/10.54097/hset.v14i.1593
9. Boisseau, P., Loubaton, B.: Nanomedicine, Nanotechnology in medicine. Comptes Rendus Hebdomadaires Des Séances de l'Académie Des Sciences (2011)
10. Corrie, S.R., Thurecht, K.J.: Nano-bio interactions: guiding the development of nanoparticle therapeutics, diagnostics, and imaging agents. Pharm. Res. (2016). https://doi.org/10.1007/S11095-016-2009-Y
11. Desai, S.: Nanomaterials and Their Distinguishing Features (2022). https://doi.org/10.1007/978-981-19-1384-6_1
12. Pelleg, J.: Deformation in Nanomaterials (2021). https://doi.org/10.1007/978-3-030-74652-0_5
13. Zhang, J., Wu, Z.-R.: The application of nanomaterials in energy materials. In: Highlights in Science Engineering and Technology (2023). https://doi.org/10.54097/hset.v58i.10059
14. Wang, Y., Yu, H., Lu, L., Tan, Y.: Molecular dynamics study on Young's modulus of silicon nanostructures at finite temperature (2009). https://doi.org/10.1117/12.836706
15. Kim, J.-Y., Kim, S., Lee, J.-S., Lee, K.-W., Kwon, D.: Mechanical characterization of nanostructured materials using nanoindentation. Metals Mater. Int. (2006). https://doi.org/10.1007/BF03027534

16. Li, H., Du, H.N., Shen, Y., Zhang, H.: The stiffening or softening of nanomaterials determined by a quantitative model (2019). https://doi.org/10.1088/2053-1591/AB3603
17. Kitamura, T., Sumigawa, T.: Experimental approach to fracture mechanics in nanometer scale. Proc. Struct. Integr. (2018). https://doi.org/10.1016/J.PROSTR.2018.12.144
18. Chicone, C.: Elasticity: Basic Theory and Equations of Motion (2017). https://doi.org/10.1016/B978-0-12-804153-6.50018-X
19. Lee, E.-H.: A study on the effect of Young's Modulus modeling on the energy conservation in elastic–plastic material computation. Int. J. Precis. Eng. Manuf. (2020). https://doi.org/10.1007/S12541-020-00384-Y
20. Nie, D., Liu, C., Yu, M., Jiang, X., Wang, N., Gan, Y.: Elasticity regulates nanomaterial transport as delivery vehicles: design, characterization, mechanisms and state of the art. Biomaterials (2022). https://doi.org/10.1016/j.biomaterials.2022.121879
21. Abbas, M., Alqahtani, M., Algahtani, A., Kessentini, A., Loukil, H., Parayangat, M., Ijyas, T., Mohammed, A.W.: Validation of nanoparticle response to the sound pressure effect during the drug-delivery process. Polymers (2020). https://doi.org/10.3390/POLYM12010186
22. Walkey, C., Chan, W.C.W.: Understanding and controlling the interaction of nanomaterials with proteins in a physiological environment. ChemInform (2012). https://doi.org/10.1002/CHIN.201227269
23. Rawat, K., Goyal, M.: Young's modulus and vibrational frequency dependence on shape and size in nanomaterials. Mater. Today Proc. (2021). https://doi.org/10.1016/J.MATPR.2020.07.188
24. Chan, S.-W., Wang, W.: Surface stress of nano-crystals. Mater. Chem. Phys. (2021). https://doi.org/10.1016/J.MATCHEMPHYS.2021.125091
25. Padilla Espinosa, I.M., Jacobs, T.D.B., Martini, A.: Atomistic simulations of the elastic compression of platinum nanoparticles. Nanoscale Res. Lett. (2022). https://doi.org/10.1186/s11671-022-03734-z
26. Yao, Y., Peng, Z., Li, J., Chen, S.: A new elastic theory of nanocomposites with incoherent interface effect based on interface energy density. J. Appl. Mech. (2020). https://doi.org/10.1115/1.4045624
27. Azizi, S., Safaei, B., Mohammadpour Fattahi, A., Tekere, M.: Nonlinear vibrational analysis of nanobeams embedded in an elastic medium including surface stress effects. Adv. Mater. Sci. Eng. (2015). https://doi.org/10.1155/2015/318539
28. Liu, M., Liu, F.: Quantum manifestation of elastic constants in nanostructures. Mater. Sci. (2012). https://doi.org/10.1088/0957-4484/25/13/135706
29. Zhang, X., Sharma, P.: Impact of size-dependent non-local elastic strain on the electronic band structure of embedded quantum dots (2005). https://doi.org/10.1243/17403499JNN57
30. Tao, J., Shi, W., Chen, K., Lu, W., Elbourne, A., Bao, L., Weng, L., Zheng, X.-D., Su, X.D., Teng, Z., Wang, L.: Elasticity of mesoporous nanocapsules regulates cellular uptake, blood circulation, and intratumoral distribution. Biomater. Sci. (2023). https://doi.org/10.1039/d2bm01701f
31. Wang, X., Meng, X., Zhu, Y., Ling, H., Chen, Y., Li, Z., Hartel, M.C., Dokmeci, M.R., Zhang, S., Khademhosseini, A.: A Sub-1-V, microwatt power-consumption iontronic pressure sensor based on organic electrochemical transistors. IEEE Electr. Device Lett. (2021). https://doi.org/10.1109/LED.2020.3042310
32. Fritea, L., Banica, F., Costea, T., Moldovan, L., Dobjanschi, L., Muresan, M., Cavalu, S.: Metal nanoparticles and carbon-based nanomaterials for improved performances of electrochemical (Bio)sensors with biomedical applications. Materials (2021). https://doi.org/10.3390/MA14216319
33. Xie, L.Y., Zhang, Z., Wu, Q., Gao, Z., Mi, G., Wang, R., Sun, H., Zhao, Y., Du, Y.: Intelligent wearable devices based on nanomaterials and nanostructures for healthcare. Nanoscale (2023). https://doi.org/10.1039/d2nr04551f
34. Zheng, Q., Zheng, Q., Lee, J., Shen, X., Chen, X., Kim, J.K.: Graphene-based wearable piezoresistive physical sensors. Mater. Today (2020). https://doi.org/10.1016/J.MATTOD.2019.12.004

35. Khosla, A., Chaudhary, V., Zhang, H.: A paradigm of microbe-mediated green nano-semiconductors and nano-metals. Nanotechnology **36**(8), 082501 (2024)
36. Jafarov, M., Ganbarov, K., Kariptaş, E., Huseynova, S., Guliyeva, S.: The impact of temperature on the synthesis of silver nanoparticles by Candida macedoniensis. Black Sea J. Health Sci. **6**(1), 163–166 (2023)
37. Anselmo, A.C., Zhang, M., Kumar, S., Vogus, D.R., Menegatti, S., Helgeson, M.E., Mitragotri, S.: Elasticity of nanoparticles influences their blood circulation, phagocytosis, endocytosis, and targeting. ACS Nano (2015). https://doi.org/10.1021/ACSNANO.5B00147
38. Chen, X., Zhang, S., Li, J., Huang, X., Ye, H., Qiao, X., Xue, Z., Yang, W., Wang, T.: Influence of elasticity of hydrogel nanoparticles on their tumor delivery. Adv. Sci. (2022). https://doi.org/10.1002/advs.202202644
39. Wu, Q., Weishou, M., Zhang, Y., Gao, H., Hui, D.: Mechanical properties of nanomaterials: a review. Nano Rev. (2020). https://doi.org/10.1515/NTREV-2020-0021
40. Khajuee, S., Memarian, N.: Hydrothermal synthesis of ultrafine SnO2 nanospheres: effect of reaction time on physical properties. Eur. Phys. J. Plus **136**(5), 1–12 (2021)
41. Hu, Y., Yu, Z., Fan, G., Tan, Z., Zhou, J., Zhang, H., Li, Z., Zhang, D.: Simultaneous enhancement of strength and ductility with nano dispersoids in nano and ultrafine grain metals: a brief review. Rev. Adv. Mater. Sci. (2020). https://doi.org/10.1515/RAMS-2020-0028
42. Vasyl, H.: Mechanics of Carbon Nanotubes and Nanomaterials (2017). https://doi.org/10.1201/9781315371795-6
43. Mei, H., Mei, X., He, X., Bie, Z., Wang, Z., Cui, J.: Chemical covalent connection of carbon nanotubes for related structural manufacturing: a molecular dynamics study. Appl. Surf. Sci. (2023). https://doi.org/10.1016/j.apsusc.2022.156296
44. Armstrong, R.W.: Crystal Engineering for Mechanical Strength at Nano-Scale Dimensions (2017). https://doi.org/10.3390/CRYST7100315
45. Stroppel, M.: Graphene: Structure, properties, preparation, modification, and applications (2022). https://doi.org/10.1016/b978-0-323-90937-2.00010-1
46. Yu, F., Li, X., Sheng, C., Li, L.: DNA nanotechnology targeting mitochondria: from subcellular molecular imaging to tailor-made therapeutics. Angew. Chem. **136**(36), e202409351 (2024)
47. Li, N., Liu, Z., Liu, G., Wang, Z., Guo, X., Guo, C., Han, J.: TiO_2 Nanocoatings with controllable crystal type and nanoscale topography on zirconia implants to accelerate bone formation. Bioinorg. Chem. Appl. (2022). https://doi.org/10.1155/2022/8650659
48. Zhao, R., Zhao, Z., Bai, P., Du, W., Tie, D., Zhang, Z., Wang, L., Wang, Y.: Research on tribological behavior of graphene with in-situ MgO nanoparticles reinforced AZ91 alloy composite. Compos. Commun. (2022). https://doi.org/10.1016/j.coco.2022.101086
49. Chen, D.X.B.: Mechanical Properties of Native Tissues and Scaffolds (2019). https://doi.org/10.1007/978-3-030-03460-3_4
50. Kang, M.S., Lee, J.-H., Song, S.-J., Shin, D.-M., Jang, J.-H., Hyon, S.-H., Hong, S.W., Lee, J.H., Han, D.-W.: Graphene oxide-functionalized nanofibre composite matrices to enhance differentiation of hippocampal neuronal cells (2020). https://doi.org/10.1039/D0MA00733A
51. Akbarpour, M.R., Torknik, F.S., Manafi, S.A.: Enhanced compressive strength of nanostructured aluminum reinforced with SiC nanoparticles and investigation of strengthening mechanisms and fracture behavior. J. Mater. Eng. Perform. (2017). https://doi.org/10.1007/S11665-017-2871-8
52. Nie, J., Liu, Y., Wang, F., Zhou, H., Yang, C., Liu, X., An, X., Liao, X., Zhu, Y., Zhu, Y., Zhao, Y.: Key roles of particles in grain refinement and material strengthening for an aluminum matrix composite. Mater. Sci. Eng. A. In: Structural Materials Properties Microstructure and Processing (2021). https://doi.org/10.1016/J.MSEA.2020.140414
53. Han, Q., Yi, X.: A unified mechanistic model for Hall–Petch and inverse Hall–Petch relations of nanocrystalline metals based on intragranular dislocation storage. J. Mech. Phys. Solids (2021). https://doi.org/10.1016/J.JMPS.2021.104530
54. Rahaman, O., Mortazavi, B., Dianat, A., Cuniberti, G., Rabczuk, T.: A structural insight into mechanical strength of graphene-like carbon and carbon nitride networks. Mater. Sci. (2017). https://doi.org/10.1088/1361-6528/28/5/055707

55. Solanke, S.G., Shinde, S.B., Raut, D., Oza, A.D., Prakash, C., Saxena, K.K., Kumar, S.: Effect of multi-walled carbon nanotubes in reinforced hydroxyapatite composite coatings for orthopedic applications. Surf. Rev. Lett. (2023). https://doi.org/10.1142/s0218625x23400097
56. Viswanathan, S., Mohan, L., Bera, P., Shanthiswaroop, S., Muniprakash, M., Barshilia, H.C., Anandan, C.: Corrosion and wear resistance properties of multilayered diamond-like carbon nanocomposite coating. Surf. Interface Anal. (2018). https://doi.org/10.1002/SIA.6353
57. Abera, B.: Nanomaterials: an overview of synthesis, classification, characterization, and applications. Nano Select (2023). https://doi.org/10.1002/nano.202300038
58. Amodeo, J., Pizzagalli, L.: Modeling the mechanical properties of nanoparticles: a review. Comptes Rendus Phys. (2021). https://doi.org/10.5802/CRPHYS.70
59. Basak, A., Gupta, A.: Simultaneous grain boundary motion, grain rotation, and sliding in a tricrystal. Mech. Mater. (2015). https://doi.org/10.1016/J.MECHMAT.2015.01.012
60. Werner, K.V., Niessen, F., Liu, W., Lu, S., Vitos, L., Villa, M., Somers, M.A.J.: Reconciling experimental and theoretical stacking fault energies in face-centered cubic materials with the experimental twinning stress. Materialia (2023). https://doi.org/10.1016/j.mtla.2023.101708
61. Kovačević, S., Mesarovic, S.D.: Diffusion-induced stress concentrations in diffusional creep. Int. J. Solids Struct. (2022). https://doi.org/10.1016/j.ijsolstr.2022.111440
62. Wu, Y., Li, K.: Deformation behavior and its effect on the mechanical properties of nanoparticles in pure nanoparticle-modulated system. J. Mater. Eng. Perform. (2022). https://doi.org/10.1007/s11665-022-07736-x
63. Suhanov, I., Ditenberg, I.A., Ditenberg, I.A., Tyumentsev, A.N., Tyumentsev, A.N.: Theoretical analysis of features of dipole and quadrupole configurations of partial disclinations in nanocrystals of metals (2016). https://doi.org/10.1088/1757-899X/116/1/012035
64. Sagdic, K., Eş, I., Sitti, M., Inci, F.: Smart materials: rational design in biosystems via artificial intelligence. Trends Biotechnol. (2022). https://doi.org/10.1016/j.tibtech.2022.01.005
65. Zhao, L., Liu, T., Li, X., Cui, Q., Wang, X., Song, K., Ge, D.: Protein adsorption on TiO_2 nanostructures and its effects on surface topography and bactericidal performance. Appl. Surf. Sci. (2022). https://doi.org/10.1016/J.APSUSC.2021.151779
66. Alcolea, L.C.: Biostable segmented thermoplastic polyurethane shape memory polymers for smart biomedical applications. ACS Appl. Polym. Mater. (2022). https://doi.org/10.1021/acsapm.1c01808
67. Garrett, S.L.: Elasticity of Solids (2017). https://doi.org/10.1007/978-3-319-49978-9_4
68. Kochmann, D.M., Bertoldi, K.: Exploiting microstructural instabilities in solids and structures: from metamaterials to structural transitions. Appl. Mech. Rev. (2017). https://doi.org/10.1115/1.4037966
69. Bhushan, B.: Nanomechanical Properties of Nanostructures and Scale Effects (2017). https://doi.org/10.1007/978-3-662-54357-3_32
70. Handy, R.D., Upton, M., Besinis, A.: Review of antimicrobial nanocoatings in medicine and dentistry: mechanisms of action, biocompatibility performance, safety, and benefits compared to antibiotics. ACS Nano (2023). https://doi.org/10.1021/acsnano.2c12488
71. Saylan, Y., Yavuz, H., Ulger, C., Denizli, A., Sağlam, N.: Introduction to Nanoscience, Nanomaterials, Nanocomposite, Nanopolymer, and Engineering Smart Materials (2019). https://doi.org/10.1007/978-3-030-16534-5_1
72. Trache, D., DeLuca, L.T.: Nanoenergetic materials: preparation, properties, and applications. Nanomaterials (2020). https://doi.org/10.3390/NANO10122347
73. Bian, J., Zhang, H., Niu, X., Wang, G.-F.: Anisotropic Deformation in the Compressions of Single Crystalline Copper Nanoparticles (2018). https://doi.org/10.3390/CRYST8030116
74. Chinh, N.Q., Racz, G., Gubicza, J., Valiev, R.Z., Valiev, R.Z., Langdon, T.G.: A possible stabilizing effect of work hardening on the tensile performance of superplastic materials. Mater. Sci. Eng. A. In: Structural Materials Properties Microstructure and Processing (2019). https://doi.org/10.1016/J.MSEA.2019.05.063
75. Arsenlis, A., Wirth, B.D., Rhee, M.: Dislocation density-based constitutive model for the mechanical behaviour of irradiated Cu. Philos. Mag. (2004). https://doi.org/10.1080/14786430412331293531

76. Volkov, I.A., Korotkikh, Yu.G.: Modeling of processes of complex plastic deformation of materials along arbitrary temperature and force loading paths. Mech. Solids (2007). https://doi.org/10.3103/S0025654407060076
77. Kadowaki, H., Liu, W.K.: A multiscale approach for the micropolar continuum model. Laser Phys. (2005)
78. Song, F., Huang, G., Park, H.S., Liu, X.: A continuum model for the mechanical behavior of nanowires including surface and surface-induced initial stresses. Int. J. Solids Struct. (2011). https://doi.org/10.1016/J.IJSOLSTR.2011.03.021
79. Zeng, G., Dirscherl, K., Garnæs, J.: Toward accurate quantitative elasticity mapping of rigid nanomaterials by atomic force microscopy: effect of acquisition frequency, loading force, and tip geometry. Nanomaterials (2018). https://doi.org/10.3390/NANO8080616
80. Mishra, R.K.: Microstructure-based modeling of metal plasticity and electron microscopy characterization of automotive structural metals at multi-scales. Microsc. Microanal. (2016). https://doi.org/10.1017/S1431927616006930
81. Endo, M., Ueki, S.: Line defect detection method and line defect detector (2012)
82. Capecchi, D.: The Foundation of Classical Mechanics (2021). https://doi.org/10.3390/ENCYCLOPEDIA1020040
83. Eastman, D.W., Shade, P.A., Uchic, M.D., Hemker, K.J.: Microscale Testing and Characterization Techniques for Benchmarking Crystal Plasticity Models at Microstructural Length Scales (2020). https://doi.org/10.1007/978-3-030-40562-5_4
84. Drapaca, C.S., Sivaloganathan, S.: Brief Review of Continuum Mechanics Theories (2019). https://doi.org/10.1007/978-1-4939-9810-4_2
85. Hénon, J.-O.: Material length scales in gradient-dependent plasticity/damage and size effects: theory and computation (2022). https://doi.org/10.31390/gradschool_dissertations.3054
86. Pastrana, J.A.: Molecular Dynamics Simulation Methods to Study Structural Dynamics of Proteins (2023). https://doi.org/10.1007/978-981-99-2079-2_5
87. Li, Z., Shen, T., Hu, X., Zhang, L., Jia, X., Li, J., Zhang, C.: The plastic deformation mechanism in nano-polycrystalline Al/Mg layered composites: a molecular dynamics study. Nanomaterials **14**(1), 114 (2024). https://doi.org/10.3390/nano14010114
88. Bourne, N.: Materials science: atomistic views of deformation. Nature (2017). https://doi.org/10.1038/550461A
89. Tilocca, A.: Molecular Dynamics Methods for Modeling Complex Interactions in Biomaterials (2012). https://doi.org/10.1007/978-1-61779-388-2_18
90. Velez-Cruz, A., Farinas-Coronado, W.: Review-bone characterization: mechanical properties based on non-destructive techniques. Athenea (2022). https://doi.org/10.47460/athenea.v3i10.45
91. Ghoshdastider, U., Wu, R., Trzaskowski, B., Mlynarczyk, K., Miszta, P., Gurusaran, M., Viswanathan, S., Renugopalakrishnan, V., Filipek, S.: Molecular effects of encapsulation of glucose oxidase dimer by graphene. RSC Adv. **5**, 13570 (2015)
92. Zhang, Y., Zhang, Y., Xu, Z., Ming, W.Q., Guo, J., Zhang, Z., Chen, J.: Atomic resolution analyses on defects in nanocrystalline Cu-based alloys generated by severe plastic deformation. Mater. Charact. (2019). https://doi.org/10.1016/J.MATCHAR.2019.109886
93. Mohamadnejad, S., Basti, A., Ansari, R.: Analyses of Dislocation Effects on Plastic Deformation (2020). https://doi.org/10.1007/S42493-020-00037-2
94. Li, J.C.M., Feng, C.R., Rath, B.B.: Emission of dislocations from grain boundaries and its role in nanomaterials. Crystals **11**(1), 41 (2021). https://doi.org/10.3390/cryst11010041
95. Zhang, Q., Huang, R., Jiang, J., Cao, T., Zeng, Y., Li, J., Xue, Y., Li, X.: Size effects and plastic deformation mechanisms in single-crystalline CoCrFeNi micro/nanopillars. J. Mech. Phys. Solids (2022). https://doi.org/10.1016/j.jmps.2022.104853
96. Luo, S., Liu, J., Lu, L., Fan, D., Bie, B., Zhao, F.: Plastic deformation equipment for materials (2019)
97. Panzarino, J.F., Pan, Z., Rupert, T.J.: Plasticity-induced restructuring of a nanocrystalline grain boundary network. Acta Mater. (2016). https://doi.org/10.1016/J.ACTAMAT.2016.08.040

98. Venkataraman, A., Linne, M.A., Daly, S., Sangid, M.D.: Criteria for the prevalence of grain boundary sliding as a deformation mechanism. Materialia (2019). https://doi.org/10.1016/J.MTLA.2019.100499
99. Danoix, F., Hoummada, K., Maugis, P., Rolland, N., Debreux, C., Blavette, D.: Grain size effect on interfacial segregation in nanomaterials. J. Phys. Chem. Solids (2022). https://doi.org/10.1016/j.jpcs.2022.110620
100. Wang, Z.L., Zhang, J., Lu, J.: Effects of crystallographic orientations and grain boundaries on nanoscratching behaviour of unique bi-crystal Cu. Wear (2022). https://doi.org/10.1016/j.wear.2022.204313
101. Atul, H.C.: Grain boundary processes in strengthening, weakening and superplasticity. Adv. Eng. Mater. **22**(1), 1900748 (2020). https://doi.org/10.1002/ADEM.201900748
102. Qishan, H., Qi, Z., Yingbin, C., Mingyu, G., Jixue, L., Ze, Z.: Twinning-assisted dynamic adjustment of grain boundary mobility. Nat. Commun. **12**(1), 6695 (2021). https://doi.org/10.1038/S41467-021-27002-3
103. Singh, C.A., Aditya, G., Anand, V., Sudhanshu, S.S., Jayant, J.: Elucidating the role of grain boundary sliding in enhanced ductility of high strain rate sensitive Mg–Ca binary alloy: an in-situ study. Mater. Sci. Eng. A **851**, 143613–143613. In: Structural Materials Properties Microstructure and Processing (2022). https://doi.org/10.1016/j.msea.2022.143613
104. Shufen, C., Panbo, L., Yin, Z., Xiaodong, W., Shuang, S., Ting, Z.: In situ atomic-scale observation of dislocation climb and grain boundary evolution in nanostructured metal. Nat. Commun. **13**(1) (2022). https://doi.org/10.1038/s41467-022-31800-8
105. Ian, C., Yuri, M.: Point-defect avalanches mediate grain boundary diffusion. Commun. Mater. **3**(1) (2022). https://doi.org/10.1038/s43246-022-00314-7
106. Zheng, F., Jianwei, X., Su, T., Chuang, D., Guofeng, W., Scott, X.M.: Atomic-scale observation of dynamic grain boundary structural transformation during shear-mediated migration. Sci. Adv. **8**(45) (2022). https://doi.org/10.1126/sciadv.abn3785
107. Qi, Z., Qishan, H., Cao, G., Xianghai, A., Scott, X.M., Wei, Y.: Metallic nanocrystals with low angle grain boundary for controllable plastic reversibility. Nat. Commun. **11**(1), 1–8 (2020). https://doi.org/10.1038/S41467-020-16869-3
108. Hyeon, K.S., Won, C.S., Deok, I.Y., Deok, I.Y., Kook, L.Y.: Grain boundary sliding during high-temperature tensile deformation in superplastic Fe–6.6Mn–2.3Al steel. Mater. Sci. Eng. A **780**, 139174. In: Structural Materials Properties Microstructure and Processing (2020). https://doi.org/10.1016/J.MSEA.2020.139174
109. Anping, H., Junhua, Z., Jian, Z., Peishi, Y., Ning, W., Wanlin, G.: Revealing the size effect mechanism of reversible grain boundary migration in nanocrystalline coppers: molecular dynamics simulations and a refined disconnection model. J. Mech. Phys. Solids **161**, 104832–104832 (2022). https://doi.org/10.1016/j.jmps.2022.104832
110. Andrew, L., Amit, S., Tammy, Y.O., Selim, E.: Quantized grain boundary states promote nanoparticle alignment during imperfect oriented attachment. Small **16**(29), 2001423 (2020). https://doi.org/10.1002/SMLL.202001423
111. Kardani, A., Montazeri, A.: Metal-matrix nanocomposites under compressive loading: towards an understanding of how twinning formation can enhance their plastic deformation. Sci. Rep. (2020). https://doi.org/10.1038/S41598-020-66696-1
112. Jose-Yacaman, M., Velazquez, J.J., Mendoza-Cruz, R., Lehr, A.: The role of twinning in multi metallic alloys at the nanoscale. J. Solid State Chem. (2022)https://doi.org/10.1016/j.jssc.2022.123641
113. Li, J., Fang, Q., Liu, B., Liu, Y., Jiang, C.: Twinning-governed plastic deformation in a thin film of body-centred cubic nanocrystalline ternary alloys at low temperature. J. Alloys Compd. (2017). https://doi.org/10.1016/J.JALLCOM.2017.08.115
114. Yang, Z., Zheng, L., Yue, Y., Lu, Z.: Effects of twin orientation and spacing on the mechanical properties of Cu nanowires. Sci. Rep. (2017). https://doi.org/10.1038/S41598-017-10934-6
115. Zhang, X., Anderoglu, O., Misra, A., Wang, H.: Influence of deposition rate on the formation of growth twins in sputter-deposited 330 austenitic stainless steel films. Appl. Phys. Lett. (2007). https://doi.org/10.1063/1.2720708

116. Yuan, F., Chen, L., Jiang, P., Wu, X.: Twin boundary spacing effects on shock response and spall behaviors of hierarchically nanotwinned fcc metals. J. Appl. Phys. (2014). https://doi.org/10.1063/1.4865738
117. Kim, S.H., Park, J.H., Kim, H.K., Ahn, J.-P., Whang, D., Lee, J.C.: Twin boundary sliding in single crystalline Cu and Al nanowires. Soc. Sci. Res. Netw. (2020). https://doi.org/10.2139/SSRN.3513123
118. Li, C., Segall, D., Xu, G.: Dislocation nucleation at a surface step by a multiscale approach. MRS Proc. (2004). https://doi.org/10.1557/PROC-841-R11.3
119. Huang, L., Zheng, F., Deng, Q., Thi, Q.H., Wong, L.W., Cai, Y., Wang, N., Lee, C.-S., Lau, S.P., Chhowalla, M., Li, J., Ly, T.H., Zhao, J.: In situ scanning transmission electron microscopy observations of fracture at the atomic scale. Phys. Rev. Lett. (2020). https://doi.org/10.1103/PHYSREVLETT.125.246102
120. Barney, C.W., Sacligil, I., Tew, G.N., Crosby, A.J.: Linking cavitation and fracture to molecular scale structural damage of model networks. Soft Matter (2022). https://doi.org/10.1039/d2sm00400c
121. Lesičar, T., Sorić, J., Tonković, Z.: On the scale-transition in multiscale modeling of ductile damage. Key Eng. Mater. (2017). https://doi.org/10.4028/WWW.SCIENTIFIC.NET/KEM.754.226
122. Thamburaja, G.T., Sarah, K., Srinivasa, A.R., Reddy, J.N.: Fracture of viscoelastic materials: FEM implementation of a non-local & rate form-based finite-deformation constitutive theory. Comput. Methods Appl. Mech. Eng. (2019). https://doi.org/10.1016/J.CMA.2019.05.032
123. Swainger, K.H.: The mathematically linear analysis of strain in continua. Meccanica (1968). https://doi.org/10.1007/BF02186948
124. Cooper, S.B., Rosatello, M., Mathis, A.T., Johnson, K., Johnson, K., Brake, M.R.W., Brake, M.R.W., Allen, M.S., Ferri, A.A., Roettgen, D.R., Pacini, B.R., Mayes, R.L.: Effect of Far-Field Structure on Joint Properties (2017). https://doi.org/10.1007/978-3-319-54930-9_7
125. Zhao, H., Li, Z., Gao, H., Lu, L.: Fracture and toughening mechanisms in nanotwinned and nanolayered materials. Mrs Bulletin (2022). https://doi.org/10.1557/s43577-022-00376-5
126. Jiang, W., Li, Y.: Multiscale analysis of microstructural effects on the fracture properties of laser repaired cracks with micro/nano composite material addition. Theor. Appl. Fract. Mech. (2023). https://doi.org/10.1016/j.tafmec.2023.103914
127. Lu, L., Pan, Q., Hattar, K., Boyce, B.L.: Fatigue and fracture of nanostructured metals and alloys. Mrs Bulletin (2021). https://doi.org/10.1557/S43577-021-00054-Y
128. Patel, Z.S., Meza, L.R.: Toughness amplification via controlled nanostructure in lightweight nano-bouligand materials. Small (2023). https://doi.org/10.1002/smll.202207779
129. Pereira Junior, M.L., de Sousa Oliveira, T., Monteiro, F.F., da Cunha, W.F., de Oliveria Neto, P.H., Ribeiro Junior, L.A.: Torsional fracture of carbon nanotube bundles: a reactive molecular dynamics study. Phys. Chem. Chem. Phys. (2022). https://doi.org/10.1039/d2cp01589g
130. Nimbagal, V., Banapurmath, N.R., Umarfarooq, M.A., Revankar, S., Sajjan, A.M., Soudagar, M.E.M., Shahapurkar, K., Alamir, M.A., Alarifi, I.M., Elfasakhany, A.: Mechanical and fracture properties of carbon nano fibers/short carbon fiber epoxy composites. Polym. Compos. (2023). https://doi.org/10.1002/pc.27371
131. Kendall, K.: Nano-samples give higher brittle strength by the Griffith energy principle. Philos. Trans. R. Soc. A (2022). https://doi.org/10.1098/rsta.2021.0349
132. Hetz, P., Kraus, M.F., Merklein, M.: Investigation of the influence by size effects on the material characterization of the uniaxial compression stress state. Key Eng. Mater. (2022). https://doi.org/10.4028/p-12whe6
133. Bortz, D.R., Heras, E.G., Gullon, I.M.: Impressive fatigue life and fracture toughness improvements in graphene oxide/epoxy composites. Macromolecules **45**(1), 238–245 (2012). https://doi.org/10.1021/ma201563k
134. Muñoz, S.H., Moreno, M.D.C.S., Domínguez, G.J.M., Rodríguez, P.A.M., Vázquez, E.: Experimental, numerical, and analytical study on the effect of graphene oxide in the mechanical properties of a solvent-free reinforced epoxy resin. Polymers **11**(12), 2115 (2019). https://doi.org/10.3390/polym11122115

135. Pogrebniak, A., Kasperska, V.: Fatigue resistance of metal construction materials and its relationship with changes in the state of thin surface layers. Mech. Adv. Technol. (2022). https://doi.org/10.20535/2521-1943.2022.6.2.261035
136. Bai, Y., Yue, H., Wang, J., Shen, B., Sun, S., Wang, S., Wang, H., Li, X., Xu, Z.P., Zhang, R., Wei, F.: Super-durable ultralong carbon nanotubes. Science (2020). https://doi.org/10.1126/SCIENCE.AAY5220
137. Zhang, Y., Li, J., Hu, Y., Ding, S., Wu, W., Xia, R.: Fatigue responses of metallic glass-based stochastic network nanomaterial: superior strain-hardening ability. Fatigue Fract. Eng. Mater. Struct. **47**(4), 1208–1218 (2024)
138. Garofalo, S., Rovale, G., Morano, C., Furgiuele, F., Minervino, M., Pagnotta, L.: On the fatigue behaviour of epoxy-based nanostructured composite materials. J. Reinf. Plast. Compos. 07316844241244499 (2024)
139. Huo, W., Fang, F., Liu, X., Tan, S., Xie, Z., Jiang, J.: Fatigue resistance of nanotwinned high-entropy alloy films. Mater. Sci. Eng. A **739**, 26–30 (2019)
140. Zhang, Y., He, C., Yu, Q., Li, X., Wang, X., Zhang, Y., Wang, J., Jiang, C., Jia, Y., Zhang, X.-C.: Nacre-like surface nanolaminates enhance fatigue resistance of pure titanium. Nat. Commun. **15**(1), 6917 (2024)
141. Chen, P., Cai, X., Min, N., Liu, Y., Wang, Z., Jin, M., Jin, X.: Enhanced fatigue resistance of nanocrystalline $Ni_{50.8}Ti_{49.2}$ wires by mechanical training. Superalloys (2023). https://doi.org/10.3390/met13020361
142. Wan, S., Xu, F., Jiang, L., Cheng, Q.: Superior fatigue resistant bioinspired graphene-based nanocomposite via synergistic interfacial interactions. Adv. Funct. Mater. (2017). https://doi.org/10.1002/ADFM.201605636
143. Pan, Q., Lu, L.: Improved fatigue resistance of gradient nanograined metallic materials: suppress strain localization and damage accumulation. Scr. Mater. (2020). https://doi.org/10.1016/J.SCRIPTAMAT.2020.06.038
144. da Costa, L.P.: Engineered nanomaterials in the sports industry (2020). https://doi.org/10.1016/B978-0-12-821381-0.00014-4
145. Greer, J.R., Jang, D., Gu, X.W.: Exploring deformation mechanisms in nanostructured materials. JOM (2012). https://doi.org/10.1007/S11837-012-0438-6
146. Naik, S.N., Walley, S.M.: The Hall–Petch and inverse Hall–Petch relations and the hardness of nanocrystalline metals. J. Mater. Sci. (2020). https://doi.org/10.1007/S10853-019-04160-W
147. Cao, L., Sengupta, A., Pantuso, D., Koslowski, M.: Effect of texture and grain size on the residual stress of nanocrystalline thin films. Model. Simul. Mater. Sci. Eng. (2017). https://doi.org/10.1088/1361-651X/AA80FB
148. Xiao, Y., Kong, X., Yao, B.N., Legut, D., Germann, T.C., Zhang, R.: Atomistic insight into the dislocation nucleation at crystalline/crystalline and crystalline/amorphous interfaces without full symmetry. Acta Mater. (2019). https://doi.org/10.1016/J.ACTAMAT.2018.09.068
149. Blum, W., Li, Y.J., Chen, J., Chen, J., Zeng, X.H., Lu, K.: On the Hall-Petch relation between flow stress and grain size. Int. J. Mater. Res. (2006). https://doi.org/10.3139/146.101398
150. Li, Y., Bushby, A.J., Dunstan, D.J.: The Hall-Petch effect as a manifestation of the general size effect. Mater. Sci. (2015). https://doi.org/10.1098/RSPA.2015.0890
151. Knipe, H.: Hall-Petch relationship and deformation mechanism of pure Mg at room temperature. J. Alloys Compd. (2022). https://doi.org/10.1016/j.jallcom.2022.165924
152. Song, Y.: Investigation of the Hall-Petch Relationship Using Strain Gradient Plasticity Model for Finite Deformation Framework (2020). https://doi.org/10.32732/JMA.2020.9.2.55
153. Morris, D.G.: The origins of strengthening in nanostructured metals and alloys. Revista De Metalurgia (2010). https://doi.org/10.3989/REVMETALM.1008
154. Liu, C., Roters, F., Raabe, D.: Finite strain crystal plasticity-phase field modeling of twin, dislocation, and grain boundary interaction in hexagonal materials. Acta Mater. (2022). https://doi.org/10.1016/j.actamat.2022.118444
155. Needleman, A.: Discrete defect plasticity and implications for dissipation. Eur. J. Mech. A Solids (2022). https://doi.org/10.1016/j.euromechsol.2023.105002

156. Li, X., Yin, S., Oh, S. H., Gao, H.: Hardening and toughening mechanisms in nanotwinned ceramics. Scr. Mater. (2017). https://doi.org/10.1016/J.SCRIPTAMAT.2017.02.003
157. Wu, B., Wang, J.: Numerical simulations of the intergranular fracture in nanocrystalline Ni. Acta Metallurgica Sinica (2009)
158. Rajulapati, K.V., Koch, C.C.: Nanocrystalline Materials: Mechanical Properties (2014). https://doi.org/10.1016/B978-0-12-803581-8.02570-4
159. Zhang, H., Jiang, C., Lu, Y.: Low-cycle fatigue testing of Ni nanowires based on a micro-mechanical device. Exp..Mech. (2017). https://doi.org/10.1007/S11340-016-0199-1
160. He, Q.: Bridging the Gap from Continuum Mechanics to Molecular Dynamics for Nanoscale Systems (2015)
161. Yang, L., Bian, J., Yuan, W., Wang, G.-F.: Statistical characterization of the yield stress of nanoparticles. Acta Mech. Solida Sinica (2021). https://doi.org/10.1007/S10338-020-00212-W
162. Duan, Y., Tian, L., Zhang, F., Qiao, H., Muyu, L., Yang, L., Guan, Y.: A Fast Identification Method of Yield Strength of Materials Based on Bending Experimental Data (2020). https://doi.org/10.3390/MET10020169
163. Ding, R., Azadehranjbar, S., Padilla Espinosa, I.M., Martini, A., Jacobs, T.D.: Separating geometric and diffusive contributions to the surface nucleation of dislocations in nanoparticles. ACS Nano **18**(5), 4170–4179 (2024)
164. Gu, X.W., Wu, Z., Zhang, Y.-W., Srolovitz, D.J., Greer, J.R.: Microstructure versus flaw: mechanisms of failure and strength in nanostructures. Nano Lett. (2013). https://doi.org/10.1021/NL403453H
165. Zhang, D.: Ultrafine grained metals and metal matrix nanocomposites fabricated by powder processing and thermomechanical powder consolidation. Progr. Mater. Sci. (2021). https://doi.org/10.1016/J.PMATSCI.2021.100796
166. Tak, T.N., Prakash, A., Samajdar, I., Guruprasad, P.J.: Role Played by Grain Boundaries in Plastic Deformation of Polycrystalline Metals: A Discrete Dislocation Dynamics Study (2021). https://doi.org/10.1007/978-981-15-8315-5_17
167. Li, Z., Zhang, Y., Zhang, Z., Cui, Y.-T., Guo, Q., Liu, P., Jin, S., Sha, G., Ding, K., Li, Z., Fan, T., Urbassek, H.M., Yu, Q., Zhu, T., Zhang, D., Wang, Y.M.: A nanodispersion-in-nanograins strategy for ultra-strong, ductile and stable metal nanocomposites. Nat. Commun. (2022). https://doi.org/10.1038/s41467-022-33261-5
168. Mahan, H.M., Konovalov, S., Panchenko, I.: Effect of heat treatment on the mechanical properties of the aluminium alloys AA2024 with nanoparticles. Int. J. Appl. Sci. Eng. (2023). https://doi.org/10.6703/ijase.202306_20(2).011
169. Hassanzadeh-Aghdam, M.K., Hassanzadeh-Aghdam, M.K., Mahmoodi, M.J., Ansari, R.: A comprehensive predicting model for thermomechanical properties of particulate metal matrix nanocomposites. J. Alloys Compd. (2018). https://doi.org/10.1016/J.JALLCOM.2017.12.232
170. Cui, Y.: Method for preparing high volume fraction particle reinforced metal matrix composites through near net shaping (2011)
171. Wang, G., Wang, R., Wang, C., Tang, C., Luo, Y., Xiao, D.: Mechanistic pathway of NP-polymer interface to engender nanoscale confinement. Compos. Commun. (2022). https://doi.org/10.1016/j.coco.2022.101186
172. VanSaders, B., Glotzer, S.C.: Pinning dislocations in colloidal crystals with active particles that seek stacking faults. Soft Matter (2020). https://doi.org/10.1039/C9SM02514F
173. Zhang, Z., Li, J.C.: Transition of plastic deformation mechanisms governed by spacing of nano-precipitates in a high entropy alloy. Scr. Mater. (2022). https://doi.org/10.1016/j.scriptamat.2022.114771
174. Mariano, P.M., Mucci, D.: Micro-Slip-Induced Multiplicative Plasticity: Existence of Energy Minimizers. Archive for Rational Mechanics and Analysis (2023). https://doi.org/10.1007/s00205-023-01867-8
175. Li, X., Li, Z., Shen, J., Zheng, Z., Liu, J.: Role of a nanoparticle network in polymer mechanical reinforcement: insights from molecular dynamics simulations. Phys. Chem. Chem. Phys. (2021). https://doi.org/10.1039/D1CP03153H

176. Jia, Y., Lopez-Pamies, O., Zhang, X.S.: Controlling the fracture response of structures via topology optimization: from delaying fracture nucleation to maximizing toughness. J. Mech. Phys. Solids (2023). https://doi.org/10.1016/j.jmps.2023.105227
177. Youssef, K., Sakaliyska, M., Bahmanpour, H., Scattergood, R.O., Koch, C.C.: Effect of stacking fault energy on mechanical behavior of bulk nanocrystalline Cu and Cu alloys. Acta Mater. (2011). https://doi.org/10.1016/J.ACTAMAT.2011.05.052

Tribological Innovations in Nanomaterials: Friction, Wear, and Lubrication for Enhanced Biomedical Performance

Arya Muraleedharan, M. Unnimaya, and Jenny Jacob

Abstract Tribological advances in nanomaterials executed dramatic advancements in biomedical engineering, significantly enhancing the performance and durability of medical implants, prosthetics, and tissue engineering scaffolds. Nanostructures like carbon nanotubes and graphene represent carbon-based materials that, along with metallic nanoparticles and polymer nanocomposites, reduce wear and friction through boundary lubrication and minimize wear particle generation. Innovative bioinspired systems, such as self-lubricating hydrogels mimicking cartilage, highlight the potential to replicate superior wear resistance and lubrication of natural tissues. Advanced surface engineering and nanoscale additives in biofluids enhance biocompatibility but face challenges like cytotoxicity and scalability. This chapter examines the role of nanomaterials in improving friction, wear, and lubrication in medical applications, addressing issues of biocompatibility and production scalability. It highlights the potential of nanomaterials to enhance biomedical devices like orthopedic implants and tissue engineering scaffolds by integrating tribology and nanopharmacology for a safer and more effective solution.

Keywords Tribology · Nanotribology · Nanomaterials · Nanopharmacology · Biomedical devices · Friction and wear · Nanolubricants

1 Introduction

Biomedical devices and implants must be carefully assessed with close attention to the tribological characteristics displayed. Using technical developments like nanostructured materials can influence the friction, wear, and lubrication characteristics of the component material. Surface changes with which friction and wear can be reduced can provide the devices with unique qualities and properties. Low friction coefficient [1] and mechanical degradation of the material are much reduced

A. Muraleedharan · M. Unnimaya · J. Jacob (✉)
School of Biosciences, Mar Athanasios College for Advanced Studies, Tiruvalla, Kerala, India
e-mail: jenny@macfast.org; jennyjosappu@gmail.com

S. N. Saud Al-Humairi et al. (eds.), *Smart Nanomaterials in Biomedical Applications*, Engineering Materials, https://doi.org/10.1007/978-981-96-8023-8_3

in nanoscales by implants covered with nanostructured materials. Hamad et al. [2] on nanofluidic ultra-thin surface coatings made of carbon nanotubes or graphene demonstrating significant reduction in wear resistance.

The inclusion of nanomaterials can minimize friction and provide antibacterial activity, which leads to improved biocompatibility. Silver nanoparticles incorporated with nanocoating have been shown to impart lubrication and antimicrobial functionality on implant surfaces [3]. Infections and debris from frictional wear on the devices can greatly reduce efficiency, while coatings with biological effects remain important. Atomic force microscopy is used for nanotribology studies to understand surface roughness, adhesion, friction, boundary lubrication, among others [1]. Study of nanomechanics and nanoscale tribology, can lead to better learning about how the nanocoating on devices interact with other surfaces.

The frictional contact between two surfaces is dependent on several factors like surface roughness, adhesion, abrasion, etc. Natural lubrication systems, such as the synovial fluid in joints, utilize boundary, fluid film, and elastohydrodynamic lubrication to prevent degeneration. Inspired by biological lubrication, biomedical engineers create hydrogels, polymeric coatings, and nanolubricants for medical uses [4]. Bioinspired technologies lower friction-generated damage in implants by increasing prosthetic efficiency and patient comfort.

Often worsened by poor lubrication, wear is the slow loss of material resulting from mechanical stress and environmental exposure. Wear manifests as abrasive, adhesive, and fatigue wear, each of which has a distinct impact on biomedical equipment [5]. Whereas adhesive wear moves material between surfaces, abrasive wear results from sharp particles scratching surfaces. Repeated stress cycles in fatigue wear cause gadget failure. Lubrication speeds up wear, leading to degradation of the device and surrounding biological tissue [6].

Improving biomedical equipment depends much on tribology, the study of friction, wear, and lubrication. It helps choose biocompatible materials with low friction and wear properties, therefore minimizing negative biological reactions [7]. In prosthetics and dental implants, tribological innovations enhance device integration and longevity. By reducing friction-induced irritation, these advancements contribute to patient comfort and treatment success [8]. Researchers continue to develop durable, high-performance medical materials to withstand physiological stresses. Improved tribological strategies lead to safer, more effective biomedical solutions, ultimately enhancing patient care.

2 Types of Nanoparticles in Tribology

Nanoparticles have transformed the field of tribology by significantly enhancing the performance of biomedical devices through reduced friction, improved wear resistance, and enhanced lubrication. Nanoparticles are minimal materials, with at least one dimension measuring 100 nm or less. They exist in various forms, including surface coatings (one-dimensional), fibers or strands (two-dimensional), and particles

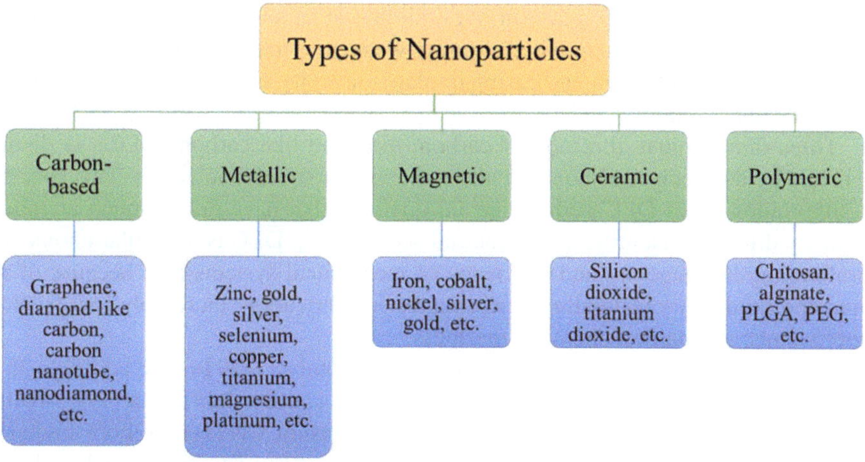

Fig. 1 Different types of nanoparticles used in tribology

(three-dimensional) [9, 10]. Nanoparticles can take on shapes such as spheres, tubes, or irregular forms. They may exist individually, combined, clustered, or grouped [11]. Nanoparticles such as quantum dots, fullerenes, dendrimers, carbon nanotubes, silver nanoparticles, and silica are widely used in nanotechnology for their unique properties. These materials can be classified based on their composition, structure, and function in tribological applications [12]. The major types of nanomaterials used in biomedical tribology are discussed below (Fig. 1).

2.1 Carbon Nanoparticles

Carbon nanomaterials have been widely used for research due to their varied structures and outstanding characteristics like corrosion resistance, thermal conductivity and mechanical strength of the nanoparticles make them valuable in fields like mechanics, optics, and electrochemistry [11]. These nanomaterials are effective solid lubricants, which can help reduce energy consumption. Research shows that solid lubricants can be used as additives or can be incorporated into coatings, in bulk materials and lubricants to improve lubricity and achieve superlubricity [13] due to their distinctive surfaces, made of mixed sp3 and sp2 bonded carbon atoms [14]. Some of the commonly used carbon nanomaterials in biomedical tribology are discussed below.

The excellent friction and wear characteristics of graphene make it a two-dimensional carbon nanomaterial primarily employed as an efficient solid lubricant. Reduced surface friction, strong mechanical strength and chemical stability make graphene the perfect component to improve the tribological properties of biological materials [15]. Benefits of graphene-based coatings that have uses in electrical contact interfaces, iron contact surfaces, sliding steel surfaces and space environments are electrical conductivity, ultralow wear, low friction, superlubricity, [16], superior surface protection, increasing implant longevity, and reducing the likelihood of wear-related failures.

Three-dimensional allotrope of carbon, diamond-like carbon (DLC), is distinguished by its special sp3 bonding arrangement among carbon atoms. This specific configuration gives DLC certain special qualities that have drawn much attention in many different scientific and industrial sectors [17]. DLC is a popular choice for prolonging the longevity and efficiency of mechanical systems [18] because of its amazing hardness, outstanding performance in lubrication, incredibly low friction coefficients, and natural chemical inertness.

Roll-up sheets of graphene called carbon nanotubes (CNTs) are additives that improve the tribological properties of both liquid and dry lubricants [19]. Mostly because of their higher surface energy and natural tension, CNTs show more tendency for winding and bundling over conventional materials [20]. This substantial tendency for bundling greatly lowers the tribological performance of the resulting tribofilm since less frictional capabilities [21] result from the discontinuity in the structure. Nanodiamonds exhibit unique properties like scalable high surface energy, biocompatibility, chemical stability and high thermal conductivity [14, 22]. Using these nanodiamonds together with conventional lubricants can reduces friction between the sliding surfaces [23, 24].

2.2 Metallic Nanoparticles

Nanomaterials are essential in modern medical science, providing significant advancements in diagnostics, imaging, and therapeutic applications. Nanoparticles can be divided into metal, metal oxide, and doped metal or metal oxide types. Metal and metal oxide nanoparticles are commonly used in tribological applications as they are able to reduce wear, improve load-bearing capacity, and increase surface durability [25]. With unique physicochemical properties, metallic nanoparticles like zinc, gold, silver, selenium, copper, titanium, magnesium, and platinum show great potential for biomedical uses. Unique optical and electrical features of gold nanoparticles include not interacting with substances. Their negative charge also makes their surfaces functionally interesting [26]. Some biosensors [27], bioimaging [28], and photothermal therapy [29] find usage for gold nanoparticles. Because gold gets easily attached to organic molecules, conjugating gold with ligands, antibodies, or medical compounds lets one use it actively or passively [30]. Applications in the electrical, photonic, antibacterial, and disinfection domains have been made possible by the good optical characteristics, chemical stability, conduction of heat and electricity, catalytic action, and chemical stability of silver nanoparticles [31]. Materials, including device coatings, wound dressings, and fabrics dyed with silver nanoparticles, are widely used in health care because of their outstanding antibacterial activity [32].

Fe–Cu, Fe–Co, Fe–Pt, Fe–Ni, and Cu–Ni are the listed bimetallic nanoparticles. High Curie temperature [33], superparamagnetism, and magnetic strength make iron cobalt (Fe–Co) nanoparticles appealing. Still, there are questions regarding their safety for living entities and their simple breakdown. Coating Fe–Co nanoparticles with a biocompatible substance is a standard solution to this problem. Treatment options for magnetic hyperthermia includes the use of Fe–Co [34] and Cu–Ni [35] nanoparticles. Ajmal et al. [36] nanoparticles can be applied as contrast agents for magnetic resonance imaging. Metal oxide nanoparticles are frequently used as nano additives in lubricant formulations due to their improved performance and efficiency [37].

2.3 *Magnetic Nanoparticles*

Magnetic nanoparticles consist mainly of Fe_3O_4, TiO_2, CeO_2, SiO_2, Al_2O_3, and so on [38]. TiO_2 nanoparticles having properties like chemical stability, and biocompatibility [39], explains their growing role in biomedical applications, including drug administration, bioimaging, photoablation treatment, and biosensors [40]. Cerium oxide (CeO_2) nanoparticles or nanoceria that have the capacity to change between several oxidation states make them prominent in the field of biomedicine as biosensor and anticancer agent [41]. Cerium (IV) and cerium (III) oxidation states coexisting on the surface of cerium oxide nanoparticles can cause many surface defects including oxygen vacancies [42].

SiO_2 and Al_2O_3 nanoparticles are widely used to enhance the hardness and durability in composite coatings. This increases the prosthetic joint and orthopaedic implant wear resistance [43]. Enhanced specific surface area, pore volume, controllable particle size, and ideal biocompatibility of porous silica (SiO_2) nanoparticles can raise their use in biomedical applications.

Magnetic hyperthermia uses iron oxide nanoparticles for various applications like contrast agents in magnetic resonance imaging (MRI) [44], targeted drug delivery to tissues [45], and cancer treatment [46]. α-Fe_2O_3, β-Fe_2O_3, γ-Fe_2O_3, and ε-Fe_2O_3 are some examples of the crystalline polymorphs of iron (III) oxide (Fe_2O_3) [47]. The magnetic properties of iron oxide nanoparticles can be increased by using metals like manganese (Mn), cobalt (Co), or nickel (Ni) [48]. Preventing agglomeration, facilitating attachment and interaction with anticancer drugs, enzymes, proteins, and antibodies are the primary functions of the shell coating of magnetic nanoparticles [49].

2.4 *Ceramic Nanoparticles*

A different subclass of advanced ceramics from traditional clay-based materials, nanoceramics were first identified in the early 1980s. Strong mechanical, chemical, physical, and magnetic properties abound in nanoceramics, composed of silicon carbide and alumina [50]. Excellent wear resistance is appreciated; orthopaedic and dental implants find usage for coatings based on zirconia [51]. Biomedical uses of ceramic nanoparticles have transformed drug delivery, bioassays, and imaging. Nanotubes, scaffolds, and nanoclays in focused drug delivery and cellular absorption yield nanosystems [52]. A perfect match for medical implants and regenerative medicine is defined by their great mechanical strength, low biodegradability, toxicity, increased biocompatibility, and long-term usage [53].

2.5 Polymeric Nanoparticles

Polymeric nanoparticles have become very common in biomedical tribology and several pharmaceutical uses, including drug administration systems, due to their flexibility, biocompatibility, and appropriateness to form lubricating films [54]. Structural form shows that polymeric nanoparticles might be mostly differentiable as either nanocapsules or nanospheres. Nanocapsules are reservoir systems whereby the drug is stored inside an oil core and then wrapped by a polymeric shell controlling its release [55, 56]. Continuous polymeric mats are either absorbed or adsorbed on the surface, and nanospheres let the drug be incorporated or adsorbed [57, 58].

Two self-lubricating coatings with less friction and more wear resistance used in biomedical implants are polytetrafluoroethylene (PTFE) and polyethylene glycol (PEG) [59, 60]. These polymers create hydrophilic surfaces that lower their biocompatibility, antimicrobial characteristics, and capacity to improve tribological performance [61] have drawn interest for chitosan-based nanoparticles. Hyaluronic acid is found in synovial fluid replacements for joint lubrication, much like natural biopolymers [62]. Fundamentally, polymer-based nanosystems are a fundamental building component used in the development of pharmaceutical nanotechnology to improve treatment stability, particle area targeting, side effect minimisation, and so on.

3 Characteristics of Nanomaterials Affecting Tribological Behaviour

Some nanomaterial characteristics can help to increase the tribological performance, or interaction under wear and friction. Encouraging interactions at contacting surfaces reduces friction coefficients and raises wear resistance with their small size, large surface area, and ideal mechanical characteristics [63]. Furthermore, adding nanoparticles to lubricants can boost viscosity and film-forming characteristics, improving performance in demanding conditions [9, 64].

3.1 Concentration of Nanoparticles

Nanoparticle effectiveness depends on their concentration, as their tribological characteristics are greatly influenced by it. Including nanoparticles in the lubricant usually helps to reduce friction and wear [65]. Moreover, multiple studies have shown that lubricants can include NPs at concentrations below 1 wt.% or beyond 2 wt.% without a single requirement for efficacy, hence, no optimal concentration fits every case [66].

More nanoparticles can first help the lubricant work better, but there is a limit to the subsequent benefits of reducing friction and wear, which do not follow correspondingly. Ideal concentration thus shows a definite range in which the reduction of friction and abrasion is maximized [67, 68]. This implies that nanoparticles will help the lubrication process, a threshold exists beyond which excessive amounts may not enhance or even work against the desired tribological enhancements. Understanding and determining this optimum concentration is crucial for harnessing the potential of nanolubricants in day-to-day applications.

3.2 Morphology of Nanoparticles

Nanoparticle dimensions can highly influence the tribological performance of nanolubricants. The smaller the particles, the greater the penetration of contacting surfaces and improved surface area-to-volume ratios [69]. Since the shape of nanoparticles greatly influences the pressure applied to the nanoparticles, synthesizing nanolubricants primarily during the loading process depends on this aspect. Spherical, granular, sheet, onion-like, and tubular are the five primary forms of nanoparticles; spherical is the most widely used form, followed by granular, sheet, onion-like, and tubular [70]. The crystalline structures of the particles develop to balance after nucleation, reducing surface energy [71]. Under isotropic surface energy conditions, nanoparticles usually take on a spherical form. Spherical nanoparticles generally foster a rolling mechanism, functioning similarly to ball bearings as they navigate between friction surfaces [72].

Spherical nanoAl_2O_3 improves load capacity, reduces friction, and minimizes wear [73], while nearly spherical nanoCu (Copper) enhances tribological properties, as shown by SEM and EDS analysis [74]. The shape of nanoparticles affects their contact with lubricated surfaces: spherical nanoparticles create point contact, while nanoplatelets offer a flat contact that can change lubrication dynamics [75]. Recent research on WS2 nanorods as additives in mineral oil demonstrated superior friction properties compared to 2H-WS2 due to the formation of a thin tribofilm promoted by adding 2H-WS2. These findings mark a significant advancement in lubrication technology [76].

3.3 Particle Stability of Nanoparticles

Nanoparticles possess a large surface area relative to their volume, which enhances surface energy and intermolecular forces, possibly resulting in agglomeration and sedimentation [77]. This reduces their effectiveness in lubrication [78]. The dispersion stability of nanoparticles can be determined by testing sedimentation rates using Stokes' law.

$$V_t = \frac{\left[2(\rho_{NP} - \rho_F)gr^2\right]}{9\eta}$$

Here, the settling velocity is described by V_t, ρ_{NP} is the mass density of the nanoparticles, ρ_F is the mass density of the fluid, g is the acceleration due to gravity, r is the radius of nanoparticles, and η is the dynamic viscosity [79, 80]. Maintaining dispersion stability ensures the lowest friction and wear. Among the techniques to increase its stability are homogeneity, high shear mixing, ultrasonic agitation, and magnetic force agitation [81]. Metallographic micrograph tests, spectral absorption tensor analysis, centrifugal testing, and zeta potential measurement are among the ways to assess the dispersion stability of nanolubricants in different uses.

4 Mechanisms of Tribological Enhancement for Nanomaterials

Effective lubricating agents, such as nanoparticles, reduce friction and wear while enhancing the load-bearing capacity of mechanical components. Understanding how nanolubricants function can help one fully appreciate their tribological behavior [4, 82]. They can be classified into two basic categories: surface-enhancing nanoparticles (polishing/mending) and nanoparticles that directly improve lubrication (ball bearing effect/protective film formation) [82].

4.1 Ball Bearing Mechanism

These days, nanoparticles are one of the main and creative options for synthesizing additives [83]. Several reliable studies on nanoparticles have been carried out, and their use yields different results, reducing the lubricant wear and tear [84]. These form additives stop wear and lessen friction between the rubbing material surfaces. An experiment [85] using diamond nanoparticles combined with liquid paraffin revealed the ball bearing mechanism of nanoparticles. It has been reported that as the duration of the experiment increases, the spherical diamond nanoparticles transform into tiny

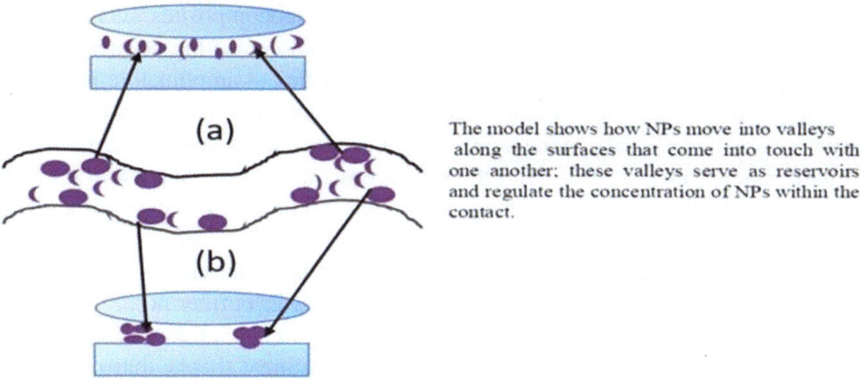

Fig. 2 Mechanisms of MoS_2 protective tribofilms: **a** Monosheet that exfoliates with a thin film formed, and **b** Aggregates separating interfacial contact with a thick film formed

rolling materials similar to ball bearings and function according to the process shown in Fig. 2 [86].

Additionally, materials with rough surfaces slide because nanoparticles deposit troughs on the rubbing surface, acting as a surface treatment medium. Although preparing graphite nanomaterial was challenging, the outcome was quite significant when it was added to an oil lubricating system [87]. Nanoscale modified graphene platelets (MGPs) work as nano-bearings between tribocouples as a lubricant. The Al_2O_3/TiO_2 nanocomposite has excellent wear and friction properties for lubrication [88].

4.2 Rolling Mechanism

This approach introduces a layer of amorphous material composed of nanoparticles to the surface of the sliding element, lowering the contact area [89]. Nanoparticles create sorbent films that can attach to the surfaces of tribo-pair entities [90]. Furthermore, filling in the generated wear tracks is the frictional pressure and heat-generating sintering of these nanoparticles [91]. Good dispersion of nanoparticles in the fluid will most certainly produce a thin layer covering the metal surface. This low-shear, dense film separates the contact points, reducing frictional effects [92]. While the system is working, the nanoparticles in the fluid naturally tend to move toward the area where the two parts rub against each other [93]. The protective film mechanism was demonstrated using magnesium borate in the 500-SN base oil. The results show an amorphous film layer forming between the sliding contact surfaces. The coefficient of friction went down when a new layer was added, and wear resistance went up simultaneously [94].

Consequently, the tribological performance of nanocomposites such as SiO_2 nanoparticles is enhanced by interacting with material-contacting surfaces, transitioning from hydrophilic or hydrophobic characteristics to amphipathic properties [95]. In the additive operation of nanoparticles, agglomeration facilitates rolling and sliding across the material contact area, influencing the tribological properties, material hardness, contact pressure, and the interactions among the nanoparticles [96]. The study of IF-MoS_2 nanoparticles demonstrates that the size and crystalline shape significantly influence their lubricating properties [97]. The coefficient of friction with the addition of 1 w% of IF-MoS_2 under boundary lubrication is lower than that of the pure PAO oil sample, as illustrated in Fig. 2. Furthermore, the performance of the COF in terms of reduction properties is not influenced by contact pressure. This tribofilm mechanism is the most prominent theory that provides a deeper understanding of lubricants' friction and wear reduction behavior [98].

4.3 Mending Mechanism

During the operation of nanofluid, nanoparticles in the lubricant tend to deposit, fill, and resurface the damaged rubbing contact by desorption and deposition, thereby facilitating the formation of a smoother and flatter interface. The self-repairing operation is predominantly observed in face-centered cubic soft metals [99]. According to Zhu et al. [100], nano-copper additives in lubricants typically exhibit properties that facilitate the separation of frictional components during friction. Nevertheless, the mending operation and self-repairing process are not confined to the accumulation of nanoparticles on material surfaces but also encompass the size of nanoparticles and their melting point characteristics [101].

At elevated temperatures beyond the melting point of the additional materials, the nanoparticles either melt or sinter in the damaged contact area, simultaneously functioning as a filler to repair the component's working surface [102]. Carbon nanotubes (CNTs) and carbon nanofibers (CNFs) have been discovered to penetrate microscopic gaps between rough surfaces in contact, generating self-lubricating thin films. It was found that the incorporation of diamond particles as a nanoparticle addition in paraffin oil resulted in a smoother surface on the rubbing contact at the commencement of the tribotest, in contrast to the surface image obtained with pure paraffin oil [103].

4.4 Polishing Mechanism

Primarily dependent on the hardness of the nanoparticles, the mechanism helps preserve or change the smoothness of the machine surface components [104]. Usually, tribological pairs grow in contact areas as the material gets smoother. This reduces the coefficient of friction, therefore raising the load-bearing capacity.

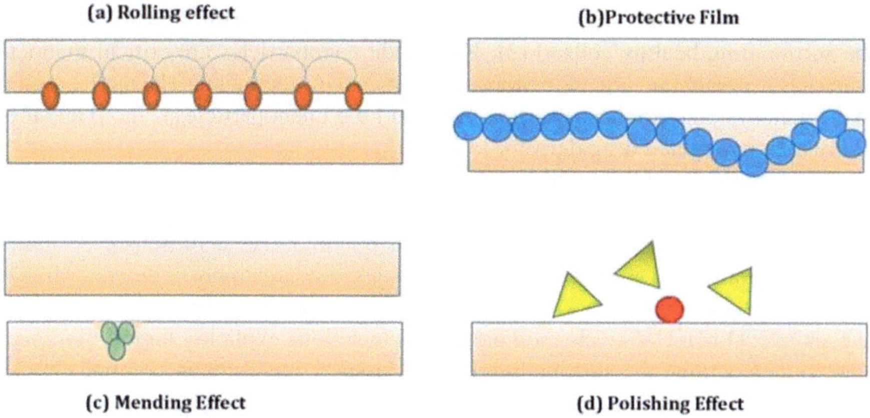

Fig. 3 Various lubrication mechanisms of nanoparticles

Lubrication employing nanofluid showed a 15% decline over standard commercial oil [105]. The polishing of diamond nanoparticles generated a smoother surface, reducing the nanofluid-type value [106]. Under this procedure, nanoparticles improve when they come into contact with surfaces that are not too rough, they don't improve the tribological features of nanofluids when they come into contact with highly rough surfaces [107]. Figure 3 shows the various lubrication mechanisms of the nanoparticles.

5 Biomedical Applications of Nanomaterial-Based Tribological Innovations

The use of nanoparticles as lubricants is a novel concept. The advantages of nanolubricants include their stability under temperature changes and reduced reactivity with other compounds compared to regular lubricant additives. These must first chemically interact with substrates before experiencing an induction period to produce the tribofilm on sliding surfaces [108]. Colloidal solutions of oils and solid nanoparticles [109] serve as nanolubricants. Because their surfaces have a lot of energy, nanoparticles tend to clump together, lowering their stability and causing precipitation [110]. There are one-step and two-step processes for synthesizing nanolubricants. The one-step process involves the simultaneous dispersion of nanoparticles and a dispersion agent in the base fluid.

In contrast, the two-step process requires the base fluid and nanoparticles to be prepared separately before mixing. Both methods necessitate surface modifications or functionalization to enhance the dispersion stability of the nanolubricant. Paul et al. [111] reported the superior performance of graphene dispersed in lubricants, which exhibited improved tribological characteristics. Drug delivery is a significant focus in medicine, as it administers cancer treatments directly to the tumor while protecting the surrounding healthy cells [112]. Iron oxide nanoparticles are critical magnetic materials for reaching target locations with anticancer treatments. The application of many nanosystems emphasizes TiO_2, Ag, and Fe–Pt nanoparticles (Fig. 4) and their likely uses in targeted drug delivery [47].

One promising field of biomedical research is biosensor development, which can detect a wide range of biomolecules [113]. Several nanoparticles, including Au, Fe–sPt, CeO_2, and TiO_2, are studied for their potential use as biosensors. Another important biomedical application of nanoparticles is the treatment of magnetic hyperthermia, as shown in Fig. 5. This therapy raises temperatures above 42 °C to destroy cancer cells [114]. Unlike chemotherapy, this technique explicitly targets malignancies while shielding the surrounding healthy tissues. The primary component used for this treatment is iron oxide (Fe_3O_4); other nanosystems under research include bimetallic nanoparticles (Fe–Co and Cu–Ni) and magnetic nanoparticles (Co–Fe_2O_4, Mn–Fe_2O_4, and Ni–Co_2O_4) [115]. Localized hyperthermia treatment also utilizes magnetic nanoparticles. This method thermally stimulates a specific limb or organ. Total body hyperthermia is usually recommended to treat metastatic cancer that spreads throughout the body.

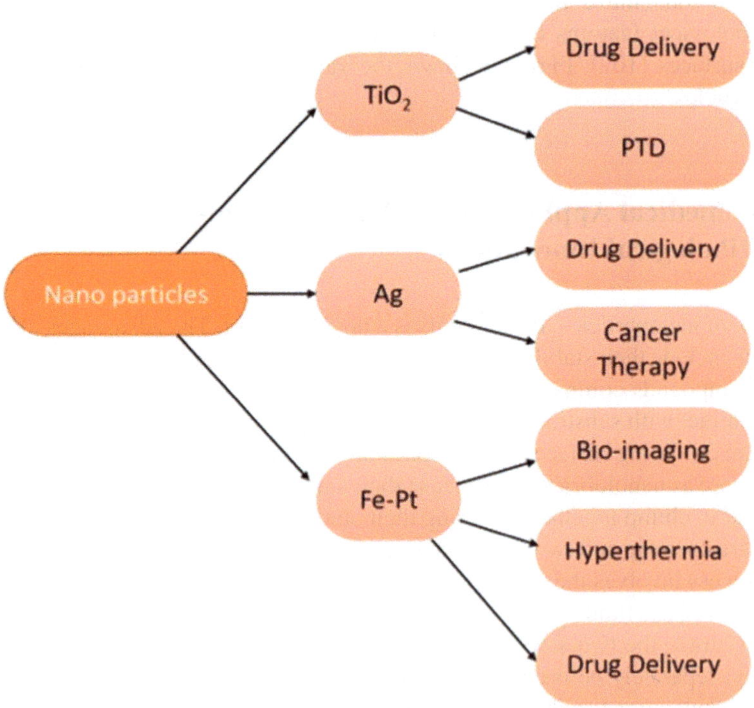

Fig. 4 Commonly used metal-based nanoparticles and their applications

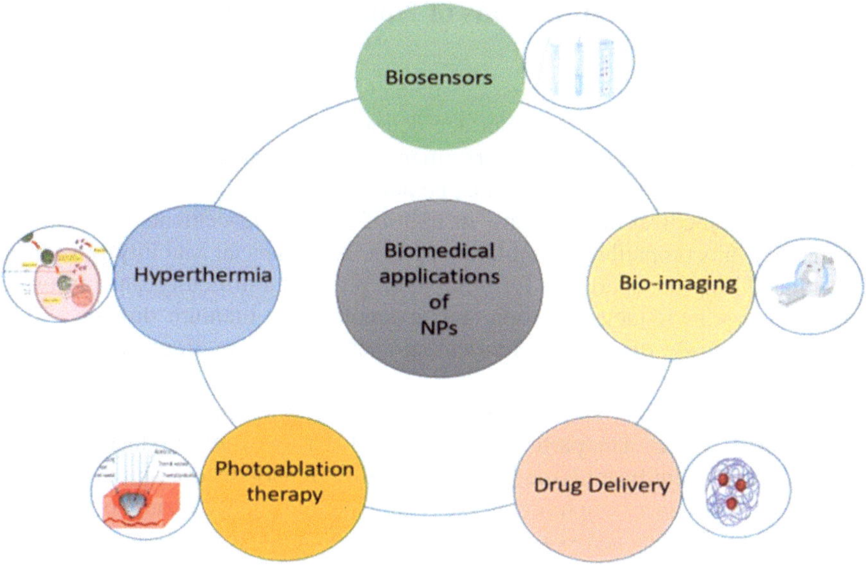

Fig. 5 Biomedical applications of nanoparticles

In bioimaging research, applications of nanoparticles define the creation of contrast agents for MRI and CT testing. Depending on friction, adhesion, and wear, these nanoparticles can influence interactions with biological tissues and improve imaging efficiency. In imaging joints, MRI contrast compounds assist in evaluating synovial fluid lubrication, a crucial factor for the diagnosis of degenerative diseases like osteoarthritis [116]. Moreover, nanoparticle tribology affects biocompatibility, retention, clearance, imaging time, and dispersion. More extraordinary contrast agents are more stable, less toxic, and more useful for therapeutic reasons can be produced by means of greater understanding of these interactions [117].

Photothermal therapy and photodynamic therapy (PDT) are two primary variants of photoablation treatment. PDT treatment mainly targets cancer cells using compounds known as photosensitizers, which are benign when not exposed to light but turn harmful at a specific wavelength [118]. It mostly aims to attack malignant cells. The PDT method exposes photosensitizers, such as TiO_2 nanoparticles, to emit light at a specific wavelength, which produces electrons and holes. Light interacts with some materials to generate electrons and positively charged creatures known as holes [119]. These can affect water or hydroxyl ions, producing single oxygen and dangerous molecules called reactive oxygen species (ROS). The development of these organisms causes unavoidable cellular harm. Targeting tumors; therapy employs a near-infrared (NIR) light source. Energy derived from this light source can cause hyperthermia and consequent cell death. Photocatalytic activity, in particular, makes it appealing for use in photothermal treatment [120].

6 Challenges and Limitations of Nanomaterials in Tribology

By mitigating friction, wear, and surface degradation, nanomaterials have significantly enhanced the tribological performance of biomedical tools. Nonetheless, various challenges and limitations must be addressed to facilitate their widespread adoption in biomedical applications, in light of these advancements [121]. Material stability, biocompatibility regulatory issues, and financial viability pose these difficulties. Using nanoparticles for lubrication presents a difficulty since it lowers their efficiency in reducing friction. For example, how titanium dioxide (TiO_2) nanoparticles cluster disturbs the lubricating coating, resulting in unequal wear on parts.

Small-scale laboratory approaches generally struggle to achieve uniform coating thickness and correct nanoparticle distribution for large-scale production; scalability becomes a significant issue [122]. Variability in particle size and distribution influences performance, emphasizing the requirement of standardizing to guarantee quality and consistency across several batches and manufacturers—an unnecessary condition for more general acceptance in many sectors [123].

The use of nanomaterials in medical implants calls for rigorous approval procedures to guarantee safety and efficacy, including FDA rules and ISO conformity. Safety, toxicity, and efficacy must all be evaluated using long-term clinical studies. Furthermore, the environmental impact of discarding devices loaded with nanoparticles raises questions about nanotoxicity and emphasizes the need for biodegradable materials [124]. The advent of nanomaterials also raises ethical and patient safety issues, so openness in risk assessments and research is essential to foster public confidence [51].

Under repeated stress, nanocoatings such as graphene face adhesion and cohesion difficulties, increasing the risk of delamination and lower durability. Diamond-like carbon (DLC) is one such material. This emphasizes how much more energetic interfacial bonding methods are needed. DLCs may become graphitic and lose lubricating qualities; friction might cause structural alterations [17]. Although nanoparticles improve load distribution, too heavy loads could cause surface fatigue and material degradation. The efficacy and lifetime of nanocoating in practical applications depend thus on ideal surface texturing and careful material choice.

7 Emerging Trends and Future Perspectives in Nanomaterial Tribology

Nanomaterials transform tribology by improving lubrication, lowering wear, and extending the life of mechanical systems and biomedical devices. Significant developments are intelligent nanolubricants and self-healing coatings that may release lubricants in response to mechanical wear. In medical uses, nature-inspired designs such as lotus leaf-like nanostructures [125] and fish scale-inspired coatings improve lubrication and lower friction [126]. Combining materials like graphene and molybdenum disulfide (MoS_2) hybrid lubricants offers extremely low friction and excellent wear resistance. Prosthetic joints and dental implants significantly benefit from these developments [127].

Offering creative answers to long-standing problems in medical devices, implants, and prostheses, recent developments in artificial intelligence (AI) driven nanomaterial design and 4D printing are changing the field of biomedical tribology [128]. Artificial intelligence and machine learning are mostly driving an acceleration in the search and optimization of nanomaterials. AI generates nanocomposites with customized features, including decreased friction, higher wear resistance, and improved biocompatibility [129]. On the other hand, 4D printing offers material design a fresh perspective by means of surfaces that could react to environmental changes, including mechanical stress, pH, or temperature fluctuations. Under many conditions, this approach produces dynamic, self-healing materials and patient-specific implants with surfaces maximizing lubrication and minimizing friction. 4D-printed cartilage-like materials helps to enhance performance and durability in joint implants, which imitate the natural lubrication mechanisms of cartilage [130].

4D printing under AI-driven design is opening extraordinary opportunities in biomedical tribology. AI enhances the ability to predict and improve material properties, while 4D printing adaptively realizes these concepts. Together, they create dynamic, intelligent, self-optimizing tribological systems for the human biomechanical environment. Future research will focus on developing new stimuli-responsive materials, expanding current technologies for therapeutic use, and integrating real-time data to facilitate predictive maintenance [8].

Additionally, bioinspired systems are being developed to replicate natural low-friction surfaces, with notable advancements including hydrophobic coatings and lubricants that mimic synovial fluid, finding significant applications in prosthetic uses [131]. These advancements promise significant impacts across various disciplines—such as health, aircraft, robotics, and renewable energy. Incorporating innovative nanomaterials is expected to enhance the lifespan and efficiency of mechanical components across sectors, fostering multidisciplinary collaboration and bridging gaps in numerous scientific domains [132]. With the urgent need to reduce environmental effects and encourage environmentally conscious innovation, sustainability is becoming the cornerstone of upcoming innovations in biomedical tribology. The production of biodegradable nanomaterials with exceptional tribological performance and low environmental effects is of greater relevance for researchers [133]. Moreover, the research of environmentally friendly lubricants is gathering momentum as non-toxic, bio-based formulations offer efficient lubrication without endangering human or environmental health. Research on ecologically acceptable substitutes for traditional petroleum-based lubricants, such as plant-based oils and biocompatible polymers, is in progress [134]. Including these green technologies meets the need for high-performance biomedical solutions and helps the industry meet world sustainability targets, assuring that ongoing developments are creative and ecologically responsible.

8 Conclusion

Nanomaterials exhibit exceptional mechanical strength and thermal stability, with self-repairing effects. Incorporating nanomaterials with lubricants can reduce friction and wear by providing surface smoothness for better action. Such nanolubricants' stability must be ensured to reduce maintenance costs and increase dispersion stability. Nanomaterial-based tribology is a transformative field with significant implications across various scientific and industrial areas. Improvements in energy-efficient tribological coatings for wind turbines, solar panels, and electric vehicles are expected to enhance efficiency and support sustainable energy efforts. By addressing current challenges and utilizing advanced technologies, future research will strengthen the development of intelligent and efficient tribological solutions. Its interdisciplinary nature promises advances in biomedical engineering, aerospace, automotive, and renewable energy. Continued innovation and collaboration will redefine friction control, wear resistance, and lubrication, contributing to a more efficient and sustainable future.

References

1. Bhushan, B.: Nanotribology and Nanomechanics: An Introduction. Springer, Berlin (2008). https://doi.org/10.1007/978-3-540-77608-6
2. Hamad, E.M., Khaffaf, A., Yasin, O., Abu El-Rub, Z., Al-Gharabli, S., Al-Kouz, W., Chamkha, A.J.: Review of nanofluids and their biomedical applications. J. Nanofluids **10**(4), 463–477 (2021). https://doi.org/10.1166/jon.2021.1806
3. Yilmaz, O., Yorgancioglu, A.: Nanocoatings: preparation, properties, and biomedical applications. In: Polymeric Nanomaterials in Nanotherapeutics, pp. 299–331. Elsevier (2019). https://doi.org/10.1016/B978-0-12-813932-5.00008-X
4. Jason, Y.J.J., How, H.G., Teoh, Y.H., Chuah, H.G.: A study on the tribological performance of nanolubricants. Processes **8**(11), 1372 (2020). https://doi.org/10.3390/pr8111372
5. Williams, J.A.: Wear and wear particles—some fundamentals. Tribol. Int. **38**(10), 863–870 (2005). https://doi.org/10.1016/j.triboint.2005.03.007
6. Myshkin, N.K., Petrokovets, M.I., Kovalev, A.V.: Tribology of polymers: adhesion, friction, wear, and mass-transfer. Tribol. Int. **38**(11), 910–921 (2005). https://doi.org/10.1016/j.triboint.2005.07.016
7. Bhushan, B., Ko, P.L.: Introduction to tribology. Appl. Mech. Rev. **56**(1), B6–B7 (2003). https://doi.org/10.1115/1.1523360
8. Kumar, A., Gajjar, P., Gudadur, K.S., Kumar, A., Gupta, A.K., Choudhari, A., Chowdhury, N., Kumar, A.: Tribological hurdles in biomedical manufacturing: a comprehensive examination. In: Kumar, A., Kumar, A., Kumar, A. (eds.): Applications of Biotribology in Biomedical Systems. Springer, Cham (2024). https://doi.org/10.1007/978-3-031-58327-8_6
9. Gulzar, M., Masjuki, H.H., Kalam, M.A., Varman, M., Zulkifli, N.W.M., Mufti, R.A., Zahid, R.: Tribological performance of nanoparticles as lubricating oil additives. J. Nanopart. Res. **18**(8), 223 (2016). https://doi.org/10.1007/s11051-016-3537-4
10. Liu, W., Qiao, X., Liu, S., Chen, P.: A review of nanomaterials with different dimensions as lubricant additives. Nanomaterials **12**(21), 3780 (2022). https://doi.org/10.3390/nano12213780
11. Raina, A., Haq, M.I.U., Victoria, P.I., Mohan, S.R.J., Anand, A. (Eds): Nanomaterials for Sustainable Tribology. CRC Press (2023). https://doi.org/10.1201/9781003306276

12. Najar, F.A., Masoodi, J.H., Banda, S., Najar, N.A.: Nanomaterials and tribology: an introduction. In: Raina, A., Ul Haq, M.I., Victoria, P.I., Mohan, S.R.J., Anand, A. (eds.): Nanomaterials for Sustainable Tribology. CRC Press (2023). https://doi.org/10.1201/9781003306276
13. Wang, H., Liu, Y.: Superlubricity achieved with two-dimensional nano-additives to liquid lubricants. Friction **8**(6), 1007–1024 (2020). https://doi.org/10.1007/s40544-020-0410-3
14. Zhai, W., Srikanth, N., Kong, L.B., Zhou, K.: Carbon nanomaterials in tribology. Carbon **119**, 150–171 (2017). https://doi.org/10.1016/j.carbon.2017.04.027
15. Penkov, O., Kim, H.J., Kim, H.J., Kim, D.E.: Tribology of graphene: a review. Int. J. Precis. Eng. Manuf. **15**(3), 577–585 (2014). https://doi.org/10.1007/s12541-014-0373-2
16. Sun, J., Du, S.: Application of graphene derivatives and their nanocomposites in tribology and lubrication: a review. RSC Adv. **9**(69), 40642–40661 (2019). https://doi.org/10.1039/C9RA05679C
17. Al Mahmud, K.A.H., Kalam, M.A., Masjuki, H.H., Mobarak, H.M., Zulkifli, N.W.M.: An updated overview of diamond-like carbon coating in tribology. Crit. Rev. Solid State Mater. Sci. **40**(2), 90–118 (2015). https://doi.org/10.1080/10408436.2014.940441
18. Katiyar, J.K., Sahu, R.K., Gupta, T.C.S.M.: Sustainable Lubrication. CRC Press (2022). https://doi.org/10.1201/9781003201199
19. Qianming, G., Dan, L., Zhi, L., Xiao-Su, Y., Ji, L.: Chapter 10—Tribology properties of carbon nanotube-reinforced composites. In: Friedrich, K., Schlarb, A.K. (eds.) Tribology Interface and Engineering Series, vol. 55, pp. 245–267. Elsevier (2008) https://doi.org/10.1016/S1572-3364(08)55010-5
20. Lim, D.S., An, J.W., Lee, H.J.: Effect of carbon nanotube addition on the tribological behavior of carbon/carbon composites. Wear **252**(5), 512–517 (2002). https://doi.org/10.1016/S0043-1648(02)00012-1
21. Zhang, F.Z., Liu, X.B., Yang, C.M., Chen, G.D., Meng, Y., Zhou, H.B., Zhang, S.H.: Insights into robust carbon nanotubes in tribology: from nano to macro. Mater. Today **74**, 203–234 (2024). https://doi.org/10.1016/j.mattod.2024.03.001
22. Neitzel, I., Mochalin, V.N., Gogotsi, Y.: Chapter 14—Nanodiamonds in composites: polymer chemistry and tribology. In: Arnault, J.C. (ed.) Nanodiamonds, pp. 365–390. Elsevier (2017). https://doi.org/10.1016/B978-0-32-343029-6.00015-5
23. Amaral, M., Abreu, C.S., Fernandes, A.J.S., Oliveira, F.J., Gomes, J.R., Silva, R.F.: Nanodiamond-based tribosystems. Surf. Coat. Tech. **204**(12), 1962–1969 (2010). https://doi.org/10.1016/j.surfcoat.2009.11.015
24. Zhao, W., Duan, F.: Friction properties of carbon nanoparticles (nanodiamond and nanoscroll) confined between DLC and a-SiO_2 surfaces. Tribol. Int. **145**, 106153 (2020). https://doi.org/10.1016/j.triboint.2019.106153
25. Wang, S., Chen, D., Hong, Q., Gui, Y., Cao, Y., Ren, G., Liang, Z.: Surface functionalization of metal and metal oxide nanoparticles for dispersion and tribological applications—a review. J. Mol. Liq. **389**, 122821 (2023). https://doi.org/10.1016/j.molliq.2023.122821
26. Akbulut, M.: Nanoparticle-based lubrication systems. J. Powder Metall. Min. **1**, e101 (2012). https://doi.org/10.4172/2168-9806.1000e101
27. Dai, W., Kheireddin, B., Gao, H., Liang, H.: Roles of nanoparticles in oil lubrication. Tribol. Int. **102**, 88–98 (2016). https://doi.org/10.1016/j.triboint.2016.05.020
28. Bakunin, V., Suslov, A., Kuzmina, G., Parenago, O., Topchiev, A.: Synthesis and application of inorganic nanoparticles as lubricant components—a review. J. Nanopart. Res. **6**, 273–284 (2004). https://doi.org/10.1023/b:nano.0000034720.79452.e3
29. Mosleh, M., Atnafu, N.D., Belk, J.H., Nobles, O.M.: Modification of sheet metal forming fluids with dispersed nanoparticles for improved lubrication. Wear **267**(5–8), 1220–1225 (2009). https://doi.org/10.1016/j.wear.2008.12.074
30. Padgurskas, J., Rukuiza, R., Prosyčevas, I., Kreivaitis, R.: Tribological properties of lubricant additives of Fe, Cu and Co nanoparticles. Tribol. Int. **60**, 224–232 (2012). https://doi.org/10.1016/j.triboint.2012.10.024
31. Yu, H., Xu, Y., Shi, P., Xu, B., Wang, X., Liu, Q.: Tribological properties and lubricating mechanisms of Cu nanoparticles in lubricant. Trans. Nonferr. Metals Soc. China **18**(3), 636–641 (2008). https://doi.org/10.1016/s1003-6326(08)60111-9

32. Ahmed, Y., Ankah, N.K., Ogunlakin, N., Toor, I.U., Farooq, W.: Exploring the future of metallic implants: a review of biodegradable and non-biodegradable solutions. Corr. Rev. (2024). https://doi.org/10.1515/corrrev-2024-0125
33. Arora, N., Thangavelu, K., Karanikolos, G.N.: Bimetallic nanoparticles for antimicrobial applications. Front. Chem. **8**, 412 (2020). https://doi.org/10.3389/fchem.2020.00412
34. Chen, L., Xu, H., Cui, H., Zhou, H., Wan, H., Chen, J.: Preparation of Cu–Ni bimetallic nanoparticles surface-capped with dodecanethiol and their tribological properties as lubricant additive. Particuology **34**, 89–96 (2017). https://doi.org/10.1016/j.partic.2016.12.006
35. Fu, X., Guo, S., Wan, Y., Sun, P., Fan, Y., Zheng, C., Li, Q.: Tribological behavior and lubrication mechanism of polyalphaolefin with added Cu–Ni bimetallic nanoparticles on a TiN coating. Appl. Surf. Sci. **679**, 161287 (2024). https://doi.org/10.1016/j.apsusc.2024.161287
36. Ajmal, M., Muslim, F., Ali, S.A., Shah, M.R.: Synthesis of bio-based nanolubricant using bimetallic nanoparticles to mitigate the surface deterioration of low carbon steel for metal forming industry. Mater. Chem. Phys. **327**, 129913 (2024). https://doi.org/10.1016/j.matchemphys.2024.129913
37. Shiv, J.K., Kumar, K., Jayapalan, S.: Recent advances in polymer using metal oxides nanocomposite and its hybrid fillers for tribological application. Adv. Mater. Process. Technol. **10**(4), 2720–2731 (2024). https://doi.org/10.1080/2374068X.2023.2171673
38. Wang, B., Qiu, F., Barber, G.C., Zou, Q., Wang, J., Guo, S., Yuan, Y., Jiang, Q.: Role of nano-sized materials as lubricant additives in friction and wear reduction: a review. Wear **490–491**, 204206 (2021). https://doi.org/10.1016/j.wear.2021.204206
39. Khoobchandani, M., Ghosh, S.: (ed.): Medicinal Applications of Phytopharmaceuticals. Springer Nature Switzerland (2024). https://doi.org/10.1007/978-3-031-63677-6
40. Aniołek, K., Kupka, M., Barylski, A., Dercz, G.: Mechanical and tribological properties of oxide layers obtained on titanium in the thermal oxidation process. Appl. Surf. Sci. **357**, 1419–1426 (2015). https://doi.org/10.1016/j.apsusc.2015.09.245
41. Pancrecious, J.K., Deepa, J.P., Jayan, V., Bill, U.S., Rajan, T.P.D., Pai, B.C.: Nanoceria induced grain refinement in electroless Ni–B–CeO_2 composite coating for enhanced wear and corrosion resistance of Aluminium alloy. Surf. Coat. Tech. **356**, 29–37 (2018). https://doi.org/10.1016/j.surfcoat.2018.09.046
42. Wu, L., Lei, X., Zhang, Y., Zhang, S., Yang, G., Zhang, P.: The tribological mechanism of cerium oxide nanoparticles as lubricant additive of poly-alpha olefin. Tribol. Lett. **68**(4), 101 (2020). https://doi.org/10.1007/s11249-020-01340-7
43. Kotia, A., Ghosh, G.K., Srivastava, I., Deval, P., Ghosh, S.K.: Mechanism for improvement of friction/wear by using Al_2O_3 and SiO_2/gear oil nanolubricants. J. Alloys Comp. **782**, 592–599 (2019). https://doi.org/10.1016/j.jallcom.2018.12.215
44. Chen, S., Wang, J., Lu, H., Xu, L.: Surfactant-modified silica nanoparticles-stabilized magnetic polydimethylsiloxane-in-water pickering emulsions for lubrication and anticorrosion. ACS Sust. Chem. Eng. **10**(33), 10816–10826 (2022). https://doi.org/10.1021/acssuschemeng.2c01753
45. Shahnazar, S., Bagheri, S., Hamid, S.B.A.: Enhancing lubricant properties by nanoparticle additives. Int. J. Hydrogen Energy **41**(4), 3153–3170 (2016). https://doi.org/10.1016/j.ijhydene.2015.12.040
46. Van Sang, N.L., Sugimura, N., Khajeh, K., Washizu, H.: Solid lubricants of combined graphene and iron nanoparticles for study of friction and stability. Langmuir **38**(5), 1860–1868 (2022). https://doi.org/10.1021/acs.langmuir.1c03075
47. McNamara, K., Tofail, S.A.M.: Nanoparticles in biomedical applications. Adv. Phys. X **2**(1), 54–88 (2017). https://doi.org/10.1080/23746149.2016.1254570
48. Guo, X., Huang, Q., Wang, C., Liu, T., Zhang, Y., He, H., Zhang, K.: Effect of magnetic field on cutting performance of micro-textured tools under Fe_3O_4 nanofluid lubrication condition. J. Mater. Process. Tech. **299**, 117382 (2021). https://doi.org/10.1016/j.jmatprotec.2021.117382
49. Liñeira del Río, J.M., López, E.R., González Gómez, M., Yáñez Vilar, S., Piñeiro, Y., Rivas, J., Gonçalves, D.E.P., Seabra, J.H.O., Fernández, J.: Tribological behavior of nanolubricants based on coated magnetic nanoparticles and trimethylolpropane trioleate base oil. Nanomater **10**(4), 683 (2020). https://doi.org/10.3390/nano10040683

50. Mukhopadhyay, A.: Tribological properties and processing challenges of bulk structural ceramic nanomaterials. Tribol. Mater. Surf. Interface **2**(3), 169–184 (2008). https://doi.org/10.1179/175158309X408351
51. Zhao, Q., Wang, L., Hu, T., Song, J., Su, Y., Hu, L.: Research on the preparation of zirconia coating on titanium alloy surface and its tribological properties. Lubricants **12**(5), 154 (2024). https://doi.org/10.3390/lubricants12050154
52. Cui, Y., Ding, M., Sui, T., Zheng, W., Qiao, G., Yan, S., Liu, X.: Role of nanoparticle materials as water-based lubricant additives for ceramics. Tribol. Int. **142**, 105978 (2020). https://doi.org/10.1016/j.triboint.2019.105978
53. Rui, L.M., Sulaiman, M.H., Ibrahim, M.S., Jamir, M.R.M.: Enhancing physical, mechanical, tribological and contact characteristics of water-based lubrication using ceramic nanomaterials MgO and SiC for metal-to-metal sliding interfaces. J. Tribol. **39**, 51–68 (2023)
54. Prabhakar, K., Debnath, S., Ganesan, R., Palanikumar, K.: A review of mechanical and tribological behaviour of polymer composite materials. IOP Conf. Ser. Mater. Sci. Eng. **344**(1), 012015 (2018). https://doi.org/10.1088/1757-899X/344/1/012015
55. Jeng, Y.R., Huang, Y.H., Tsai, P.C., Hwang, G.L.: Tribological properties of carbon nanocapsule particles as lubricant additive. J. Tribol. **136**(4), 041801 (2014). https://doi.org/10.1115/1.4027994
56. Sun, J., Wang, Y., Li, N., Tian, L.: Tribological and anticorrosion behavior of self-healing coating containing nanocapsules. Tribol. Int. **136**, 332–341 (2019). https://doi.org/10.1016/j.triboint.2019.03.062
57. Luo, T., Chen, X., Wang, L., Wang, P., Li, C., Zeng, H., Cao, B.: Green laser irradiation-stimulated fullerene-like MoS_2 nanospheres for tribological applications. Tribol. Int. **122**, 119–124 (2018). https://doi.org/10.1016/j.triboint.2018.02.040
58. Wu, B., Zhu, L., Sun, Y., Wu, S.: Tribological performance of carbon nanospheres as lubricant additives: insights from molecular dynamics simulation and experimental analysis. J. Tribol. **147**, 042201 (2025). https://doi.org/10.1115/1.4067691
59. Deshwal, D., Belgamwar, S.U., Bekinal, S.I., Doddamani, M.: Role of reinforcement on the tribological properties of polytetrafluoroethylene composites: a comprehensive review. Polym. Compos. **45**(16), 14475–14497 (2024). https://doi.org/10.1002/pc.28802
60. Ghaednia, H., Hossain, M.S., Jackson, R.L.: Tribological performance of silver nanoparticle-enhanced polyethylene glycol lubricants. Tribol. Transac. **59**(4), 585–592 (2016). https://doi.org/10.1080/10402004.2015.1092623
61. Li, X., Wang, Z., Dong, G.: Preparation of nanoscale liquid metal droplet wrapped with chitosan and its tribological properties as water-based lubricant additive. Tribol. Int. **148**, 106349 (2020). https://doi.org/10.1016/j.triboint.2020.106349
62. Ghosh, S., Choudhury, D., Das, N.S., Pingguan-Murphy, B.: Tribological role of synovial fluid compositions on artificial joints—a systematic review of the last 10 years. Lubr. Sci. **26**(6), 387–410 (2014). https://doi.org/10.1002/ls.1266
63. Ali, Z.A.A.A., Takhakh, A.M., Al-Waily, M.: A review of use of nanoparticle additives in lubricants to improve its tribological properties. Mater. Today Proc. **52**, 1442–1450 (2022). https://doi.org/10.1016/j.matpr.2021.11.193
64. Loo, D.L., Teoh, Y.H., How, H.G., Le, T.D., Nguyen, H.T., Rashid, T., Pottmaier, D., Sher, F.: Effect of nanoparticles additives on tribological behaviour of advanced biofuels. Fuel **334**, 126798 (2023). https://doi.org/10.1016/j.fuel.2022.126798
65. Wu, H., Zhao, J., Cheng, X., Xia, W., He, A., Yun, J.H., Huang, S., Wang, L., Huang, H., Jiao, S., Jiang, Z.: Friction and wear characteristics of TiO_2 nano-additive water-based lubricant on ferritic stainless steel. Tribol. Int. **117**, 24–38 (2018). https://doi.org/10.1016/j.triboint.2017.08.011
66. Kannan, K.T., Babu, S.R.: Tribological behavior of modified jojoba oil with graphene nanoparticle as additive in SAE20W40 oil using pin on disc tribometer. Energy Sour. Part A Recovery Utilization Environ. Eff. **39**(17), 1842–1848 (2017). https://doi.org/10.1080/15567036.2017.1376006

67. Alves, S.M., Barros, B.S., Trajano, M.F., Ribeiro, K.S.B., Moura, E.: Tribological behavior of vegetable oil-based lubricants with nanoparticles of oxides in boundary lubrication conditions. Tribol. Int. **65**, 28–36 (2013). https://doi.org/10.1016/j.triboint.2013.03.027
68. Koshy, C.P., Rajendrakumar, P.K., Thottackkad, M.V.: Evaluation of the tribological and thermo-physical properties of coconut oil added with MoS_2 nanoparticles at elevated temperatures. Wear **330–331**, 288–308 (2015). https://doi.org/10.1016/j.wear.2014.12.044
69. Uflyand, I.E., Zhinzhilo, V.A., Burlakova, V.E.: Metal-containing nanomaterials as lubricant additives: state-of-the-art and future development. Friction **7**(2), 93–116 (2019). https://doi.org/10.1007/s40544-019-0261-y
70. Calestani, D.: Chapter 2—Characterization of the physical and chemical properties of engineered nanomaterials. In: Marmiroli, E., White, J.C., Song, J. (eds.) Exposure to Engineered Nanomaterials in the Environment, Elsevier, pp. 31–57 (2019). https://doi.org/10.1016/B978-0-12-814835-8.00002-9
71. Yang, G., Zhang, Z., Zhang, S., Yu, L., Zhang, P., Hou, Y.: Preparation and characterization of copper nanoparticles surface-capped by alkanethiols. Surf. Interface Anal. **45**(11–12), 1695–1701 (2013). https://doi.org/10.1002/sia.5309
72. Azam, S., Park, S.S.: Sonochemical synthesis of CuO nanoplatelets and their tribological properties as an additive in synthetic oil using reciprocating tribometer. Lubricants **11**(4), 185 (2023). https://doi.org/10.3390/lubricants11040185
73. Luo, T., Wei, X., Huang, X., Huang, L., Yang, F.: Tribological properties of Al_2O_3 nanoparticles as lubricating oil additives. Ceram. Int. **40**(5), 7143–7149 (2014). https://doi.org/10.1016/j.ceramint.2013.12.050
74. Viesca, J.L., Hernández Battez, A., González, R., Chou, R., Cabello, J.J.: Antiwear properties of carbon-coated copper nanoparticles used as an additive to a polyalphaolefin. Tribol. Int. **44**(7), 829–833 (2011). https://doi.org/10.1016/j.triboint.2011.02.006
75. Tevet, O., Von-Huth, P., Popovitz-Biro, R., Rosentsveig, R., Wagner, H.D., Tenne, R.: Friction mechanism of individual multilayered nanoparticles. Proc. Natl. Acad. Sci. **108**(50), 19901–19906 (2011). https://doi.org/10.1073/pnas.1106553108
76. Zhang, L.L., Tu, J.P., Wu, H.M., Yang, Y.Z.: WS_2 nanorods prepared by self-transformation process and their tribological properties as additive in base oil. Mater. Sci. Eng. A **454**, 487–491 (2007). https://doi.org/10.1016/j.msea.2006.11.072
77. Ghosh, G.K., Kotia, A., Kumar, N., Ghosh, S.K.: Multi-response optimization of tribological characteristics for Graphene-Gear oil nanolubricants using Grey-Taguchi methodology followed by scrutinization of lubrication mechanisms. J. Mater. Eng. Perform. (2024). https://doi.org/10.1007/s11665-024-10365-1
78. Yu, W., Xie, H.: A review on nanofluids: preparation, stability mechanisms, and applications. J. Nanomater. **2012**(1), 435873 (2012). https://doi.org/10.1155/2012/435873
79. Arakawa, K.: Dynamic sliding friction and similarity with Stokes' law. Tribol. Int. **94**, 77–81 (2016). https://doi.org/10.1016/j.triboint.2015.08.007
80. Chen, Y., Zhang, Y., Zhang, S., Yu, L., Zhang, P., Zhang, Z.: Preparation of nickel-based nanolubricants via a facile in situ one-step route and investigation of their tribological properties. Tribol. Lett. **51**(1), 73–83 (2013). https://doi.org/10.1007/s11249-013-0148-4
81. Ahirwar, B.K., Kumar, A.: Comparative analysis of CuO–water and ZnO–water nanofluids in the turbulent regime for enhanced performance in double-pipe heat exchanger. J. Thermal Anal. Calor. **149**(23), 14213–14240 (2024). https://doi.org/10.1007/s10973-024-13623-5
82. Singh, A., Chauhan, P., Mamatha, T.: A review on the tribological performance of lubricants with nanoparticle additives. Mater. Today Proc. **25**, 586–591 (2019). https://doi.org/10.1016/j.matpr.2019.07.245
83. Gegner, J. (ed.): Tribology: Fundamentals and Advancements. InTech, Croatia (2013). https://doi.org/10.5772/3427
84. Liu, X., Li, S., Yang, P., Yang, P.: On the lubricating mechanism of roller skew in cylindrical roller bearings. Tribol. Trans. **56**(6), 929–942 (2013). https://doi.org/10.1080/10402004.2013.812760

85. Zhang, C., Liu, T., Wang, Y., He, H., Zhang, C.: Mechanism study of symmetric mechanism-rolling bearing dynamic system considering the influence of temperature factor. Symmetry **15**(12), 2166 (2023). https://doi.org/10.3390/sym15122166
86. Cann, P., Lubrecht, A.A.: An analysis of the mechanisms of grease lubrication in rolling element bearings. Lubr. Sci. **11**(3), 227–245 (1999). https://doi.org/10.1002/ls.3010110303
87. Grebe, M., Molter, J., Schwack, F., Poll, G.: Damage mechanisms in pivoting rolling bearings and their differentiation and simulation. Bearing World J. **3**, 71–86 (2018)
88. Ebert, F.: Fundamentals of design and technology of rolling element bearings. Chin. J. Aeronaut. **23**(1), 123–136 (2010). https://doi.org/10.1016/s1000-9361(09)60196-5
89. Aizawa, T., Mitsuo, A., Yamamoto, S., Sumitomo, T., Muraishi, S.: Self-lubrication mechanism via the in situ formed lubricious oxide tribofilm. Wear **259**(1–6), 708–718 (2005). https://doi.org/10.1016/j.wear.2005.02.025
90. Gosvami, N.N., Bares, J.A., Mangolini, F., Konicek, A.R., Yablon, D.G., Carpick, R.W.: Mechanisms of antiwear tribofilm growth revealed in situ by single-asperity sliding contacts. Science **348**(6230), 102–106 (2015). https://doi.org/10.1126/science.1258788
91. Gustavsson, F., Jacobson, S.: Diverse mechanisms of friction induced self-organisation into a low-friction material—an overview of WS2 tribofilm formation. Tribol. Int. **101**, 340–347 (2016). https://doi.org/10.1016/j.triboint.2016.04.029
92. Zhang, J., Spikes, H.: On the mechanism of ZDDP antiwear film formation. Tribol. Lett. **63**(2) (2016). https://doi.org/10.1007/s11249-016-0706-7
93. Sharma, V., Gabler, C., Doerr, N., Aswath, P.B.: Mechanism of tribofilm formation with P and S containing ionic liquids. Tribol. Int. **92**, 353–364 (2015). https://doi.org/10.1016/j.tri boint.2015.07.009
94. Zhou, Y., Leonard, D.N., Guo, W., Qu, J.: Understanding tribofilm formation mechanisms in ionic liquid lubrication. Sci. Rep. **7**(1) (2017). https://doi.org/10.1038/s41598-017-09029-z
95. Xu, D., Wang, C., Espejo, C., Wang, J., Neville, A., Morina, A.: Understanding the friction reduction mechanism based on molybdenum disulfide tribofilm formation and removal. Langmuir **34**(45), 13523–13533 (2018). https://doi.org/10.1021/acs.langmuir.8b02329
96. Lei, X., Zhang, Y., Zhang, S., Yang, G., Zhang, C., Zhang, P.: Study on the mechanism of rapid formation of ultra-thick tribofilm by CeO_2 nano additive and ZDDP. Friction **11**(1), 48–63 (2022). https://doi.org/10.1007/s40544-021-0571-8
97. Wood, R.J.K.: Tribo-corrosion of coatings: a review. J. Phys. D Appl. Phys. **40**(18), 5502–5521 (2007). https://doi.org/10.1088/0022-3727/40/18/s10
98. Yang, K., Ma, H., Cao, S., Li, X., Li, A., Cao, Z.: Analysis of friction interfaces with sinusoidal microchannels and the hybrid lubrication mechanisms of a tribo-film. Appl. Surf. Sci. **525**, 146502 (2020). https://doi.org/10.1016/j.apsusc.2020.146502
99. Wang, B., Zhong, Z., Qiu, H., Chen, D., Li, W., Li, S., Tu, X.: Nano serpentine powders as lubricant additive: tribological behaviors and self-repairing performance on worn surface. Nanomaterials **10**(5), 922 (2020). https://doi.org/10.3390/nano10050922
100. Zhu, L., Wang, S., Yu, H., Yin, Y., Xing, Z., Wang, H., Kang, J.: Friction-triggered self-repairing behaviors and mechanism of natural serpentine mineral/Fe composites. Appl. Surf. Sci. **688**, 162406 (2025). https://doi.org/10.1016/j.apsusc.2025.162406
101. Huang, Q., Shi, X., Xue, Y., Zhang, K., Wu, C.: Wear-triggered self-repairing behavior of bionic textured AISI 4140 steel filled with multi-solid lubricants. Wear **504–505**, 204416 (2022). https://doi.org/10.1016/j.wear.2022.204416
102. Guo, Z.C., Chen, M., Zhao, D.B., Zhang, F.M.: Tribology and self-repairing nature of RE nanophase composite Material. Adv. Mater. Res. **659**, 3–6 (2013). https://doi.org/10.4028/www.scientific.net/amr.659
103. Chu, H.Y., Hsu, W.C., Lin, J.F.: The anti-scuffing performance of diamond nano-particles as an oil additive. Wear **268**(7–8), 960–967 (2009). https://doi.org/10.1016/j.wear.2009.12.023
104. Runnels, S.R., Eyman, L.M.: Tribology analysis of chemical-mechanical polishing. J. Electrochem. Soc. **141**(6), 1698–1701 (1994). https://doi.org/10.1149/1.2054985
105. Lee, H.: Tribology research trends in chemical mechanical polishing (CMP) process. Tribol. Lubr. **34**(3), 115–122 (2018). https://doi.org/10.9725/kts.2018.34.3.115

106. Datta, D., Rai, H., Singh, S., Srivastava, M., Sharma, R.K., Gosvami, N.N.: Nanoscale tribological aspects of chemical mechanical polishing: a review. Appl. Surf. Sci. Adv. **11**, 100286 (2022). https://doi.org/10.1016/j.apsadv.2022.100286
107. Lin, J.F., Chern, J.D., Chang, Y.H., Kuo, P.L., Tsai, M.S.: Analysis of the tribological mechanisms arising in the chemical mechanical polishing of copper-film wafers. J. Tribol. **126**(1), 185–199 (2004). https://doi.org/10.1115/1.1631010
108. Azman, N.F., Samion, S.: Dispersion stability and lubrication mechanism of nanolubricants: a review. Int. J. Precis. Eng. Manuf. Green Tech. **6**(2), 393–414 (2019). https://doi.org/10.1007/s40684-019-00080-x
109. Tobar, M.G., Urgiles, R.W.C., Cordero, B.J., Matute, J.G.: Nanotechnology in lubricants: a systematic review of the use of nanoparticles to reduce the friction coefficient. Lubricants **12**(5), 166 (2024). https://doi.org/10.3390/lubricants12050166
110. Akl, S., Elsoudy, S., Abdel-Rehim, A.A., Salem, S., Ellis, M.: Recent advances in preparation and testing methods of engine-based nanolubricants: a state-of-the-art review. Lubricants **9**(9), 85 (2021). https://doi.org/10.3390/lubricants9090085
111. Paul, G., Hirani, H., Kuila, T., Murmu, N.C.: Nanolubricants dispersed with graphene and its derivatives: an assessment and review of the tribological performance. Nanoscale **11**(8), 3458–3483 (2019). https://doi.org/10.1039/c8nr08240e
112. Apostolou, M., Fatokun, A.A., Assi, S., Khan, I.: Targeted lipid-based drug delivery systems for lung cancer therapy. Appl. Sci. **14**(15), 6759 (2024). https://doi.org/10.3390/app14156759
113. Dykman, L., Khlebtsov, N.: Gold nanoparticles in biomedical applications: recent advances and perspectives. Chem. Soc. Rev. **41**(6), 2256–2282 (2011). https://doi.org/10.1039/c1cs15166e
114. Thanh, N.T., Green, L.A.: Functionalisation of nanoparticles for biomedical applications. Nano Today **5**(3), 213–230 (2010). https://doi.org/10.1016/j.nantod.2010.05.003
115. Tran, N., Webster, T.J.: Magnetic nanoparticles: biomedical applications and challenges. J. Mater. Chem. **20**(40), 8760 (2010). https://doi.org/10.1039/c0jm00994f
116. Jiang, J., Pi, J., Cai, J.: The advancing of zinc oxide nanoparticles for biomedical applications. Bioinorg. Chem. Appl. **2018**, 1–18 (2018). https://doi.org/10.1155/2018/1062562
117. De Crozals, G., Bonnet, R., Farre, C., Chaix, C.: Nanoparticles with multiple properties for biomedical applications: a strategic guide. Nano Today **11**(4), 435–463 (2016). https://doi.org/10.1016/j.nantod.2016.07.002
118. Veg, E., Hashmi, K., Raza, S., Joshi, S., Rahman Khan, A., Khan, T.: The role of nanomaterials in diagnosis and targeted drug delivery. Chem. Biodivers. **22**(1), e202401581 (2025). https://doi.org/10.1002/cbdv.202401581
119. Sauvage, F., Fraire, J.C., Remaut, K., Sebag, J., Peynshaert, K., Harrington, M., Van De Velde, F.J., Xiong, R., Tassignon, M., Brans, T., Braeckmans, K., De Smedt, S.C.: Photoablation of human vitreous opacities by light-induced vapor nanobubbles. ACS Nano **13**(7), 8401–8416 (2019). https://doi.org/10.1021/acsnano.9b04050
120. Wu, C., Yang, Y., Hsu, Y., Wu, T., Hung, C., Huang, J., Chang, C.: Nanoparticle-induced intraperitoneal hyperthermia and targeted photoablation in treating ovarian cancer. Oncotarget **6**(29), 26861–26875 (2015). https://doi.org/10.18632/oncotarget.4766
121. Syu, W., Huang, C., Hsiao, J., Lee, Y., Huang, Y., Venkatesan, P., Lai, P.: Co-precipitation synthesis of near-infrared iron oxide nanocrystals on magnetically targeted imaging and photothermal cancer therapy via photoablative protein denaturation. Nanotheranostics **3**(3), 236–254 (2019). https://doi.org/10.7150/ntno.24124
122. Demirer, G.S., Okur, A.C., Kizilel, S.: Synthesis and design of biologically inspired biocompatible iron oxide nanoparticles for biomedical applications. J. Mater. Chem. B **3**(40), 7831–7849 (2015). https://doi.org/10.1039/c5tb00931f
123. Arkhipova, V.I., Mochalova, E.N., Nikitin, M.P.: Au-based bimetallic nanoparticles: current biomedical applications. J. Nanopart. Res. **26**(9), 214 (2024). https://doi.org/10.1007/s11051-024-06122-z
124. Zhang, Q., Pan, J., Zhang, X., Lu, J., Yan, Q.: Tribological behavior of 6H–SiC wafers in different chemical mechanical polishing slurries. Wear **472–473**, 203649 (2021). https://doi.org/10.1016/j.wear.2021.203649

125. Latthe, S.S., Terashima, C., Nakata, K., Fujishima, A.: Superhydrophobic surfaces developed by mimicking hierarchical surface morphology of lotus leaf. Molecules **19**(4), 4256–4283 (2014). https://doi.org/10.3390/molecules19044256
126. Chatterjee, D.K., Gnanasammandhan, M.K., Zhang, Y.: Small upconverting fluorescent nanoparticles for biomedical applications. Small **6**(24), 2781–2795 (2010). https://doi.org/10.1002/smll.201000418
127. Elsabahy, M., Wooley, K.L.: Design of polymeric nanoparticles for biomedical delivery applications. Chem. Soc. Rev. **41**(7), 2545 (2012). https://doi.org/10.1039/c2cs15327k
128. Zhu, X., Li, Y., Gu, N.: Application of artificial intelligence in the exploration and optimization of biomedical nanomaterials. Nano Biomed. Eng. **15**(3), 342–353 (2023). https://doi.org/10.26599/NBE.2023.9290035
129. Mazumdar, H., Khondakar, K.R., Das, S., Halder, A., Kaushik, A.: Artificial intelligence for personalized nanomedicine; from material selection to patient outcomes. Expert Opin. Drug Deliv. **22**(1), 85–108 (2025). https://doi.org/10.1080/17425247.2024.2440618
130. Miao, W.J., Wang, S.Q., Wang, Z.H., Wu, F.B., Zhang, Y.Z., Ouyang, J.H., Wang, Y.M., Zou, Y.C.: Additive manufacturing of advanced structural ceramics for tribological applications: principles, techniques. Microstruct. Prop. Lubr. **13**(3), 112 (2025). https://doi.org/10.3390/lubricants13030112
131. Truong, T.T., Mondal, S., Doan, V.H.M., Tak, S., Choi, J., Oh, H., Nguyen, T.D., Misra, M., Lee, B., Oh, J.: Precision-engineered metal and metal-oxide nanoparticles for biomedical imaging and healthcare applications. Adv. Colloid Interface Sci. **332**, 103263 (2024). https://doi.org/10.1016/j.cis.2024.103263
132. Kuchur, O., Tsymbal, S., Shestovskaya, M., Serov, N., Dukhinova, Shtil, A.: Metal-derived nanoparticles in tumor theranostics: potential and limitations. J. Inorg. Biochem. **209**, 111117 (2020). https://doi.org/10.1016/j.jinorgbio.2020.111117
133. Malik, M.A.I., Kalam, M.A., Mujtaba, M.A., Almomani, F.: A review of recent advances in the synthesis of environmentally friendly, sustainable, and nontoxic bio-lubricants: Recommendations for the future implementations. Environ. Tech. Innov. **32**, 103366 (2023). https://doi.org/10.1016/j.eti.2023.103366
134. Prasannakumar, P., Sankarannair, S., Prasad, G., HariKrishna, P.H., Pranav, S., Vivek, P., Sidharth, S., Shanmugam, R.: Bio-based additives in lubricants: addressing challenges and leveraging for improved performance toward sustainable lubrication. Biomass Conv. Bioref. 1–29 (2025). https://doi.org/10.1007/s13399-025-06563-z

Understanding Phase Transformations in Nanometals: Shape-Memory Alloys in Medicine

Treesa Varghese, V. H. Haritha, and Jenny Jacob

Abstract Shape memory alloys (SMAs) are innovative materials that allow a reversible martensitic phase transformation. They have a wide range of biomedical applications due to their pseudoelasticity and the ability to restore and retain a 'memorized' shape. In the biomedical industry, their macroscopic and microdevice properties offer a significant advantage in manufacturing components for microelectronic devices, including resistors, capacitors, and transistors. Furthermore, using biodegradable SMAs may enhance rehabilitation efficiency and decrease patients' discomfort compared to non-degradable SMAs. The development of nanocomposite shape memory alloys provides an additional benefit to the scientific community by overcoming the size limitations of traditional alloys. It thus opens new avenues for the use of nanoscaled biomedical devices. This chapter presents the latest findings and advancements in the mechanics and functionalization of nanotechnology-based shape memory alloys.

Keywords Shape memory alloys · Biomedical applications · Pseudoplasticity · Martensitic phase transformation · Smart materials · Nanocomposites · Nanotechnology · NiTi alloys

1 Introduction

Shape memory alloys (SMAs), also known as memory or smart alloys, are unique materials capable of returning to a pre-programmed shape when exposed to a thermal stimulus. These materials have garnered significant research interest over the past few decades. SMAs exhibit two key properties: Shape Memory Effect (SME) and

T. Varghese · J. Jacob (✉)
School of Biosciences, Mar Athanasios College for Advanced Studies, Tiruvalla, Kerala, India
e-mail: jenny@macfast.org

V. H. Haritha
National Institute of Plant Science and Technology, Mahatma Gandhi University, Kottayam, Kerala, India

S. N. Saud Al-Humairi et al. (eds.), *Smart Nanomaterials in Biomedical Applications*, Engineering Materials, https://doi.org/10.1007/978-981-96-8023-8_4

Super Elasticity (SE). SME is achieved through temperature-induced phase transformation, while SE occurs through stress-induced martensitic phase transformation in the materials [1].

1.1 Shape Memory Effect (SME)

The shape memory effect (SME) is a unique characteristic of martensitic alloys that allows them to remember and restore their original shape after being deformed in response to an external stimulus. This distinctive effect occurs in materials that display a reversible phase transformation between austenite (the high-temperature phase) and martensite (the low-temperature phase). In summary, when cooled, the stable high-temperature austenitic phase transforms into a soft, less stable, and easily deformable metastable phase. However, the austenite form is regained upon heating, as evidenced by its rigidity [2], highlighting the significance of thermal activation in the transition from martensite to austenite. This property of NiTi (Nickel–Titanium) alloys has been thoroughly examined by Jani et al. [1], who outlined the potential for using these shape memory alloys in various medical applications by taking advantage of their smart reversible property in response to heat, even at body temperature [1, 3].

1.1.1 Pseudoelasticity

Pseudoelasticity, also known as superelasticity, is a defining property of shape memory alloys. Under stress conditions, these alloys can undergo temporary deformation [2, 4]. Unlike conventional materials, which may deform plastically under stress, shape memory alloys (SMAs) experience stress-induced phase transformations [5]. Specifically, when subjected to sufficient stress, SMAs in the austenite phase undergo elastic deformation, transitioning to a new phase called the martensite state. This deformation is reversible; under stress-free conditions, they can return to their original state without any thermal activation [4, 5]. The pseudoelastic behavior of NiTi alloys at room temperature has been observed, showing up to eight percent temporary strain during the phase transformation from austenite to martensite [4]. The superelastic nature of this metal makes it an ideal material for biomedical applications, including the construction of advanced devices like vascular stents, where high elasticity and recovery after deformation are critically important.

2 Phase Transformations

2.1 *Martensitic Transformation*

In the martensitic transformation, the atoms within the material reposition themselves cooperatively without long-range atomic diffusion, resulting in a new crystalline structure [6]. This change in crystal structure during the phase transition arises from the coordinated movement of atoms over distances typically smaller than interatomic lengths [6]. The martensitic phase exhibits a shear mechanism, leading to a distorted crystal structure with lower symmetry compared to its parent phase, making it harder and more brittle [7]. Furthermore, twin boundaries serve as sites for stress, rendering them easily deformable [8].

2.1.1 Temperature-Dependent Transformation

Martensitic transformation in steel alloys occurs during rapid cooling from elevated temperatures. This transformation is associated with two critical temperatures: the martensite starts temperature and the martensite finish temperature. The martensite start temperature is defined as the temperature during cooling that initiates the martensitic transformation, while the martensite finish temperature is the temperature at which the process is concluded [9].

2.1.2 Stress-Induced Martensitic Transformation

The second type of martensitic transformation is induced by external stress or strain. In this type, the initiation and propagation of the transformation are random, without specifying or predicting the location. The disparity in free energy between the austenitic and martensitic phases acts as a driving force for spontaneous martensitic transformation, which arises from the difference in chemical free energy between the parent and martensite phases [10].

2.2 *Austenitic Transformation*

Austenitic transformation, also known as austenitization, is a phase transformation in metal alloys that occurs in response to thermal stimuli. This process is characterized by the reversible transition of parent alloys from the martensite phase to the high-symmetry shape memory austenite phase, facilitated by thermal energy [11]. It is a diffusionless process in which the initial and final phases differ in their crystal symmetry [2, 12] and involves cooperative shifts in atomic positions [2], commonly seen in steel and NiTi alloys [12]. This unique structural arrangement in the austenitic phase enhances its ductility, toughness, and resistance to deformation [7], as well as increases stiffness and mechanical stability [13]. The transformation from martensite to austenite is regulated by the austenite start temperature, which marks the beginning of the austenite phase, and the austenite finish temperature, indicating when the transformation is complete [2]. These critical temperature shifts depend on the alloy compositions and are crucial parameters that determine the material's ability to recover its original shape [14].

2.3 *Factors Affecting Phase Transformation*

2.3.1 Hysteresis

Hysteresis in shape memory alloys (SMAs) refers to the delay between the forward (martensitic) and reverse (austenitic) transformations. This delay results from energy dissipation during phase changes that occur in temperature- and stress-driven transformations [2, 11]. The transformation characteristics, internal stress, and alloy composition influence the extent of the hysteresis [15]. In pseudoelastic SMAs, mechanical loading triggers a stress-induced martensite phase transformation, while unloading causes a reverse transformation, thus creating a hysteresis loop in the stress–strain curve [16, 17]. The overall extent of the hysteresis depends on alloy composition, grain size, and processing conditions. Alloys of nickel, copper, and aluminum can affect their memory effect by modifying the energy barrier and transformation temperatures. Differences in microstructures within alloys can also significantly influence their pseudoelastic properties [15, 18].

2.3.2 Thermomechanical Processing

The microstructure, phase stability, and mechanical properties of shape memory alloys (SMAs) are influenced by thermomechanical processing, which plays a crucial role in reshaping their transformational behavior. Additionally, transformation temperatures, hysteresis width, and the overall shape memory effect are regulated by various processing techniques, including heat treatment, cold working, and precipitation hardening. Moreover, phase stability and mechanical performance have been shown to be affected by grain refinement [13, 17].

2.3.3 Heat and Cold Treatment

Heat treatment can influence phase transformation behavior by affecting the distribution of defects, dislocations, and precipitates, thereby altering the microstructure. Additionally, depending on the alloy systems, solution treatment and aging temperature can stabilize the austenite or martensite phases by regulating the transformation temperatures [11]. However, cold treatment initiates strain hardening and elevates dislocation density, thus hindering martensitic transformation, shifting transformation temperatures [16], and increasing internal stresses, which modifies the stress-induced martensitic transformation and pseudoelastic properties [17].

2.3.4 Precipitation Hardening

Precipitation hardening, also known as age hardening, in heat treatment involves forming fine, uniformly distributed precipitates within the metal matrix, which restricts dislocation movement and enhances mechanical properties and strength, leading to altered transformation behavior. For example, the formation of Ni_4Ti_3 in Ni–Ti alloys effectively adjusts martensitic transformations (MTs), resulting in an increased transformation temperature and a highly stable austenite phase [15]. The characteristics related to the shape, size, and distribution of precipitates in alloy materials and their volume fraction are essential for predicting thermal, chemical, and mechanical performance, ultimately determining whether the transformation is acceptable or rejected [18]. Furthermore, precipitation hardening may also improve fatigue strength, enabling alloys to endure repeated cycles of strain or stress without failing by increasing their strength and hardness, which boosts resistance to cyclic loading and failure [2].

2.3.5 Grain Refinement

The stability of phases and the optimization of transformation characteristics in shape memory alloys (SMAs) are critically enhanced by grain refinement. The homogeneous nucleation of martensite can reduce transformation hysteresis and improve cyclic stability, a property attributed to the fine-grained microstructure of SMAs [19]. Plastic deformation or controlled thermomechanical treatments result in a reduction in grain size, thereby enhancing mechanical properties such as strength and ductility without compromising functional properties like superelasticity and the shape memory effect [16] while also improving fatigue resistance by minimizing crack initiation sites and thus enhancing transformation reversibility [17].

3 Types of SMAs and Their Composition

Heating and cooling cycles facilitate the transformation of shape memory alloys (SMAs) between a martensitic state (low temperature) and an austenitic state (high temperature). This transformation enables SMAs to recover from strain and generate force by exhibiting two shape memory effects: the one-way shape memory effect (SME) and the two-way SME [20]. As seen in NiTi alloys, one-way SME materials regain their original shape only when heated [1]. Conversely, two-way SME materials retain high- and low-temperature shapes, transforming accordingly during heating and cooling [20], allowing for bi-directional movement. This effect is noted in Cu-based and Fe-based shape memory alloys, and this property is utilized in actuators and robotics [4]. In contrast to two-way SME, which possesses a lower recovery force and less pronounced phase transformation [4], one-way SME exhibits stronger shape recovery. This characteristic makes one-way SME more valuable for high-force applications.

3.1 Nickel–Titanium (NiTi) Shape Memory Alloys

NiTi SMAs demonstrate excellent biocompatibility, corrosion resistance, superior mechanical properties, and a unique ability for reversible martensitic transformation. This makes them the most commonly used SMAs crucial in medical, aerospace, and actuator applications [2]. A slight change in the concentration of nickel or titanium in near-equiatomic NiTi alloys (Nitinol) can significantly impact the phase transition characteristics by affecting their mechanical behavior and transformation temperature [3].

3.1.1 Effects of Composition and Alloying on NiTi SMAs Transformation Behavior

The martensitic transformation in NiTi alloys occurs between the austenite and martensite phases through the formation of an intermediate phase, significantly influenced by the Ni to Ti ratio. In other words, Ni-dominant NiTi alloys (greater than 50% Ni) may exhibit lower transformation temperatures due to the enhanced stability of their austenite phase. Conversely, alloys with less than 50% Ni (Ti–rich alloys) have higher transformation temperatures because of their increased rigidity in the martensite phase [21]. Even a minor increase (0.5%) in Ni content can reduce the transformation temperature of the martensite phase by 10–15 °C [5]. Ni-rich alloys will likely form Ni_4Ti_3 precipitates, which can influence their mechanical properties and transformation hysteresis [16]. Additionally, the heat treatment temperature can affect precipitation, thereby adjusting the martensitic transformation and related properties of fabricated NiTi SMAs [5, 22].

3.1.2 Impact of Minor Alloying Elements on NiTi Transformation

i. Copper (Cu) Addition

The B2–B19′ transformation in NiTi alloys refers to the transition from the high-temperature austenite phase (B2) to the low-temperature martensite phase (B19′).

This phase change gives NiTi its shape memory and superelastic properties. In standard NiTi, an intermediate R-phase may occasionally appear during cooling. Adding copper to NiTi alloys (NiTiCu) modifies the transformation process, often resulting in a direct B2–B19 transformation without the intermediate R-phase. Copper enhances actuation speed and reduces energy loss by stabilizing stress-induced martensite. However, excessive copper (beyond 15%) can render the alloy brittle and diminish its superelasticity [5].

ii. Hafnium (Hf) Addition

Hafnium (Hf) is a critical element in NiTi-based high-temperature shape memory alloys (SMAs). Substituting Ni with Hf (e.g., Ni_{50}–xTi_{50}Hfx) significantly raises the transformation temperature. When 10% of the total atoms in the NiTi alloy are replaced with hafnium (Hf), it can increase the martensite start temperature by approximately 100 °C, making it suitable for high-temperature actuators [23]. Hf enhances the alloy's strength and stability while reducing flexibility [24].

iii. Platinum (Pt) Addition

Substituting Platinum (Pt) in NiTi enhances stability at high temperatures by forming an oxide layer, significantly improving corrosion resistance. It has been reported that a 10% replacement of Pt in NiTi alloys will raise the transformation temperature, with the martensite start temperature increasing by nearly 200 °C. The high-temperature stability, strength, and better flexibility retention make NiTiPt alloys particularly suitable for aerospace applications [25].

3.2 Copper-Based SMAs

Lower costs and higher transformation temperatures have made copper-based shape memory alloys (SMAs) more dominant than NiTi alloys. These include Cu–Zn–Al and Cu–Al–Ni alloys [2] and are the most widely studied types among Cu-SMAs. In addition to exhibiting shape memory effect (SME) and superelasticity, they offer lower costs, higher transformation temperatures, and easier processing. This makes them particularly suitable for aerospace and vibration-damping applications compared to NiTi alloys [13, 15]. However, the aging effects on copper-based SMAs can degrade shape memory performance over time, and some copper SMAs demonstrate lower ductility, brittleness, and intergranular fracture [18, 26]. Furthermore, their lower fatigue resistance compared to NiTi limits long-term durability in cyclic applications [27].

3.2.1 Cu–Zn–Al Alloys

Cu–Zn–Al alloys exhibit high transformation temperatures and moderate mechanical properties. Depending on composition and processing conditions, these alloys undergo a reversible martensitic transformation at temperatures between 100 and 300 °C [19]. Although they are susceptible to aging effects that lead to the degradation of the shape memory effect over time [18], their relatively high damping capacity allows Cu–Zn–Al alloys to be widely used in vibration control and seismic damping applications [27].

3.2.2 Cu–Al–Ni Alloys

Among Cu-SAMs alloys, Cu–Al–Ni alloys are preferred over Cu–Zn–Al alloys due to their superior thermal stability [13]. These alloys demonstrate transformation temperatures ranging from 150 to 400 °C, making them suitable for high-temperature applications such as aerospace and automotive actuators [26]. However, a significant limitation of Cu–Al–Ni alloys is their low ductility, which makes them susceptible to brittleness and intergranular fracture, particularly under repeated thermal cycling [15].

3.3 Iron-Based SMAs

Iron-based shape memory alloys (Fe-SMAs) have attracted considerable attention due to their cost-effectiveness, high strength, and superior workability compared to traditional Ni–Ti alloys. Among Fe-SMAs, Fe–Mn–Si alloys are the most extensively studied due to their excellent shape memory effect and superior fatigue resistance, making them suitable for seismic dampers and structural reinforcements [28, 29]. However, the shape memory effect in Fe–Mn–Si is non-thermoelastic, which means that superelasticity cannot be achieved. Another promising Fe-SMA, Fe–Mn–Al–Ni, has shown significant superelastic behavior across a broad temperature range, from −263 to 240 °C, with minimal temperature sensitivity [30]. Unlike conventional Fe-based-SMAs, Fe–Mn–Al–Ni undergoes a martensitic transformation from a body-centered cubic (BCC) ferrite phase to an austenite phase. Well-organized fine-sized precipitates within a disordered matrix of alloys are crucial for accelerating the thermoelastic transformation [31, 32]. The Fe–Mn–Al–Ni alloy exhibits a comparatively broader range of operational temperatures, superior cold functionality, and minimal stress sensitivity, thus providing an efficient composition and excellent mechanical performance [33, 34]. Fe-SMA-based rebars and strips have already been used in numerous structural rehabilitation projects to strengthen heat-critical concrete T-beams and offer a sustainable solution for developing reinforced structures [35, 36]. Due to their exceptional biocompatibility, Fe-incorporated Fe–Mn–Si-based SMAs can also be explored for various biomedical applications, particularly in developing scaffolds for tissue engineering [34]. Although they have a lower shape memory effect than Ni–Ti alloys, Fe–Mn–Si alloys represent an effective, cost-efficient, and corrosion-resistant alternative for many structural and functional materials in large-scale real-world applications [37].

4 Emerging and High-Performance SMAs

The use of remarkably high-performing shape memory alloys has attracted attention for their suitability in various advanced industrial applications, as shown in Fig. 1. These materials exhibit high thermal stability and superior performance at elevated temperatures when responding to external stimuli, such as high temperatures or a magnetic field. The main categories of these high-performance shape memory alloys can be classified as (1) high-temperature shape memory alloys (HTSMAs) and (2) magnetic shape memory alloys (MSMAs), along with nanostructured options SMAs.

4.1 High-Temperature SMAs (HTSMAs)

High-temperature shape memory alloys (HTSMAs) are a class of shape memory alloys that can resist structural deformation induced by temperatures above 100 °C without altering functional properties. Compared to conventional nickel–titanium (NiTi) metal alloys, platinum (Pt) and hafnium (Hf) incorporated high-temperature shape memory alloys exhibit better thermal stability. At high temperatures, NiTiPt and NiTiHf alloys display comparatively higher phase transformation stability than

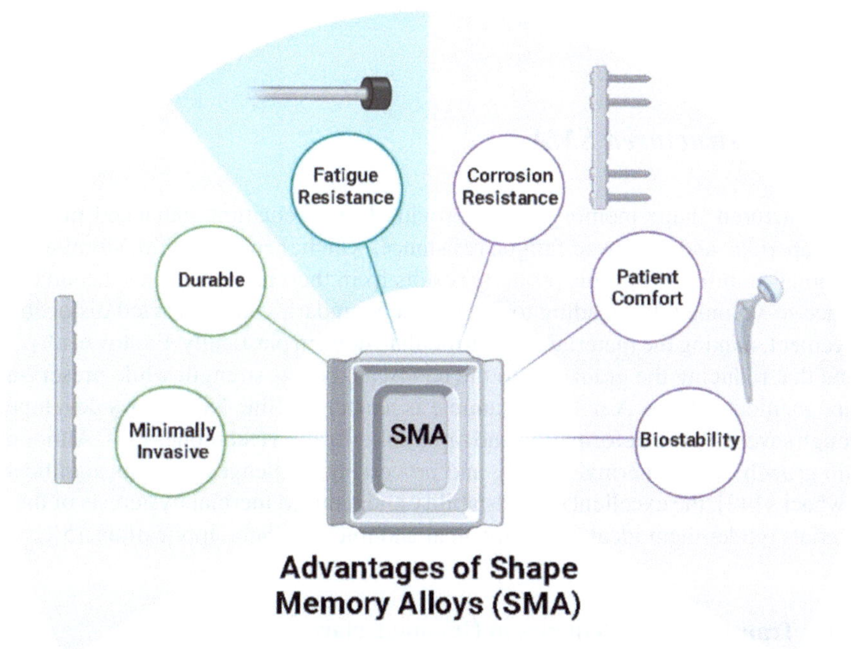

Fig. 1 Advantages of shape memory alloys

conventional shape memory alloys, making them good candidates for biomedical industries [13]. Furthermore, these high-performance alloys may experience martensitic phase transformations at high temperatures above 100 °C while maintaining their functionality up to 200 °C [38]. Hence, these HTSMAs can be potential candidates in automotive, aerospace, and nuclear power plants, operating at high temperatures while maintaining relatively stable performance [11].

4.2 Magnetic SMAs (MSMAs)

Magnetic shape memory alloys (MSMAs) are a promising type of shape memory alloy (SMA) that can undergo significant temperature or mechanical deformation-induced shape memory changes with reversibility in a magnetic field. Due to this phase transformation phenomenon responding to magnetic fields rather than thermal stimuli, they are also called ferromagnetic shape memory alloys (FSMAs). It has been reported that Ni–Mn–Ga functions as a magnetic SMA in an external magnetic field and is suitable for applications requiring magnetic actuation [39]. The ability of MSMAs to control phase transformation in response to a magnetic field enables them to function as devices such as precision instruments, sensors, and actuators. These properties can also be explored in biomedical engineering to develop modern medical devices like artificial muscles, which can provide well-regulated high-force output [40].

4.3 Nanostructured SMAs

Nanostructured shape memory alloys provide faster actuation, enhanced mechanical properties, and improved fatigue resistance, which increase overall strength and functional stability [41]. This property results from their refined grain size and high surface-to-volume ratio, leading to more grain boundaries and restricted dislocation movement, making the material more difficult to deform plastically. Firstov et al. [24] noted that reducing the grain size to the nanoscale boosts strength while preserving shape memory effects. A notable example is nanocrystalline NiTi alloys developed through severe plastic deformation and rapid solidification techniques [24]. Although grain growth during thermal cycling and processing challenges presents significant drawbacks [42], the excellent cyclic stability and reduced thermal hysteresis of these materials render them ideal for biomedical and microactuator applications [5].

4.3.1 Transformation Kinetics in Nanomaterials

Increased interfaces and refined grain sizes could alter phase transformation kinetics, enhancing the functional properties, as illustrated in steel, AI alloy, and Mg alloy.

This effect is also observable in nanostructured SMAs, where a higher nucleation rate at grain boundaries accelerates transformation, thus enhancing response time in actuation applications [5]. Additionally, refined grains reduce internal stresses and improve fatigue resistance and durability in cyclic applications [42]. Control over phase transformation behavior in nanostructured SMAs allows for a better shape memory effect and superelasticity, making them ideal for precise applications in medical devices and actuators [24].

4.3.2 Conventional and Nanostructured SMAs

Microstructural factors, particularly grain size, have significantly influenced phase transformation behaviors in shape memory alloys (SMAs). The synthesis of nanostructured SMAs from conventional materials involves grain refinement and the formation of a high density of grain boundaries. This phenomenon dramatically impacts shifts in transformation temperatures, reduces mechanical hysteresis, and enhances fatigue resistance, making them suitable for advanced biomedical applications [43, 44]. Smaller grain sizes in nanostructured SMAs inhibit martensitic transformation because the high density of grain boundaries imposes constraints on the nucleation and growth of the martensitic phase [45, 46]. Reducing grain size and increasing the fraction of grain boundaries leads to rapid phase transformation kinetics and raises the energy barrier for phase transformation, thereby decreasing the martensite start temperature. This property is exhibited by nanostructured SMAs, which display lower transformation temperatures compared to conventional SMAs [2, 47]. The martensitic transformation can be inhibited by reducing the grain size below a critical threshold, where the austenite phase temperature stabilizes, while martensite eventually forms into coarse-grained models of shape memory alloys [48, 49]. Additionally, during phase transformation cycles, nanocrystalline shape memory alloys exhibit minimal mechanical hysteresis due to their high surface-to-volume ratio of grain boundaries, which alters transformation energetics and reduces energy dissipation [50, 51]. Consequently, these materials demonstrate narrower hysteresis loops and improved energy efficiency, making them ideal solutions for essential devices with superior mechanical properties and durability [52].

5 SMAs Used in Biomedical Applications

Shape memory alloys, such as NiTi, exhibit unique properties, including superior corrosion resistance and superelasticity. They are biocompatible and suitable for biomedical applications, particularly in non-invasive devices that can return to their original shape [1]. Similarly, copper alloys are also biocompatible and demonstrate shape memory effects, but operate at lower transformation temperatures and exhibit poor mechanical performance. Iron SMAs possess pseudoelastic properties.

These SMAs are biodegradable and used in orthodontics, endodontics, and surgical instrument sutures.

5.1 Biocompatibility and Corrosion Resistance of Nano-SMAs

NiTi alloys are biocompatible and do not cause any toxicity reactions in biological systems, making them suitable for use as bone fixation devices, such as screws [53]. Unlike conventional devices, they can maintain proper posture during tissue healing and remain minimally invasive [54]. NiTi was also the first alloy used as an orthodontic device for teeth alignment, which reduces patient discomfort. Orthodontic archwires with SMPs are also biocompatible, offering a more aesthetic experience for patients [55]. NiTi alloys exhibit superior corrosion resistance to iron-based SMAs [54]. They are less toxic due to very low concentrations [56] and are biocompatible with osteoblasts and fibroblasts. However, oxygen in the biological system and physiological temperatures can cause the material to corrode [57].

5.2 Superelasticity and Shape Memory Effect in Medical Applications

Due to SMAs' superelasticity, reversible shape deformation in response to external stimuli occurs [11]. Due to their flexibility, stability, and resorbability, these materials can be utilized in surgical cardiovascular stents, catheters, and sutures [58]. NiTi alloys are the most commonly used medical devices and can undergo dynamic structural deformations at in situ body temperatures to adapt to the required shape. Figure 2 shows the various medical applications for shape memory alloys.

5.3 Thermomechanical Properties Enhancing Biomedical Device Performance

The thermomechanical properties of shape memory alloys (SMAs) involve the phase transitions the alloys undergo from martensite to austenite and vice versa. These temperature-controlled phase transformations help the material maintain structural integrity in biological systems [11]. Orthopedic devices must exhibit shape-changing characteristics at physiological temperatures while maintaining structural integrity to reduce bone and shear stress at the tissue-implant interfaces [59]. Other examples

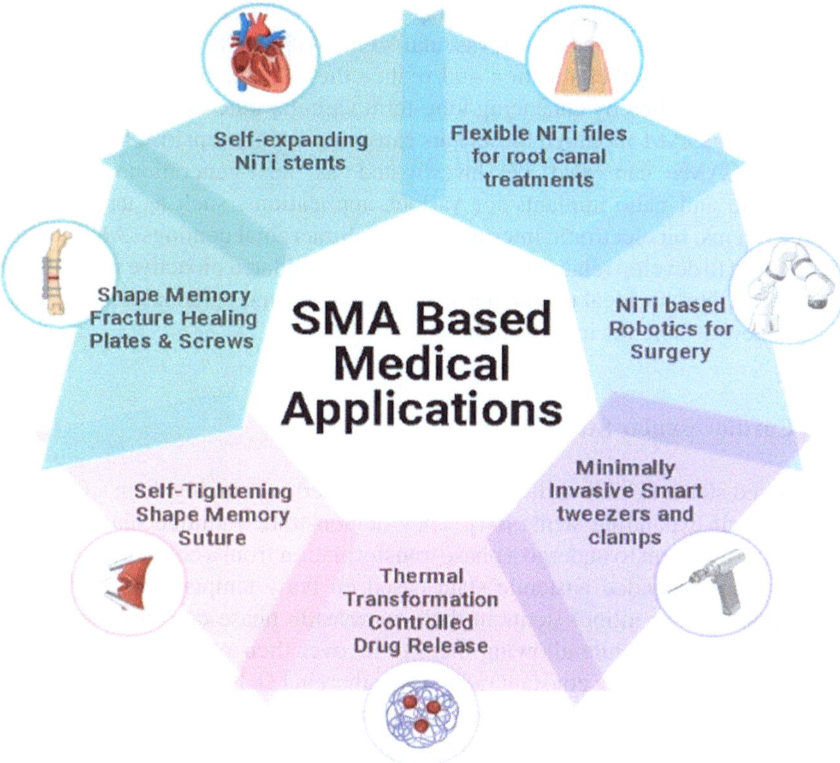

Fig. 2 Medical applications of shape memory alloys

include micro-actuators in robotic surgeries [60], bone clips/staples [61], and self-expanding aneurysm clips [62], each of which is considered for its efficacy and precision biocompatibility.

5.4 Biomedical Devices Using SAMs and Nano-SAMs

Self-assembled monolayers (SAMs) and their nanomaterials (nano-SAMs) on NiTi alloys (nitinol) represent an emerging and effective surface modification technology in the biomedical industry aimed at altering biofunctionality, biocompatibility, and material synergism. SAMs are organic molecules that spontaneously assemble into highly organized structures as thin single layers on the nanometer scale over the surfaces of materials, forming specialized tissue-implant modifications [4]. SAMs enhance devices' capacity in biomedical applications to improve biocompatibility, immunoregulation, microbial activity, and controlled drug delivery systems. The integration of nanotechnology for SAM formation may offer additional advantages in developing biomedical devices, including implants, stents, drug delivery systems, and biosensors. For instance, cardiovascular stents with SAM modifications could improve vascular endothelialization and reduce the likelihood of thrombosis and in-stent restenosis, thereby enhancing long-term vascular outcomes [63].

Furthermore, SAM-modified biosensors can detect highly sensitive biomolecules [46]. Nano-SAMs can also be investigated as next-generation biomedical nanomedicine and nano-implants for various applications, such as targeted drug delivery systems, bioelectronic interfaces, and antimicrobial coatings. Additionally, SAMs' ability to develop relatively stable and highly regulated bioactive surfaces aids in fine-tuning them as ideal candidates for regenerative and personalized medicine, opening a new avenue for improved therapy efficacy.

5.4.1 Cardiovascular Stents

Nitinol-based stents exhibit multiple functional properties, making them suitable for developing self-expanding stents [64]. They demonstrate a unique shape-memory effect that allows them to undergo a phase transformation from a compressed martensite state to an expanded austenite state based on body temperature. It has been reported that cooling nitinol stents in their martensitic phase converts them into a more crumpled form while allowing them to recover their preset expanded shape within the vessel, applying constant radial force therein [3]. Furthermore, generating stress points at stent junctions triggers local phase transformations, thus regulating their flexibility while keeping the vessel fully functional [65]. The superelastic nature of NiTi provides an added advantage by supporting blood vessels for an extended period with minimal damage to the arterial wall.

In addition to their mechanical properties, stents developed using nanostructured self-assembled monolayers (SAMs) can also demonstrate biocompatibility and functionality. The formation of self-assembled monolayers over nitinol stents enhances their performance by lowering thrombogenicity, minimizing in-stent restenosis, and improving endothelialization [66]. Various bioactive compounds with endothelial cell adhesion, antimicrobial, or antioxidant properties can be candidates for creating these monolayers [4]. Moreover, the corrosion-resistant properties of nitinol-SAMs provide additional advantages and are essential for maintaining stent stability under normal physiological conditions [20, 67]. Nano-SAMs-coated NiTi stents also enhance biological and mechanical properties, resulting in better performance as self-expanding stents [1].

5.4.2 Orthodontic Archwires

Thermally activated NiTi archwires can achieve optimal and efficient tooth movement, which provides light and continuous forces during their temperature-dependent phase transitions. A study conducted by Bradley et al. [68] characterized the transformation temperatures of various commercially available NiTi archwires using differential scanning calorimetry. This study reported that NiTi archwires with an austenite finish (Af) temperature near body temperature transitioned from the flexible martensitic phase at room temperature to a stiffer austenitic phase intraorally, allowing dynamic reshaping. This transition helps exert a gentle force that minimizes periodontal stress and reduces the risk of root resorption. It offers significant clinical insight, indicating that even minor variations can impact force delivery and subsequently influence the rate of tooth alignment. Therefore, archwires can be optimized for more predictable and efficient orthodontic outcomes by adjusting thermal properties during manufacturing.

Nano Self-Assembled Monolayers (nano-SAMs) can be implemented to enhance the performance and biocompatibility of NiTi archwires. Additionally, nano-SAM coatings facilitate surface functionalization and reduce friction, thereby improving sliding mechanics for smoother and more efficient tooth movement [66]. Functionalized nano-SAM layers can prevent bacterial adhesion, plaque formation, and gingival inflammation [4]. Nano-SAMs also possess the ability to gradually release bioactive ions such as silver or fluoride, promoting oral health and decreasing the risk of enamel demineralization [65].

5.4.3 Bone Fixation Devices

Nitinol SMA staples are highly significant in orthopedic applications as they provide continuous compressive support over fracture sites, facilitating primary bone healing. It has been reported that NiTi staples, in their soft martensitic state, are used to fix small bone fractures [69]. When exposed to moderate temperatures, the martensitic staples regain their preshape, aiding interfragmentary compression and stabilizing fractures by stimulating callus formation, which promotes faster recovery. Hoon et al. [61] reported that NiTi's superelastic and shape memory properties significantly improve implant fixation compared to conventional methods. Nano-SAM coatings and surface modifications can enhance cell adhesion and osteogenic differentiation for quicker bone integration [66] and also prevent the release of nickel ions, reducing the potential for immune rejection [4]. Silver nanoparticles or antimicrobial peptides can also be incorporated into nano-SAMs to mitigate post-surgical infections and improve wound healing [65]. Nano-SAM can also enable controlled drug delivery to the fracture site, accelerating healing and decreasing complications [20].

6 Challenges and Limitations

Several challenges, such as toxicity issues, microstructural changes affecting stability, cost constraints, etc., need to be addressed while using nano-SAMs to ensure their safety, efficacy, and long-term clinical success. However, nano-SMAs have been reported to be used for various biomedical applications, especially medical implants and drug delivery systems.

6.1 Biocompatibility and Toxicity Issues

The release of nickel ions and their potential toxicity raise concerns when using NiTi alloy in medical devices. Prolonged exposure to nickel has been linked to allergic reactions and cytotoxic effects. Therefore, further studies are required to assess the biocompatibility of this material before clinical application [70, 71].

6.2 Fatigue and Long-Term Stability

While medical devices are designed, repetitive mechanical loading is unavoidable, making fatigue resistance a critical factor. SMAs, such as NiTi alloys, have been widely used in medical devices. However, many authors have reported structural and functional fatigue in SMAs. They experience functional degradation over time due to cyclic loading and a loss of shape memory properties, which can lead to device failure. Moreover, Harumi et al. reported that microstructural changes during repeated use affect the long-term stability of these devices' materials [72].

6.3 Precision Manufacturing and Cost Constraints

The high cost of SMAs and nanomaterials is challenging due to the need for purity in the raw materials and complex manufacturing processes. Moreover, more specifications are needed to manufacture medical devices. Attaining the desired micro/nanostructure and mechanical properties in clinical settings is also challenging.

7 Emerging Trends and Future Research

Biomedical nanotechnology focuses on tailoring highly ordered nanostructured materials that provide distinct responses when exposed to specific stimuli. The field prioritizes nanoscale alloy models, advanced characterization techniques, and surface modification, and it incorporates these materials into smart electronics, medical devices, and personalized medicine. Compared to coarse-grained counterparts, improved mechanical strength, enhanced corrosion resistance, and superior biocompatibility make nanocrystalline metallic materials suitable for implants and prosthetics [73]. In-depth knowledge of complex phase transformations at the nanoscale is essential for generating materials with desired properties. High-resolution transmission electron microscopy (HR-TEM) and synchrotron X-ray scattering can characterize and comprehensively analyze microstructural changes in nanoscale materials during phase transitions [74]. Smart materials like shape memory alloys (SMAs) can undergo dynamic changes, such as reversible transformations in response to physiological conditions, making them ideal for self-expanding stents and responsive orthopedic implants, thus enhancing therapeutic outcomes. The application of advanced surface modification methods on SMAs leads to better osseointegration and reduced microbial infection with high survival and success rates. As technology advances, the shift toward personalized medicine utilizing shape memory alloys fosters the customization of patient-specific implants and devices to enhance the efficacy of the treatment treatments.

8 Conclusion

Shape memory alloys (SMAs) and nanomaterials in biomedical applications have become a top priority for developing medical devices, including orthopedic staples, orthodontic archwires, and cardiovascular stents, by leveraging their unique phase transformation properties to enable controlled drug delivery and protection against infectious organisms. Key findings indicate that thermally induced phase transformations in NiTi-based SMAs are critical for ensuring optimal performance, flexibility, and biocompatibility. At the same time, nano-SMAs present promising opportunities for enhanced surface modification, reduced ion release, and improved mechanical durability. However, challenges related to biocompatibility, fatigue resistance, large-scale manufacturing, and regulatory approval must be addressed to maximize clinical safety and effectiveness. A deeper understanding of phase transformations, advanced nanotechnology, and computational modeling will be vital for designing next-generation smart medical implants that provide superior functionality, longevity, and patient outcomes.

References

1. Jani, J.M., Leary, M., Subic, A., Gibson, M.A.: A review of shape memory alloy research, applications and opportunities. Mater. Des. (1980–2015) **56**, 1078–1113. https://doi.org/10.1016/j.matdes.2013.11.084
2. Otsuka, K., Wayman, C.M. (eds.): Shape Memory Materials. Cambridge University Press (1999)
3. Duerig, T.W., Pelton, A.R., Stöckel, D.: An overview of nitinol medical applications. Mater. Sci. Eng. A **273–275**, 149–160 (1999). https://doi.org/10.1016/S0921-5093(99)00294-4
4. Feninat, F.E., Laroche, G., Fiset, M., Mantovani, D.: Shape memory materials for biomedical applications. Adv. Eng. Mater. **4**(3), 91–104 (2002). https://doi.org/10.1002/1527-2648(200203)4:3%3c91::AID-ADEM91%3e3.0.CO;2-B

5. Otsuka, K., Ren, X.: Physical metallurgy of Ti–Ni-based shape memory alloys. Progr. Mater. Sci. **50**(5), 511–678 (2005). https://doi.org/10.1016/j.pmatsci.2004.10.001
6. Bhadeshia, H.K.D.H.: Developments in martensitic and bainitic steels: role of the shape deformation. Mater. Sci. Eng. A **378**(1–2), 34–39 (2004). https://doi.org/10.1016/j.msea.2003.10.328
7. Christian, J.W., Mahajan, S.: Deformation twinning. Prog. Mater. Sci. **39**(1–2), 1–157 (1995). https://doi.org/10.1016/0079-6425(94)00007-7
8. Mayer, S., Petersmann, M., Fischer, F.D., Clemens, H., Waitz, T., Antretter, T.: Experimental and theoretical evidence of displacive martensite in an intermetallic Mo-containing γ-TiAl based alloy. Acta Mater. **115**, 242–249 (2016). https://doi.org/10.1016/j.actamat.2016.06.006
9. Kelly, P.M.: Crystallography of martensite transformations in steels. In: Phase Transformations in Steels, pp. 3–33. Woodhead Publishing (2012). https://doi.org/10.1533/9780857096111.1.3
10. Dlouhy, A., Khalil-Allafi, J., Eggeler, G.: Multiple-step martensitic transformations in Ni-rich NiTi alloys–an in-situ transmission electron microscopy investigation. Philos. Mag. **83**(3), 339–363 (2003). https://doi.org/10.1080/0141861021000034531
11. Duerig, T.W., Melton, K.N., Stöckel, D.W.C.M.: Engineering Aspects of Shape Memory Alloys. Butterworth-Heinemann (2013)
12. Bhattacharya, K.: Microstructure of Martensite: Why it Forms and How it Gives Rise to the Shape-Memory Effect. Oxford University Press, Oxford (2003). https://doi.org/10.1093/oso/9780198509349.001.0001
13. Otsuka, K., Ren, X.: Martensitic transformations in nonferrous shape memory alloys. Mater. Sci. Eng. A **273**, 89–105 (1999). https://doi.org/10.1016/S0921-5093(99)00291-9
14. Gall, K., Sehitoglu, H.: The role of texture in tension–compression asymmetry in polycrystalline NiTi. Int. J. Plast. **15**(1), 69–92 (1999). https://doi.org/10.1016/S0749-6419(98)00060-6
15. Torra, V., Martorell, F., Lovey, F.C., Sade, M.: Remarks on the particular behavior in martensitic phase transition in Cu-based and Ni–Ti shape memory alloys. Shape Mem. Superelast. **4**(2), 272–284 (2018). https://doi.org/10.1007/s40830-018-0178-8
16. Gall, K., Sehitoglu, H., Chumlyakov, Y.I., Kireeva, I.V.: Tension–compression asymmetry of the stress–strain response in aged single crystal and polycrystalline NiTi. Acta Mater. **47**(4), 1203–1217 (1999). https://doi.org/10.1016/S1359-6454(98)00432-7
17. Villa, F., Villa, E., Nespoli, A., Passaretti, F.: Internal friction parameter in shape memory alloys: correlation between thermomechanical conditions and damping properties in NiTi and NiTiCu at different temperatures. J. Mater. Eng. Perf. **30**(4), 2605–2616 (2021). https://doi.org/10.1007/s11665-021-05609-3
18. Qin, Q., Peng, H., Fan, Q., Zhang, L., Wen, Y.: Effect of second phase precipitation on martensitic transformation and hardness in highly Ni-rich NiTi alloys. J. Alloys Comp. **739**, 873–881 (2018). https://doi.org/10.1016/j.jallcom.2017.12.128
19. Pushin, V.G., Kuranova, N.N., Marchenkova, E.B., Belosludtseva, E.S., Kourov, N.I., Kuntsevich, T.E., Uksusnikov, A.N.: Thermoelastic martensitic transitions and shape memory effects: classification, crystal and structural mechanisms of transformations, properties, production and application of promising alloys. Mater. Sci. Found. **81**, 174–206 (2015). https://doi.org/10.4028/www.scientific.net/MSFo.81-82.174
20. Yoneyama, T., Miyazaki, S.: (eds.): Shape Memory Alloys for Biomedical Applications. Woodhead Publishing (2008)
21. Elahinia, M.H.: Shape Memory Alloy Actuators: Design, Fabrication, and Experimental Evaluation. Wiley & Sons (2016)
22. Li, B., Wang, B., Wang, L., Oliveira, J.P., Cui, R., Wang, Y., Zhu, G., Yu, J., Su, Y.: Effect of post-heat treatments on the microstructure, martensitic transformation and functional performance of EBF3-fabricated NiTi shape memory alloy. Mater. Sci. Eng. A **871**, 144897 (2023). https://doi.org/10.1016/j.msea.2023.144897
23. Lin, B., Gall, K., Maier, H.J., Waldron, R.: Structure and thermomechanical behavior of NiTiPt shape memory alloy wires. Acta Biomater. **5**(1), 257–267 (2009). https://doi.org/10.1016/j.actbio.2008.07.015

24. Firstov, G.S., Van Humbeeck, J., Koval, Y.N.: High-temperature shape memory alloys: some recent developments. Mater. Sci. Eng. A **378**(1–2), 2–10 (2004). https://doi.org/10.1016/j.msea.2003.10.324
25. Bigelow, G.S., Padula, S.A., Garg, A., Gaydosh, D., Noebe, R.D.: Characterization of ternary NiTiPd high-temperature shape-memory alloys under load-biased thermal cycling. Metall. Mater. Trans. A **41**, 3065–3079 (2010). https://doi.org/10.1007/s11661-010-0365-5
26. Kaya, I., Karaca, H.E., Nagasako, M., Kainuma, R.: Effects of aging temperature and aging time on the mechanism of martensitic transformation in nickel-rich NiTi shape memory alloys. Mater Charact **159**, 110034 (2020). https://doi.org/10.1016/j.matchar.2019.110034
27. Bertacchini, O.W., Lagoudas, D.C., Calkins, F.T., Mabe, J.H.: Thermomechanical cyclic loading and fatigue life characterization of nickel rich NiTi shape-memory alloy actuators. In: Proceedings of the SPIE 6929, Behavior and Mechanics of Multifunctional and Composite Materials 692916 (2008). https://doi.org/10.1117/12.776502
28. Sato, A., Chishima, E., Soma, K., Mori, T.: Shape memory effect in $\gamma \rightleftarrows \epsilon$ transformation in Fe-30Mn-1Si alloy single crystals. Acta Metall. **30**(6), 1177–1183 (1982). https://doi.org/10.1016/0001-6160(82)90011-6
29. Zhang, J., Fang, C., Yam, M.C., Lin, C.: Fe–Mn–Si alloy U-shaped dampers with extraordinary low-cycle fatigue resistance. Eng. Struct. **264**, 114475 (2022). https://doi.org/10.1016/j.engstruct.2022.114475
30. Ozcan, H., Ma, J., Wang, S.J., Karaman, I., Chumlyakov, Y., Brown, J., Noebe, R.D.: Effects of cyclic heat treatment and aging on superelasticity in oligocrystalline Fe–Mn–Al–Ni shape memory alloy wires. Scr. Mater. **134**, 66–70 (2017). https://doi.org/10.1016/j.scriptamat.2017.02.023
31. Morales-Rivas, L., Garcia-Matereo, C., Sourmail, T., Kuntz, M., Rementeria, R., Caballero, F.G.: Ductility of nanostructured bainite. Metals **6**(12), 302 (2016). https://doi.org/10.3390/met6120302
32. Omori, T., Kainuma, R.: Martensitic transformation and superelasticity in Fe–Mn–Al-based shape memory alloys. Shape Mem. Superelast. **3**(4), 322–334 (2017). https://doi.org/10.1007/s40830-017-0129-9
33. Ghafoori, E., Hosseini, E., Leinenbach, C., Michels, J., Motavalli, M.: Fatigue behavior of a Fe–Mn–Si shape memory alloy used for prestressed strengthening. Mater. Des. **133**, 349–362 (2017). https://doi.org/10.1016/j.matdes.2017.07.055
34. Walnsch, A., Kriegel, M.J., Motylenko, M., Korpala, G., Prahl, U., Leineweber, A.: Thermodynamics of martensite formation in Fe–Mn–Al–Ni shape memory alloys. Scr. Mater. **192**, 26–31 (2021). https://doi.org/10.1016/j.scriptamat.2020.10.003
35. Abdulridha, A., Palermo, D., Foo, S., Vecchio, F.J.: Behavior and modeling of superelastic shape memory alloy reinforced concrete beams. Eng. Struct. **49**, 893–904 (2013). https://doi.org/10.1016/j.engstruct.2012.12.041
36. Code, P.: Eurocode 2: design of concrete structures-part 1–1: general rules and rules for buildings, vol. 668, pp. 659–668. British Standard Institution, London (2005)
37. Cladera, A., Weber, B., Leinenbach, C., Czaderski, C., Häupl, P., Motavalli, M.: Iron-based shape memory alloys for civil engineering structures: an overview. Constr. Build. Mater. **63**, 281–293 (2014). https://doi.org/10.1016/j.conbuildmat.2014.04.032
38. Kainuma, R., Takahashi, S., Ishida, K.: Thermoelastic martensite and shape memory effect in ductile Cu–Al–Mn alloys. Metall. Mater. Trans. A **27**, 2187–2195 (1996). https://doi.org/10.1007/BF02651873
39. Karaca, H.E., Karaman, I., Basaran, B., Ren, Y., Chumlyakov, Y.I., Maier, H.J.: Magnetic field-induced phase transformation in NiMnCoIn magnetic shape-memory alloys—a new actuation mechanism with large work output. Adv. Func. Mater. **19**(7), 983–998 (2009). https://doi.org/10.1002/adfm.200801322
40. Moya, X., Kar-Narayan, S., Materhur, N.D.: Caloric materials near ferroic phase transitions. Nat. Mater. **13**(5), 439–450 (2014). https://doi.org/10.1038/nmat3951
41. Eggeler, G., Hornbogen, E., Yawny, A., Heckmann, A., Wagner, M.F.: Structural and functional fatigue of NiTi shape memory alloys. Mater. Sci. Eng. A **378**(1–2), 24–33 (2004). https://doi.org/10.1016/j.msea.2003.10.327

42. Khalil-Allafi, J., Dlouhy, A., Eggeler, G.: Ni_4Ti_3-precipitation during aging of NiTi shape memory alloys and its influence on martensitic phase transformations. Acta Mater. **50**(17), 4255–4274 (2002). https://doi.org/10.1016/S1359-6454(02)00257-4
43. Ashbli, S., Menzemer, C.: On the fatigue behavior of nanocrystalline NiTi shape memory alloys: a review. J. Nanomed. Nanotechnol. **10**(529), 2 (2019). https://doi.org/10.35248/2157-7439.19.10.529
44. Pushin, V.G., Stolyarov, V.V., Valiev, R.Z., Lowe, T.C., Zhu, Y.T.: Nanostructured TiNi-based shape memory alloys processed by severe plastic deformation. Mater. Sci. Eng. A **410**, 386–389 (2005). https://doi.org/10.1016/j.msea.2005.08.071
45. Ahluwalia, R., Quek, S.S., Wu, D.T.: Simulation of grain size effects in nanocrystalline shape memory alloys. J. Appl. Phys. **117**(24) (2015). https://doi.org/10.1063/1.4923044
46. Li, Y., Li, J.Y., Liu, M., Ren, Y.Y., Chen, F., Yao, G.C., Mei, Q.S.: Evolution of microstructure and property of NiTi alloy induced by cold rolling. J. Alloys Comp. **653**, 156–161 (2015). https://doi.org/10.1016/j.jallcom.2015.09.056
47. Waitz, T., Antretter, T., Fischer, F.D., Karnthaler, H.P.: Size effects on martensitic phase transformations in nanocrystalline NiTi shape memory alloys. Mater. Sci. Tech. **24**(8), 934–940 (2008). https://doi.org/10.1179/174328408X3026
48. Chen, J., Yin, H., Sun, Q.: Effects of grain size on fatigue crack growth behaviors of nanocrystalline superelastic NiTi shape memory alloys. Act Mater. **195**, 141–150 (2020). https://doi.org/10.1016/j.actamat.2020.05.008
49. Sun, Q., Aslan, A., Li, M., Chen, M.: Effects of grain size on phase transition behavior of nanocrystalline shape memory alloys. Sci. China Tech. Sci. **57**, 671–679 (2014). https://doi.org/10.1007/s11431-014-5505-5
50. Chowdhury, P.B., Sehitoglu, H., Rateick, R.G.: Predicting fatigue resistance of nano-twinned materials: Part I-role of cyclic slip irreversibility and Peierls stress. Int. J. Fatigue **68**, 277–291 (2014). https://doi.org/10.1016/j.ijfatigue.2014.05.014
51. Li, S., Cong, D., Xiong, W., Chen, Z., Zhang, X., Nie, Z., Wang, Y.: A low-cost Ni–Mn–Ti–B high-temperature shape memory alloy with extraordinary functional properties. ACS Appl. Mater. Interface **13**(27), 31870–31879 (2021). https://doi.org/10.1021/acsami.1c07619
52. Battaglia, M., Sellitto, A., Giamundo, A., Visone, M., Riccio, A.: Advanced material thermomechanical modelling of shape memory alloys applied to automotive design. Shape Mem. Superelast. **10**(3), 297–313 (2024). https://doi.org/10.1007/s40830-024-00494-w
53. Shabalovskaya, S.A.: On the nature of the biocompatibility and on medical applications of NiTi shape memory and superelastic alloys. Biomed. Mater. Eng. **6**(4), 267–289 (1996). https://doi.org/10.3233/BME-1996-6405
54. Sabahi, N., Chen, W., Wang, C.H., Kruzic, J.J., Li, X.: A review on additive manufacturing of shape-memory materials for biomedical applications. Adv. Manuf. Biomater. Biol. Mater. **72**, 1229–1253 (2020). https://doi.org/10.1007/s11837-020-04013-x
55. Petrini, L., Migliavacca, F.: Biomedical applications of shape memory alloys. J. Metall. **1**, 501483 (2011). https://doi.org/10.1155/2011/501483
56. Ryhänen, J., Kallioinen, M., Tuukkanen, J., Junila, J., Niemelä, E., Sandvik, P., Serlo, W.: In vivo biocompatibility evaluation of nickel-titanium shape memory metal alloy: muscle and perineural tissue responses and encapsule membrane thickness. J. Biomed. Mater. Res. **41**(3), 481–488 (1998). https://doi.org/10.1002/(SICI)1097-4636(19980905)41:3%3c481::AID-JBM19%3e3.0.CO;2-L
57. Vojtěch, D., Fojt, J., Joska, L., Novák, P.: Surface treatment of NiTi shape memory alloy and its influence on corrosion behavior. Surf. Coat. Tech. **204**(23), 3895–3901 (2010). https://doi.org/10.1016/j.surfcoat.2010.05.010
58. Ryhänen, J., Niemi, E.U.O.O., Serlo, W., Niemelä, E., Sandvik, P., Pernu, H., Salo, T.: Biocompatibility of nickel–titanium shape memory metal and its corrosion behavior in human cell cultures. J. Biomed. Mater. Res. **35**(4), 451–457 (1997). https://doi.org/10.1002/(SICI)1097-4636(19970615)35:4%3c451::AID-JBM5%3e3.0.CO;2-G
59. Mariappan, N.: Recent trends in nanotechnology applications in surgical specialties and orthopedic surgery. Biomed. Pharmacol. J. **12**(3), 1095–1127 (2019). https://doi.org/10.13005/bpj/1739

60. Ikuta, K., Nokata, M.: Two-lead-wire drive for multi-micro actuators [medical micro robots]. In: Proceedings 1999 IEEE International Conference on Robotics and Automation (Cat. No. 99CH36288C) (Vol. 3, pp. 2378–2384). IEEE (1999, May)
61. Hoon, Q.J., Pelletier, M.H., Christou, C., Johnson, K.A., Walsh, W.R.: Biomechanical evaluation of shape-memory alloy staples for internal fixation—an in vitro study. J. Exp. Orthop. **3**, 1–11 (2016). https://doi.org/10.1186/s40634-016-0055-3
62. Puffer, A.J.: Design and Testing of a Minimally Invasive Blood Clot Removal Device Constructed With Elements of Superelastic Nitinol. The University of Toledo (2014)
63. Mao, L., Shen, L., Chen, J., Zhang, X., Kwak, M., Wu, Y., Fan, R., Zhang, L., Pei, J., Yuan, G., Song, C., Ge, J., Ding, W.: A promising biodegradable magnesium alloy suitable for clinical vascular stent application. Sci. Rep. **7**(1), 46343 (2017). https://doi.org/10.1038/srep46343
64. Shabalovskaya, S., Anderegg, J., Van Humbeeck, J.: Critical overview of Nitinol surfaces and their modifications for medical applications. Acta Biomater. **4**(3), 447–467 (2008). https://doi.org/10.1016/j.actbio.2008.01.013
65. Sun, F., Feng, X., Wang, H., Cao, X., Liu, X., Liu, Y., Li, Y., Liu, Z., Yi, X.: Brief overview of functionally graded NiTi-based shape memory alloys. Adv. Eng. Mater. **25**(9), 2201434 (2023). https://doi.org/10.1002/adem.202201434
66. Chiu, T.C., Huang, C.C.: Aptamer-functionalized nano-biosensors. Sensor **9**(12), 10356–10388 (2009). https://doi.org/10.3390/s91210356
67. Pelton, A.R., Dicello, J., Miyazaki, S.: Optimization of processing and properties of medical-grade Nitinol wire. Min. Invas. Ther. Allied Technol. **9**(1), 107–118 (2009)
68. Bradley, T.G., Brantley, W.A., Culbertson, B.M.: Differential scanning calorimetry (DSC) analyses of superelastic and nonsuperelastic nickel-titanium orthodontic wires. Am. J. Orthod. Dentofac. Orthop. **109**(6), 589–597 (1996). https://doi.org/10.1016/S0889-5406(96)70070-7
69. Wu, J.C., Mills, A., Grant, K.D., Wiater, P.J.: Fracture fixation using shape-memory (ninitol) staples. Ortho. Clin. **50**(3), 367–374 (2019). https://doi.org/10.1016/j.ocl.2019.02.002
70. Es-Souni, M., Es-Souni, M., Fischer-Brandies, H.: Assessing the biocompatibility of NiTi shape memory alloys used for medical applications. Anal. Bioanal. Chem. **381**(3), 557–567 (2005). https://doi.org/10.1007/s00216-004-2888-3
71. Ramazanzadeh, B.A., Ahrari, F., Sabzevari, B., Habibi, S.: Nickel ion release from three types of nickel-titanium-based orthodontic archwires in the as-received state and after oral simulation. J. Dent. Res. Dent. Clin. Dent. Prosp. **8**(2), 71 (2014). https://doi.org/10.5681/joddd.2014.013
72. Yokokawa, H., Tu, H., Iwanschitz, B., Mai, A.: Fundamental mechanisms limiting solid oxide fuel cell durability. J Power Sources **182**(2), 400–412 (2008). https://doi.org/10.1016/j.jpowsour.2008.02.016
73. Yang, C., Cao, W., Yang, Z., Wang, M., Jing, X., Tian, Y.: The study on the anti-corrosion performance of NiTi alloy in human body solution with the fabricating processes of laser irradiation and PDMS modification. J. Bionic Eng. **18**, 77–91 (2021). https://doi.org/10.1007/s42235-021-0011-5
74. Peng, Y., Miao, K., Sun, W., Liu, C., Wu, H., Geng, L., Fan, G.: Recent progress of synchrotron X-ray imaging and diffraction on the solidification and deformation behavior of metallic materials. Acta Metall. Sinica (English Letters) **35**(1), 3–24 (2022). https://doi.org/10.1007/s40195-021-01311-4

Bioactive Nanomaterials: Revolutionizing Biomedical Science

Safaa Najah Saud Al-Humairi and Kritanjali Singh

Abstract This chapter seeks to thoroughly investigate the developing discipline of bioactive nanomaterials and their transformative effect on biomedical research. We examined the definition and importance of these advanced materials, explored the various bioactive nanomaterials currently investigated, metallic, ceramic, polymeric, and carbon-based, and discussed their unique features supporting biomedical applications. Moreover, we will delve into the complex interactions between these nanomaterials and biological systems at both the molecular and cellular levels. A review of the extensive uses of bioactive nanoparticles in critical areas such as targeted drug delivery, tissue engineering, and diagnostics will constitute a significant section of this chapter. Essential for their safe and effective use in medicine, we will also address the salient aspects of their biocompatibility and potential toxicity. Lastly, we will discuss the prospects of this exciting topic, emphasizing the challenges that must be addressed to fully exploit the transformative potential of bioactive nanomaterials in revolutionizing biomedical research and enhancing human health.

Keywords Nanoparticles · Diagnostics · Drug delivery · Biomedical applications · Nanotoxicity

1 Introduction

The growing worldwide prevalence of crippling illnesses calls for constant research and creative development of new approaches for diagnosis and treatment. Although traditional medical techniques have reached vital benchmarks, numerous difficulties still exist, especially in treating complex diseases such as cancer, neurological

S. N. S. Al-Humairi (✉)
Faculty of Information Sciences and Engineering, Management and Science University, 40100 Shah Alam, Malaysia
e-mail: safaa_najah@msu.edu.my

K. Singh
Subharti Medical College and Research and Development Cell, Swami Vivekanand Subharti University, Meerut, U.P. 250005, India

S. N. Saud Al-Humairi et al. (eds.), *Smart Nanomaterials in Biomedical Applications*, Engineering Materials, https://doi.org/10.1007/978-981-96-8023-8_5

disorders, and significant tissue damage. Emerging as a transformative discipline with unprecedented possibilities to transcend limitations and usher in a new era of precision medicine is the intersection of nanotechnology and biomedical research [1].

Initially, traditional biomaterials extensively utilized in medical applications, such as implants and drug delivery systems, were primarily developed to be inert within the biological environment [2]. This focus on passivity aimed to enhance biocompatibility and minimize adverse reactions. However, this inherent inertness limited the materials' capacity to engage actively in the complex biological processes essential for tissue healing and regeneration [3]. For instance, while a standard hip implant provides mechanical joint support, it typically does not stimulate natural bone growth or integrate biologically beyond mechanical fixation. This lack of bioactivity posed significant challenges for applications requiring active tissue repair, targeted drug delivery, or precise cellular function modulation.

Recognition of these limitations catalyzed a transformative shift towards bioactive materials. Pioneering research by Larry Hench in the 1960s on bioactive glass was pivotal, demonstrating for the first time that materials could persist within living tissue and actively integrate with it, forming strong interfacial bonds with bone [4]. This groundbreaking finding overturned the conventional requirement for biomaterial inertness and launched a new field focused on designing materials that actively elicit specific biological responses for therapeutic gain. This conceptual evolution signified that materials could be engineered to participate proactively in the body's healing mechanisms [5].

Building on this foundation, the development of nanotechnology has provided bioactive materials with a new level of control and utility. Typically ranging from 1 to 100 nm in at least one dimension, scientists have created bioactive nanomaterials with unprecedented control over their physicochemical characteristics and interactions with biological systems through nanoscale material manipulation [6]. This miniaturization to the nanoscale has produced a synergistic effect, whereby the unique characteristics of nanomaterials, such as their very high surface area to volume ratio and the emergence of quantum effects, significantly enhance their bioactivity and offer new opportunities for interacting with biological entities at a molecular level. Bioactive nanomaterials are smaller counterparts of their macroscopic versions and exhibit unique behaviors and functions that may be absent in bulk materials, facilitating more precise and effective treatments in biological systems [7].

This chapter seeks to thoroughly investigate the developing discipline of bioactive nanomaterials and their transformative effect on biomedical research. We will examine the definition and importance of these advanced materials, explore the various bioactive nanomaterials currently under investigation metallic, ceramic, polymeric, and carbon-based and discuss their unique features supporting biomedical applications. Moreover, we will delve into the complex interactions between these nanomaterials and biological systems at both the molecular and cellular levels. A review of the extensive uses of bioactive nanoparticles in critical areas such as targeted drug delivery, tissue engineering, and diagnostics will constitute a significant section of this chapter. Essential for their safe and effective use in medicine, we will also address the salient aspects of their biocompatibility and potential toxicity. Lastly, we will discuss the prospects of this exciting topic, emphasizing the challenges that must be addressed to fully exploit the transformative potential of bioactive nanomaterials in revolutionizing biomedical research and enhancing human health. This shift focuses on designing materials that actively stimulate specific biological responses for therapeutic benefit. The implications of this transition are profound, suggesting that materials can be engineered to play a more active role in the body's natural healing mechanisms.

Building upon this foundation, the advent of nanotechnology has ushered in a new dimension of control and functionality in bioactive materials. By manipulating materials at the nanoscale, typically ranging from 1 to 100 nm in at least one dimension, scientists have created bioactive nanomaterials with unprecedented control over their physicochemical properties and interactions with biological systems. Miniaturization to the nanoscale imbues materials with unique properties, such as exceptionally high surface-area-to-volume ratios and emergent quantum effects. These characteristics synergistically enhance bioactivity, creating novel possibilities for molecular-level interactions with biological systems. Bioactive nanomaterials are distinct from their macroscopic precursors; they exhibit behaviours and functionalities often absent in bulk materials, enabling more precise and effective interventions within biological systems.

This chapter comprehensively explores the burgeoning field of bioactive nanomaterials and their significant impact on biomedical science. We will define these advanced materials, discuss their importance, and survey the diverse classes under investigation (including metallic, ceramic, polymeric, and carbon-based). We will also examine the unique properties driving their biomedical applications and the intricate mechanisms governing their interactions with biological systems at the cellular and molecular levels.

A substantial portion reviews key applications like targeted drug delivery, tissue engineering, and diagnostics. Critical considerations of biocompatibility and potential toxicity, essential for safe and effective deployment, are also addressed. Finally, the chapter explores future directions, highlighting challenges that must be overcome to fully realize the transformative potential of bioactive nanomaterials in advancing biomedical science and improving human health.

2 Understanding Bioactive Nanomaterials and Their Significance

Bioactive materials are characterized by their ability to elicit a specific biological response at the interface with surrounding living tissues. A strong interfacial bond between the material and tissue enhances bioactivity, promoting favorable biological outcomes [6]. The advent of bioactive materials marked a significant departure from the earlier biomaterials paradigm, which emphasized inertness primarily to prevent adverse physiological reactions. The scope of biomedical applications broadened considerably when it was recognized that materials could be engineered to actively participate in biological processes, such as inducing healing or promoting tissue formation [8].

Extending this principle, bioactive nanomaterials are engineered explicitly at the nanoscale to trigger desired biological reactions upon contact with proteins, cells, or tissues within the body. Their distinct characteristics—small size, high surface area, and precisely engineered structures—lead to unique interactions with biological systems and often result in novel biological activity compared to their bulk counterparts [9]. These nanoscale features, such as size effects, specific surface area, and controlled nanostructure, are critical determinants of their pronounced bioactivities and interaction mechanisms.

The bioactivity of these nanoparticles is dictated by their physical and chemical properties, including crystal structure, surface charge, specific surface functional groups, overall shape, and size. These features can be precisely controlled during synthesis to achieve specific biological goals. Consequently, bioactive nanoparticles are essential tools in biomedical research, enabling applications such as disease detection, targeted treatment, and tissue regeneration [10]. They have the potential to revolutionize healthcare, offering more precise and efficient treatments for various conditions compared to traditional biomaterials or conventional pharmaceutical formulations. Their nanoscale dimensions provide a unique platform for specialized, individualized therapies by enabling interaction at the molecular level. Bioactive nanomaterials can, therefore, overcome the limitations of "one-size-fits-all" approaches by allowing properties and bioactivities to be tailored to specific patient needs and conditions, leading to more effective and potentially less invasive therapeutic strategies [7].

Furthermore, these cutting-edge materials have significant potential to address critical global health challenges, such as treating various malignant cancers and severe chronic illnesses, including nervous system diseases. Bioactive nanoparticles frequently work by direct contact with biological components, circumventing the typical drug-release mechanisms found in conventional nanomedicines [11]. This direct biological action represents substantial development in the area and can result in more rapid and possibly more successful therapy outcomes.

3 The Diverse Landscape of Bioactive Nanomaterials: Types and Properties

Figure 1 shows that bioactive nanomaterials are classified into three types: metallic nanoparticles, ceramic nanoparticles, polymeric nanoparticles, and carbon-based nanomaterials [12]. Each class has unique characteristics that make it suited for specific biomedical applications. Metallic nanoparticles, which are commonly made from noble metals such as gold, silver, and platinum, as well as transition metals such as iron and copper, exhibit exceptional properties that vary greatly depending on their size, shape, and surface features [13]. Metallic nanoparticles are notable for their size-dependent bioactivity. Smaller silver nanoparticles, for example, have demonstrated more significant antibacterial activity than larger ones.

Furthermore, specific metallic nanoparticles exert effects via ion release; for example, iron ions released from iron-based nanoparticles can be used in cancer therapy, while silver ions confer potent antibacterial properties [14]. Many metallic nanoparticles also exhibit enzyme-mimicking catalytic activity, proving valuable in therapeutic and diagnostic applications. The surface charge of these nanoparticles critically influences their biological interactions by modulating protein adsorption and cellular uptake. Noble metal nanoparticles, noted for their distinct optical properties like localized surface plasmon resonance (LSPR), are particularly useful. LSPR enables the absorption and manipulation of light at specific wavelengths, making these nanoparticles advantageous for bioimaging and photothermal therapy [15]. Magnetic metallic nanoparticles, especially iron oxides, possess a wide spectrum of biomedical applications. These include therapeutic uses (such as antimicrobial

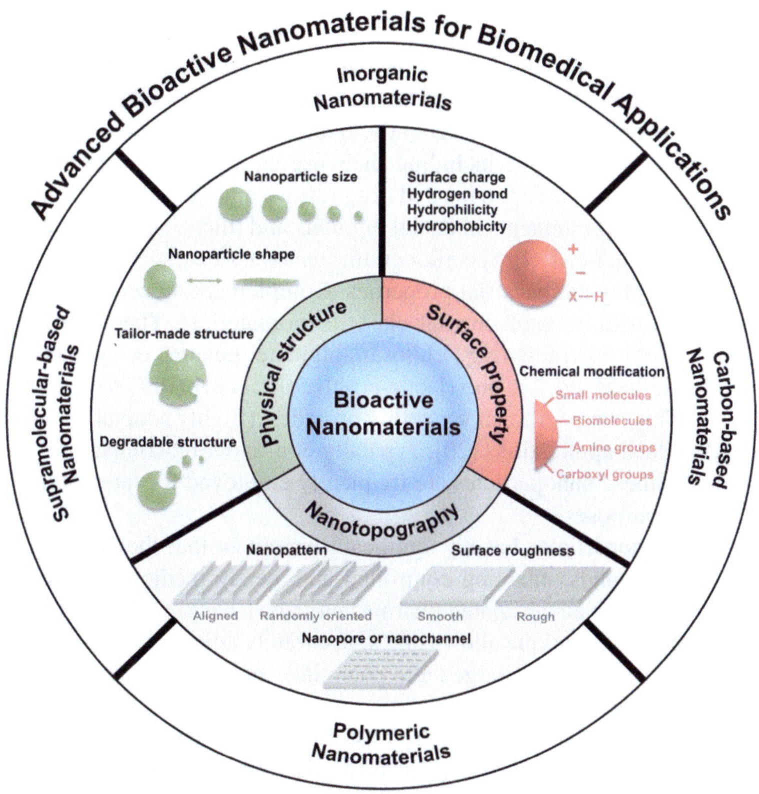

Fig. 1 Biomedical applications of advanced bioactive nanomaterials [12]

agents, cancer therapy, and inhibiting protein aggregation associated with neurodegeneration), diagnostic tools (like immunoassays and contrast agents for imaging), and roles in regenerative medicine (e.g., bone tissue engineering) [16]. Typical applications leverage their magnetic properties for hyperthermia therapies, targeted drug delivery, and magnetic resonance imaging (MRI). Additionally, the development of environmentally benign "green synthesis" methods, employing resources like plant extracts or microorganisms, offers sustainable routes for the production of these metallic nanoparticles [17].

Ceramic nanoparticles composed of alumina, zirconia, hydroxyapatite, and bioactive glasses are characterized by high stiffness, strength, hardness, and resilience to elevated temperatures and chemical degradation. Ceramic biomaterials are categorized based on their interaction with biological tissues into three main groups: inert materials (e.g., alumina, zirconia) that exhibit low reactivity; surface-active materials (e.g., bioactive glasses) capable of forming direct bonds with tissues; and bioresorbable materials (e.g., hydroxyapatite, tricalcium phosphate) which progressively dissolve and are replaced by new tissue [18]. The specific chemical composition of the ceramic dictates its category. Owing to their characteristic blend of mechanical robustness and biocompatibility, bioceramics are well-suited for load-bearing tissues and bone regeneration applications.

Furthermore, ceramic nanocomposites—wherein ceramic nanoparticles are integrated into a matrix—can demonstrate enhanced mechanical strength, biocompatibility, and osteoconductivity. Consequently, ceramic nanoparticles have found extensive biomedical applications, including their use as bioactive coatings for orthopedic implants, platforms for controlled drug release and gene delivery, scaffolds in tissue engineering, components in dental implants and fillers, and materials for bone grafts and reconstruction [19]. Specific ceramic nanoparticles, such as hydroxyapatite and alumina, display antibacterial properties. Complementing ceramics, polymeric nanoparticles constitute another vital class of biomaterials. These can be formulated from natural polymers (e.g., chitosan, alginate, gelatin) or synthetic options (e.g., polylactic acid (PLA), poly(lactic-co-glycolic acid) (PLGA), polycaprolactone (PCL)). This compositional versatility provides a highly adaptable platform for diverse biomedical applications [20]. Typically non-toxic, biocompatible, and often biodegradable, these nanoparticles are frequently employed for internal therapeutic and diagnostic purposes.

Polymeric nanoparticles have a significant benefit in that they can encapsulate therapeutic compounds, enabling controlled and steady medication release and a higher half-life of the encapsulated pharmaceuticals. Changing the type of polymer and the preparation technique allows one to accurately adjust their physicochemical characteristics size, surface charge (zeta potential), and encapsulation effectiveness [21]. Moreover, the surface of polymeric nanoparticles can be easily functionalized with molecules like polyethylene glycol (PEG) to lengthen their circulation time or target ligands to improve their absorption by particular cells. Particularly helpful in solubilizing hydrophobic medicines, some polymeric materials can also self-assemble into nanostructures such as micelles or polymersomes [22]. Polymeric nanoparticles thus find extensive use in drug delivery, including targeted therapies for cancer and arthritis, antimicrobial delivery, gene therapy, tissue engineering as scaffolds in nanofibers, wound healing, and even in theranostics, so combining treatment and diagnostics [23].

Another important class of bioactive nanomaterials based on carbon is fullerenes, carbon nanotubes (single-walled and multi-walled), graphene, graphene oxide, carbon dots, and nanodiamonds. Highly promising for biomedical uses, these materials show a surprising spectrum of structural, electrical, mechanical, and optical characteristics [24]. For example, carbon nanotubes and graphene have extraordinary mechanical strength and electrical conductivity, while carbon dots show unique photoluminescent qualities. Many carbon-based nanomaterials show good biocompatibility when adequately functionalized. Their enormous surface area-to-ratio makes possible high drug-loading capabilities and simplicity of surface modification with several functional groups and biomolecules [25]. Consequently, carbon-based nanomaterials are under significant research for drug administration, gene therapy, bioimaging, and biosensing as scaffolds in tissue engineering, wound healing applications, photothermal therapy, and even immunization techniques [26].

4 Comparison of Bioactive and Traditional Nanomaterials

Nanomaterials are solid materials composed of nanoparticles with a size of no more than 100 nm. In the material revolution, nanotechnology has been widely used in energy and environment, electronics and information, medicine, and health [27], as shown in Fig. 2. The rapid development of nanotechnology and the synthesis of diverse functional nanomaterials have provided a broad space for innovation in medical technology. Drug delivery, biological detection, and disease diagnosis and prevention have become the research hotspots in the nanoscience and technology healthcare field.

Bioactive nanomaterials are solid materials in vitro or implanted, consisting of nanoparticles exhibiting biological activities and interacting with proteins, cells, or tissues in vivo and causing biological reactions depending on their physical and chemical properties, showing their superiority over traditional materials [28]. Bioactive nanomaterials have been developed with nanoscale size and precise structure, enabling them to regulate the interaction between materials and biological systems, exhibiting unique biological activities. Researchers in the biomedical field have favored bioactive nanomaterials in recent years due to their specific induction and response to particular pathology and sites [9]. As the study of mechanisms of bioactive materials and biomedicine deepens, biodegradable and Bioactive-supported materials are expected to more effectively treat challenging diseases that could not be treated using conventional methods. Challenging nervous system diseases and malignant tumors can expect better therapeutic effects through their development [29].

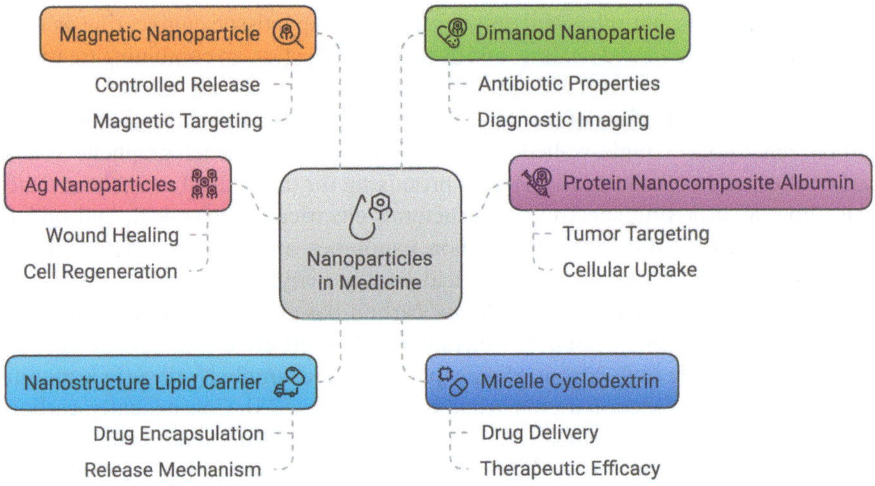

Fig. 2 Diagrammatic illustration of the use of nanoparticles for biomedical applications

5 Synthesis Methods of Bioactive Nanomaterials

Recent advances in nanotechnology have allowed researchers to develop devices with promising potential for use in environmental science, pharmaceuticals, biology, and medicine, and synthesizing biogenic nanoparticles has become an area of intense focus and interest. Due to their small size, the NPs possess unique physicochemical properties that offer advantages such as a high surface-volume ratio, thermal stability improvement, optical properties such as fluorescence or absorbance of light, and a size-dependent ability to interact with other molecules [30]. There is, however, a growing strain on current chemical processes synthesizing materials on the nanoscale due to environmental pollution brought about by the toxic nature of the chemical compounds routinely used as a part of these processes [31]. Since nanoparticles are generated, handled, and manipulated with their large surface area in mind, methods of nanoparticle creation need to have a diameter-controlling element, a functionality-generating element, or both. The commonly known physical methods for NP synthesis do not retain this small-scale control in the particle product and require a complexity and overhead that excludes the average biological laboratory from entry. The traditional chemical counterparts can more readily overcome this obstacle. "Bio-NPs" can be synthesized in only one step under a wide range of conditions, and with the provided template, large control of the size and crystallinity can be achieved [32]. Additionally, standard laboratory operations in any biological laboratory can be assumed, and the required time and overhead for this bio-mimetic NP synthesis are minimal.

Green synthesis to preserve biological activity chemistry offers a variety of strategies to create complex structures from simple building blocks [33]. Soft solution-based chemistry in the presence of suitable linkers permits nano-structuring and self-assembly of molecular building blocks to fabricate small defined clusters. In contrast to purely chemical approaches, especially when using biological materials, bio-inorganic synthesis allows the introduction of functional biomolecules [34]. However, the self-assembly and de-novo synthesis of large inorganic structures is a synthetic feat that demands considerable effort and still largely eludes current scientific capabilities. Antagonistically, the growth of high-quality inorganic crystallites from solution upon nucleation and growth centers is a process that nature has long since perfected. The green synthesis method is gaining popularity due to its cost-effectiveness, non-toxic nature, and ease of scalability, making it suitable for large-scale production without the need for high pressure, temperature, or toxic chemicals [35]. Chemical processes such as co-precipitation, hydrothermal, sol–gel, and chemical vapor deposition enable precise control over the size, shape, and surface properties of nanomaterials, allowing them to be tailored for specific applications [36]. Mechanical milling, laser ablation, and lithography are all used to create unique nanomaterials. Biological procedures occasionally necessitate additional resources and energy [37]. The combination of these approaches has resulted in various nanomaterials with applications in health, agriculture, and environmental remediation [38]. Despite the considerable potential of these synthesis approaches, difficulties such as scalability, biocompatibility, and purification remain, necessitating additional research and invention to optimize these procedures for more general industrial and commercial use [39]. The future of bioactive nanomaterials depends on continued research into bio-based technologies, particularly those that use plant resources and microbial nano factories to create more sustainable and efficient synthesis pathways [40].

6 Advantages of Bioactive Nanomaterials in Biomedical Applications

Bioactive nanomaterials have attracted much attention worldwide due to their interactions with biological systems compared with traditional nanomaterials [28]. The rapid development and applications of nano-synthesis and nanotechnology result in many kinds of nanomaterials being used in various fields, such as industry, electronics, cosmetics, pharmaceuticals, and biomedicine, due to specific physical and chemical properties derived from their nanoscale structure and physicochemical properties. There are a number of advantages of biomaterials that can be described as follows:

6.1 Enhanced Targeting

Nanomaterials with an exciting and exceptional surface area refer to a vast spectrum of nanoscale materials with at least a 1–100 nm dimension limit. Nanomaterials' first and foremost merit is to precisely adjust and direct the final nanomaterial's characteristics from the initial steps by focusing on the synthesis methods, proper functionalization and modification, and monitoring the size and shape [41]. Hence, various nanomaterials are synthesized considering the mentioned variables. Because nanomaterials have a feature size in the biomolecule range, it is widely accepted that nanomaterial surface chemistry is the primary factor for cytotoxic effects. In order to achieve a stable and high-performance outcome in nanomaterial-biological interactions, it is essential to modify their surfaces with a biocompatible coating or ligand [42]. Such modifications can sometimes mitigate the immune system's toxic response and help nanomaterials achieve the intended goal, accumulating in targeted organs or lesions. Various nanomaterials are functionalized with appropriate characteristics to diminish cytotoxicity and access the desired position in the organism [43]. Aside from using a variety of ways to define NBCs and their constituents, as well as numerous biological targeting systems, the significant benefits of NBCs are highlighted. According to Li et al., the development of self-assembled nanomaterials employs bioactivated in vivo assembly (BIVA) approaches to create nanostructures that increase medication transport and targeting by responding to natural stimuli like enzymes and pH changes. These nanostructures have beneficial biological benefits such as improved targeting and retention at the lesion site, promising them for successful medication administration and therapy optimization [44]. Nanomaterials are used in cancer treatment to boost targeting efficiency through size, shape, and surface modifications, often in conjunction with ligands. This technique improves treatment efficiency by ensuring that anticancer medications are delivered directly to tumor sites, reducing systemic toxicity and bypassing biological barriers such as drug resistance [45]. Stimuli-responsive systems, which release drugs in reaction to environmental changes, have been proven in clinical trials to improve outcomes and reduce side effects [46]. Carbon nanomaterials and iron oxide/gold core–shell nanoparticles are being studied for their unique physicochemical properties and ability to be functionalized for targeted delivery, which has therapeutic applications including cancer and neurodegenerative diseases [47]. Bioactive nanomaterials such as silk fibroin and gold nanorods have been shown to accelerate wound healing by modulating immune responses and encouraging tissue regeneration, highlighting nanomaterials' adaptability in improving therapeutic targeting across many medical disciplines [48]. With ongoing research targeted at overcoming challenges such as immune system evasion and scalability, the incorporation of nanotechnology in drug delivery systems represents a transformative strategy for improving the targeting and efficacy of bioactive compounds in general [49].

6.2 Reduced Side Effects

Nanotechnology is a new technology that controls the construction of materials and devices in the nanometer range. As an emerging interdisciplinary science, nanotechnology has broad application prospects in various fields. The nanomaterials developed by nanotechnology will be applied to many fields, and nanomaterials will impact medicine at multiple levels and have different effects [50]. It will bring revolutionary changes to the future development of medicine, so it is one of the development directions of modern medicine. With the precise control of nanoparticles and nanodevices, much research has focused on bioactive nanomaterials. Compared with traditional materials, bioactive nanomaterials have unique nanoscale properties [28]. A bioactive nanomaterial is a kind of material that interacts with biological systems and can cause biological reactions. Bioactive nanomaterials function as crucial interfaces, modulating interactions between materials and biological systems through their unique biological activities [51]. As a material's biological behaviour is predominantly determined by its surface properties, bioactive nanomaterials represent a highly active research field with extensive application prospects. Significant research efforts have been directed towards preparing low-dimensional materials with distinct surface structures and properties [52].

Foundational research continues to explore the biomedical potential of bioactive nanomaterials, particularly for molecular diagnostics and therapeutic interventions. Compared to traditional biomaterials, low-dimensional bioactive nanomaterials offer distinct advantages for biological detection, including enhanced signal amplification and superior molecular recognition capabilities [53]. Their capacity to amplify biological signals or regulate biological systems underpins their growing importance in the scientific community [54]. In therapeutics, the high surface-area-to-volume ratio and molecular-level interaction capabilities of nanomaterials make them ideal vectors for targeted drug delivery, aiming to improve treatment efficacy while minimizing adverse effects [55]. Examples include surface-modified carbon-based nanomaterials demonstrating improved safety profiles for drug delivery [56], and biologically active black phosphorus nanoparticles showing potential to mitigate chemotherapy resistance and side effects in pancreatic cancer (cited appropriately if source available).

Furthermore, nanoformulation strategies address limitations of conventional drugs, such as poor solubility and bioavailability, thereby enhancing therapeutic potential and reducing side effects in applications like wound healing [57]. Bioactive nanocarriers, exemplified by systems incorporating oleanolic acid, have also shown promise in reducing chemotherapy-induced organ toxicity [58]. Key strategies like chemical modification and surface activation are employed to mitigate inherent nanomaterial toxicity, improving overall biocompatibility and safety [59]. Collectively, these advancements highlight the transformative potential of bioactive nanoparticles to provide novel and improved therapeutic alternatives.

7 Mechanisms of Bioactive Nanomaterial Interactions in Different Implementations

7.1 Bioactive Nanomaterial Interactions with Biological Systems

The biomedical utility of bioactive nanomaterials stems from their interactions with biological systems occurring at the cellular and molecular scales. Therefore, a fundamental understanding of these interaction systems is essential for designing and applying these materials in diagnostics, therapeutics, and regenerative medicine [60]. Key cellular interaction processes include nanoparticle uptake (internalization), subsequent intracellular trafficking, and the resulting cellular reactions. The initial uptake phase can proceed via multiple pathways: phagocytosis, various forms of endocytosis (clathrin-mediated, caveolin-mediated, clathrin/caveolae-independent), macropinocytosis, or passive membrane diffusion [61]. Which uptake mechanism prevails depends primarily on the nanoparticle's physicochemical characteristics (size, shape, surface charge, hydrophobicity) and the specific cell type involved (see Fig. 3). Factors such as smaller size, often correlating with a larger relative surface area, typically enhance internalization efficiency [62]. Nanoparticle surface charge also critically impacts interaction with the anionic cell membrane. Following uptake, nanoparticles are transported within the cell through intracellular trafficking pathways.

Following cellular uptake, nanoparticles internalized via endocytosis typically become entrapped within endosomal vesicles. These vesicles progressively mature into lysosomes, acidic organelles containing degradative enzymes, which can compromise the integrity and function of therapeutic payloads, particularly macromolecular cargoes like genes or proteins. Consequently, facilitating the escape of nanoparticles from these endosomal compartments to reach their intended cytosolic or nuclear destinations is a critical challenge in nanomedicine delivery [63]. Several strategies have been developed to promote endosomal escape, including the modification of nanoparticles with cationic polymers that induce membrane disruption via the "proton sponge effect," leading to osmotic swelling and rupture [64]. Other mechanisms involving membrane fusion or pore formation are also being explored.

Beyond facilitating escape into the general cytoplasm, precise therapeutic action often requires targeting specific subcellular locations. Functionalizing nanoparticles with specialized targeting ligands—such as peptides or aptamers that recognize receptors on organelles like mitochondria or the nucleus—enables directed delivery. This approach can concentrate therapeutic agents at the site of action, enhancing efficacy while potentially minimizing off-target effects, or allowing intervention in specific cellular processes, such as modulating mitochondrial metabolism or directly influencing gene expression [65]. It is crucial to recognize that bioactive nanoparticle interactions extend beyond simple uptake and trafficking, profoundly influencing a wide spectrum of cellular behaviors. Nanomaterials can modulate cell proliferation,

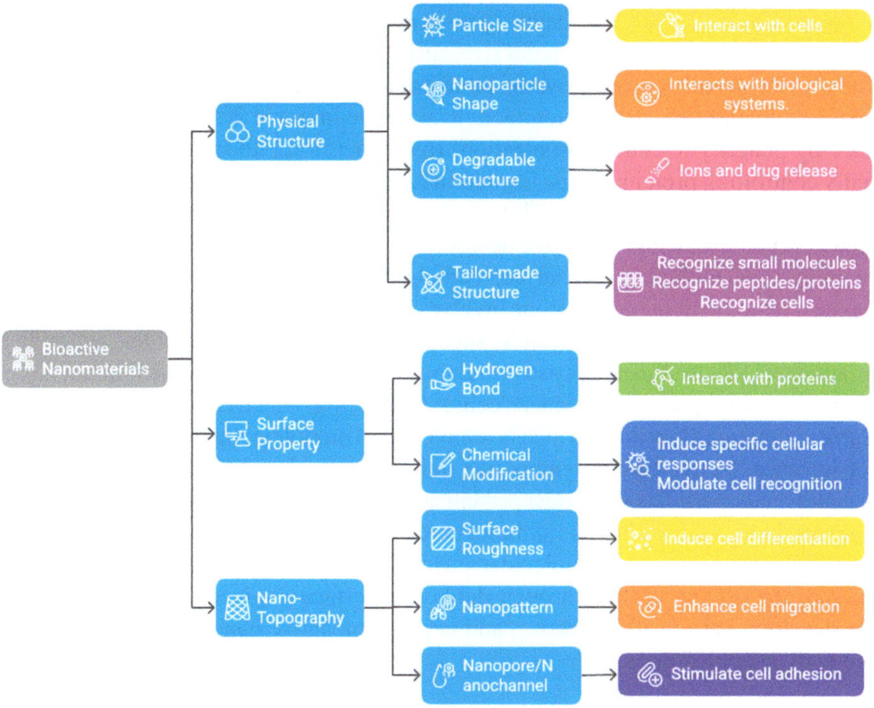

Fig. 3 The elements that influence bioactivities

differentiation pathways, induce apoptosis (programmed cell death), alter cell adhesion and migration dynamics, and trigger modifications to the cytoskeleton [66]. The physicochemical and topographical characteristics of the nanoparticle surface, including roughness, nano-topography, charge, and hydrophilicity/hydrophobicity balance, are key determinants dictating the nature and magnitude of these cellular responses. For instance, substrate stiffness and nanotopography influence cell adhesion, spreading, and lineage specification in stem cells.

The interaction landscape becomes even more complex at the molecular level upon introduction into biological environments. When nanoparticles encounter biological fluids, such as blood plasma or interstitial fluid, they are rapidly coated by a layer of biomolecules, primarily proteins, forming what is known as the "protein corona" [67]. This corona is not a static entity but a dynamic structure with layers of varying affinities, tightly bound proteins form a "hard corona," while more loosely associated proteins constitute a "soft corona," with ongoing exchange between the surface and the bulk fluid [68]. The composition of this corona (including albumin, fibrinogen, immunoglobulins, complement factors, etc.) effectively dictates the nanoparticle's biological identity.

This acquired biological identity profoundly influences subsequent interactions, altering the nanoparticle's apparent size, surface charge, and hydrophobicity. Consequently, the protein corona governs biodistribution patterns, cellular uptake mechanisms, potential toxicity, and interactions with the immune system, including clearance by phagocytic cells (opsonization) [69]. Therefore, understanding and potentially controlling protein corona formation is paramount for predicting and designing nanoparticle behavior in vivo. While the protein corona often mediates interactions, bioactive nanoparticles can also be designed for direct engagement with other specific biomolecules, including cell surface carbohydrates, lipids, or nucleic acids [70]. A targeted nanomedicine cornerstone involves the nanoparticle surface's strategic functionalization with bioactive ligands, such as small molecules, peptides, antibodies, or aptamers. These ligands possess high affinity for specific biological targets (e.g., receptors overexpressed on cancer cells), facilitating selective binding and uptake. Surface functionalization can also be employed to modulate the protein corona itself, for instance, by grafting polymers like polyethylene glycol (PEGylation) to create "stealth" nanoparticles that resist protein adsorption, thereby prolonging circulation time and potentially reducing immunogenicity. Ultimately, navigating these intricate intracellular pathways and molecular interactions, particularly managing the influence of the protein corona through deliberate surface engineering, is essential for the development of safe, effective, and targeted bioactive nanomaterials for diverse biomedical applications.

7.2 Bioactive Nanomaterials in Drug Delivery

Targeted drug delivery systems, particularly those utilizing bioactive nanomaterials, represent a pivotal advancement in therapeutic strategies. Directing pharmacological agents preferentially to diseased cells or tissues aims to significantly enhance treatment efficacy while concurrently mitigating the systemic adverse effects often associated with conventional therapies. Key methodologies employed to achieve this targeted delivery include passive targeting, active targeting, and the development of stimuli-responsive nanomaterials [71]. Passive targeting primarily leverages the Enhanced Permeability and Retention (EPR) effect, a pathophysiological feature commonly observed in tumor microenvironments and sites of inflammation. This phenomenon arises from aberrant, leaky vasculature and impaired lymphatic drainage, which facilitates the preferential accumulation of nanoparticles within these specific locations, typically within a size range below 200 nm. This mechanism effectively increases the local concentration of the therapeutic payload at the disease site without relying on specific molecular recognition elements [72]. Although the EPR effect has been extensively exploited, especially in cancer nanomedicine, its inherent efficacy exhibits considerable variability, influenced by factors such as tumor heterogeneity, disease stage, and the specific physicochemical properties of the nanoparticles utilized.

In contrast, active targeting strategies enhance delivery specificity through the surface functionalization of nanomaterials with targeting ligands. These ligands, which can encompass antibodies, peptides, aptamers, or small molecules, are selected for their high binding affinity towards receptors or antigens that are distinctly over-expressed on the surface of target cells, such as malignant or inflamed cells [73]. This specific molecular interaction promotes the selective adhesion, accumulation, and subsequent internalization of the nanoparticle-drug conjugate at the intended site of action. Consequently, this approach significantly improves therapeutic selectivity and minimizes detrimental off-target effects on healthy tissues. The ultimate success of active targeting is critically dependent on the rational selection of appropriate ligands that guarantee efficient recognition and interaction with the designated cellular targets [74].

Further refining spatial and temporal control over drug release, stimuli-responsive drug delivery systems constitute another sophisticated strategy. This approach utilizes bioactive nanoparticles engineered to release their encapsulated therapeutic cargo predominantly in response to specific triggers within the target microenvironment. These triggers can be endogenous, such as alterations in pH or enzyme concentrations, or exogenous, like the application of light, heat, or ultrasound. By restricting drug release to the precise site and time of action, stimuli-responsive systems offer the potential for even greater therapeutic precision and reduced systemic exposure. Changes in pH, commonly lower in tumor microenvironments and within cellular endosomes, particularly in sick tissues, or changes in the redox potential within cells can all be internal stimuli [75]. Light (used in photothermal and photodynamic therapies), temperature (for nanoparticles that release their payload above a specified temperature), magnetic fields (for magnetically triggered release from magnetic nanoparticles), and ultrasound are among the external stimuli that might set off medication release. These stimuli-responsive devices minimize systemic medication exposure, employing exact spatial and temporal control over drug release, enabling on-demand drug delivery at the target site and further improving therapeutic efficacy [76].

Additionally, it successfully improves the efficacy of different therapeutic medicines by including bioactive nanoparticles in drug delivery systems. When combined with nanoparticles, many poorly soluble medications in water show increased solubility and bioavailability. By shielding medications from breakdown by enzymes and other bodily components, nanoparticles also help to ensure that a greater percentage of the given dosage reaches the target site [77]. Moreover, surface modification of nanoparticles helps to lengthen their circulation time in the bloodstream, enabling continuous drug release over a lengthy period and higher accumulation at the target site [78]. Nanoparticles' small size and unusual surface characteristics can also improve cell absorption, enabling medication administration into the intracellular region. Especially in cancer treatment, nanoparticle-based drug delivery systems have shown promise in overcoming drug resistance, a major obstacle in treating many kinds of cancer, by focusing on efflux transporters or apoptotic pathways inside cancer cells [79].

7.3 Bioactive Nanomaterials in Tissue Engineering and Regenerative Medicine

Bioactive nanomaterials are increasingly integral to the progression of regenerative medicine and tissue engineering. Their prominence arises significantly from a unique capacity to emulate the nanoscale features and biological cues inherent to the natural extracellular matrix (ECM). This biomimicry facilitates the design of sophisticated scaffolds that actively support cellular proliferation, directed differentiation, and organized tissue regeneration [80]. The native ECM provides more than just structural support, it constitutes a dynamic microenvironment rich in biochemical and biophysical signals that meticulously regulate cellular functions. Nanomaterial-based scaffolds endeavor to replicate critical aspects of this complex milieu at the nanoscale, thereby aiming to guide the formation and maturation of functional tissues [81]. The development of effective nanomaterial scaffolds involves diverse strategies. Electrospinning techniques, for example, fabricate nanofibrous meshes from natural or synthetic polymers. These structures offer high surface-area-to-volume ratios and porosity resembling native tissue architecture, which promotes efficient cell adhesion and proliferation [82]. Another prevalent strategy involves incorporating inorganic (e.g., ceramic, metallic) or polymeric nanoparticles within hydrogel networks. This enhances the hydrogel's mechanical properties, introduces specific bioactivity, and enables the controlled release of embedded growth factors or other therapeutic molecules crucial for tissue repair. Furthermore, advanced nanocomposites are being engineered by combining different types of nanomaterials or integrating nanomaterials within traditional biomaterial frameworks. This approach yields scaffolds with synergistic characteristics tailored to meet the complex demands of regenerating particular tissues [80].

Beyond serving as structural components, the intrinsic properties of bioactive nanoparticles are fundamental to directing cellular responses. For instance, the specific surface topography and nanoscale roughness of these materials profoundly influence cell adhesion dynamics, morphology, and the activation of distinct intracellular signaling pathways controlling differentiation [83]. Moreover, bioactive nanoparticles function as highly effective carriers for the localized and sustained delivery of potent signaling molecules—including growth factors, cytokines, and genetic material—directly to the regenerating tissue microenvironment, thereby enabling precise spatial and temporal control over the healing cascade. Bioactive glass nanoparticles have demonstrated considerable success in enhancing the osteogenic differentiation of mesenchymal stem cells, culminating in improved bone regeneration outcomes [84]. Similarly, specialized DNA-peptide nanotubes are being investigated for their ability to promote the differentiation of neural stem cells into functional neurons. Unique approaches also utilize magnetic nanoparticles, where external magnetic fields can be applied to remotely influence stem cell behavior and direct tissue development. Collectively, accumulating research underscores the versatility and efficacy of bioactive nanoparticles in promoting the regeneration of a wide spectrum of tissues, including bone, cartilage, neural structures, skin, and cardiac muscle [85].

7.4 *Bioactive Nanomaterials in Diagnostics and Medical Imaging*

Bioactive nanomaterials represent a significant frontier in diagnostics and medical imaging, presenting compelling advantages over established conventional methods. Their distinct physicochemical characteristics are fundamental to engineering highly efficacious contrast agents tailored for diverse imaging modalities and developing sophisticated biosensors that enable accurate and early disease detection [86]. In the domain of biosensing, the integration of bioactive nanomaterials is increasingly central to modern sensor design. The inherently high surface-area-to-volume ratio of nanoparticles dramatically enhances their interaction efficiency with target analytes compared to traditional platforms. This translates directly into notable gains in analytical sensitivity, specificity, and rapidity of detection. The development of advanced biosensors leverages various nanomaterial classes: metallic nanoparticles, such as gold and silver, are prized for their unique optical and electrochemical properties [87]. Carbon-based nanomaterials encompassing carbon nanotubes, graphene, and carbon dots are widely employed owing to their exceptional electrical conductivity, expansive surface area, and the relative ease with which their surfaces can be functionalized for selective analyte binding. Additionally, due to their remarkable and tunable optical characteristics, semiconductor quantum dots are valuable as fluorescent reporters in biosensing applications [88].

Concurrently, bioactive nanomaterials substantially augment medical imaging capabilities through their application as advanced contrast agents. Key benefits conferred by nanomaterial-based agents include enhanced image resolution and sensitivity, prolonged systemic circulation times permitting extended imaging windows, and the capacity for targeted imaging, achieved by decorating the nanoparticle surface with specific biorecognition ligands [89]. Specific material types are exploited based on the imaging modality: Iron oxide nanoparticles, for example, serve extensively as contrast agents in Magnetic Resonance Imaging (MRI), altering local water proton relaxation times to improve image contrast [90]. Nanomaterials composed of high atomic number elements, like gold or bismuth, function as efficient contrast agents for Computed Tomography (CT) owing to their significant X-ray attenuation properties. In nuclear medicine, radiolabeled nanoparticles permit highly sensitive visualization of biological processes using Positron Emission Tomography (PET) [91]. Furthermore, fluorescent nanomaterials, such as quantum dots and carbon dots, enable high-resolution optical imaging of cells and tissues [92].

Beyond enhancing single-modality imaging, nanomaterials facilitate the design of multimodal contrast agents. These sophisticated agents integrate the capabilities of multiple imaging techniques within a single platform, thereby providing richer and more comprehensive diagnostic information. Advancing this integration further is the field of theranostics, where diagnostic and therapeutic functionalities are combined within the same nanoparticle construct. Theranostic nanoparticles offer the transformative potential for simultaneous disease detection, targeted therapy delivery, and real-time monitoring of treatment response, heralding a significant move towards more personalized and effective healthcare strategies [93].

8 Biocompatibility and Potential Toxicity of Bioactive Nanomaterials

While bioactive nanomaterials hold considerable promise for advancing diverse areas of biomedical science, their successful and ethical translation into practical medical applications hinges upon the rigorous evaluation of their biocompatibility and potential toxicity. Ensuring patient safety necessitates a meticulous understanding of how these novel materials interact with complex biological systems. Biocompatibility, in this context, refers to the capacity of a material to perform its intended function within a specific biological environment without eliciting unacceptable local or systemic adverse effects in the host organism [94]. Evaluating the biocompatibility profile of nanomaterials is a multifaceted process requiring a thorough assessment across several critical parameters. These include potential cytotoxicity (harm to cells), immunogenicity (capacity to provoke an immune response), hemocompatibility (interactions with blood components), and the propensity to induce inflammatory or other adverse tissue reactions upon administration.

The assessment framework typically follows a hierarchical approach, commencing with comprehensive in vitro studies. These initial assays evaluate the effects of nanomaterials on various cell types in culture, probing endpoints such as cell viability, proliferation, and specific cellular stress responses. Subsequently, promising candidates progress to in vivo investigations in relevant animal models. These studies are designed to elucidate the nanomaterial's biodistribution, pharmacokinetic profile (absorption, distribution, metabolism, excretion—ADME), persistence within tissues, and overall physiological impact, including any potential organ toxicity [95].

It is crucial to recognize that the toxicological profiles of bioactive nanomaterials are not uniform; they exhibit significant variability influenced by a complex interplay of factors. Key determinants include the nanoparticle's intrinsic physicochemical properties—such as size, shape, surface area, composition, crystal structure, surface charge, and functionalization as well as extrinsic factors like dose, route, and frequency of administration, and the specific biological milieu encountered. For example, certain types of metallic nanoparticles have been implicated in generating elevated levels of oxidative stress, which can subsequently damage vital biomacromolecules, including DNA. Furthermore, some nanomaterials demonstrate a propensity for bioaccumulation within particular organs (e.g., liver, spleen), raising pertinent questions regarding potential long-term adverse health consequences [96]. Depending on their specific characteristics (like aspect ratio in fibrous nanomaterials), other materials may trigger persistent inflammatory responses or granuloma formation.

Therefore, a comprehensive characterization and cautious toxicological assessment, considering both acute and chronic effects, are indispensable for the responsible development and clinical implementation of nanomaterial-based technologies. This underscores the ongoing need for robust, standardized testing protocols and continued research to delineate the long-term biological implications of these innovative materials fully. Mostly inhaled, ceramic nanoparticles have generated questions about their possible ability to induce lung inflammation. The inherent characteristics of the polymers utilized, the release of toxic monomers or degradation products, or the inclusion of possibly hazardous stabilizers in their formulation could cause polymeric nanoparticles to show toxicity [97]. Additionally, under close examination are carbon-based nanomaterials, whose studies suggest possible long-term toxicity, inflammation, and genotoxicity. Still, the particular type of carbon nanomaterial and its surface functionalization will significantly affect these results [98]. Several factors are known to influence the biocompatibility and toxicity of nanomaterials, including their size (with smaller nanoparticles often exhibiting higher toxicity due to increased surface area and cellular uptake), shape (e.g., rod-shaped nanoparticles can have different toxicological profiles compared to spherical ones), surface charge (cationic nanoparticles can sometimes be more cytotoxic), chemical composition (the inherent toxicity of the constituent material), the administered dosage (overdosing can lead to adverse effects), and the tendency of nanoparticles to aggregate or agglomerate [99].

With regulatory authorities like the Food and Drug Administration (FDA) in the United States and the European Medicines Agency (EMA) striving to create clear rules, the environment for bioactive nanomaterials in biomedical applications is still changing. The approval process is complicated in some areas where particular regulatory definitions and thorough nanoparticle rules are lacking [100]. Nonetheless, the current legislative framework stresses a product-oriented, science-based approach whereby the safety and efficacy of medical devices based on nanotechnology are assessed case-by-case, considering their unique qualities and intended application [101]. Demonstrating the biocompatibility of nanomaterials depends on following global guidelines for the biological evaluation of medical devices, such as the ISO 10993 series. Reflecting a risk-based approach to their control, the Medical Device Regulation (MDR) in the European Union classifies medical devices, including nanomaterials, based on their possible internal exposure [100].

9 Directions and Challenges in Bioactive Nanomaterial Research and Translation

Numerous promising avenues for future research and application exist in the dynamic and swiftly evolving field of bioactive nanomaterials. Dealing with the scalability of bioactive nanomaterial manufacture to satisfy the needs of extensive application presents one of the main difficulties. Traditional nanoparticle synthesis techniques often involve time-consuming, labor-intensive batch-by-batch operations with batch-to-batch variability in quality and repeatability [102]. Increasing attention is being paid to creating scalable, continuous manufacturing methods to overcome these constraints. Following good manufacturing practice (GMP) rules, recent developments in microfluidic-based platforms present intriguing possibilities for the fast and large-scale manufacture of nanoparticles with highly regulated structures and features [103].

The translation of bioactive nanomaterials involves another substantial challenge in navigating the regulatory channels leading to clinical approval. Particularly with regard to their definition, characterization, safety testing, and quality control, the lack of clear, explicit regulatory rules for nanomaterials in biomedical uses can cause ambiguity for producers and impede clearance [104]. Still, significant difficulties are the complexity of characterizing nanomaterials and the necessity of consistent toxicity testing procedures to evaluate the several spectra of nanomaterials with different physicochemical properties. More international cooperation and harmonization of regulatory frameworks are required to assure patient safety and enable the worldwide translation of nanomedicines [105].

Usually, overcoming the obstacles to clinical translation is a complex task. The complexity of biological systems sometimes impedes the translation of bioactive nanoparticles to successful therapeutic outcomes despite promising preclinical results; this can lead to unexpected in vivo behavior and possibly off-target effects. Preclinical models that replicate human physiology and disease conditions more accurately are needed [106]. Furthermore, it is essential to conduct comprehensive long-term safety assessments to ensure the potential benefits of bioactive nanoparticles outweigh any risks associated with their use in human subjects [107].

Looking ahead, bioactive nanomaterials are beginning to show various interesting lines of investigation. Developing multifunctional bioactive nanomaterials that combine therapeutic and diagnostic properties onto a single platform, thereby enabling theranostics for individualized medicine, is attracting increasing attention [51]. Additionally, under investigation are combination treatments using bioactive nanomaterials to provide several therapeutic agents concurrently to overcome obstacles, including drug resistance, and provide synergistic benefits. More individualized nanomedicine approaches where nanomaterials and their payloads are customized to the particular demands of individual patients and their distinct illness characteristics—are also under appeal in the field [108]. Moreover, by allowing the prediction of their properties and interactions with biological systems, artificial intelligence and machine learning are starting to influence the design, production, and use of bioactive nanomaterials, hence quickening their development.

10 Conclusion

Bioactive nanomaterials are at the forefront of a scientific revolution and hold immense transformative potential for biomedical science. Their unique ability to interact with biological systems at the nanoscale has opened up unprecedented opportunities for diagnostics, therapeutics, and regenerative medicine advancements. Bioactive nanomaterials are poised to overcome many of the limitations of traditional medical approaches, from enabling highly sensitive biosensors for early disease detection to delivering drugs with remarkable precision and promoting the regeneration of damaged tissues.

This chapter has explored the diverse landscape of bioactive nanomaterials, highlighting the key properties and applications of metallic, ceramic, polymeric, and carbon-based nanomaterials. We have delved into the intricate mechanisms governing their interactions with cells and biomolecules, underscoring the importance of understanding these fundamental processes for effective utilization. The chapter has also examined the critical role of bioactive nanomaterials in targeted drug delivery and their contribution to the exciting field of tissue engineering and regenerative medicine. Furthermore, we have addressed their biocompatibility and potential toxicity, emphasizing the need for rigorous evaluation and adherence to evolving regulatory guidelines.

While the progress in the field of bioactive nanomaterials has been remarkable, several challenges remain that need to be addressed to fully realize their potential. Scalable and cost-effective manufacturing, the establishment of clear and harmonized regulatory pathways, and the successful translation of promising preclinical findings into clinically viable therapies are crucial steps that lie ahead. Overcoming these barriers will require continued interdisciplinary research, fostering innovation in nanomaterial design and synthesis, and a strong commitment to ensuring the safety and efficacy of these advanced materials for human use.

In conclusion, the future of bioactive nanomaterials in biomedical science is auspicious. With ongoing research and development, we can expect to see even more sophisticated and targeted applications that will significantly improve human health and transform the way we diagnose, treat, and prevent a wide range of diseases. The continued exploration of these fascinating materials holds the key to unlocking a new era of precision and personalized medicine, ultimately leading to better outcomes and an enhanced quality of life for patients worldwide.

References

1. Hicks, A.: The Emerging Field of Precision Medicine—The New Paradigm for Healthcare. Advancing Global Bioethics (2023). https://doi.org/10.1007/978-3-031-28593-6_2
2. Biobaku-Mutingwende, B.: Introduction to Biomaterials (2021). https://doi.org/10.1007/978-981-33-6888-0_1
3. Crawford, L., Wyatt, M., Bryers, J.D., Ratner, B.D.: Biocompatibility evolves: phenomenology to toxicology to regeneration. Adv. Healthc. Mater. (2021). https://doi.org/10.1002/ADHM.202002153
4. Shearer, A., Montazerian, M., Sly, J.J., Hall, R., Mauro, J.C.: Trends and perspectives on the commercialization of bioactive glasses. Acta Biomater. (2023). https://doi.org/10.1016/j.actbio.2023.02.020
5. Nour, S., Baheiraei, N., Imani, R., Rabiee, N., Khodaei, M., Alizadeh, A., Moazzeni, S.M.: Bioactive materials: a comprehensive review on interactions with biological microenvironment based on the immune response. J. Bionic Eng. (2019). https://doi.org/10.1007/S42235-019-0046-Z
6. Li, R., Liu, K., Huang, X., Li, D., Ding, J., Liu, B., Chen, X.: Bioactive materials promote wound healing through modulation of cell behaviors. Adv. Sci. (2022). https://doi.org/10.1002/advs.202105152
7. Pandya, M.: Application of Nanoparticles in Medicine (2022). https://doi.org/10.59143/isas.jisas.1.2.mvsb9110
8. Kumar, V., Varghese, S., Kumar, D.K., Khatri, N., Singh, G.: Biomaterials in medical applications. Curr. Mater. Sci. (2023). https://doi.org/10.2174/2666145416666230420094148
9. Khalid, U.A., Tariq, R., Zahra, S.T., Khalid, T.A.: Nanomaterials: an overview with respect to biological applications. J. Appl. Emerg. Sci. (2020). https://doi.org/10.36785/JAES.102401
10. Alves de Almeida, F., Srinivasan, R., Vijayakumar, S.: Editorial: biosynthesis of bio-inspired nanoparticles/nanomaterials and evaluation of their therapeutic potential in the medical field. Front. Nanotechnol. (2023). https://doi.org/10.3389/fnano.2023.1198994
11. Bhagat, D.S., Gurnule, W.B., Bumbrah, G.S., Koinkar, P., Chawla, P.: Recent advances in biomedical applications of biogenic nanomaterials. Curr. Pharm. Biotechnol. (2022). https://doi.org/10.2174/1389201023666220513101628
12. Zhao, Y., Zhang, Z., Pan, Z., Liu, Y.: Advanced bioactive nanomaterials for biomedical applications. In: Exploration, vol. 1, no. 3, p. 20210089 (2021). https://doi.org/10.1002/EXP.20210089
13. Jamkhande, P.G., Ghule, N.W., Bamer, A.H., Kalaskar, M.G.: Metal nanoparticles synthesis: an overview on methods of preparation, advantages and disadvantages, and applications. J. Drug Deliv. Sci. Technol. (2019). https://doi.org/10.1016/J.JDDST.2019.101174
14. Tortella, G., Rubilar, O., Fincheira, P., Pieretti, J.C., Duran, P., Lourenço, I.M., Seabra, A.B.: Bactericidal and virucidal activities of biogenic metal-based nanoparticles: advances and perspectives. J. Antibiot. (2021). https://doi.org/10.3390/ANTIBIOTICS10070783
15. Plascencia-Villa, G., Mendoza-Cruz, R.: Gold nanoparticles for biocatalysis (2022). https://doi.org/10.1016/b978-0-12-824436-4.00013-7
16. Zasońska, B.A., Horák, D.: Magnetic iron oxide particles for theranostics (2022). https://doi.org/10.1007/978-3-030-76235-3_4
17. Kaur, M., Gautam, A., Guleria, P., Singh, K., Kumar, V.: Green synthesis of metal nanoparticles and their environmental applications. Curr. Opin. Environ. Sci. Health (2022). https://doi.org/10.1016/j.coesh.2022.100390
18. Punj, S., Singh, J., Singh, K.: Ceramic biomaterials: properties, state of the art and future prospectives. Ceram. Int. (2021). https://doi.org/10.1016/J.CERAMINT.2021.06.238
19. Feng, J., Thian, E.S.: Applications of nanobioceramics to healthcare technology. Nano Rev. (2013). https://doi.org/10.1515/NTREV-2012-0065
20. Ying, B., Coull, J.M., Sim, D.S.: Polymer-based nanoparticles, related formulation methods, and apparatus (2019)

21. Nölke, A.: Nanopolymers in drug delivery system. Mater. Today Proc. (2022). https://doi.org/10.1016/j.matpr.2022.05.019
22. Leach, A.B., Finnegan, M., Ferguson, L.B., Furniss, J., Smyth, P., Porter, A.J., Barelle, C.J., Themistou, E., Scott, C.J.: Functionalization of polymeric nanoparticles with targeting VNAR ligands using vinyl sulfone conjugation. J. Mater. Chem. B (2023). https://doi.org/10.1039/d2tb01985j
23. Begines, B., Ortiz, T., Pérez-Aranda, M., Martínez, G., Merinero, M., Argüelles-Arias, F., Alcudia, A.: Polymeric nanoparticles for drug delivery: recent developments and future prospects. Nanomaterials (2020). https://doi.org/10.3390/NANO10071403
24. Owida, H., Turab, N.M., Al-Nabulsi, J.: Carbon nanomaterials advancements for biomedical applications. Bull. Electr. Eng. Inform. (2023). https://doi.org/10.11591/eei.v12i2.4310
25. Sarkar, P., Ghosal, K., Chakraborty, D., Sarkar, K.: Biocompatibility and biomedical applications of various carbon-based materials (2021). https://doi.org/10.1016/B978-0-12-821996-6.00015-4
26. Victor-Solana, K., Kharira, S.: Carbon nanomaterials for biomedicine applications. Exp. Theor. Nanotechnol. (2023). https://doi.org/10.56053/6.2.117
27. Akther, W.: Nanomaterials (2022). https://doi.org/10.1016/b978-0-323-90968-6.50005-9
28. Abbas, H.S., Mahmoud, A.M., Wahed, R.A., Elsantawy, M.A.A., Hamdy, N.M., Ismail, S.E.S., Nabil, M.A.: Prospects of using bioactive compounds in nanomaterials surface decoration and their biomedical purposes. Int. Nano Lett. (2021). https://doi.org/10.1007/S40089-021-00355-9
29. Ramos-Zúñiga, R., Segura-Durán, I., Gonzalez-Castaneda, R.E., González Rios, J.A.: The challenges of the bioactive scaffolds in nervous system: from their molecular conformation to their therapeutic efficiency. Neurol. Perspect. (2022). https://doi.org/10.1016/j.neurop.2021.07.007
30. Tasic, L., Stanisic, D., Martins, L.G., Cruz, G., Savu, R.: Insights of green and biosynthesis of nanoparticles (2022). https://doi.org/10.1016/b978-0-12-824547-7.00020-5
31. Nel, A.E., Parak, W.J., Chan, W.C.W., Xia, T., Hersam, M.C., Brinker, C.J., Zink, J.I., Pinkerton, K.E., Baer, D.R., Weiss, P.S.: Where are we heading in nanotechnology environmental health and safety and materials characterization. ACS Nano (2015). https://doi.org/10.1021/ACSNANO.5B03496
32. Carstens, N., Mirigliano, M., Strunskus, T., Faupe, F., Lupan, O., Milan, P., Vahl, A.: Nanoparticles as building units for bio-inspired electronics—switching and sensing (2022). https://doi.org/10.52326/ic-ecco.2021/el.01
33. Jyoti, A., Kushwaha, S., Baranwal, J., Singh, S.: A review on green synthesis of biologically active compounds. Curr. Green Chem. (2022). https://doi.org/10.2174/221334611066622121 3092734
34. Mayen, L., Jensen, N.D., Laurencin, D., Marsan, O., Bonhomme, C., Gervais, C., Smith, M.E., Coelho, C., Laurent, G., Trébosc, J., Gan, Z., Chen, K., Rey, C., Combes, C., Soulié, J.: A soft-chemistry approach to the synthesis of amorphous calcium ortho/pyrophosphate biomaterials of tunable composition. Acta Biomater. (2020). https://doi.org/10.1016/J.ACTBIO.2019.12.027
35. Sharma, M., Sharma, M.K.: Fundamental and Principles of Green Synthesis (2021). https://doi.org/10.1007/978-3-030-67884-5_14
36. Auerbach, S.M.: An Overview of the Synthesis of Nanomaterials. Materials Research Foundations (2023). https://doi.org/10.21741/9781644902370-2
37. Laurens, L.M.: Synthesis of Nanomaterials by Biological Route (2022). https://doi.org/10.1007/978-981-16-6819-7_5
38. Islam, T., Ali, I., Naz, I., Kayshar, S.: Appraisal of nanotechnology for sustainable environmental remediation (2022). https://doi.org/10.1016/b978-0-12-824547-7.00010-2
39. Peng, Q., Zang, B., Xiong, T., Huang, C.-F., Xu, T., Zhang, C., Ren, J., Ji, F., Jia, L.: Single and dual functionalization of proteins using site-specific nucleophilic carbon ligations. Chem. Commun. (2022). https://doi.org/10.1039/d2cc01630c

40. Zanbili, F., Poursattar Marjani, A.: Innovative green and bio-based approaches for photosensitive nanoparticle synthesis: a review on methodologies, characterization, and applications. Micro Nano Syst. Lett. **13**(1), 3 (2025)
41. Muñoz, S.H., Moreno, M.D.C.S., Domínguez, G.J.M., Rodríguez, P.A.M., Vázquez, E.: Experimental, numerical, and analytical study on the effect of graphene oxide in the mechanical properties of a solvent-free reinforced epoxy resin. Polymers **11**(12), 2115 (2019). https://doi.org/10.3390/polym11122115
42. Sancak, S., Yazgan, I., Ugurlu Bayarslan, A., Ayna, A., Evecen, S., Taşdelen, Z., Gümüş, A., Sönmez, H.A., Demir, M.A., Demir, S., Bakar, F., Tepe, H.D., Kasemets, K., Otsus, M., Çeter, T.: Surface chemistry dependent toxicity of inorganic nanostructure glycoconjugates on bacterial cells and cancer cell lines. J. Drug Deliv. Sci. Technol. (2023). https://doi.org/10.1016/j.jddst.2022.104054
43. Rabiee, N., Ahmadi, S., Iravani, S., Varma, R.: Functionalized silver and gold nanomaterials with diagnostic and therapeutic applications. Pharmaceutics (2022). https://doi.org/10.3390/pharmaceutics14102182
44. Li, R., Ren, H.N., Liu, X.-M., Chen, Z., Li, L., Wang, H.: Peptide-based bioactivated in vivo assembly nanomaterials and its biomedical applications: a review (2022). https://doi.org/10.13345/j.cjb.210282
45. Kumar, S., Kumar Shukla, M., Sharma, A., Jayaprakash, G.K., Tonk, R.K., Chellappan, D.K., Singh, S.K., Dua, K., Ahmed, F., Bhattacharyya, S., Kumar, D.: Metal-based nanomaterials and nanocomposites as promising frontier in cancer chemotherapy. MedComm (2023). https://doi.org/10.1002/mco2.253
46. Parvin, N., Joo, S.W., Mandal, T.K.: Biodegradable and stimuli-responsive nanomaterials for targeted drug delivery in autoimmune diseases. J. Funct. Biomater. **16**(1), 24 (2025)
47. Lee, S.S., Paliouras, M., Trifiro, M.: Functionalized carbon nanoparticles as theranostic agents and their future clinical utility in oncology. Bioengineering (2023). https://doi.org/10.3390/bioengineering10010108
48. Wang, J., Jin, N., Xie, Z., Cheng, Q., Jiang, B., Shuai, Y., Xu, Z., Wan, Q., Chen, Y., Mao, C.: Gold nanorods coated with self-assembled silk fibroin for improving their biocompatibility and facilitating targeted photothermal—photodynamic cancer therapy. Nanoscale (2025)
49. Chavda, V.P., Patel, A.B., Mistry, K.J., Suthar, S.F., Wu, Z.-X., Chen, Z.-S., Hou, K.: Nano-drug delivery systems entrapping natural bioactive compounds for cancer: recent progress and future challenges. Front. Oncol. **12**, 867655 (2022)
50. Shinde, R.K., Naik, S.V.: Nano technology. Int. J. Adv. Res. Sci. Commun. Technol. (2022). https://doi.org/10.48175/ijarsct-5125
51. Ramburrun, P., Khan, R.A., Choonara, Y.E.: Design, preparation, and functionalization of nanobiomaterials for enhanced efficacy in current and future biomedical applications. Nanotechnol. Rev. (2022). https://doi.org/10.1515/ntrev-2022-0106
52. Zheng, Y.: Biological design of materials (2019). https://doi.org/10.1016/B978-0-12-814843-3.00002-8
53. Yadav, K., Kumar, P., Teja, D., Chakraborty, S., Chakraborty, M., Mohapatra, S.S., Sahoo, A., Chou, M.M.C., Liang, C.-T., Hang, D.-R.: A review on low-dimensional nanomaterials: nanofabrication, characterization and applications. Nanomaterials (2022). https://doi.org/10.3390/nano13010160
54. Heydari-Badrooei, E., Ensafi, A.A.: Nanomaterials-based biosensing strategies for biomarkers diagnosis, a review (2022). https://doi.org/10.1016/j.biosx.2022.100245
55. Manzari-Tavakoli, A., Babajani, A., Tavakoli, M.M., Safaeinejad, F., Jafari, A.: Integrating natural compounds and nanoparticle-based drug delivery systems: a novel strategy for enhanced efficacy and selectivity in cancer therapy. Cancer Med. **13**(5), e7010 (2024)
56. Alexander, A., Agrawal, M., Yadav, P., Jeswani, G., Verma, V.S., Siddique, S.: Targeted delivery through carbon nanomaterials: applications in bioactive delivery systems (2020). https://doi.org/10.1016/B978-0-12-819666-3.00017-1
57. Kant, V., Kumari, P., Kamothi, D., Ahuja, M., Kumar, V.: Nanomaterials of natural bioactive compounds for wound healing: novel drug delivery approach. Curr. Drug Deliv. (2021). https://doi.org/10.2174/1567201818666210729103712

58. Wang, Y., Li, G., Zhu, S., Jing, F.-C., Liu, R.-D., Li, S., He, J., Lei, J.: A self-assembled nanoparticle platform based on amphiphilic oleanolic acid polyprodrug for cancer therapy. Chin. J. Polym. Sci. (2020). https://doi.org/10.1007/S10118-020-2401-2
59. Lemus, M.A., Monroy, H., Lopez, T., de la Cruz Hernández, E.N., López-González, R.: Effect of surface modification on the bioactivity of sol–gel TiO_2-based nanomaterials. J. Chem. Technol. Biotechnol. (2016). https://doi.org/10.1002/JCTB.4915
60. Lazurko, C., Jacques, E., Ahumada, M., Alarcon, E.I.: Nanomaterials for its use in biomedicine: an overview (2019). https://doi.org/10.1007/978-3-030-31261-9_1
61. Augustine, R., Hasan, A.: Cellular response to nanobiomaterials (2020). https://doi.org/10.1016/B978-0-08-102967-1.00022-0
62. Zhang, X., Ma, G., Wei, W.: Simulation of nanoparticles interacting with a cell membrane: probing the structural basis and potential biomedical application. NPG Asia Mater. (2021). https://doi.org/10.1038/S41427-021-00320-0
63. Voltà-Durán, E., Parladé, E., Serna, N., Villaverde, A., Vázquez, E., Unzueta, U.: Endosomal escape for cell-targeted proteins. Going out after going in. Biotechnol. Adv. (2023). https://doi.org/10.1016/j.biotechadv.2023.108103
64. Lee, D.U., Park, J.-Y., Kwon, S., Park, J.Y., Kim, Y.H., Khang, D., Hong, J.H.: Apoptotic lysosomal proton sponge effect in tumor tissue by cationic gold nanorods. Nanoscale (2019). https://doi.org/10.1039/C9NR04323C
65. Kashaw, S., Tomar, S., Rajoriya, V., Sahu, P., Agarwal, S., Vyas, S.P.: Multifunctional nanoparticles for organelle-specific targeted drug delivery in cancer therapy. Curr. Nanomed. (2022). https://doi.org/10.2174/2468187313666221219150315
66. Bodey, B., Siegel, S.E.: Mechanisms and markers of carcinogenesis and neoplastic progression. Exp. Opin. Biol. Therapy (2005). https://doi.org/10.1517/14712598.5.10.1317
67. Akhter, H., Khalilullah, H., Gupta, M., Alfaleh, M.A., Alhakamy, N.A., Riadi, Y.: Impact of protein corona on the biological identity of nanomedicine: understanding the fate of nanomaterials in the biological milieu. Biomedicines (2021). https://doi.org/10.3390/BIOMEDICINES9101496
68. Hosseinpour, N., Abdollahi, M.: Characterization of protein corona formation on nanoparticles via the analysis of dynamic interfacial properties: Bovine serum albumin—silica particle interaction. Coll. Surf. A Physicochem. Eng. Aspects (2022). https://doi.org/10.1016/j.colsurfa.2022.128273
69. Cui, G., Su, W., Tan, M.: Formation and biological effects of protein corona for food-related nanoparticles. Compr. Rev. Food Sci. Food Saf. (2021). https://doi.org/10.1111/1541-4337.12838
70. Hajipour, M.J., Safavi-Sohi, R., Sharifi, S., Mahmoud, N.N., Ashkarran, A.A., Voke, E., Serpooshan, V., Ramezankhani, M., Milani, A.S., Landry, M.P., Mahmoudi, M.: An overview of nanoparticle protein corona literature. Small (2023). https://doi.org/10.1002/smll.202301838
71. Cheng, X., Xie, Q., Sun, Y.: Advances in nanomaterial-based targeted drug delivery systems. Front. Bioeng. Biotechnol. (2023). https://doi.org/10.3389/fbioe.2023.1177151
72. Wu, J.: The enhanced permeability and retention (EPR) effect: the significance of the concept and methods to enhance its application. J. Pers. Med. (2021). https://doi.org/10.3390/JPM11080771
73. Azandaryani, A.H., Kashanian, S., Jamshidnejad-Tosaramandani, T.: Recent insights into effective nanomaterials and biomacromolecules conjugation in advanced drug targeting. Curr. Pharm. Biotechnol. (2019). https://doi.org/10.2174/1389201020666190417125101
74. Yoo, J., Park, C., Yi, G., Lee, D.H., Koo, H.: Active targeting strategies using biological ligands for nanoparticle drug delivery systems. Cancers (2019). https://doi.org/10.3390/CANCERS11050640
75. Sethuraman, V., Janakiraman, K., Krishnaswami, V., Kandasamy, R.: Recent progress in stimuli-responsive intelligent nano scale drug delivery systems: a special focus towards pH-sensitive systems. Curr. Drug Targets (2021). https://doi.org/10.2174/1389450122999210128180058

76. Armenia, I., Cuestas Ayllón, C., Torres Herrero, B., Bussolari, F., Alfranca, G., Grazú, V., Martinez de la Fuente, J.: Photonic and magnetic materials for on-demand local drug delivery. Adv. Drug Deliv. Rev. (2022). https://doi.org/10.1016/j.addr.2022.114584
77. Karthick, G., Akiladevi, D., Ahamed, M.I.: A comprehensive review of a new nanosuspension for improving the oral bioavailability of poorly soluble drugs. J. Pharm. Res. Int. (2022). https://doi.org/10.9734/jpri/2022/v34i20b35828
78. Osman, N., Osman, N., Devnarain, N., Omolo, C.A., Omolo, C.A., Fasiku, V.O., Jaglal, Y., Govender, T.: Surface modification of nano-drug delivery systems for enhancing antibiotic delivery and activity. Wiley Interdisc. Rev. Nanomed. Nanobiotechnol. (2021). https://doi.org/10.1002/WNAN.1758
79. Yao, Y., Zhou, Y., Liu, L., Xu, Y., Chen, Q., Wang, Y., Wu, S., Deng, Y., Zhang, J., Shao, A.: Nanoparticle-based drug delivery in cancer therapy and its role in overcoming drug resistance. Front. Mol. Biosci. (2020). https://doi.org/10.3389/FMOLB.2020.00193
80. Panigrahi, B., Suryakanta, U., Mishra, S., Singh, R.K., Mandal, D.: Role of nanotechnology in tissue engineering and regenerative medicine (2022). https://doi.org/10.1201/9781003161158-24
81. Singla, R., Singla, R., Abidi, S.M.S., Abidi, S.M.S., Dar, A.I., Acharya, A., Acharya, A.: Nanomaterials as potential and versatile platform for next generation tissue engineering applications. J. Biomed. Mater. Res. Part B (2019). https://doi.org/10.1002/JBM.B.34327
82. Cui, J., Yu, X., Shen, Y., Sun, B., Guo, W., Liu, M., Chen, Y., Wang, L., Zhou, X., Shafiq, M., Mo, X.: Electrospinning inorganic nanomaterials to fabricate bionanocomposites for soft and hard tissue repair. Nanomaterials (2023). https://doi.org/10.3390/nano13010204
83. Cortez-Jugo, C., Czuba-Wojnilowicz, E., Tan, A., Caruso, F.: A focus on "bio" in bionanoscience: the impact of biological factors on nanomaterial interactions. Adv. Healthc. Mater. (2021). https://doi.org/10.1002/ADHM.202100574
84. Toosi, S., Naderi-Meshkin, H., Esmailzadeh, Z., Behravan, G., Ramakrishna, S., Behravan, J.: Bioactive glass-collagen/poly (glycolic acid) scaffold nanoparticles exhibit improved biological properties and enhance osteogenic lineage differentiation of mesenchymal stem cells. Front. Bioeng. Biotechnol. (2022). https://doi.org/10.3389/fbioe.2022.963996
85. Mohsin, A.I., Hussain, M., Mohsin, M.Z., Zaman, W.Q., Aslam, M.S., Shan, A., Dai, Y., Khan, I., Niazi, S., Zhang, Y., Guo, M.: Recent advances of magnetic nanomaterials for bioimaging, drug delivery, and cell therapy. ACS Appl. Nano Mater. (2022). https://doi.org/10.1021/acsanm.2c02014
86. Sobhanan, J., Anas, A., Biju, V.: Nanomaterials for fluorescence and multimodal bioimaging. Chem. Rec. (2023). https://doi.org/10.1002/tcr.202200253
87. Tripathi, A., Bonilla-Cruz, J.: Review on healthcare biosensing nanomaterials. ACS Appl. Nano Mater. (2023). https://doi.org/10.1021/acsanm.3c00941
88. Pourmadadi, M., Rahmani, E., Rajabzadeh-Khosroshahi, M., Samadi, A., Behzadmehr, R., Rahdar, A., Romanholo Ferreira, L.F.: Properties and application of carbon quantum dots (CQDs) in biosensors for disease detection: a comprehensive review. J. Drug Deliv. Sci. Technol. (2023). https://doi.org/10.1016/j.jddst.2023.104156
89. Patel, D.K., Kesharwani, R., Kumar, V.: Nanoparticles: an emerging platform for medical imaging (2021). https://doi.org/10.1016/B978-0-12-821163-2.00007-8
90. Marashdeh, M.W., Ababneh, B., Lemine, O.M., Alsadig, A., Omri, K., Mir, L., Sulieman, A., Mattar, E.: The significant effect of size and concentrations of iron oxide nanoparticles on magnetic resonance imaging contrast enhancement. Results Phys. (2019). https://doi.org/10.1016/J.RINP.2019.102651
91. Yu, H., Guo, H., Wang, Y., Cao, Y., Wang, B., Zhang, L.W.: Bismuth nanomaterials as contrast agents for radiography and computed tomography imaging and their quality/safety considerations. Wiley Interdiscip. Rev. Nanomed. Nanobiotechnol. (2022). https://doi.org/10.1002/wnan.1801
92. Kim, T., Kim, T., Jokerst, J.V.: Inorganic Fluorescent Nanomaterials (2019). https://doi.org/10.1007/7355_2019_85

93. Chakraborty, A., Selomulya, C.: Functionalized nanoparticles in pulmonary disease diagnosis (2020). https://doi.org/10.1016/B978-0-12-820658-4.00014-5
94. Sobirjon o'g'li, T.M., Okunev, R.: (2023).Biocompatibility. https://doi.org/10.1016/b978-0-12-824315-2.01099-x
95. Domb, A.J., Sharifzadeh, G., Nahum, V., Hosseinkhani, H.: Safety evaluation of nanotechnology products. Pharmaceutics (2021). https://doi.org/10.3390/PHARMACEUTICS13101615
96. Sengul, A.B., Asmatulu, E.: Toxicity of metal and metal oxide nanoparticles: a review. Environ. Chem. Lett. (2020). https://doi.org/10.1007/S10311-020-01033-6
97. You, D.J., Bonner, J.C.: Susceptibility factors in chronic lung inflammatory responses to engineered nanomaterials. Int. J. Mol. Sci. (2020). https://doi.org/10.3390/IJMS21197310
98. Soltani, R., Guo, S., Bianco, A., Ménard-Moyon, C.: Carbon nanomaterials applied for the treatment of inflammatory diseases: preclinical evidence (2020). https://doi.org/10.1002/ADTP.202000051
99. Egbuna, C., Egbuna, C., Parmar, V.K., Jeevanandam, J., Ezzat, S.M., Patrick-Iwuanyanwu, K.C., Adetunji, C.O., Khan, J., Onyeike, E.N., Uche, C.Z., Akram, M., Ibrahim, M.S., El Mahdy, N.M., Awuchi, C.G., Saravanan, K., Tijjani, H., Odoh, U., Messaoudi, M., Ifemeje, J.C., Ibeabuchi, C.G., et al.: Toxicity of nanoparticles in biomedical application: nanotoxicology. J. Toxicol. (2021). https://doi.org/10.1155/2021/9954443
100. Rehman, N., Moore, S.: An overview of the state of the regulatory and preclinical requirements for nanomaterials including medical devices. Integr. Environ. Assess. Manage. (2021). https://doi.org/10.1002/IEAM.4426
101. Rickerby, D.G.: Nanotechnological medical devices and nanopharmaceuticals: the European regulatory framework and research needs. J. Nanosci. Nanotechnol. (2007). https://doi.org/10.1166/JNN.2007.009
102. Mughal, B., Zaidi, S.Z.J., Zhang, X., Hassan, S.: Biogenic nanoparticles: synthesis, characterisation and applications. Appl. Sci. (2021). https://doi.org/10.3390/APP11062598
103. Osouli-Bostanabad, K., Puliga, S., Serrano, D.R., Bucchi, A., Halbert, G., Lalatsa, A.: Microfluidic manufacture of lipid-based nanomedicines. Pharmaceutics (2022). https://doi.org/10.3390/pharmaceutics14091940
104. Mühlebach, S.: Regulatory challenges of nanomedicines and their follow-on versions: a generic or similar approach? Adv. Drug Deliv. Rev. (2018). https://doi.org/10.1016/J.ADDR.2018.06.024
105. Faunce, T.: Toxicological and public good considerations for the regulation of nanomaterial-containing medical products. Exp. Opin. Drug Saf. (2008). https://doi.org/10.1517/14740338.7.2.103
106. Kleynhans, J., Sathekge, M., Ebenhan, T.: Obstacles and recommendations for clinical translation of nanoparticle system-based targeted alpha-particle therapy. Materials (2021). https://doi.org/10.3390/MA14174784
107. Tirumala, M.G., Anchi, P., Raja, S., Rachamalla, M., Godugu, C.: Novel methods and approaches for safety evaluation of nanoparticle formulations: a focus towards in vitro models and adverse outcome pathways. Front. Pharmacol. (2021). https://doi.org/10.3389/FPHAR.2021.612659
108. Buljan, M., Wick, P.: Tailoring design of nanomaterials and systems to individualize patient treatments. Chimia (2022). https://doi.org/10.2533/chimia.2022.236

Innovative Coating Technologies for Nanomaterials: Enhancing Functionality and Performance in Biomedical Devices

Saman Iqbal, Farheen Nusrat, Muhammad Noman, Saman Mehmood, and Nida Iqbal

Abstract Innovative coating techniques have significantly improved and enhanced nanomaterials' biocompatibility, bioactivity, and corrosion resistance. This chapter focused on the most recent and cost-effective coating techniques, such as Physical Vapor Deposition (PVD, Chemical Vapor Deposition, Plasma Spraying, Electrospinning, and Sol–Gel for medical device applications. It provides a detailed analysis of nanocoating applications, including improvements due to smart and self-healing coatings by responding to environmental stimuli, preventing infections, and enabling controlled drug release. Researchers explore the role of nanomaterial-based coatings in improving patient care. Furthermore, it discusses the use of these technologies in important biomedical areas, including orthopedic implants, cardiovascular stents, and dental implants. This chapter also highlights the relationship between nanotechnology and advanced nanomaterial coatings for biomedical device applications, offering insights into the future of biomedical engineering.

Keywords Innovative coating technologies · Physical vapor deposition (PVD) · Chemical vapor deposition (CVD) · Plasma spraying · Electrospinning · Sol–gel

---

S. Iqbal (✉)
Department of Physics, University of the Punjab, Lahore, Pakistan
e-mail: samaniqbal.physics@pu.edu.pk

F. Nusrat · M. Noman · S. Mehmood · N. Iqbal
Department of Biomedical Engineering, University of Engineering and Technology, Lahore, Pakistan
e-mail: 2024msbme1@student.uet.edu.pk

M. Noman
e-mail: 2024msbme6@student.uet.edu.pk; muhammad.noman@ahs.uol.edu.pk

S. Mehmood
e-mail: 2024msbme7@student.uet.edu.pk

N. Iqbal
e-mail: nidaiqbal@uet.edu.pk

M. Noman
Centre for Healthcare Simulation and Innovation, The University of Lahore, Lahore, Pakistan

S. N. Saud Al-Humairi et al. (eds.), *Smart Nanomaterials in Biomedical Applications*, Engineering Materials, https://doi.org/10.1007/978-981-96-8023-8_6

coatings · Smart coatings · Self-healing coatings · Orthopedic implants · Cardiovascular stents · Antimicrobial coatings · Eco-friendly coatings · Biodegradable materials

1 Introduction

The field of biomedical engineering has undergone a paradigm shift with the advent of nanotechnology, leading to groundbreaking advancements in medical devices, implants, and therapeutic approaches. Nanomaterials, the structural components at the nanoscale (typically below 100 nm), exhibit exceptional physicochemical properties distinct from their bulk counterparts. One of the unique properties of graphene and graphene-like materials include a high surface area to volume ratio, improved strength, quantum effects, and tunable surface chemistry, empowering unprecedented applications in healthcare. Integration of nanomaterials into biomedical devices has dramatically enhanced their function, longevity, and interaction with biological environments [1]. Nanocoating is a staged development that has gained considerable influence as advanced surface modifying exhibits are essential in overcoming the shortcomings of the current industrial and medical devices to achieve better performance, biocompatibility, and continuous stability in the human body.

The revolution of nanomaterials in the design of medical devices rested on controllable surface properties. Biocompatibility of traditional biomedical materials such as stainless steel, titanium alloys, and ceramics often fails because they suffer from immune rejection, bacterial infections, and implant failure. Additionally, they are inherently susceptible to degradation, i.e., corrosion, wear, and surface degradation, and they can lose functionality over time. Applying nanomaterial-based coatings to the material increases surface properties while keeping the underlying mechanical properties of the material intact. This work has designed these coatings to be bioactive and have wear resistance, corrosion protection, and antibacterial properties superior to bare metals. Designing new biomedical implants through the ability to tailor surface interactions at the nanoscale has created ways to address the lack of interaction with human tissue yet avoid adverse reactions [2]. The enhancing biocompatibility of nanocoatings in biomedical applications is one of the key advantages of nanocoatings for biomedical applications. Introducing foreign material into the body often results in the immune system viewing the item as an invader, which triggers inflammatory and encapsulating fibrotic responses, which can result in implant rejection. Changes in the implant surface with bifunctional nanomaterials can inhibit immune system activation and enhance cell adhesion, proliferation, and tissue integration. Nanocoatings have also had a significant impact on both antimicrobial and aerospace coatings. With FDA rate publicizing tied to the number of deaths associated with medical devices alone, risk of bacterial infection and subsequent severe complications, prolonged stays in the hospital, and device failure during implantation are some of its most significant challenges. Bacterial biofilms formed on medical implants are incredibly resistant to antibiotics and immune defenses, which are hard

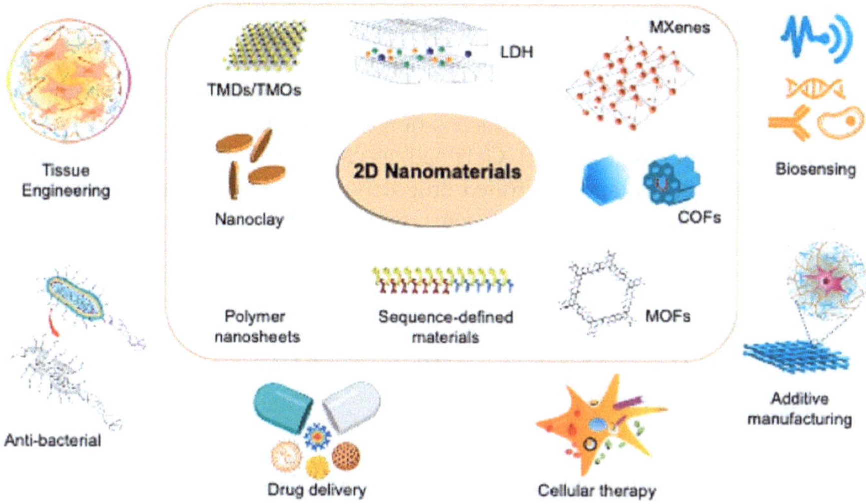

Fig. 1 Different types of 2D nanomaterials for various biomedical applications [4]

to remove. Systemic antibiotics used in conventional antimicrobial treatments cause systemic toxicity and can lead to antibiotic resistance [3]. Figure 1 shows the different types of 2D nanomaterials involved in biomedical applications.

The chapter explores the diverse applications of nanomaterials in biomedical coatings, focusing on coating techniques for medical implants to improve their durability, biocompatibility, and performance.

A discussion of nanomaterials used in coatings with nanoparticles, nanotubes, and nanofilms will be discussed in detail, followed by a description of the biosensing capabilities of various materials used for coatings and their coatings efficacy. Then, it will discuss innovative coating technologies with increased surface properties that have functional properties, including antimicrobial resistance and controlled drug release. Also, the importance of coatings in dealing with significant challenges encountered by biomedical devices, such as corrosion, wear, and immune compatibility, will be examined. Experts will address future directions of nanocoating innovations, regulatory impacts, and translational perspectives. They will provide critical perspectives on nanotechnology in modern medicine. The chapter aims to provide a detailed comprehension of how nanocoating is restructuring the future of biomedical engineering to produce better patient outcomes and improve medical device performance through a comprehensive study.

2 Innovative Coating Technologies

Innovating coating techniques make Biomedical materials more functional, durable, and biocompatible. Physical Vapor Deposition (PVD), Chemical Vapor Deposition (CVD), Plasma Spraying, and Sol–Gel coatings can provide better wear resistance, antibacterial properties, and controlled drug delivery. Advanced coatings are vital in medical implants, cardiovascular stents, and drug delivery systems to increase performance and patient outcomes.

2.1 Physical Vapor Deposition (PVD) and Chemical Vapor Deposition (CVD)

To improve biomedical implants' performance, durability, and biocompatibility, surface engineering modifies the surface characteristics to address specific clinical requirements. An implant surface's roughness, chemical composition, and wettability significantly affect the implant's interaction with biological tissues.

A well-engineered surface enhances implant integration, reduces the risk of rejection, and extends the device's lifespan. Traditional surface modification techniques, such as mechanical polishing and acid etching, impart limited properties on surfaces at the micro and nanoscale. The adoption of advanced coating technologies such as Physical Vapor Deposition (PVD) and Chemical Vapor Deposition (CVD) has even resulted in the control of film thickness, composition, and structural integrity [5].

Despite its wide use, surface engineering techniques have not been able to achieve high-purity thin films with high adhesion and uniformity in various surface engineering techniques; PVD and CVD have emerged as leading methods that can effectively modify the surface functional properties of biomedical implants. These techniques' deposition techniques make biocompatible materials, such as titanium dioxide (TiO_2) or hydroxyapatite, which help promote osteointegration and improve implant performance in orthopedic and dental applications. In addition, PVD and CVD coatings make metallic implants more resistant to corrosion so that the harmful metal ions of the implants do not leak into surrounding tissues. Due to the propensity of titanium and cobalt-chromium-based implants to degradation in the physiological environment, this is especially important. These techniques help create protective coatings that provide protective armor against corrosion and wear of medical implants [6].

2.1.1 Physical Vapor Deposition (PVD) Process

Physical Vapor Deposition (PVD) is one of the most advanced coating techniques; it is a material transfer pattern from a solid phase to a vapor phase and again to a solid phase onto a substrate. The process occurs in a high vacuum, so the deposited coatings remain exceptionally pure and uniform. Unlike electroplating and other traditional coating methods, such as thermal spraying, which involve chemical reactions or high processing temperatures, PVD relies solely on physical processes to deposit thin films with precise thickness, stoichiometry, and microstructure. As these surface properties are critical for biomedical applications, the ability to tailor coatings at the nanoscale makes PVD highly valuable [5]. PVD mainly offers to deposit thin films without destroying the substrate integrity. This process has three main stages: material vaporization, vapor transport, and condensation. First, let us consider a target material, typically a metal or ceramic, and vaporize it utilizing high-energy sources such as ion bombardment, an electron beam, or an electric arc. This vaporized atom passes through a prearranged sequence through the vacuum chamber, minimizing the contaminant's interaction. Last, the vapor condenses onto the substrate surface, depositing an ultra-uniform and adherent coating. This sequential process guarantees that PVD coatings have excellent adhesion, high density, and low porosity to increase the durability and performance of medical implants [7].

Another critical feature of PVD is that it allows the creation of multi-layered and composite coatings with specific properties. The coating's functionalities, such as enhanced hardness, corrosion resistance, and bioactivity, can be engineered by adjusting deposition parameters such as temperature, deposition rate, and gas composition. PVD effectively deposits on various materials, including metals, nitrides, oxides, and carbides, making it highly versatile for biomedical applications. PVD deposits titanium dioxide (TiO_2) coatings, the most common coating that promotes osseointegration in orthopedic and dental implants. Additionally, the controlled deposition process allows these coatings to have a very crystalline, bioactive, and long-term stable structure within the human body [6].

2.1.2 PVD Deposition Techniques

Physical Vapor Deposition (PVD) is a series of advanced coating methods employing the deposition of high-purity thin films onto various substrates. Such techniques are widely used in biomedical applications to enhance the function properties of medical implants, including their biocompatibility, wear resistance, and corrosion protection. The main PVD deposition methods are sputtering electron beam evaporation and cathodic arc deposition. These techniques run under vacuums to produce clean, uniform coatings that adhere well. A particular deposition method is chosen based on the film composition needed, thickness, and application in the medical field. As the surface characteristics of implants improve with the help of PVD coatings, PVD coatings are used in orthopedic, dental, and cardiovascular applications [5].

- *Sputtering Deposition*

Sputtering is one of the most commonly used PVD techniques to produce highly uniform and well-adhesion coatings. The process of bombarding a target material with high-energy ions, usually from an inert gas such as argon, occurs. When the impact occurs, atoms on the target dislodge, and the atoms travel in a vaporized state, reaching the substrate where they adhere to it to form a thin film. It is a highly effective coating method that imparts precise thickness control and excellent surface smoothness, which is well-suited for biomedical applications. One of the advantages of sputtering in medical implantology is the deposition of biocompatible materials, for example, titanium dioxide (TiO_2) and hydroxyapatite, that promote osseointegration (the bone integration of the implant). For instance, better bioactivity and antibacterial properties of TiO_2 coatings have reduced the risk of post-implantation infectionsReactive sputtering involves adding gases such as oxygen or nitrogen into the chamber to form compound coatings, such as titanium nitride (TiN) or titanium oxide (TiO_2), which enhance the corrosion and wear resistance of implants [6]. Besides biomedical applications, sputtering provides superior film density and low porosity, improving the implant's long-term stability. As a result, it enables uniform thickness coating of complex geometries, including complex implant designs used in cardiovascular stents or orthopedic prostheses, which is beneficial toward improved surface properties. Additionally, the advances in magnetron sputtering have enabled higher deposition rates and greater control of the composition of the coating, which is one of the most reliable PVD techniques used in modern medical applications [1].

- *Electron Beam Evaporation*

Elect electron beam evaporation is another widely used PVD technique to deposit thin films with high purity and chemical stability. Here, an intense electron beam is directed onto the target material to vaporize it. The atoms are then vaporized on the substrate and condense into a thin film of excellent adhesion and structural integrity. This method is very convenient for coating deposition from degassed ceramic and oxide solutions since high temperatures are needed to achieve the material properties required. Electron beam evaporation produces contamination-free coatings by operating in an ultra-high vacuum environment. It is, therefore, very suitable for medical applications concerning purity and biocompatibility. For instance, this technique can effectively deposit hydroxyapatite coatings that are like the mineral composition of bone. It has been demonstrated that these coatings accelerate the integration of the implant with adjacent bone tissues for improved healing and reduce the risk of implant failure. In addition, it is possible to deposit multi-layered coatings with their tailored properties, like improved mechanical strength and controlled porosity, with electron beam evaporation. One of the most significant advantages of this new material is its ability to control porosity, which is very important in drug-eluting implants, where drug release aspires to prevent infection and promote tissue regeneration. Finally, the high deposition rate and scalability of the method make it, in addition to its use in the biomedical industry for orthopedic and dental implants that need excellent wear resistance and long-term durability [7], a highly used technique.

2.1.3 Advantages of PVD Coatings in Biomedical Applications

One of the most essential benefits of PVD coatings in the biomedical sector is its ability to improve the wear resistance of implants, minimize surface shrinkage, and extend the entire life of medical devices. PVD coatings have extraordinary challenges. This is especially crucial for orthopedic implants subjected to continuous mechanical stress, such as hip and knee prosthetics [6]. Additionally, PVD coatings offer excellent corrosion resistance, a vital property for implants exposed to body fluids. For instance, titanium-based coating foams, including protective oxide layers, prevent metal ions from leaching into surrounding tissues, reducing the risk of inflammatory and rejection responses. This mainly benefits cardiovascular implants, where material degradation could lead to severe complications [8]. Materials such as titanium dioxide (TIO) and hydroxyapatite promote PVD-mediated Oseo integration, which is responsible for the implant surface by osteocytes. This ensures stable fixation of implants in the body and accelerates healing. In particular, hydroxyapatite coatings have demonstrated prominent biological activity as they mimic the natural mineral composition of bone and promote cell binding and proliferation [1].

Additionally, the PVD coating can be adjusted to suit the antibacterial properties. This is extremely important for preventing postoperative infections. A recent study examined the placement of silver nanoparticles on PVD coatings to create an antibacterial surface that inhibits the sedimentation of bacteria in implants. This innovation is up-and-coming for dental and orthopedic implants, where infectious disease-related complications can lead to implant failure and revision surgery [5].

2.2 Chemical Vapor Deposition (CVD) Process

Chemical Vapor Separation (CVD) is a sophisticated thin film separation technique that uses chemical reactions to create high-quality coatings on different substrates. CVD utilizes reactive chemical precursors in a controlled environment to form solid thin films, unlike PVD, which relies on the physical transfer of vapor-phase materials. This method allows the production of coatings with excellent responsibility, uniformity, and high purity, making them particularly valuable for biomedical applications [9]. The CVD process usually involves several steps, including pioneer steam, transport to the reaction chamber, chemical reactions at the substrate interface, and removal of by-products, as shown in Fig. 2. With its ability to control deposition parameters such as temperature, pressure, and gas composition, CVD ensures that it creates coatings with the made's mechanical, chemical, and biological properties.

An essential advantage of CVD compared to other separation techniques is that it can achieve compliant coatings even in complex shapes. Furthermore, the versatility of CVD allows for various materials, such as ceramics, polymers, and diamond-like carbon (DLC). This process can be altered by techniques such as plasma-enhanced CVD (PECVD) and CVD (LPCVD) at low pressure (PECVD) to improve membrane properties, lower separation temperatures, and enhance coating adhesion of temperature-dependent substrates [7]. CVD coatings are highly suitable for biomedical applications due to their widespread recognition for their excellent mechanical strength, chemical resistance, and biocompatibility. With advances in nanotechnology and materials science, CVD plays an essential role in the next

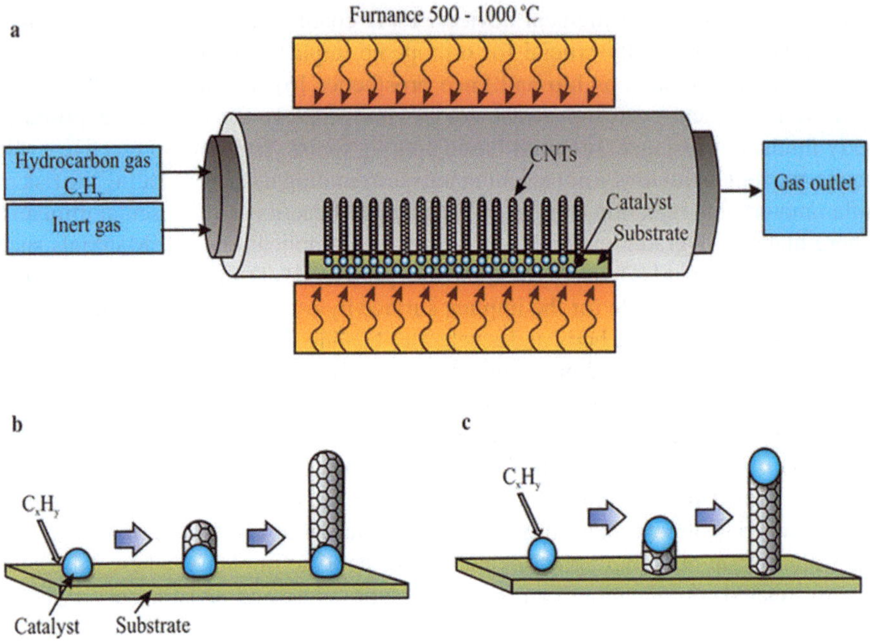

Fig. 2 Schematic representation of chemical vapor deposition (CVD) process. **a** Simplified scheme of a CVD reactor for CNTs synthesis; **b** base growth (Zaytseva and Neumann 2016)

generation of biomedical coatings, providing innovative solutions to improve the performance of medical devices and patient care.

2.2.1 Types of CVD Techniques Used in Biomedical Applications

The various separation methods of Chemical Vapor Separation (CVD) require strict attention for specific biomedical applications. The techniques display differences concerning operating requirements, energy supplies, and storage modes. The method can adapt to different implant and medical device applications in prosthetics. PECDV is the most common CVD technique for biomedical applications, along with LPCVD and thermal CVD. The medical field benefits from these coating techniques because they provide separate benefits regarding their ability to achieve specific coating thicknesses while ensuring strong adhesion, desired chemical content, and biocompatibility properties.

- *Plasma-Enhanced Chemical Vapor Deposition (PECVD)*

A wide CVD process modification known as Plasma-Enhanced Chemical Vapor Decomposition (PECVD) activates precursor energy using plasma-generated energy to enable low-temperature film deposition. The method of CVD must maintain high temperatures to achieve molecule reduction and initiate reactions between substrates. PECVD achieves temperature reduction by operating in gas-phase reactors where plasma fields enable efficient film development at low temperatures [8]. PECVD technology proves beneficial for working with temperature-fragile biomedical polymers that typically serve cardiovascular and orthopedic medical applications. PECVD stands out for biomedical technology applications because it enables the creation of functionalized surface coatings that exhibit enhanced biocompatibility. PECVD operates as a predominant method to deposit diamond-like carbon coating (DLC) in cardiovascular implants, including stents and artificial heart valves. DLC coatings applied by PECVD methods excel in blood compatibility, stop blood clot development, and lower the possibility of thrombus development [7]. PECVD enables the creation of antibacterial coatings that stop bacterial sedimentation in catheters, thus reducing the likelihood of hospital-acquired infections. PECVD coatings are essential in improving orthopedic implants' durability and wear resistance. By preserving thin hydroxyapatite and titanium oxide films, researchers enhanced the osseointegration of joint prosthetics and dental implants. These coatings mimic natural bone mineral structures, make fabric integration more flexible, and reduce the risk of implant rejection [9]. Additionally, PECVD allows for accurate control of film composition, allowing for the development of drug-delivery coatings that release therapeutically active ingredients in a controlled manner to prevent infection and promote healing in postoperative applications.

- *Low-Pressure Chemical Vapor Deposition (LPCVD)*

Chemical vapor separation with low pressure (LPCVD) is a wide range of CVD technology in biomedical applications. In LPCVD, deposition occurs under reduced pressure, improving vapor phase diffusion and promoting coating formation. This technology is highly effective in coating medical devices with complex structures, such as MEMS biosensors and diagnostic chips (microelectromechanical systems) [8]. One of the main applications of LPCVD in the biomedical sector is the deposition of layers on silicon-based biosensors and CHIP laboratory devices. These coatings improve the sensitivity and durability of diagnostic equipment that detects disease, glucose monitoring, and portable health tracking systems. LPCVD coatings strengthen the reliability and accuracy of medical sensors by providing protective barriers to environmental losses and maintaining excellent electrical conductivity [7]. LPCVD is also used to produce biocompatible coatings for orthopedic and dental implants. Titan-nitride coatings deposited via LPCVD are known for their excellent hardness, corrosion resistance, and biocompatibility. These coatings protect the underlying metal from failure and improve the implant's life by reducing wear and friction. Furthermore, LPCVD deposits bioactive glass coatings that promote bone regeneration and accelerate healing processes in orthopedic operations [9].

2.2.2 Advancements and Future Perspectives in CVD for Biomedical Engineering

As biomedical research progresses, the potential for CVD for medical applications is growing. Emerging trends focus on developing intelligent coatings that actively interact with the biological environment. For example, drug-loaded CVD coatings enable the controlled release of therapeutic agents and localized drug delivery directly at implant sites. These coatings can revolutionize the treatment of chronic diseases such as postoperative infections, inflammatory diseases, and osteoporosis [9]. Nanostructured CVD coatings also pay attention to the ability to mimic the nanoscale architecture of natural tissues. By including nanoscale properties on implant surfaces, researchers hope to improve cell adhesion, accelerate tissue regeneration, and improve the overall biocompatibility of medical devices. These nanoscale coatings can be tuned to enhance antibacterial resistance, bone introduction properties, and mechanical stability, making them extremely valuable for orthopedic, dental, and cardiovascular applications [7]. Researchers will explore a hybrid approach in which CVD coatings are applied to 3D-printed implants and scaffolds to create highly personalized, functionalized medical devices. This approach allows for producing patient-specific implants with precisely developed surfaces that optimize biological integration and mechanical performance. Furthermore, advances in environmentally friendly precursors and energy-efficient separation technologies promote the development of sustainable CVD processes, reducing environmental impact and the maintenance of strong coatings [8].

2.3 Plasma Spraying Technology

Plasma spray is an energy-oriented thermal spray process often used to store protection and functional coatings on various substrates. It is an essential technique in aerospace, biomedical engineering, manufacturing, and other industries where surface modification is required to improve properties such as abrasion resistance, corrosion protection, and biocompatibility. This process produces a plasma agent as a heat source for accelerating trace powder particles to the target surface. According to the effect, these particles solidify quickly and form a dense, well-classified coating. Plasma spray is particularly advantageous as it is versatile for processing various materials, including ceramics, metals, and composites.

i. **Generation of Plasma Jet and Energy Source**

The plasma generation process begins with high-temperature Plasma bursts in a specially designed plasma chamber. This process involves a cathode and an anode that create an electric arc to initiate plasma formation. Arrangements are usually inert working gases that convert argon, nitrogen, helium, or an inert working gas that converts it into a highly energetic plasma state. The temperature within the plasma CAD can exceed 10,000 °C, and even refractory materials can melt. The plasma jet is released from the torch nozzle at an over-sound speed, providing the energy needed to quickly melt and drive the coating material. The composition of the Plus Magazine affects a variety of parameters, including radiation temperature, energy density, and velocity. All of these are extremely important to achieving optimal coating quality. Argon is typically used as the primary gas due to its stability.

In contrast, auxiliary gases like hydrogen and helium are introduced to modify plasma jets' enthalpy and acceleration characteristics. The electrical parameters of the Plasma Grapel, including arch currents and voltages, play an essential role in determining the overall energy output. Higher arc currents produce more intensive plasma jets, facilitating efficient melting of coated particles. However, excessive energy input can lead to unwanted effects such as oxidation and fleeting evaporation of coating components. Therefore, accurate control of electrical parameters is crucial to maintain consistency and reliability in the plasma germination process.

ii. **Powder Injection and Particle Behavior in Plasma Jet**

When the plasma jet is set up, a customarily powdered starting material is inserted into the plasma current. The powder injection method depends on the specific application requirements and the coating properties of your choice. The powder can be injected inside and introduced directly into the hottest region of the plasma or outside and is fed from the periphery of the plasma jet. The choice of injection method affects the melting degree and trajectory of the particles in the beam. As the powder particles pass through the plasma, they experience fast heating and melting. The degree to which a particle melts depends on the plasma's size, thermal conductivity, and exposure time. Particles tend to melt faster due to their higher proportions from the surface range. Larger particles can melt, producing coatings containing melted and semi-molten particles. Particle velocity plays a vital role in determining the final properties of the coating. High speeds lead to more compact and bonding coatings, but lower speeds increase porosity and weaken responsibility. Optimization of particle size distribution, injection rate, and feed rate are of great importance in the production of coatings with uniform, high density, and desirable mechanical properties.

iii. **Substrate Interaction and Coating Deposition Mechanism**

Once the melted particles reached the substrate, they affected the surface rapidly and were distributed in thin layered structures called "splashes." These splats quickly solidify for contact and form the essential components of the coating. The interaction between the melted particles and the substrate is vital to determine the deposition layer's adhesion strength and overall integrity. The substrate temperature, surface roughness, and kinetic energy of affected particles all affect the binding mechanism for coating the substrate. The substrate is learned to improve responsibility before the plasma sprays extensive surface preparation. This includes grinding jets with materials such as aluminum oxide or carbide silicon to create rough surfaces where mechanical locking potential increases. The parameters that may influence the quality of nanostructure coating are illustrated in Table 1.

In some cases, substrates are required to minimize thermal loading and improve binding efficiency. The formation of a well-classified coating relies on balancing

Table 1 Process parameters influencing coating quality

Process parameter	Description	Effect on coating quality
Plasma gas composition and flow rate	Determines the thermal energy and velocity of the plasma jet	Affects particle melting, deposition efficiency, and coating adhesion
Arc current and voltage	Regulates the energy output of the plasma torch	Influences particle heating, coating uniformity, and density
Spray distance	Distance between the plasma torch and the substrate	Ensures particles reach the substrate fully molten, preventing excessive cooling or oxidation
Powder feed rate and distribution	Controls the amount and consistency of powder delivery	Affects coating morphology, thickness, porosity, and adhesion strength
Substrate temperature	Temperature of the substrate during spraying	It influences residual stress, reduces thermal expansion mismatches, and prevents delamination or cracking
Cooling and heating cycles	Controlled heating and cooling during deposition	Reduces internal stresses and improves mechanical stability

thermal and kinetic energy to ensure that the melted particles spread evenly without excessive oxidation or decomposition. One of the defining properties of plasma germination is the rapid consumption of particles deposited on impact. High cooling rates lead to unique microstructural features such as fine grain size, amorphous phase, and controlled porosity. The cooling rate also affects the crystallinity of the coating. This is particularly important in biomedical applications where bioactive materials such as hydroxyapatite must maintain structural integrity to promote cell interaction and osseointegration.

2.3.1 Enhancement of Surface Properties Through Plasma-Sprayed Nanocoating

i. *Improved Adhesion and Mechanical Strength*

One of the most critical factors in biomedical implant-ology is ensuring that implants achieve strong adhesion with the surrounding biological tissue to maintain stability and long-term functionality. Plasma-sprayed coatings significantly enhance adhesion strength by creating a robust interface between the implant substrate and the applied coating. During the plasma spraying process, the high velocity of molten particles ensures deep penetration into the substrate surface, forming a mechanical interlock that prevents delamination or premature coating failure. This strong bonding mechanism is particularly beneficial for titanium-based implants, where plasma-sprayed hydroxyapatite (HA) coatings have excellent adhesion properties, improving osseointegration and long-term implant success [10]. Beyond adhesion, plasma-sprayed coatings contribute to the overall mechanical strength of implants. Medical implants, particularly those for orthopedic applications such as hip and knee replacement, are exposed to continuous mechanical tension. Repeated periodic stress can lead to material fatigue and potential failure if the implant is not sufficiently abrasion-resistant. Plasma spray technology addresses this issue by strengthening the implant surface with a ceramic or metal coating with excellent mechanical durability. The microstructure refinement achieved by plasma spray ensures that these coatings have high hardness and fracture toughness, making them resistant to wear, fatigue, and mechanical degradation. Using nanostructured coatings further improves toughness and strength, as their fine grain size contributes to improved capacity and resistance to crack spreading [11].

ii. *Corrosion Resistance and Longevity of Implants*

Corrosion, which physically happens when an implant material slowly degrades by interacting with bodily fluids, is often the reason behind a metallic implant's short lifespan. The corrosion, along with corrosion products, not only weakens the structural integrity of the implant but also releases metal ions that can induce inflammatory responses, tissue damage, and, eventually, rejection of the implant. Coatings applied by plasma spray have the advantage of suppressing corrosion and preventing direct contact between the metal substrate and the physiological environment. The exceptional corrosion resistance of titanium and plasma-sprayed coatings (i.e., plasma-sprayed coatings based on titanium and hydroxyapatite) has been extensively studied for orthopedic and dental applications. These coatings have a dense well, adhered microstructure, which restrains penetration of corrosive agents and decreases the risk of material degradation with the years [12]. In addition, plasma-sprayed coatings engineered with specific porosity afford enhancements of the protective function with retention of biocompatibility. In addition to preventing the infiltration of corrosive fluid, porous coatings allow bone cells to infiltrate, to this end providing for stable fixation and integration with the surrounding tissue.

Recently, more attention has been generated concerning using advanced ceramic materials like zirconia and alumina, partly due to recent advancements in plasma spraying technology to improve corrosion resistance. They possess high chemical stability and play defensive roles for the metal as protective layers by protecting the underlying metal from oxidative degradation. The application of corrosion-resistant coatings enhances the lifespan of the implants as well as preserving the mechanical properties of implants, such that they remain structurally sound within their lifespan [13]. One of the breakthroughs in implant technology is the continuous evolution of plasma spraying techniques in a corrosion-resistant coating that can last with the patient and those more or less permanent biomedical implants.

iii. *Antibacterial Properties for Infection Prevention*

Post-surgically acquired infections are one of the most serious complications associated with biomedical implants, such as the need for additional surgeries, prolonged hospitalization, and implant failure due to such infections. Bacterial colonization of implants that leads to biofilm formation on implant surface leads to infections resistant to conventional antibiotics. With plasma spraying technology, infection prevention has been achieved by developing antimicrobial coatings that deter the adhesion and proliferation of bacteria. In the case of antibacterial plasma sprayed coatings, integrating antibacterial agents like silver, zinc, or titanium dioxide (TiO_2) has shown to be very effective in preventing microbial contamination. It is shown that both silver and gold nanoparticles, particularly colloidal silver nanoparticles, have potent antibacterial activity due to their ability to disrupt bacterial cell membranes and metabolic processes. Coatings of Ag–TiO_2 plasma sprayed onto orthopedic materials have also been shown to have dual functionality in blockage of bacterial colonization and promote cell proliferation and tissue regeneration. This multifunctional capacity guarantees that implants are infection-resistant and associated with quick recuperation and coordination with the host tissue [14].

2.4 Sol–Gel Coating: A Multifunctional Approach for Biomedical Applications

It has been widely reported here that the sol–gel technique is of extreme technological interest in materials science since it allows various targeted advanced coatings to be obtained with controlled physicochemical properties. This method involves the hydrolysis and condensation reactions of a sol (liquid precursor solution) to turn it into a gel-like network, followed by controlled drying and heat treatment to create a stable and functional coating [15]. The sol–gel process is a widely studied technique for modifying material surface at the micro- and nanoscale. It permits coatings synthesis with tunable porosity, chemical composition, and surface morphology [16]. Due to this adaptability, sol–gel coatings have been an area of extensive research, especially where surface properties, including adhesion, wettability, and mechanical durability, are essential [17].

One of the key reasons for using the sol–gel method is that it can produce homogeneous, high-purity coatings with controlled microstructures. Unlike conventional coating techniques, sol–gel synthesis does not need high processing temperatures to be applied to heat-sensitive materials [18]. Further, the sol–gel process introduces diverse bioactive compounds to improve mechanical strength, corrosion resistance, and surface bioactivity, like hydroxyapatite, metal oxide nanoparticles, and polymeric additives [19]. Its applications are further extended to introducing functional agents into the sol–gel matrix to modify metallic, ceramic, and polymeric substrates and improve their stability and functional properties [17]. An additional important characteristic of sol gel-derived coatings is the formation of porous structures with tailored pore size distribution. Surface porosity and hydraulic conductivity are significant for applications for which a certain amount of surface porosity is needed for fluid permeability, molecular diffusion, or biological reactions [16]. Sol–gel films may be tailored for specific functional needs by adjusting precursor concentration, reaction kinetics, and drying conditions, and the porosity of the films can be fine-tuned [15]. Besides, the sol–gel method for coating deposition also uses an environmentally friendly deposition process, as it usually works with water or alcohol-based precursors to minimize or eliminate hazardous solvents [18]. More recently, the work has concentrated on optimizing the sol–gel formulations to improve the coating adhesion and mechanical durability and protect them from environmental degradation. Advances in hybrid organic–inorganic sol–gel systems [16] are providing new opportunities for developing more flexible, more rigid, and stimulus-responsive multifunctional coatings, as shown in Fig. 3. To enhance the optical, electrical, or antimicrobial properties of sol–gel coatings, nanomaterials such as silica, titanium dioxide, or bioactive glass can be incorporated to improve these characteristics.

With continued research on the sol–gel process, an expansion of reaction mechanisms, structural evolution, and processing parameters will be vital in improving performance over various entities and disciplines, as shown in Table 2. The sol–gel-based coating is an emerging technology in materials science through ongoing development. It makes it significant and provides a versatile platform for designing coatings with various properties to multifunctional capabilities [17].

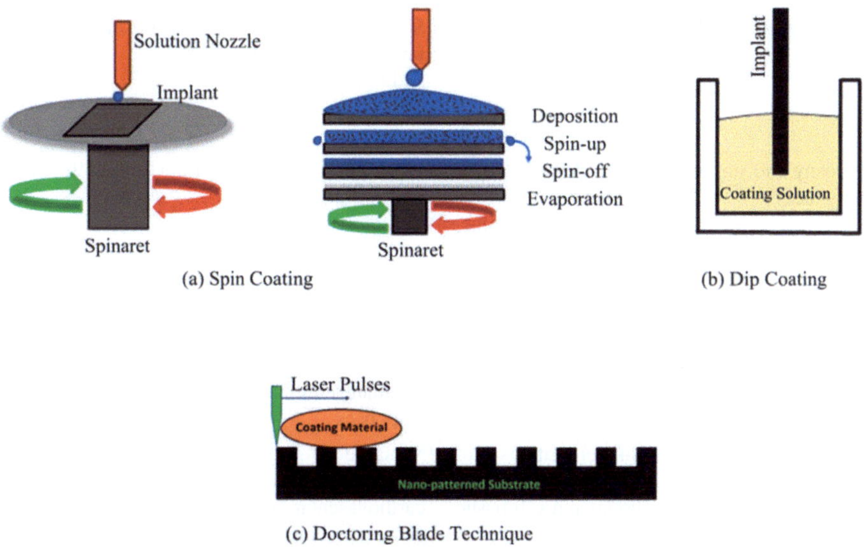

Fig. 3 Types of sol–gel deposition on implant substrates [20]

Table 2 Biocompatibility and antimicrobial properties of sol–gel coatings

Feature	Description	Key benefits	References
Biocompatibility	Sol–gel coatings exhibit excellent biocompatibility, ensuring compatibility with biological tissues and preventing adverse immune reactions	Reduces foreign body reactions inflammation and supports implant integration	Mohammad et al. [16], Fernández-Hernán et al. [17]
Bio integration	Controlled surface chemistry enhances the biointegration of implants, especially in orthopedic, dental, and cardiovascular applications	Promotes long-term stability and function of implants	Mohammad et al. [16]
Cell adhesion and proliferation	Sol–gel coatings support cell adhesion, proliferation, and differentiation, which is crucial for implant success	Enhances healing and tissue regeneration	Murari and Chauhan [15]
Hydroxyapatite (HA) coatings	HA-based sol–gel coatings mimic natural bone mineral and promote osseointegration	Strengthens implant-bone bonding and improves bioactivity	Catauro and Ciprioti18
Porosity and surface roughness	Controlled porosity and nanoscale roughness mimic the extracellular matrix (ECM)	Enhances protein adsorption, cellular attachment, and fluid exchange	Tahir et al. [19]
Bioactive agent incorporation	Coatings can be loaded with peptides, growth factors, or bioresorbable polymers	Promotes wound healing, reduces fibrosis, and modulates immune responses	Catauro and Ciprioti [18]
Soft and hard tissue applications	Sol–gel coatings can be applied in hard (bone) and soft tissue engineering	Versatile for orthopedic, dental, cardiovascular, and neural implants	Murari and Chauhan [15]
Antimicrobial properties	Sol–gel coatings prevent bacterial adhesion and infection by incorporating antimicrobial agents	Reduces implant-associated infections and biofilm formation	Tahir et al. [19]

(continued)

Table 2 (continued)

Feature	Description	Key benefits	References
Organic antimicrobial agents	Chitosan, quaternary ammonium compounds, and antimicrobial peptides enhance bacterial inhibition	Provides biocompatible and non-toxic antimicrobial activity	Fernández-Hernán et al. [17]
Hybrid coatings (organic + inorganic)	Combination of organic and inorganic agents for synergistic antimicrobial effects	Effective against multidrug-resistant bacterial strains	Tahir et al. [19]
Smart antimicrobial coatings	Coatings that respond to pH changes or bacterial presence to release antimicrobial agents	Targeted infection prevention with minimal toxicity	Mohammad et al. [16]

2.5 *Self-Healing and Smart Coatings*

Innovative coatings have recently attracted great interest in biomedical applications to impart innovative properties to improve the durability and functionality of coatings on surfaces. These coatings aim to offer superior properties, including self-healing, corrosion resistance, and regulated drug release by including responsive materials that respond to surrounding conditions. Recently, effective, innovative coatings, including those developed by sol–gel techniques, can be tailored to respond to temperature variations, acid/base changes, mechanical damage, and biological effects for obvious biomedical and implant applications [2]. Innovative coatings are one of the primary advantages because they improve the longevity and stability of biomedical implants. However, traditional coatings degrade with time because of mechanical wear, bacterial colonization, or physiological conditions, which decrease the performance or failure of the implant. On the contrary, innovative coatings can respond to surrounding stimuli and provide superior protection and functionality. However, recent research has concentrated on inserting micro- and nanocontainers in these coatings with the controlled release of healing agents or drugs based on environmental cues [21]. Furthermore, nanotechnology has been integrated into innovative coatings that provide their capabilities to control drug delivery with a prior desired amount and self-repair mechanisms.

i. **Self-Healing Mechanisms in Coatings**

Surface protection technology for biomedical applications requires long-term coating stability, which self-healing coatings can afford. They sense these minor damages autonomously without forming cracks and corrosion and maintain structural integrity. The self-healing process can be grouped as intrinsic and extrinsic. Chemical bonds, such as hydrogen bonds or dynamic covalent interactions, form reversibly, and the material self-heals after mechanical damage owing to the inherent drive to reform the existing reversible bond network. In contrast, extrinsic self-healing relies on the inclusion of healing agent(s) encapsulated inside micro- or nanocontainers that release their contents in response to external stimuli such as pH, temperature, and mechanical stress [3]. Extensive exploration of sol–gel-derived self-healing coatings has been conducted due to their potential for long-lasting metallic implant protection. These coatings can enhance the longevity of implants by embedding functional nanoparticles or polymeric microcapsules into a sol–gel matrix, which can autonomously repair microcracks and defects [22]. A promising method is using mesoporous silica nanoparticles impregnated with healing agents released under mechanical stress for microcrack sealing and protective function restoration of the coating. The investigations have shown that incorporating coatings with these self-healing capabilities significantly reduces the risk of implant degradation and failure, making them very useful for orthopedic and dental implants [21].

An innovative approach to introducing biomimetic materials used in self-healing coatings is based on mimicking the natural healing mechanisms of biological systems. For example, the coatings bearing polydopamine, which mimics the mussel adhesion proteins, boast excellent self-repairing capabilities because they can form strong interface bonds with implant surfaces. Moreover, these biomimetic coatings are biofouling resistant, antimicrobial, and biocompatible, desirable for biomaterials [22]. In particular, the continuous advancements in self-healing coating technology assert that future biomedical implants will be far more resistant to many uses during physiological conditions. Figure 4 shows the self-healing mechanisms.

ii. **Stimuli-Responsive Smart Coatings**

Intelligent coatings (also called stimuli-sensitive coatings) are designed to elicit changes in their properties in response to external environmental stimuli such as pH,

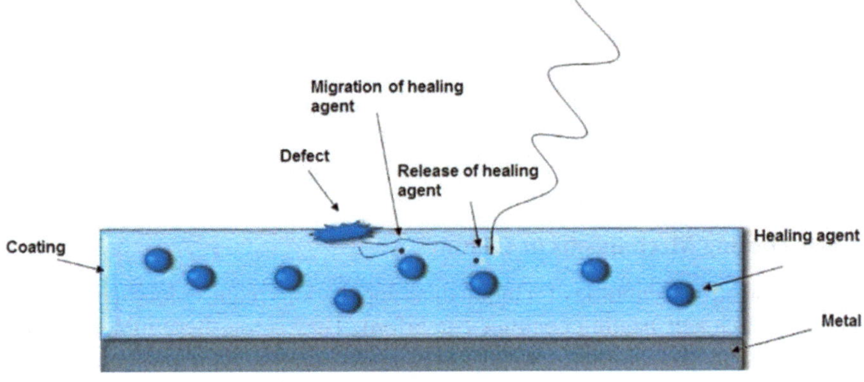

Fig. 4 Schematic of the self-healing coating mechanism on metallic substrates [23]

temperature, light, or mechanical stress. These coatings hold great promise for bio applications in controlled drug delivery and implant protection. pH sensitivity is one of the most widely studied stimuli–responsive mechanisms, allowing the coatings to release the therapeutic agent only under specific pH conditions, such as those found in infected or inflamed tissue [2]. Regarding temperature, temperature-responsive coatings also show potential to control drug release based on fluctuations in body temperature. These coatings are typically made from thermo-responsive polymers that undergo phase transitions at physiological temperatures to activate drug release when needed. In particular, such coatings are highly applicable to cancer therapy, where localized hyperthermia can trigger drug release from these systems while avoiding systemic toxicity and improving treatment outcomes [3].

Furthermore, near-infrared (NIR) light-absorbing innovative coatings exhibit pH and temperature responsiveness and light responsivity. Controlled release of drugs can be achieved upon exposure to specific wavelengths of light, allowing for a noninvasive therapy. For example, superhydrophobic coatings with NIR and pH dual responsivity can be prepared based on mesoporous polydopamine microspheres, which can be used to accurately control drug release and surface protective properties [22]. These innovations represent new possibilities for stimuli-responsive coatings in personalized medicine and advanced implant technologies. The second class of smart coatings in the making are electro-responsive materials whose properties can be changed mechanically in response to electrical stimuli. Coatings of this type would be helpful, especially in neural implants or other similar bioelectronic devices that use electrical signals to control drug release or improve tissue integration. With recent efforts, electroactive polymers are employed as coatings that respond to changes in voltage (external control factor) to control the diffusion of drugs and cellular interactions [21]. This technology allows us to produce implantable devices that are real-time adaptive to certain physiologic conditions, which constitutes the development of the next generation of implantable devices.

3 Applications in Biomedical Devices

Nanomaterial coatings are the cornerstone of advanced coatings types in various biomedical devices for enhancements and improvements in the device results to provide the best assistive solutions. Some of the applications are discussed below.

3.1 Nanomaterial Coatings for Dental Implants

Dental implants have become the most reliable solution for oral healthcare and a long-term remedy for patients with dental issues or missing teeth. However, these implants also have various drawbacks, including bacterial infections, poor osseointegration, and corrosion-related degradation. To address such problems, titanium-based [24]

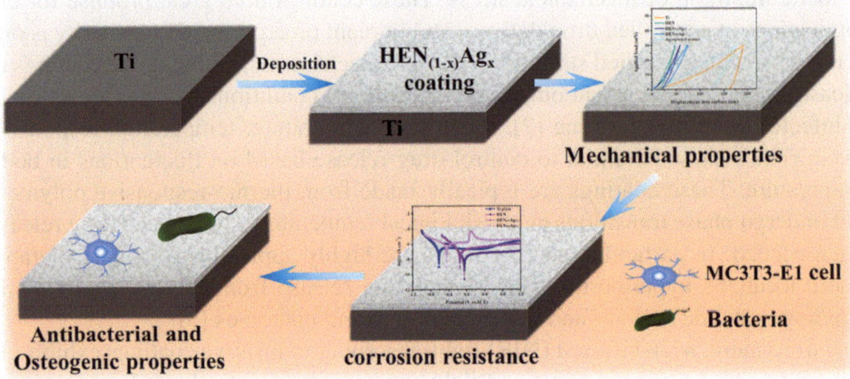

Fig. 5 Schematic diagram of the entire experimental route [25]

implants are widely utilized for their biocompatibility, lower antibacterial properties, and reduced susceptibility to biofilm formation. Significant advancements have been made to enhance the performance of dental implants by increasing antibacterial resistance, corrosion protection, and osseointegration. To achieve this, the application of nanomaterial coatings has been investigated. Recent research has demonstrated that high-entropy nitride (HEN) coatings embedded with silver nanoparticles (AgNPs) offer a promising solution, as shown in Fig. 5.

Development of High-Entropy Nitride (HEN) Coatings for Dental Implants can be summarized as the HEN coatings for dental implants comprising multiple transition metals such as Ti, Zr, Nb, Ta, and Mo combined with nitrogen [25]. Then, Ag nanoparticles are added to enhance antibacterial functionality further. These coatings are synthesized through reactive sputter deposition using a double cathode glow discharge plasma system. The face-centered cubic (FCC) crystal structure of HEN coatings has ensured high mechanical strength and wear resistance. Silver nanoparticles are used to enhance antibacterial efficacy and maintain mechanical integrity. After developing HEN coatings, potentiodynamic polarization tests will be done in simulated body fluid to check its corrosion resistance. Alkaline Phosphatase (ALP) Activity will check early osteoblast differentiation, indicating accelerated bone formation. Thus, mechanical durability, corrosion resistance, and antibacterial efficacy make HEN-Ag coatings a viable option for dental implants that offer long-term stability in the oral environment.

3.2 *Orthopedic Implants*

Orthopedic implants are the most frequently used fracture fixation technique, and they are highly effective in facilitating bone growth and developmentas, of well as treating various bone diseases and defects. However, these commonly used orthopedic implants also experience multiple drawbacks and complications, such as bacterial adhesion, poor cell proliferation, and limited corrosion resistance [26]. Many significant disadvantages of currently available orthopedic implants stem from their osseointegration at the tissue-implant interface, which can lead to loosening due to immunological rejection, wear debris formation, low mechanical fixation, and implant-related infections. Nevertheless, incorporating nanomaterials as a coating for these implants offers exciting potential due to their outstanding tribological qualities, high resistance to wear, ability to sustain drug release, capacity for osseointegration, and capability to regenerate tissue, as shown in Fig. 6. Additionally, these materials can mimic the features and hierarchical structure of native bones [26] They also promote cell proliferation, reduce infection rates, and prevent biofilm formation, among other various functions. Nanomaterials play a crucial addrerole in scaffolds as they can support tissue structure, provide mechanical strength, and regulate biodegradability. Before discussing the design and formation of scaffolds using nanomaterials, let us outline the requirements for an optimal scaffold. It should meet the criteria of shape, volume, and mechanical strength in three dimensions; feature a porous surface essential for facilitating cell infiltration; minimize;inflammatory responses by ensuring biocompatibility, and ultimately, possess a precise degradation rate that aligns with the regenerative rate of the affected tissue.

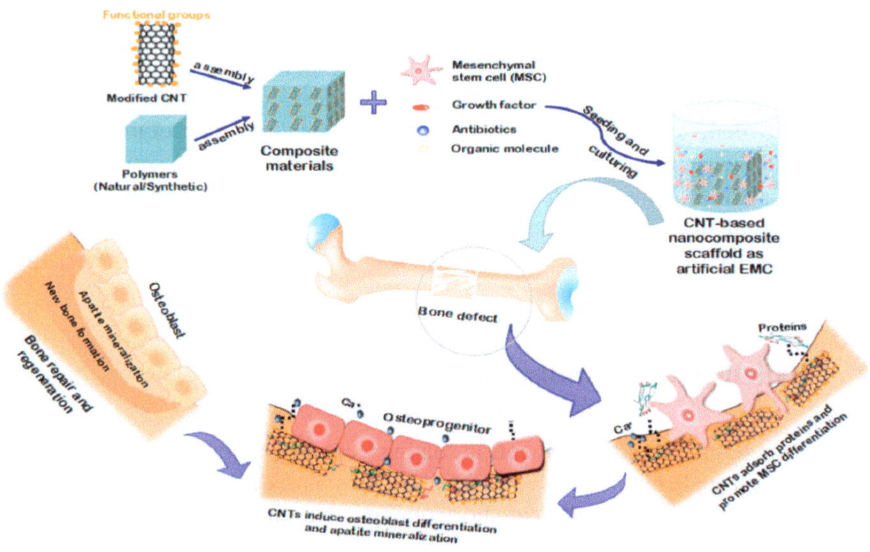

Fig. 6 Nanoparticles widely used in orthopedic implants as scaffold composites in BTE and regeneration [27]

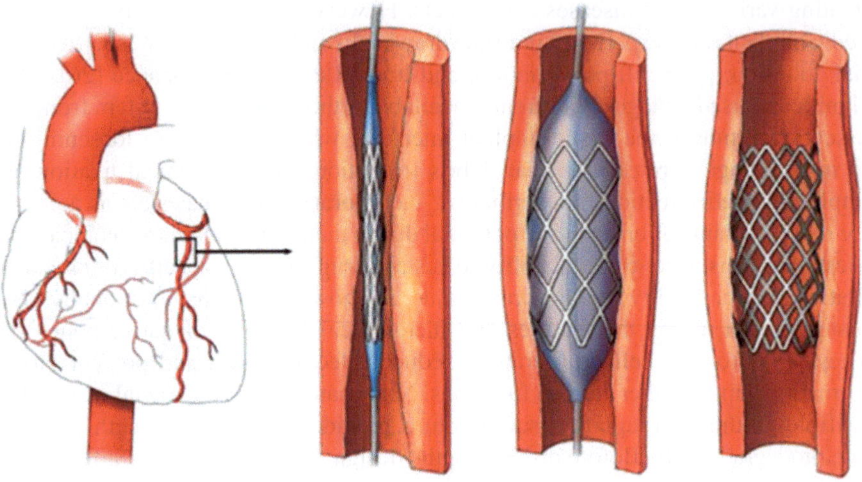

Fig. 7 Cardiovascular dilation by stent

3.3 *Cardiovascular Stents and Catheters*

Cardiovascular diseases (CVDs) are becoming the center of attention nowadays, being the leading cause of premature death and disability in humans worldwide. The most abundantly used treatment directs the insertion of a stent at the lesion site [28]. Their clinical success depends on different parameters, including the degradability of the polymers, drug release profiles, stent platforms, coating polymers, and the metals and their alloys employed as metal frames in the stents, as shown in Fig. 7. Thus, it is difficult to investigate and solve problems with the inflammatory response, sub-acute stent thrombosis, and corrosion resistance caused by these stents [29]. One of the proposed solutions to all of these problems is the use of nanoparticle coatings.

4 Challenges and Limitations

Despite the significant advancements in coating technologies, their practical adoption in biomedical applications faces substantial barriers. These include financial burdens, regulatory complexities, technical shortcomings, and environmental sustainability concerns. Addressing these challenges is pivotal for adopting innovative coatings in healthcare.

The cost of implementing advanced coating techniques is a critical limitation. Methods like Physical Vapor Deposition (PVD), Chemical Vapor Deposition (CVD), and Atomic Layer Deposition (ALD) involve costly materials and infrastructure:

- **Material Expenses**: High-performance materials such as hydroxyapatite (HAp), graphene, and carbon nanotubes (CNTs) are costly, as are specialized chemicals like silica precursors for Sol–Gel coatings.
- **Specialized Equipment**: Techniques like plasma spraying require high-vacuum chambers, plasma generators, and controlled environments, significantly increasing costs.
- **Training Requirements**: Operating these advanced systems necessitates skilled professionals, further increasing financial strain.

The biomedical industry is highly regulated, as coatings used on medical devices must meet stringent safety and efficacy standards. Agencies such as the **FDA (U.S.)** and **EMA (Europe)** require:

- extensive biocompatibility testing to assess and eliminate risks of cytotoxicity, genotoxicity, and hemocompatibility
- Proof of long-term stability and non-toxicity under physiological conditions.

These processes can take years, significantly delaying market availability. Compounding this is the lack of global standardization, which complicates cross-border product approvals. Several technical challenges remain unaddressed:

i. **Degradation in Biological Environments**: Coatings such as polymer-based films or bioactive ceramics degrade in physiological fluids, leading to corrosion, wear, or toxicity due to nanoparticle release.
ii. **Adhesion and Durability**: Coatings like titanium nitride on orthopedic implants may delaminate under mechanical stress, compromising device longevity.
iii. **Unpredictable Biodegradation Rates**: Biodegradable coatings (e.g., chitosan) degrade inconsistently across patients due to physiological variations.

The environmental impact of coating fabrication and disposal is often overlooked. Challenges include:

i. **Toxic By-Products**: Coating processes like CVD may involve hazardous chemicals, while improperly disposed nanomaterials can pollute ecosystems.
ii. **Energy-Intensive Production**: High energy demands for vacuum environments and plasma generation contribute to the carbon footprint.

Developing eco-friendly alternatives and biodegradable coatings is essential for sustainable adoption.

5 Future Trends and Opportunities

Innovative coating technologies are evolving rapidly, driven by nanotechnology, AI, and sustainability-focused research advancements. Innovative coatings are transformative innovations that respond dynamically to environmental stimuli such as pH, temperature, or mechanical forces. Applications include:

i. **pH-Responsive Coatings**: These coatings are used in cancer therapy and release chemotherapeutic agents selectively in acidic tumor microenvironments.
ii. **Temperature-Sensitive Polymers**: These coatings release drugs during localized inflammation.
iii. **Self-Healing Coatings**: These coatings, specifically polymeric, are repaired when scratched or worn, ensuring long-term stability.

AI and Machine Learning (ML) are revolutionizing coating (Fig. 8) design by:

- Predicting long-term performance under different environmental conditions.
- Optimizing material composition to balance durability and biocompatibility.
- Automating production processes, reducing costs.

Sustainability is becoming a key focus in biomedical coatings. Researchers are exploring biodegradable materials such as:

- **Chitosan**: Used in wound dressings for antimicrobial protection.
- **Silk Fibroin**: Ideal for corneal implants, offering transparency and biocompatibility.
- **Natural Polymers**: Reduce environmental impact and are metabolized naturally by the body.

6 Conclusion and Future Prospects

The intersection of nanotechnology, biomaterials, and AI is driving groundbreaking advancements in coatings, paving the way for smarter, safer, and more sustainable solutions. Recent innovations include graphene-based coatings that enhance signal transmission in neural implants and superhydrophobic coatings that prevent biofilm formation, crucial for catheters. Bioresorbable materials such as magnesium stents naturally dissolve, eliminating the need for surgical removal. Meanwhile, coatings inspired by lotus leaves prevent protein adhesion and enhance the longevity of implants. Smart coatings hold immense market potential in emerging regions like Asia and Latin America, fueled by improving healthcare infrastructure and the rise of chronic diseases. It is essential to address ethical concerns regarding affordability and environmental safety. Future research should also focus on wearable medical devices, biosensors, and coatings for regenerative medicine.

Fig. 8 Workflow of AI-driven coating development

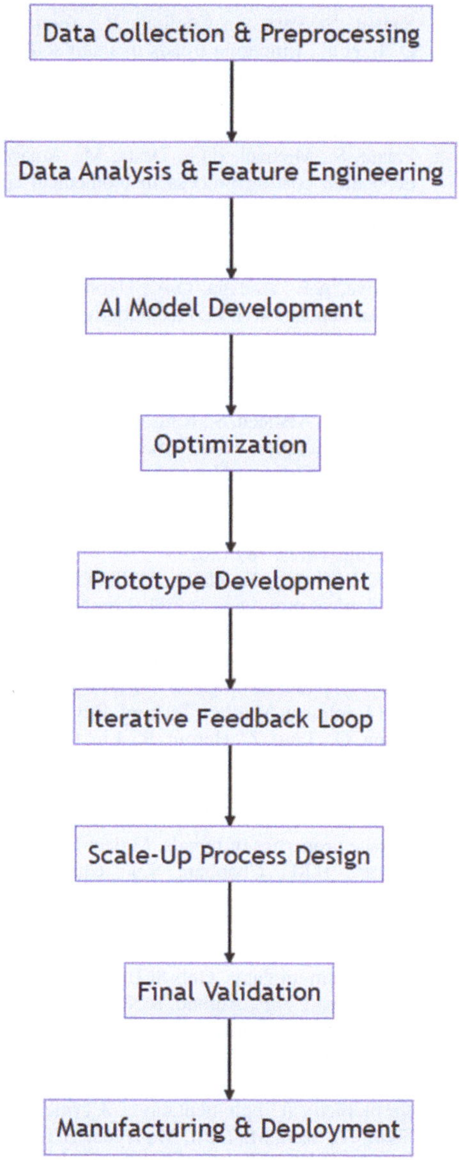

References

1. Saba, T., Saad, K.S.K., Rashid, A.B.: Precise surface engineering: leveraging chemical vapor deposition for enhanced biocompatibility and durability in biomedical implants. Heliyon **10**(18) (2024)
2. Zhao, Y., Bai, J., Xue, F., Zeng, R., Wang, G., Chu, P.K., Chu, C.: Smart self-healing coatings on biomedical magnesium alloys: a review. Smart Mater. Manuf. **1**, 100022 (2023)

3. Sanyal, S., Park, S., Chelliah, R., Yeon, S.J., Barathikannan, K., Vijayalakshmi, S., Oh, D.H., et al.: Emerging trends in smart self-healing coatings: a focus on micro/nanocontainer technologies for enhanced corrosion protection. Coatings **14**(3), 324 (2024)
4. Murali, A., Lokhande, G., Deo, K.A., Brokesh, A., Gaharwar, A.K.: Emerging 2D nanomaterials for biomedical applications. Mater. Today **50**, 276–302 (2021)
5. Palani, S., Michael, E.G., Desta, M., Atnaw, S.M., Banoth, R., Kolanji, S.: Physical vapor deposition coating process in biomedical applications: an overview. Sustain. Adv. Manuf. Mater. Process. 67–93 (2022)
6. Saad, K.S.K., Saba, T., Rashid, A.B.: Application of PVD coatings in medical implantology for enhanced performance, biocompatibility, and quality of life. Heliyon **10**(16) (2024)
7. Adeoye, A.E., Adeaga, O.A., Ukoba, K.: Chemical vapour deposition (CVD) and physical vapour deposition (PVD) techniques: advances in thin film solar cells. Niger. J. Technol. **43**(3), 479–489 (2024)
8. Adesina, A.Y., Ogunlakin, N.: Physical and chemical vapour deposition coatings. In: Advances in Corrosion Control of Magnesium and its Alloys, pp. 305–324. CRC Press (2023)
9. Awan, T.I., Afsheen, S., Kausar, S.: Chemical vapor deposition technique. In: Thin Film Deposition Techniques: Thin Film Deposition Techniques and Its Applications in Different Fields, pp. 65–96. Springer Nature Singapore, Singapore (2025)
10. Heimann, R.B.: Plasma-sprayed osteoconductive hydroxylapatite coatings for endoprosthetic hip implants: phase composition, microstructure, properties, and biomedical functions. Coatings **14**(7), 787 (2024)
11. Gao, R., Huang, Y., Zhang, J., Jin, G., Ma, G., Xing, Z., Chen, S., et al.: Microstructure and tribological properties of nickel-based coatings by plasma enhanced high-speed arc spraying: non-transfer arc, transfer arc modes. Surf. Coat. Technol. 131823 (2025)
12. Alontseva, D., Yantsen, Y., Voinarovych, S., Obrosov, A.: Microplasma-sprayed titanium and hydroxyapatite coatings on Ti6Al4V Alloy: in vitro biocompatibility and corrosion resistance: Part I. Johnson Matthey Technol. Rev. **69**(1), 45–58 (2025)
13. Wang, Q., Tang, Z., Herout, R., Liu, C., Yu, K., Lange, D., Wang, R., et al.: Axial suspension plasma sprayed Ag–TiO_2 coating for enhanced photocatalytic and antimicrobial properties. Surf. Interfaces **45**, 103856 (2024)
14. Benčina, M., Resnik, M., Starič, P., Junkar, I.: Use of plasma technologies for antibacterial surface properties of metals. Molecules **26**(5), 1418 (2021)
15. Murari, B.M., Chauhan, S.S.: Characterisation of TEOS-based sol–gel-hydroxyapatite composite coating on titanium implants. Biomed. Mater. Devices 1–8 (2024)
16. Fernández-Hernán, J.P., Torres, B., López, A.J., Rams, J.: The role of the sol-gel synthesis process in the biomedical field and its use to enhance the performance of bioabsorbable magnesium implants. Gels **8**(7), 426 (2022)
17. Mohammad, N.F., Ahmad, R.N., Rosli, N.M., Manan, M.A., Marzuki, M., Wahi, A.: Sol–gel deposited hydroxyapatite-based coating technique on porous titanium niobium for biomedical applications: a mini-review. Mater. Today Proc. **41**, 127–135 (2021)
18. Catauro, M., Ciprioti, S.V.: Characterization of hybrid materials prepared by sol-gel method for biomedical implementations. A critical review. Materials **14**(7), 1788 (2021)
19. Tahir, H., Saad, M., Latif, M., Hyder, S.T., Ahmed, R.: Synthesis of metal oxide nano particles by sol–gel method and investigation of its biomedical applications. Nanochem. Res. **9**(1), 19–27 (2024)
20. Priyadarshini, B., Rama, M., Chetan, Vijayalakshmi, U.: Bioactive coating as a surface modification technique for biocompatible metallic implants: a review. J. Asian Ceram. Soc. **7**(4), 397–406 (2019)
21. Udoh, I.I., Shi, H., Daniel, E.F., Li, J., Gu, S., Liu, F., Han, E.H.: Active anticorrosion and self-healing coatings: a review with focus on multi-action smart coating strategies. J. Mater. Sci. Technol. **116**, 224–237 (2022)
22. Ni, X., Gao, Y., Zhang, X., Lei, Y., Sun, G., You, B.: An eco-friendly smart self-healing coating with NIR and pH dual-responsive superhydrophobic properties based on biomimetic stimuli-responsive mesoporous polydopamine microspheres. Chem. Eng. J. **406**, 126725 (2021)

23. Anjum, M.J., Yasin, G., Tabish, M., Malik, M.: A review on self-healing coatings applied to Mg alloys and their electrochemical evaluation techniques. Int. J. Electrochem. Sci. **15**, 3040–3053 (2020)
24. Jiang, P., et al.: Advanced surface engineering of titanium materials for biomedical applications: from static modification to dynamic responsive regulation. Bioact. Mater. Italy (2023)
25. Zhao, Y.: Corrigendum to "(TiZrNbTaMo)N nanocomposite coatings embedded with silver nanoparticles: imparting mechanical, osteogenic and antibacterial traits to dental implants". J. Alloy. Compd. Australia **972** (2024)
26. Gautam, S.B.: Advancements in nanomaterials. Biomed. Eng. Adv. Zhejiang **3**, 100029 (2022)
27. Pei, B., Wang, W., Dunne, N., Li, X.: Applications of carbon nanotubes in bone tissue regeneration and engineering: superiority, concerns, current advancements, and prospects. Nanomaterials **9**(10), 1501 (2019)
28. Mutuk, T., et al.: Surface modification of Ni–Ti stents by biodegradable binary PVA/propolis electrospun nano fibers. Arab. J. Sci. Eng. Samsun **48**, 3391–3402 (2023)
29. Zhang, B., et al.: A polyphenol-network-mediated coating modulates inflammation and vascular healing on vascular stents. Yunbing Wang ACS Nano **16**(4), 6585–6597 (2022)

Thermal Properties in Polymer-Nanoparticle Composites: Applications in Therapeutic and Biomedical Engineering

Soufiene Ilahi and Dhekra Loubiri

Abstract Enhancing the thermal properties of polymers presents a significant problem across multiple applications. This chapter explores the impact of ZnSe and CdSe nanoparticles on the thermal conductivity and diffusivity of Poly(3-hexylthiophene) (P3HT) thin films. Using the photothermal deflection technique, PTD called the "Mirage effect," thermal conductivity and diffusivity were assessed. As nanoparticle concentration in CdSe-based composites increased, thermal conductivity rose from 1.78 to 3.18 W/m·K, while thermal diffusivity decreased from 3.52 to 0.23 m^2/s. A poly (3-hexylthiophene) with 40% CdSe nanoparticles (NPs) concentration produced the highest conductivity and lowest diffusivity. Similarly, ZnSe-doped P3HT films exhibited a redshift in sub-bandgap energy and decreased thermal properties at higher nanoparticle concentrations. These variations in opto-thermal parameters are attributed to nanoparticle concentration and their interaction with the polymer matrix. The findings emphasize the role of nanoparticle concentration in tailoring polymer nanocomposite properties, offering the potential for advanced therapeutic technologies, including pain management and controlled heat therapy for biomedical applications.

Keywords P3HT · Nanoparticles · Thermal properties · Biomedical · Photothermal deflection technique PTD

1 Introduction

In the 1970s, Heeger, Mac Diarmid, and Shirakawa, whose work earned them the Nobel Prize in Chemistry in 2000 as a recognition of their pioneering contributions [1], first discovered the electrical conductivity properties of polymers. During this period, significant research also focused on π-conjugated polymers, particularly in

S. Ilahi (✉) · D. Loubiri
Pysique, MathématiqueModélisation Quantique, IPEIN, Carthage University, Carthage, Tunisia
e-mail: soufiene.ilehi@fsm.rnu.tn

Physics Department, Monastir Faculty of Sciences, Monastir University, Monastir, Tunisia

S. N. Saud Al-Humairi et al. (eds.), *Smart Nanomaterials in Biomedical Applications*, Engineering Materials, https://doi.org/10.1007/978-981-96-8023-8_7

plastic electronics, where their unique properties were applied. These materials offer several benefits over their inorganic counterparts, such as the ability to precisely control their properties through functionalization, ease of processing in liquid form, the possibility of fabricating devices on flexible substrates, and lower manufacturing costs. In their undoped state, π-conjugated oligo- or polymers are organic semiconductors, with their optical (absorption and photoluminescence) and electronic properties primarily influenced by their chemical structure, precisely the conjugation length of the carbon chain. In contrast, semiconductor nanocrystals (NCs) are nanoscale objects ranging from 2 to 10 nm, consisting of hundreds to thousands of atoms arranged in a crystalline pattern. Nanomaterials have become a rapidly advancing area of research, driven by substantial progress in the tools available for synthesizing and characterizing materials at the nanometer scale. Over later decades, these materials have expanded consideration in crucial inquiries about mechanical applications. Warm conductivity is an inborn property of materials that decides their capacity to exchange warm.

This property is affected by the material's nuclear or atomic structure. Warm is effectively conducted through free electrons and grid vibrations in materials with high conductivity, such as metals [2]. In differentiation, materials with low conductivity, such as polymers, conduct warm less successfully, with values regularly extending from 0.1 to 0.5 W/m·K [3]. Three main methods are used to improve the thermal conductivity of polymers: the first involves stretching or aligning polymer chains through techniques like electrospinning, mechanical stretching, or external fields [4]. The second focuses on modifying intermolecular and intramolecular interactions, while the third method enhances crystallinity and strengthens intramolecular bonds in bulk polymers [5]. These methods differ in their approach, with the first focusing on uniaxial or 2D alignment, while the second and third target 3D structural improvements [6].

Polymer composites are widely utilized in medical applications, such as drug delivery systems and wearable devices, due to their flexibility, lightweight nature, ease of fabrication, and low cost. Their relatively low thermal conductivity is advantageous for reducing heat buildup, which is crucial for the safe and efficient operation of medical devices [7]. These composites are also utilized to create practical thermal management components, such as cooling plates and conductive cushions, which help regulate the temperature of medical equipment, ultimately improving device reliability and lifespan and ensuring patient safety [8]. One such polymer, Poly(3-hexylthiophene) (P3HT), is a synthetic organic semiconductor composed of repeating 3-hexylthiophene units. P3HT is a promising material for various applications, including medical use, due to its biocompatibility. It has shown potential in areas such as drug delivery and tissue engineering, where materials must be safe for biological systems without causing adverse effects. The versatility and biocompatibility of P3HT-based materials position them as a significant contributor to improving human health and well-being in biomedical fields. This chapter emphasizes the role of nanoparticle concentration (CdSe and ZnSe) in the thermal properties of P3HT [9–11].

2 Theory of Photothermal Deflection Technique PTD

The photothermal deflection, also known as "mirage detection," utilizes two beams: one that creates a disturbance at the surface of the sample (referred to as the pump beam) and a second (referred to as the probe beam). The probe beam travels through the gradual variation in the refractive index near the material's heated surface. By analyzing the deflection of this beam, it is possible to assess the thermal characteristics of the sample [12, 13]. This technique is employed in various fields, such as spectroscopy [14] and in the thermal and optical characterization of solids and liquids. This method is highly sensitive and easy to interpret, but requires a flat sample surface. Two types of configurations are distinguished: the first, in which the pump and probe beams are parallel, is termed the collinear configuration; the second, where the two beams are perpendicular, is known as the perpendicular configuration. By utilizing this technique, a modulated laser beam evenly illuminates the sample, creating a thermal gradient that travels through the material and the surrounding medium. A probe laser beam passing close to the material surface is redirected due to the variation in the fluid's refractive index distribution induced by the heat wave. The heat equation for the different media involved is solved to determine the probe beam's deflection. In addition to the sample's optical characteristics, this deflection also depends on the substrate, the surrounding fluid, and the sample's temperature attributes. The relationship between the sample's optical and thermal characteristics and the deflection has been covered in earlier research [15].

$$\Psi = \frac{L}{n}\frac{dn}{dT}\sigma_f T_0 \exp(-\sigma_f x_0) \tag{1}$$

The temperature profile is obtained by computing the material's one-dimensional heat equation, supporting substrate, and surrounding medium. The heat equations for each medium are shown in Fig. 1.

$$\frac{\partial^2 T_f}{\partial x^2} - \sigma_f^2 T_f = 0 \quad h \le x \le h + h_f \tag{2}$$

$$\frac{\partial^2 T_c}{\partial x^2} - \sigma_c^2 T_c = A_c \quad 0 \le x \le h \tag{3}$$

$$\frac{\partial^2 T_s}{\partial x^2} - \sigma_s^2 T_s = A_s \quad -l \le x \le 0 \tag{4}$$

$$\frac{\partial^2 T_b}{\partial x^2} - \sigma_b^2 T_b = 0 \quad -l_b - l \le x \le -l \tag{5}$$

$\sigma_f = (1 + j)/\mu_f$, where μ_f is the thermal diffusion length in the fluid given by $\mu_f = (D/F)^{1/2}$. Here, D stands for thermal diffusivity, and F is the modulation frequency of the heating beam. Similarly, $\sigma_i = (1 + j)/\mu_i$ and $\mu_i = (D_i/F)^{1/2}$ corresponds to

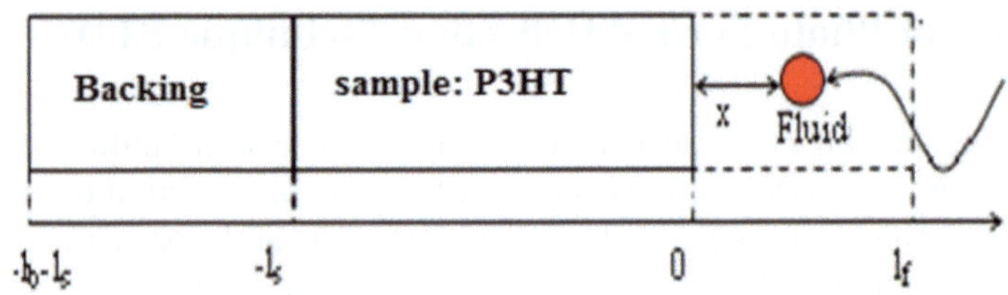

Fig. 1 Schematic of different media

the i-medium number, $A_c = \frac{-dI_c}{2K_c dx}$, $A_s = \frac{-dI_s}{2K_s dx}$. The quantities Ai represent the thermal sources generated by the absorption of light within the material, where $I_c(x) = I_0 \exp[-\alpha_c(h - x)]$ in the layer and $I_s(x) = I_0 \exp(\alpha_s x) \exp(-\alpha_c x)$ in the substrate.

The solutions to the heat equation with periodic conditions can be expressed as:

$$T_f = T_0 \exp(-\sigma_f (x - h)) \quad h \le x \le h + h_f \tag{8}$$

$$T_c = U_1 \exp(-\sigma_c(h - x) + V_1 \exp(\sigma_c(h - x)) - Z \exp(-\alpha_c(h - x)) : 0 \le x \le h \tag{9}$$

$$T_s = U \exp(\sigma_s x) + V \exp(-\sigma_s x) - E \exp(\alpha_s x) \quad -l \le x \le 0 \tag{10}$$

$$T_b = W \exp(\sigma_b(x + l)) \quad -l_b - l \le x \le -l \tag{11}$$

T_0 denotes the temperature at the surface of the material (x = h), whereas E and Z are the specific solutions to Eqs. (3) and (4), respectively. The constants U1, V1, and W are determined by applying the boundary conditions related to temperature and heat flux across the different interfaces.

$$T_f(h) = T_s(h) \tag{12}$$

$$T_c(0) = T_s(0) \tag{13}$$

$$T_s(-l) = T_b(-l) \tag{14}$$

$$K_f \frac{\partial T_f}{\partial x}(h) = K_c \frac{\partial T_c}{\partial x}(h) \tag{15}$$

$$K_c \frac{\partial T_c}{\partial x}(0) = K_s \frac{\partial T_s}{\partial x}(0) \tag{16}$$

$$K_s \frac{\partial T_s}{\partial x}(-l) = K_b \frac{\partial T_b}{\partial x}(-l) \tag{17}$$

In practice, the thickness of the layer h is shorter than the thermal diffusion length (h <<< μc). Therefore, we can make the approximation, leading to a simplified analytical expression for T_0, as outlined in Ilahi et al. [16].

$$T_0 = \frac{\left\{\begin{array}{l} E\big[(1-r_s)(b+1)\exp(\sigma_s l) - (1-b)(1+r_s)\exp(-\sigma_s l) - 2(b-r_s)\exp(-\alpha_s l)\big] \\ +Z\big[(1-r_c)(b+1)\exp(\sigma_s l) - (1-b)(1+r_s)\exp(-\sigma_s l)\big]\big[1-\exp(-\alpha_c h)\big] \end{array}\right\}}{\big[(1-g)(1-b)\exp(-\sigma_s l) + (1+g)(1+b)\exp(\sigma_s l)\big]}$$

$E = \frac{\alpha_s I_0 \exp(-\alpha_c h)}{2K_s(\alpha_s^2-\sigma_s^2)}$, $Z = \frac{\alpha_c I_0}{2K_c(\alpha_c^2-\sigma_c^2)}$, $g = \frac{K_f\sigma_f}{K_s\sigma_s}$, $b = \frac{K_b\sigma_b}{K_s\sigma_s}$, $r_s = \frac{\alpha_s}{\sigma_s}$ and $r_c = \frac{\alpha_c K_c}{\sigma_s K_s}$.
The deflection is given by:

$$\Psi = \frac{L}{n}\frac{dn}{dT}\sigma_f T_0 \exp(-\sigma_f x_0) \tag{19}$$

Let n be the refractive index of the fluid, x_0 the separation between the probe beam and the sample's surface, and L the length of the sample.

The principle and experimental setup of the Mirage effect technique are detailed elsewhere [17, 18]. A monochromatic light beam, manipulated by a mechanical chopper, is directed onto the sample's surface. Both the material and the surrounding air absorb the incident light, generating a heat wave that causes the probe laser beam (a He–Ne laser with a 632 nm wavelength) to bend as it moves across the surface of the sample. A lock-in amplifier (EG&G 5210) measures the electrical signal produced after a photodetector detects this deflection. As the modulation frequency F changes, the amplitudes and phases of the photothermal signal are noted.

Understanding the thermal properties of nanocomposite thin films is essential, as the thermal properties of polymers are susceptible to the arrangement of their molecules. One key parameter, thermal diffusivity, defines how quickly heat spreads through the material. The inverse of this value indicates the time it takes for a system to reach thermal equilibrium following a transient temperature change. However, research on the thermal properties of these materials and the effects of nanoparticles remains relatively scarce. The fitting procedure used for the simulations involves a multiparameter approach, implemented through the least squares method. To ensure accuracy and uniqueness in the extracted parameters, the root mean square (RMS) value of the deviation is computed, as detailed in previous studies [2, 19–21]. The method for calculating the RMS variance using the PTD technique is also discussed in other literature [22].

3 CdSe Nanoparticles' Impact on P3HT's Thermal Characteristics

Experimental and theoretical variations in the amplitude and phase of the photothermal deflection (PTD) signal, corresponding to different thermal diffusivity values for a thin P3HT layer, are shown elsewhere [9]. The thermal conductivity and diffusivity can be determined by contrasting the theoretical predictions with the experimental results.

To highlight the observed variations, we plotted the thermal conductivity and diffusivity against the concentration of CdSe nanoparticles in P3HT, as shown in Fig. 2. The thermal conductivity increased to 3.18 W/m·K for P3HT with 40% TBPO-capped CdSe, but decreased to 2.21 W/m·K when the CdSe concentration was raised to 60%. On the other hand, diffusivity initially decreased, reaching 0.23×10^{-7} W/m·K for P3HT with 40% CdSe, but showed a slight increase at the 60% CdSe concentration. These observed changes in thermal conductivity and diffusivity are consistent with the expected trends.

Optical microscopy images of both pure P3HT and the nanocomposite films reveal the uneven distribution of nanocrystals within the P3HT matrix [11]. In the sample containing 10% CdSe, no aggregation of the nanocrystals is observed. However, as the CdSe concentration increases, a noticeable tendency for aggregation emerges, likely due to electrostatic interactions between nanoparticles of varying sizes. These aggregates can grow to several micrometers in size, forming larger nanocrystal regions, where organic ligands are unable to fully cover the nanoparticle surfaces during synthesis, which leads to reduced roughness.

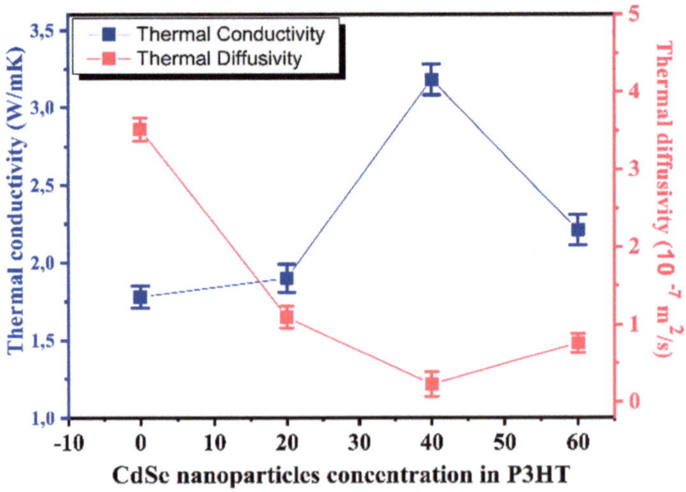

Fig. 2 Role of CdSe nanoparticle concentration in modifying the thermal conductivity and diffusivity of P3HT [9]

Based on this, it can be inferred that a nanoparticle concentration of around 40% results in the most uniform dispersion, which corresponds well with the thermal conductivity results. As is well known, thermal conductivity is governed by phonon behavior and scattering processes within a material. In crystals with an identical structure, a higher density often correlates with a reduced number of defects, leading to improved thermal conductivity. On the other hand, thermal diffusivity is the ratio between a material's capacity for heat conduction and its capacity to store it, and it typically decreases with increasing density. These observations have been confirmed by Feng et al. [23] and Feng and Wang [24].

An optical study [10, 11] also showed similar results for the same samples. We observed that incorporating higher concentrations of CdSe-TBPO improved optical absorption by up to 40% TBPO-capped CdSe. However, the absorption decreased for the 60% CdSe-TBPO sample compared to the others. Several factors could explain the observed improvements in thermal parameters. Different interfacial interactions can influence heat transport in nanocomposites. Specifically, functionalizing CdSe nanoparticles with the TBPO capping agent may help minimize the thermal resistance at the interfaces, thereby improving thermal conductivity and diffusivity. Functionalized nanoparticles can form covalent bonds with the surrounding structure. This hypothesis is supported by experimental findings from Hu et al. [25].

On the other hand, any morphological imperfections, such as dislocations or grain boundaries, hinder the propagation of thermal energy, leading to a decrease in thermal conductivity [26]. More specifically, thermal resistance is influenced by the mean free path between collisions, which is typically caused by carrier collisions. When a material's characteristic dimensions are shorter than an average free path, the interface plays a more dominant role than the carriers' interactions—potentially resulting in a substantial reduction in thermal conductivity. Moreover, the formation of aggregates and the increase in surface roughness at higher concentrations of nanoparticles may contribute to these effects.

Optical microscopy images of comparable samples reveal that as the concentration of TBPO-capped CdSe increases to 60%, nanoparticle aggregation intensifies, and the bonding between the nanoparticles and P3HT weakens [10]. Thus, the ideal interaction and connectivity between P3HT and TBPO-capped CdSe nanoparticles is likely achieved at around 40% CdSe concentration [10]. These observations align with the findings of Chen [27], who attributed the reduction in thermal conductivity to the development of amorphous areas around crystalline domains in highly ordered P3HT. These areas, which exhibit elevated thermal impedance, contribute to the observed decline in thermal conductivity.

4 ZnSe Nanoparticles' Impact on P3HT's Thermal Characteristics

There is limited information on the thermal properties of P3HT: ZnSe hybrid structures [11]. Although the thermal properties of each component (organic and inorganic) present in the mixture are well-known, it remains challenging to estimate the properties of the composite as a whole. The thermal properties of P3HT: ZnSe nanocomposites were evaluated in the same manner described in the first section of this chapter. Indeed, Figs. 3 and 4 illustrate the experimental and theoretical analysis of the photothermal signal in relation to the modulation frequency for P3HT with different concentrations of TBPO-capped ZnSe, respectively.

The relationship between thermal diffusivity D and thermal conductivity K can be expressed as $\frac{K}{\rho C_p}$, where ρ is the density and Cp is the specific heat capacity. Using this equation, we can calculate the nanomaterial's density and heat capacity, presuming the mixture's volume remains constant for various nanoparticle addition rates. The diffusivity (D), conductivity (K), density (ρ), and specific heat capacity (Cp) values for P3HT composites with varying concentrations of TBPO-capped CdSe nanoparticles. The table shows that as the CdSe concentration increases, the values

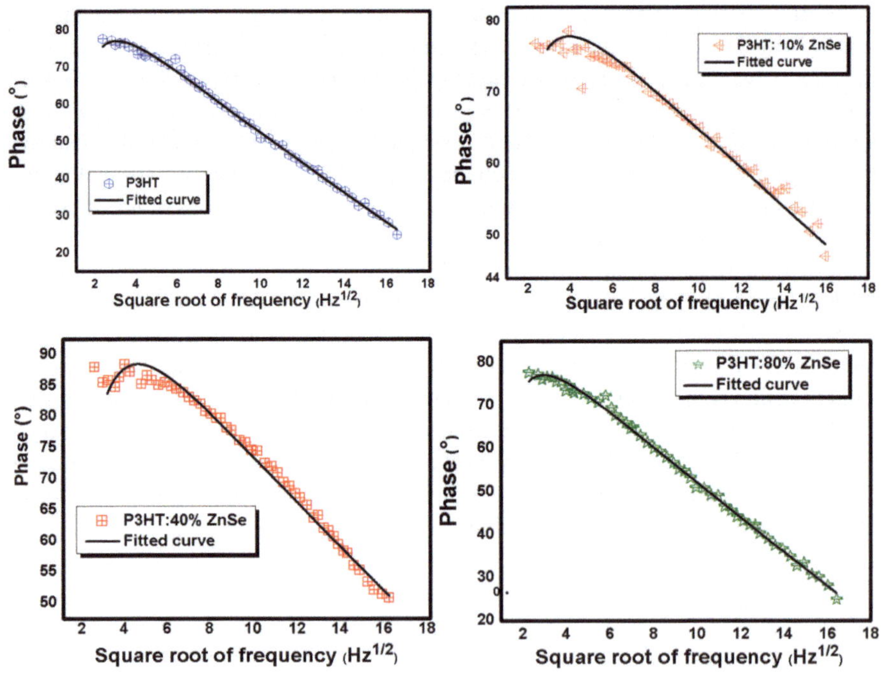

Fig. 3 Analysis of photothermal phase as a function of square root frequency for P3HT with varying TBPO-Capped ZnSe concentrations [11]

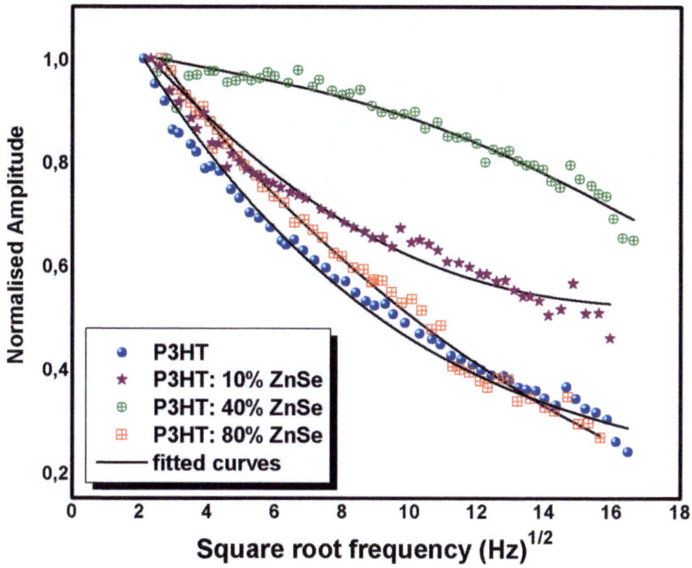

Fig. 4 Analysis of normalized photothermal amplitude as a function of square root frequency for P3HT with various TBPO-Capped ZnSe concentrations [11]

of these thermal properties change accordingly, reflecting the impact of nanoparticle incorporation on the composite's thermal behavior.

For pure P3HT, the thermal diffusivity D is 3.522 × 10^{-7} m^2/s, K is 1.782 W/m·K, the density ρ is 1.33 g/cm^3, and the specific heat capacity Cp is 0.3807 × 10^4 J/kg·K. For the P3HT composite with 10% ZnSe, the thermal diffusivity decreases to 0.987 × 10^{-7} m^2/s, and the thermal conductivity drops to 1.043 W/m·K. The density increases to 1.46 g/cm^3, while the specific heat capacity rises to 0.7243 × 10^4 J/kg·K. At 40% ZnSe concentration, the thermal diffusivity decreases further to 0.179 × 10^{-7} m^2/s, with the thermal conductivity dropping significantly to 0.250 W/m·K. The density increases to 1.86 g/cm^3, and the specific heat capacity is 0.7575 × 10^4 J/kg·K. For the P3HT composite with 80% ZnSe, the thermal diffusivity rises to 0.249 × 10^{-7} m^2/s, the thermal conductivity enhances to 1.964 W/m·K, and the density is 2.39 g/cm^3. The specific heat capacity increases substantially to 3.3008 × 10^4 J/kg·K. These data reflect the changes in the thermal properties of P3HT as the ZnSe content increases in the nanocomposite. It is observed that pure P3HT exhibits higher thermal diffusivity and thermal conductivity than the nanocomposites. In the initial two mass ratios (10 and 40 wt.%), incorporating ZnSe nanoparticles into the polymer matrix decreased these thermal properties. This behavior can be attributed to an interfacial thermal contact resistance between the different phases of the nanocomposite, as well as their incompatible thermal expansion [28]. The presence of such thermal barriers leads to a reduction in the thermal properties of the nanocomposite. However, our hybrid films' thermal properties are lower than those of pure polymer, which is likely related to the topology of the mixture, making interpreting the experimental results more challenging.

In contrast, the nanocomposite with an 80% ZnSe content shows thermal conductivity close to that of pure P3HT. The crystallinity of the mixture may play a significant role in thermal transport, particularly for high nanoparticle concentrations. Moreover, density and specific heat increase is observed as the ZnSe content increases. However, thermal resistances primarily arise from collisions between charge carriers and are characterized by the mean free path between these collisions. When a characteristic dimension of the material is reduced to less than the average free path, the influence of interactions at the interfaces surpasses that of the interactions between the carriers, leading to a significant decrease in thermal conductivity. More generally, any morphological disorder, such as dislocations, precipitates, or grain boundaries [26], impedes the propagation of elastic waves and reduces thermal conductivity.

5 Conclusion

In conclusion, the generation of substantial heat during the operation of medical devices underscores the need for effective cooling solutions to ensure both safety and optimal functionality. Thermal conductivity is critical in determining how well a material manages heat. Thus, the careful selection of materials is essential for maximizing the performance and reliability of medical devices. This is particularly true for biomedical devices, where efficient thermal management is crucial to maintain optimal operating conditions and safeguard patient safety. In this regard, the thermal characterization of P3HT with various concentrations of CdSe-TBPO nanomaterial revealed a moderate increase in thermal conductivity with higher CdSe nanoparticle concentrations. This improvement is attributed to the enhanced interfaces between the nanoparticles and the polymer and interactions between the nanoparticles themselves, which facilitate better heat transfer within the active layer.

In contrast, the thermal properties of P3HT with various concentrations of ZnSe nanocomposites showed only a slight modification compared to the pure polymer. This could be due to the mixture's topology, which complicates the interpretation of the experimental results, as well as the crystallinity of the composite, which significantly impacts thermal transport—especially at higher nanoparticle concentrations. These findings emphasize the importance of selecting materials with optimized thermal conductivity to enhance heat dissipation and improve the overall performance of biomedical devices, ultimately contributing to their safety and efficiency in medical applications.

References

1. MacDiarmid, A.G.: "Synthetic metals": a novel role for organic polymers (Nobel lecture). Angew. Chem. Int. Ed.. Chem. Int. Ed. **40**, 2581 (2001)
2. Jaiswal, S., Jaiswal, A.: Thermal conductivity of materials: implications for heat management in medical devices. ShodhKosh J. Vis. Perform. Arts **5**(7), 131–138 (2024)
3. Choy, C.L.: Thermal conductivity of polymers. Polymer **18**, 984 (1977)
4. Choy, C.L., Luk, W.H., Chen, F.C.: Thermal conductivity of highly oriented polyethylene. Polymer **19**, 155–162 (1978)
5. Liu, J., Yang, R.: Tuning the thermal conductivity of polymers with mechanical strains. Phys. Rev. B **81**, 174122 (2010)
6. Meng, X., Yu, H., Wang, L., Wu, X., Amin, B.U.: Recent progress on fabrication and performance of polymer composites with thermal conductivity. Macromol. Mater. Eng. **306**, 2100434 (2021)
7. Zhang, C., Li, Y., Kang, W., Liu, X., Wang, Q.: Current advances and future perspectives of additive manufacturing for functional polymeric materials and devices. SusMat **1**, 127–147 (2021)
8. Bahru, R., Hamzah, A.A., Mohamed, M.A.: Thermal management of wearable and implantable electronic healthcare devices: perspective and measurement approach. Int. J. Energy Res. **45**, 1517–1534 (2021)
9. Loubiri, D., Ilahi, S.: Effects of TBPO-capped CdSe nanoparticles concentration on thermal conductivity and diffusivity in Poly (3-hexylthiophene) thin films. Appl. Phys. A **130**, 723 (2024)
10. Loubiri, D., Hamed, Z.B., Ilahi, S., Sanhoury, M.A., Kouki, F., Yacoubi, N.: Effect of TBPO-capped CdSe nanoparticles concentration on sub-bandgap absorption in poly (3-hexylthiophene) thin films studied by photothermal deflection spectroscopy. Synth. Met. **206**, 1–7 (2015)
11. Loubiri, D., Ben Hamed, Z., Ilahi, S., Sanhoury, M.A., Kouki, F., Yacoubi, N.: Photothermal investigation of poly (3-hexylthiophene): ZnSe nanocomposites. Appl. Phys. A **119**, 581–588 (2015)
12. Ilahi, S., Yacoubi, N., Genty, F.: Thermal annealing effects on AlGaAsSb/GaSb laser structure: bandgap energy blueshift and thermal conductivity enhancement. Opt. Mater. **69**, 226–229 (2017)
13. Ilahi, S., Yacoubi, N., Genty, F.: Photothermal deflection technique investigation of thermal annealing effects of AlGaAsSb/GaSb laser structure: non-radiative recombination parameters enhancement. Mater. Res. Bull. **106**, 332–336 (2018)
14. Ilahi, S., Saidi, F., Hamila, R., Yacoubi, N., Maaref, H., Auvray, L.: Shift of the gap energy and thermal conductivity in BGaAs/GaAs alloys. Phys. B **421**, 105–109 (2013)
15. Ilahi, S., Khalfaoui, A., Genty, F., Yacoubi, N.: Significant improvement of thermal conductivity and bandgap energy blue shift in annealed GaInAsSb/GaSb. Opt. Laser Technol. **137**, 106819 (2021)
16. Ilahi, S., Saadalah, F., Yacoubi, N.: Photothermal deflection investigation of bulk Si and GaSb transport properties. Appl. Phys. A **110**, 459–464 (2013)
17. Khalfaoui, A., Ilahi, S., Abdel-Rahman, M., Zia, M.F., Alduraibi, M., Ilahi, B.: Photothermal deflection technique investigation of annealing temperature and time effects on optical and thermal conductivity of V/V_2O_5 alternating layers structure. Phys. B Condens. Matter **522**, 26–30 (2017)
18. Ilahi, S., Yacoubi, N.: Interface recombination velocity measurement in SiO_2/Si. Semiconductors **48**(3), 302–306 (2014)
19. Ilahi, S., Cornet, C., Yacoubi, N., Durand, O.: Improvement of carriers diffusion length and mobility in annealed GaAsPN materials for intermediate band solar cells. Sol. Energy Mater. Sol. Cells **215**, 110622 (2020)
20. Ilahi, S.: Doping effects on minority carrier parameters in bulk GaAs. Phys. B **652**, 414612 (2023)

21. Ilahi, S., Baira, M., Saidi, F., Yacoubi, N., Auvray, L., Maaref, H.: Non-radiative recombination process in BGaAs/GaAs alloys: two layer photothermal deflection model. J. Alloy. Compd. **581**, 358–362 (2013)
22. Bouagila, S., Ilahi, S., Baira, M., Mandelis, A., Yacoubi, N.: Quantitative photothermal investigation of nonradiative recombination parameters in GaAs/InAs (QD)/GaAs quantum dot structures using a three-layer laser beam deflection model. J. Appl. Phys. **136**, 16 (2024)
23. Feng, X., Liu, G., Shen, X., Lin, H., Wang, X.: 3-dimensional anisotropic thermal transport in microscale poly (3-hexylthiophene) thin films. Polymer **54**(7), 1887–1895 (2013)
24. Feng, X., Wang, X.: Thermophysical properties of free-standing micrometer-thick Poly(3-hexylthiophene) films. Thin Solid Films **5700**, 5705 (2011)
25. Hu, Y., Shen, J., Li, N., Ma, H., Shi, M., Yan, B., Huang, W., Wang, W., Ye, M.: Comparison of the thermal properties between composites reinforced by raw and amino-functionalized carbon materials. Compos. Sci. Technol. **70**(15), 2176–2182 (2010)
26. Gerl, M., Issi, J.-P.: Conductivité thermique. In: Physique des Matériaux, chapitre 7. Presses Polytechniques et Universitaires Romandes, Lausanne, pp. 347–405 (1997)
27. Chen, G.: Nanoscale Energy Transport and Conversion: A Parallel Treatment of Electrons, Molecules, Phonons, and Photons. Oxford University Press (2005)
28. Mendoza, P.J., Mandlies, A., Nicolaides, L., Huerata, J., Rodriguez, E.M.: Combined photothermal and photoacoustic characterization of silicon-epoxy, and the existence of a particle percolation threshold. Anal. Sci. 17.29 (2001)

Revolutionizing Diagnostics: Biosensors, Nanoprobes, and Imaging Agents Powered by Nanomaterials

Milan Hait, Deepika Patel, Pradeep Sahu, Sarita Diwakar, Nand Kumar Kashyap, Benvikram Barman, Manoj Kumar Banjare, and Amit Kumar Chaturwedi

Abstract Biomedical diagnostics have been radically altered by the introduction of nanomaterials, which have allowed for the development of detection technologies that are very sensitive, quick, and accurate. This chapter explores the ways in which imaging agents, biosensors, and nanoprobes are revolutionizing customized medicine and disease diagnostics. Nanomaterials have special qualities like being highly conductive, able to change the way light bends, and having a very large surface area. These latest instruments use these behaviors to find biomarkers, infections, and molecular changes on a scale that has never been seen before. This integration biosensors with nanomaterials enables very precise early illness detection and real-time point-of-care diagnostics. Nanoprobes designed to find molecules improve targeted imaging and therapy monitoring to help us learn more about diseases developing. There are advancements in contrast, resolution, and the ability to accurately identify cellular abnormalities. This medical imaging has been reimagined with the help of advanced imaging agents like quantum dots and gold nanoparticles. Despite their revolutionary promise, these breakthroughs must overcome obstacles such as scalability, obtaining regulatory clearances, and addressing biocompatibility issues. In order to integrate diagnostics based on nanomaterials into standard healthcare, it is crucial to address these challenges through multidisciplinary research, ethical considerations, and robust clinical validation. In the coming years, nanotechnologies like imaging agents, nanoprobes, and biosensors will performance a bigger part in primary disease finding, personalized therapy, and non-invasive diagnostics. This will eventually have an effect on the coming of personalized medication. This chapter explored the age of healthcare innovations powered by nanotechnology.

M. Hait · D. Patel · P. Sahu · S. Diwakar · N. K. Kashyap · A. K. Chaturwedi (✉)
Department of Chemistry, Dr. C. V. Raman University, Bilaspur, Chhattisgarh 495113, India
e-mail: amitchaturwedi1@gmail.com

B. Barman
Department of Chemistry, Govt Naveen College, Thelkadih, K.C.G., Chhattisgarh 491444, India

M. K. Banjare
Department of Chemistry, State Forensic Science Laboratory, Raipur, C.G. 492001, India

S. N. Saud Al-Humairi et al. (eds.), *Smart Nanomaterials in Biomedical Applications*, Engineering Materials, https://doi.org/10.1007/978-981-96-8023-8_8

Keywords Biosensors · Nanoprobes · Imaging agents · Nanomaterials · Precision medicine · Real-time diagnostics

Abbreviations

NMs	Nanomaterials
NPs	Nanoparticles
ML	Machine learning
CNTs	Carbon nanotubes
CVD	Chemical vapor deposition
SPR	Surface plasmon resonance
SPIO	Superparamagnetic iron-oxide
PNPs	Polymer nanoparticles
ZP	Zeta potential
TEM	Transmission Electron Microscopy
FTIR	Fourier transformation infrared spectroscopy
XRD	X-ray diffraction
SEM	Scanning Electron Microscopy
CBs	Cancer biomarkers
Pp	Porphyrin
Pc	Phthalocyanine
TNF	Tumor necrosis factor
MRI	Magnetic Resonance Imaging
CAs	Carbon Atoms
MO-NPs	Metal Oxide Nanoparticles

1 Introduction

When it comes to early illness identification, reliable monitoring, and individualized treatment methods, advanced diagnostics are essential. Patients benefit from nanotechnology-enabled imaging agents, biosensors, and nanoprobes because of the enhanced sensitivity, specificity, and speed of findings. Decreased reliance on traditional laboratory testing is made possible by these advancements in non-invasive, quick, and inexpensive diagnostics [1]. Precision medicine and tailored treatments are both aided by the improved visibility of illnesses brought about by nano-enhanced imaging agents. Improved and more predictive diagnostics made possible by AI integration pave the way for proactive management of healthcare. Ultimately, the revolutionary advancements in diagnostics are saving people's lives by enabling faster treatments and improved disease management [2].

The usage of nanomaterials has the possible to dramatically improve medical diagnostics by facilitating the quick, accurate, and very sensitive identification of illness. They are better biosensors, nanoprobes, and agents for imaging because of their biocompatibility, bulky external extent, and adjustable electrical, magnetic, and optical properties [3, 4]. Various NPs contrast representatives have been advanced to improve magnetic resonance imaging, positron emission tomography, fluorescent image, and biomarker detection. These developments pave the way for more precise diagnosis, more precise medication distribution, and more tailored approaches to patient care. Miniaturized, wearable diagnostic gadgets made possible by nanomaterials improve healthcare accessibility, efficiency, and proactivity, leading to better clinical results and patient care [5].

1.1 History of Nanotechnology

In the ninth century, in Mesopotamia, scientists found evidence of the earliest usage of nanotechnology, which means that the technology has deep roots in the first civilization of humans. Copper and silver nanoparticles were included in the glassy matrix of the pottery by the artisans to provide the glittering look desired on the pottery's surface. The Lycurgus cup, which has nanoparticles made of gold-silver alloy, Damascus steel swords that have tiny particles incorporated into them, and Maya Blue pigment are all examples of nanocomposites that have been found in ancient artifacts [6]. Following the establishment of pioneering microscopy advancements and imaginative conceptual underpinnings, the discipline rose in the latter part of the twentieth century. Stephen Papell and colleagues produced magnetic colloidal suspensions in the early 1960s. These suspensions utilize magnetite nanoparticles disseminated in paraffin with oleic acid. We prepared similar suspensions using a variety of pure metal nanoparticles immersed in a variety of carrier fluids [7].

Biologically active nanoparticles (NPs) are present in many different kinds of creatures, including bacteria, algae, birds, insects, and mammals. In 1975, Blakemore succeeded in discovering biogenic magnetite NPs (Fe_3O_4). In the 1980s, new methods of microscopy allowed for the direct observation and manipulation of nanostructures, a major stage onward in nanotechnology. The discovery of buckyballs and carbon nanotubes in 1985 and 1991, respectively, significantly enhanced the utility of nanomaterials in technology, materials, and medicine [8].

The usage of materials such as CNs, nanowires, and NPs in chemical and biosensing applications was at the forefront of nanosensor research in the early 2000s. In the early 1990s, magnetic resonance force microscopy was able to pick up a single electron spin. In the late 1990s, micromechanical cantilever sensors were created, and in 2004, graphene showed that it could pick up very small biomolecules [9]. These were all important steps in the field's development. Nanosensor proof-of-concept demonstrations in optical, electrochemical, magnetically and scanning probe modalities were born out of these and other innovations. With continuous growth and development, nanotechnology has expanded into several interdisciplinary sensory domains, including biology, electronics, energy, materials, medicine, consumer goods, artificial intelligence, and consumer goods [10].

1.2 An Overview of Nanotechnology

Nanotechnology has facilitated the advance of nanoscale strategies that utilize NMs to cooperate with analytes or biomolecules. When contrasted with conventional biosensors, these novel devices provide improved electrical conductivity, heightened sensitivity, a short reaction time, and tailored features [11]. Bioengineering uses them for medication administration, while medicine uses them for illness diagnosis, among other things. Some of the benefits of using nanotechnology in biosensing include improved electrochemical characteristics, a higher surface-to-volume ratio, and enhanced biological processing and signaling processes. Several bioengineering applications stand to benefit greatly from this development [12].

New developments in actuators, nanotechnology, and signal amplification have significantly advanced biosensors. Although these biosensors are very accurate and reliable, they nevertheless encounter problems including inconsistent signal noises, unstable sensing, and a lack of commercialization opportunities [13]. Scientists are attempting to find solutions to these problems by improving biosensor performance via the routine of chemometrics and machine learning (ML). With the help of ML, biosensors may evolve into intelligent devices that can use decision algorithms to forecast the concentration or species of analytes. One of the most popular ways to evaluate chemical data is via chemometrics, which uses mathematical and statistical approaches to find the best way to quantify chemicals. In general, scientists are making efforts to address these issues and enhance the functionality of biosensors [14].

When dealing with large, complicated matrices or samples of big data from sensors, ML is an invaluable tool. Data rendering from overlapping, low-resolution, noisy sensor data is one of its benefits. On a large scale, ML can reveal hidden connections between bio-events and signals, as well as analyte parameters and signals [15]. ML frequently interprets raw sensing data from biosensors in the following ways:

- **Categorization**

The type of target analyte determines the method used for signal classification based on sensor data.

- **Anomaly detection**

Biosensor performance is highly dependent on operating circumstances and sample matrix, and contaminations at the target location might impede sensing signal transmission. In such a situation, the ML process verifies the accuracy of the acquired indication and, if it is erroneous, corrects the biofouling-impacted signals to enhance the sensor's performance [16].

- **Reduction of noise**

Noise is often included in sensed signals. Electrical noise disrupts the perceived signal in seconds or minutes, shifting it to a sub-second timescale. It is possible to greatly enhance the accuracy of the sensor signal by training ML models to identify noise.

- **Pattern recognition and object identification**

Then, with the help of ML algorithms, the signals collected by the biosensors may be readily and efficiently evaluated by uncovering hidden patterns and objects (Serey 2023).

2 An Overview of Nanomaterials

An overview of NMs discusses them, covering their properties, characterization, types, and preparation. Additionally, subsequent sections go into excellent depth on the NMs often used in nanosensors.

2.1 Types of Nanoparticles

There are many categories for NPs (Fig. 1.) that are based on their size, shape, and chemical possessions. According to their physical and chemical properties, the following NP classes are among the most famous [17].

2.1.1 Carbon-Based NPs

Some examples of NPs based on carbon are fullerenes and CNTs.

i. **Fullerenes**

The sp^2 hybridization of carbon atoms into fullerenes (C60) forms spherical carbon molecules. Fullerenes are spherical structures composed of 28–1500 CAs, with single-layer widths of 8.2 nm and multi-layer diameters of 4–36 nm [18]. Fullerenes are nanomaterials that consist of globular porous cages (Fig. 2), like allotropic carbon. Businesses are interested in them because of their conductivity to electricity, structure, flexibility, and outstanding strength [19].

ii. **Graphene**

Graphene is an allotrope of carbon. Graphene is a carbon-based honeycomb lattice network with two dimensions and a flat, hexagonal surface (Fig. 3). Graphene sheets typically are thicker than about 1 nm [20].

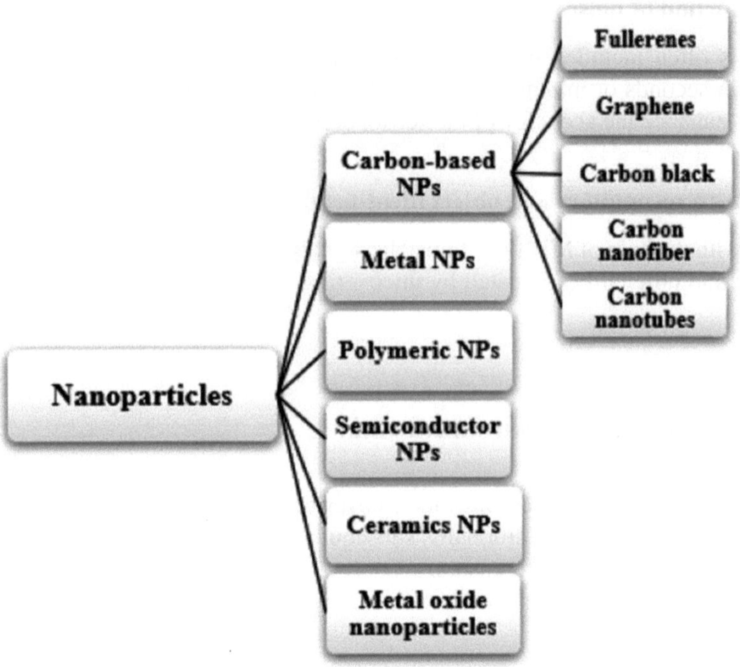

Fig. 1 Types of nanoparticles

Fig. 2 Fullerene carbon nanoparticles

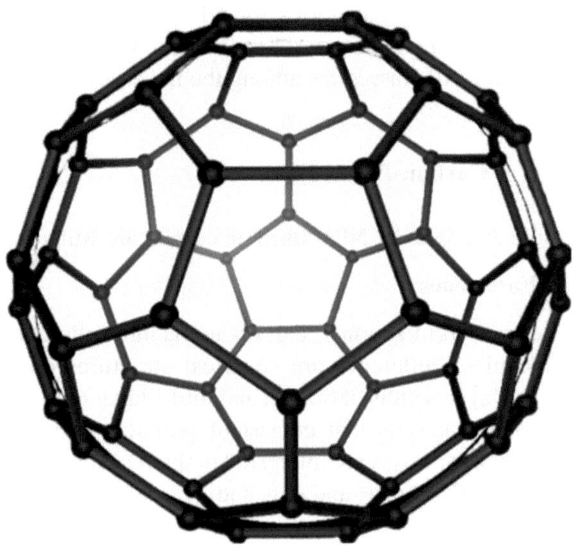

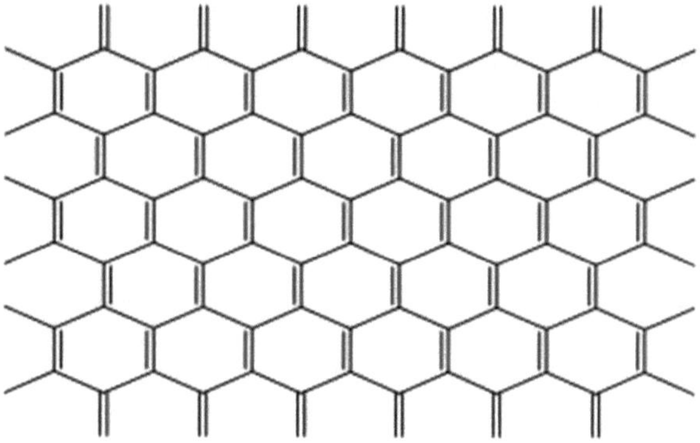

Fig. 3 Graphene carbon nanostructure

iii. Carbon Nanotubes (CNT)

Nanotubes made of carbon, known as carbon nanotubes (Fig. 4), may be as small as 0.7 nm for a single-layer CNT and 100 nm for multiple-layer CNTs. Their lengths typically vary from just a few μm to a few mm, and their honeycomb arrangement of atoms is what makes them unique. Half fullerene molecules may seal the ends, or they can be empty [21]. A graphite sheet crumpling up on itself is the best way to describe their structure [22]. We refer to the rolled sheets as single-walled, double-walled, or multi-walled CNTs because they can have one, two, or more walls. It is usual practice to create carbon precursors via deposition, especially atomic carbon. Graphite vaporization achieves the laser or electric arc deposition of carbon onto metal particles. Their most recent manufacturing process is chemical vapor deposition (CVD) [23].

iv. Carbon Nanofiber

Graphene nanofoil, carbon nanotubes, and carbon nanofiber all share the same manufacturing process. The key difference is that, instead of being cylinder-shaped, it was warped into a funnel form [24].

v. Carbon Black

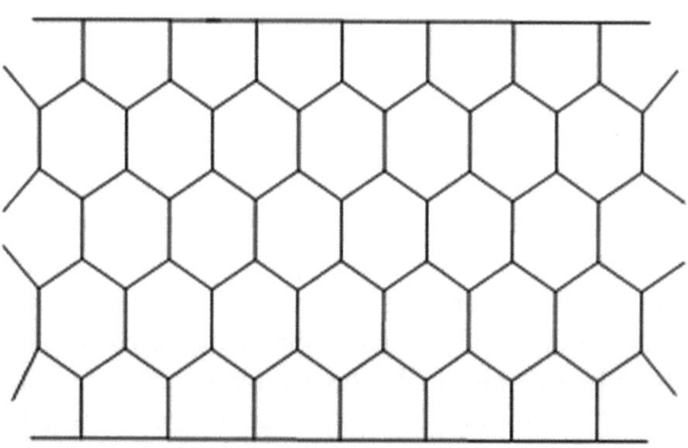

Fig. 4 Carbon nanotube

Amorphous carbon is typically spherical and ranges in size from 20 to 70 nm in diameter. Due to the fast interaction between the particles, it comes together in clumps of around 500 nm in size [25].

2.1.2 Metal NPs

There are destructive and constructive methods for synthesizing metals into nanoparticles [26]. Zn, Al, Cd, Co, Au, Cu, Fe, Pb, and Ag are among the most common metals used. Features such as size, surface features, and color set these nanoparticles apart. Metal precursors synthesize these nanoparticles using restricted surface plasmon resonance (SPR), which results in their distinctive optoelectrical properties [27, 28]. There is a solar electromagnetic spectrum absorption band that noble metal and alkali NPs like Cu, Au, and Ag exhibit. Modern materials rely on the combination of metal NPs with regulated proportions and forms [29].

2.1.3 Metal Oxide Nanoparticles (MO-NPs)

Metals with a biocidal effect, such as Cu and Ag, are often used as antimicrobials. Researchers have investigated metal-based additives for the electrochemical detection of biomolecules [30]. These additions include MO-NPs such as ZnO, NiO, MnO_2, TiO_2, Fe_2O_3, and Co_3O_4 [31]. CuO-NPs are very useful in many areas because they have unique properties that make them useful as antibacterial agents, catalysts, ultra-strong materials, and sensors [32]. Due to their polymer coating, paints and textiles often use them as antibacterial agents. The photolytic properties of TiO_2 and ZnO make them other popular materials [33, 34].

Iron oxide nanoparticles (IO-NPs) must have high magnetic induction ethics, consistent excellence, and virtuous biocompatibility in living settings. They also need to dissolve in water and spread out evenly. Medical applications currently use two forms of superparamagnetic IONP-based materials: superparamagnetic iron oxide (SPIO) and ultra-small SPIO nanoparticles [35, 36]. The mean crystal size of these nanoparticles ranges from 50 to 100 nm, and they are smaller than 50 nm. The medical profession has paid a lot of attention to these two categories, particularly since they might be the next big thing in MRI contrast agents and because they could be vectors for the transfer of genes and medications [37].

The correct surface chemistry of SPIO-NPs permits them to be utilized in several in vivo claims, such as improving MRI contrast, mending tissues, performing immunoassays, detoxifying biological fluids, delivering drugs, and separating cells. For SPIO-NPs to be useful in biology, their surfaces need to be changed with compounds that are safe and break down naturally. Adding multidentate ligands to inorganic nanoparticles can improve their electrical, optical, and magnetic properties. These ligands also make the nanoparticles more stable in solution [38, 39].

Gadolinium-based contrast agents are quickly flushed out of the body by the kidneys. Because SPIO-NPs have higher relaxation values, they are better for imaging

[40]. Four SPIO-NPs—Feridex, Endorem, Combidex, and Sinerem—have dextran coatings that are safe for magnetic resonance imaging (MRI) in patients. Their core diameters range from three to six nanometers. By tracking their biodispersion and guiding them with an outside magnetic field, we can direct the drug-loaded SPIO-NPs to the right place. This feature makes them useful for both diagnostic and therapeutic purposes [41].

2.1.4 Ceramics NPs

The boiler and cooling process produces ceramic NPs, which are non-metallic mineral particles. They may be solid, semi-solid, dense, porous, or hollow, among other possible forms and sizes. Research into these NPs is booming because of their possible usages in grounds as miscellaneous as imaging, photocatalysis, photodegradation of dyes, and catalysis [42].

2.1.5 Semiconductor NPs

Semiconductor materials have extensive applications in the literature due to their unique combination of metal and nonmetal characteristics. As a result of semiconductor NPs' huge band gaps, bandgap tuning made significant changes to their features. Their essential role in photonics, electrical devices, and photocatalysis is due to these qualities. Some semiconductor NPs work well as water splitters because of where their band gaps and band edges are located [43].

2.1.6 Polymeric NPs

Organic NPs, known as polymer nanoparticles (PNPs), usually come in nanospherical or nano-capsular shapes. They have several literary applications and are simple to functionalize [44, 45]. A subfield of nanotechnology, known as lipid nanotechnology, studies and develops methods for producing lipid nanoparticles for cancer treatment and drug delivery, among other areas. Because of their versatility and ease of functionalization, PNPs have several potential uses [46, 47].

2.2 Properties of Nanomaterials

NMs denote a captivating and developing area of research at the intersection of science and technology. Their nanoscale size gives them special qualities and makes them useful in many ways. To quickly review, NMs are materials whose structures or components are on the nm scale, frequently among 1 and 100 nm. Compared to their bulk-state counterparts, materials at this size often display unique characteristics, such as quantum effects and a rise in surface area [48]. At the nanoscale, objects with very high surface area to volume ratios are best understood via the lens of thermodynamics. Due to their increased molecule count, nanostructured materials have unique optical, electronic, magnetic, physical, and chemical properties [49].

2.3 *Characterization of Nanomaterials*

Characterizing nanomaterials involves using specialized techniques to measure and understand their unique physical and chemical properties, which are heavily dependent on their nanoscale dimensions (typically 1–100 nm). Because their size, shape, surface characteristics, and internal structure dictate their behaviour, precise characterization is crucial for both fundamental research and quality control in applications. Figure 5 displays the results of the nanoparticle characterization.

i. **UV–visible Spectrometry**

The use of UV–visible spectroscopy made it possible to study the formation of different nanoparticles in various ways. The unique peak steadily rises with the response period and the increasing concentration of plant extract and salt ions, showing that nanoparticles are forming. The particles on the nanoscale show peaks in their UV–vis absorption spectra that are indicative of surface plasmon resonance [50].

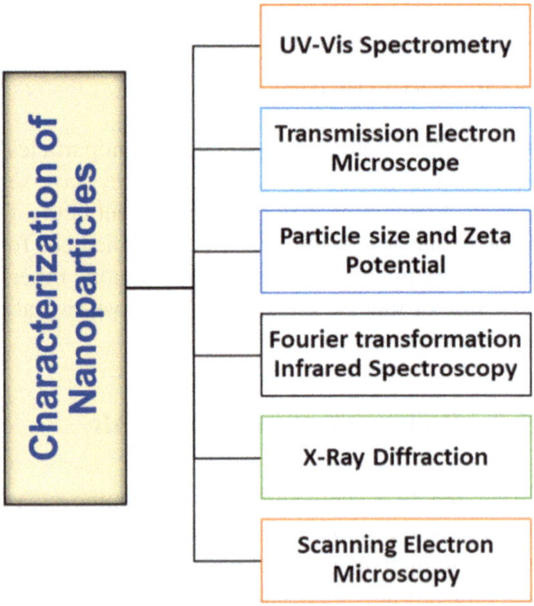

Fig. 5 Nanoparticle characterizations

ii. **Transmission Electron Microscopy (TEM)**

A transmission electron microscope (TEM) was used to record the shape of nanoparticles. TEMs use the idea of electron penetration to provide information about the primary material at different magnifications. Nanoparticles were found to have a four-polar hollow shell shape, which sheds light on multi-layered materials [51].

iii. **Particle size and zeta potential (ZP)**

ZP analyzer detected the surface charge that is possible to use the zeta potential to figure out how particles behave in water, whether you're trying to guess how stable they will be or look into how they settle in water-cooling equipment. The charging of the surface and biodistribution of nanoparticles (NPs) affect how stable the particles will be. It can get a useful idea of how a surface group behaves when it comes into contact with an electrolyte solution by measuring the electrical field at the point where two layers of suspended nanoparticles (NPs) meet. High zeta potential values, caused by electrostatic interaction, indicate dispersion-stable NPs. This method relies on density-functional theory (DFT) and is crucial for figuring out how nanoparticles' sizes are distributed [52].

iv. **Fourier Transformation Infrared Spectroscopy (FTIR)**

FTIR is a method that is often utilized for the purpose of identifying functional groups in mixtures and pure substances, as well as for comparing compounds. There is a connection between infrared analysis and the vibratory motion of the molecules and atoms [53].

v. **X-ray Diffraction (XRD)**

The existence and type of crystalline phase may be revealed by the non-destructive application of X-ray fluorescence, which can be used to investigate chemical diffraction patterns and the elemental composition of materials. Information about the sample may be gleaned from the X-ray beam diffraction (XRD) data, which includes the sample's location (angle) and intensities [54].

vi. **Scanning Electron Microscopy (SEM)**

An electron SEM can observe MNPs in three dimensions by capturing the secondary electrons generated when an electron beam strikes a sample. This allows researchers to analyze their size and shape [55].

2.4 Preparation of Nanomaterials

i. **Carbon Nanotubes**

The hexagonal arrangement of carbon nanotubes makes them desirable in nanotechnology, electronic devices, and materials research owing to their conductivity of electricity, flexibility, and strength, especially mechanical strength. As a material, they are among the strongest known, having a tensile strength higher than steel [56]. Their ability to reinforce composite materials and produce lightweight but sturdy structures is what makes them so special. Carbon nanotubes' electrical conductivity is another important feature. CNTs are highly adaptable materials with a high aspect ratio, electrical conductivity, and exceptional mechanical strength, making them ideal for use in nanotechnology and electronics. They have been used in conductive films, sensors, and high-performance transistors. Their high surface area-to-volume ratio makes them effective in energy storage, catalysis, and medication delivery systems [57].

ii. **Graphene**

A two-dimensional hexagonal carbon lattice, graphene is renowned for its high electrical and thermal conductivity as well as its remarkable mechanical strength. Its electrical conductivity is a key feature of graphene. Graphene outperforms copper and other conventional conductive materials when it comes to electrical conductivity. Due to this quality, graphene is much sought after for use in the creation of transparent conductive films, sensors, and high-speed electronics. The mechanical strength of graphene is another important feature. Graphene is ultra-strong and malleable despite having a thickness of only one atom. Due to its higher tensile strength compared to steel, it may be used to reinforce composite materials and construct lightweight yet sturdy buildings [58]. In addition to its exceptional thermal conductivity, graphene is an excellent heat dissipator. Graphene has a lot of promise in several areas, including thermal management, medicinal devices, storage of energy, and catalysis, because of its huge surface area and specific electrical characteristics. Finally, graphene is a very useful material that can be used in many different fields because it is strong, doesn't carry electricity poorly, doesn't melt in hot temperatures, and has special electronic qualities [59].

iii. **Quantum Dots**

Nanoscale semiconductor particles known as "quantum dots" have extraordinary electrical and optical characteristics brought forth by quantum confinement processes. Common components of these microscopic formations include indium arsenide, cadmium selenide, and lead sulfide. Quantum dots' size-dependent optical characteristics are an essential aspect to note. The wavelength of light emitted by a quantum dot is defined by its bandgap, which is in turn affected by its size. Quantum dots' adaptability makes them useful in fields including lighting, displays, and biological imaging. The high photoluminescence efficiency of quantum dots is another important feature. Quantum dots have limited emission spectra and produce intense, multicolored light when activated by electric or light fields. This quality is useful for sensors and medical imaging, and it also makes for bright and energy-efficient screens. The electrical properties of quantum dots allow them to be modified for specific purposes [60]. Their versatile engineering capabilities make them a potential material for quantum computers, solar cells, and transistors due to features like strong charge carrier mobility. Furthermore, quantum dots' quantum coherence and capacity to produce entangled photon pairs have made them promising candidates for use in quantum information processing and quantum cryptography. Considering their size-dependent optical characteristics, high photoluminescence efficiency, and tunable electronic properties, quantum dots are an incredibly interesting and potentially useful material with many potentials uses in fields like quantum computing, displays, lighting, and sensors [61].

iv. **Metal Nanoparticles (MNPs)**

Nanoscale particles ranging in size from 1 to 100 nm are known as metal nanoparticles. These particles are made up of metal atoms. Their chemical and physical characteristics set them apart from the competition and make them useful in many fields. They have a lot of surface area compared to their volume; metal nanoparticles are good at speeding up chemical reactions that are used to make medicines and compounds [48]. Additionally, they have unique optical properties that make them useful for sensing, imaging, and photothermal therapy (such as localized surface plasmon resonance) [62]. Electronics, sensors, and batteries may all benefit from their excellent thermal and electrical conductivity. Researchers are developing new, highly useful high-tech materials due to their diminutive size and remarkable versatility. Biomedical uses for metal nanoparticles include cancer treatment, imaging, and medication administration due to their antibacterial characteristics. Because of their little size and unusual features, metal nanoparticles provide a plethora of desirable qualities and practical uses [63].

v. **Nanowires**

Nanowires are often composed of insulators, metals, or semiconductors and have lengths and diameters on the nanoscale. They possess exceptional characteristics, such as a large aspect ratio, which enables them to have huge surface areas in comparison to their volume. As a result, they are well-suited for applications such as sensors, catalysts, and storage devices for energy. The malleability of nanowires, together with their electrical and optical characteristics, makes them very promising in the fields of optoelectronics, photovoltaics, and nanoelectronics [64]. The ideal for use in nanomechanical and flexible electronics, they are both mechanically flexible and forceful. Nanoscale devices have a high surface-to-volume ratio and are very tiny, which allows them to convert and store power effectively. Nanowires have the potential to be the building blocks of next-generation electronics, including transistors, sensors, and nanolasers [65].

vi. **Nanoparticles**

Nanoparticles are microscopic particles that possess unique properties that make them useful in imaging, medicine delivery, and catalysis. These particles also have a high surface area-to-volume ratio [66]. They have applications in imaging, drug administration, and catalysis due to their size-dependent behavior, which enables them to undergo substantial changes as their size decreases. Quantum processes at the nanoscale have resulted in the creation of materials with exceptional optical, electrical, and magnetic characteristics, these make them well-suited for a wide range of uses, including drug delivery, sensing, imaging, demonstrations, and biological labeling. Medicine, electronics, environmental cleanup, and energy storage are just a few of the many areas that might benefit from nanoparticles due to their malleability in terms of form, composition, and surface alterations. When it come with, technology, and business, nanoparticles are very adaptable materials [67].

3 Biosensors

A biosensor is an analytical tool that can detect chemical or biological compounds by converting a biological reaction into a quantifiable signal. Bioreceptors may be enzymes, antibodies, or DNA, whereas transducers can be electrochemical, optical, or piezoelectric. A signal processor is the third and final component [68].

The basic idea is that the bioreceptor will attach to the analyte of interest in a certain way, setting off a chain reaction. The transducer amplifies and displays this reaction-generated electrical or optical signal. Biosensors are extensively utilized for quick and accurate detection in biotechnology, food safety, environmental monitoring, and medical diagnostics [69]. The speed, sensitivity, precision, real-time detection, and reasonable price tag are just a few of the desirable qualities of a reliable sensing system, as seen in Fig. 6.

3.1 Types of Biosensors

The ability to measure biological interactions classifies biosensors based on their transducer type. Biosensors based on nanomaterials, electrochemical, optical, piezo-electric, or thermal technology are among the most prevalent, and each has its set of benefits in contemporary medicine [70]. Figure 7 shows in the types of biosensors and explain the following are:

i. **Electrochemical Biosensors**

Biosensors that use electrochemistry to detect biological interactions measure impedance, voltage, or current changes. They find extensive use in devices that measure glucose levels (such as blood glucose meters), identify pathogens, and track environmental conditions. These biosensors are perfect for use as point-of-care diagnostics since they are small, inexpensive, and very sensitive [71].

ii. **Optical Biosensors**

When a biomolecule binds to a bioreceptor, an optical biosensor may measure the resulting change in the light's absorption, fluorescence, or refractive index. Methods that increase sensitivity include SPR (surface plasmon resonance) and detection based on fluorescence. Because of their great selectivity and ability to be monitored in realtime, these biosensors find widespread use in medical imaging, DNA sequencing, and biomarker identification [72].

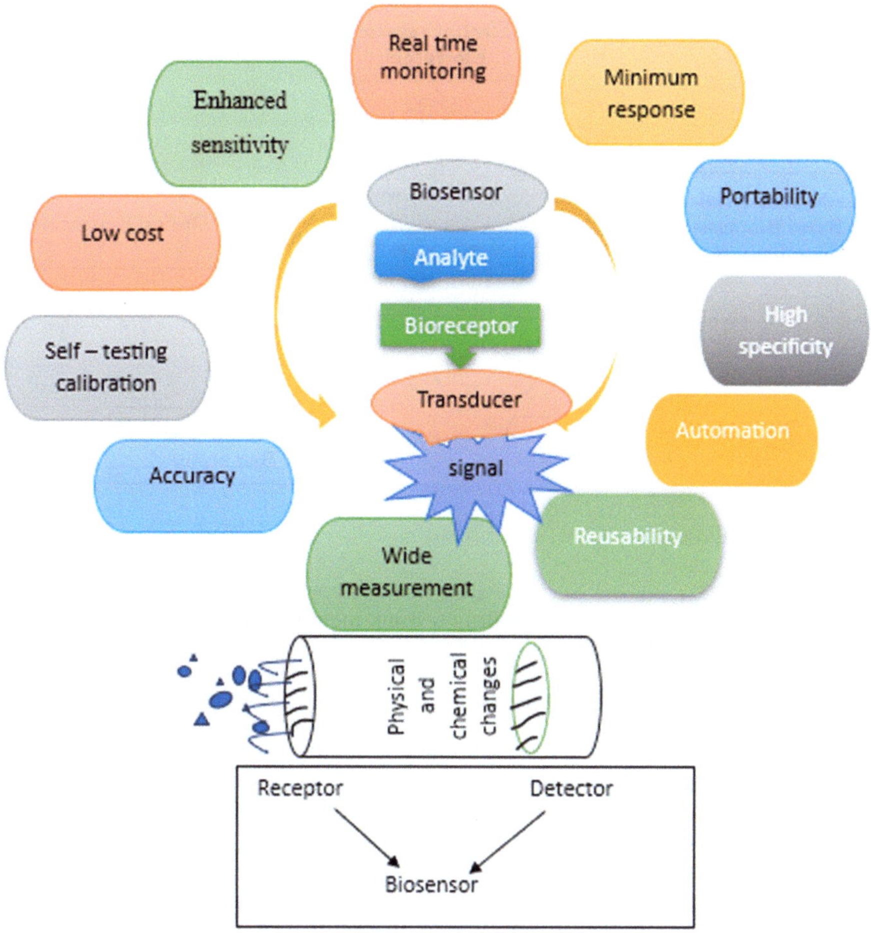

Fig. 6 The desirable qualities of a reliable sensing system

iii. **Piezoelectric Biosensors**

Biosensors that use piezoelectric technology may detect changes in mass by translating vibrations from machinery into electrical impulses. Biomolecule binding to the surface of the sensor changes the vibration frequency, allowing for accurate measurement. With their label-free, real-time detection capabilities, these biosensors find infections, poisons, and medication interactions [73].

iv. **Thermal Biosensors**

Whenever a biological process causes a change in temperature, thermal biosensors can detect it. Thermosensors pick up on the heat produced or absorbed by target molecules when they bind to bioreceptors. Because of their great accuracy,

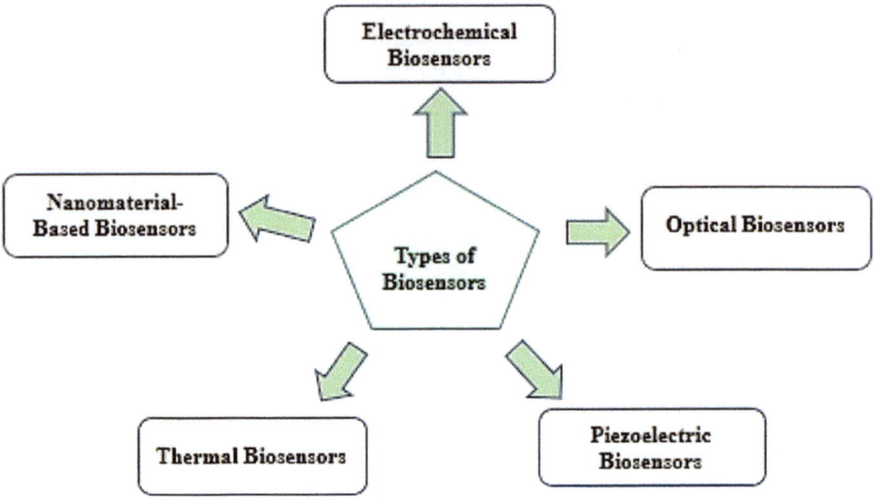

Fig. 7 Types of Biosensors

researchers use these biosensors for drug testing, metabolic monitoring, and enzyme function research [74].

v. **Nanomaterial-Based Biosensors**

The use of quantum dots, carbon nanotubes, and nanoparticles has transformed biosensors, raising their detection limits, stability, and sensitivity to a whole new level. Environmental toxin monitoring, cancer detection, and infectious disease diagnoses are some of the many applications of nanomaterial-based biosensors, which provide very sensitive and fast findings [75].

Biosensors are essential in contemporary medicine for the quick, precise, and economical credentials of a wide variation of living organisms and chemical compounds. Their variety of uses in healthcare, ecology, and industry will determine how customized treatment and fast diagnostics are developed in the years to come. Biosensors are rapidly developing in response to advances in artificial intelligence and nanotechnology, which are enhancing the efficacy and accessibility of strength repair [76].

3.2 Applications of Biosensors

Biomedical research, food processing, and healthcare have all been profoundly affected by the development of biosensors [70]. A deeper comprehension of materials' basic characteristics has been made possible by the development of nanotechnologies, which can manage and govern substances at atomic and molecular levels. Dimensionality determines the physicochemical and biological characteristics of materials, making them applicable to a wide range of fields. Biomedical diagnostics make use of nanobiosensors to track and detect physicochemical events in inaccessible regions and very low concentrations of analytes. Bioreceptors and transducers are two components that help classify biosensors [77, 78].

Four main types of biosensors exist: enzymatic, immunological, aptamer, and microbial ones. Bioreceptors are the main component of biosensors [79]. Biosensors can also be categorized by the type of transducer they use: electrochemical [80], thermal [81], electronic [82], optical [83], or gravimetric [84]. Additionally, we can categorize them into four types: electrometers [85], biosensors-on-chip [86], surface plasma resonance biosensors, and nano-biosensors [87]. Nanotechnology has advanced significantly, and in the upcoming years, biosensors will reach their peak in terms of compatibility, selection, sensitivity, usability, and LOD [88]. The following sections provide more detailed explanations of a few biosensor applications.

i. **Biomedical Applications**

Biosensors are essential in biomedicine because they allow for the accurate tracking of disease development and the reduction of death rates. Their primary function is to provide treatment plans for patients' recuperation in implanted and wearable medical devices. For the purpose of curing chronic diseases, Shin et al. created a pressure sensor that is bioresorbable and protected by a SiO_2 layer [89]. An elastic condenser biosensor was developed by Sheng et al. for use in wearable medical devices; it measures pressure, provides motion feedback, and can be controlled remotely, among other biological features [90]. Recently, biodegradable polymers, a novel category of biomaterials for sensing, have emerged. Glucose catalytic reactions may be detected within a range of 0–10 mM using an implanted biosensor. We will discuss some of the many ways biosensors have changed the biomedical industry in the sections that follow [91].

ii. **Cancer and Bone Disease**

It is impossible to screen patients for malignant cells and implement point-of-care treatment without cancer biomarkers (CBs) [92]. Traditional methods need the development of new, more practically priced technologies due to their inherent technological limits. A vital tool, biomarker detection may help diagnose cancer at an earlier stage, which in turn lowers death rates. Because of their remarkable characteristics, biosensors have recently gained support for use in cancer detection. Optic, mass-sensitive, and electrochemical cancer biosensors are classified by transducer type and biological reaction. Transducer and biological reaction are the two main criteria for classifying biosensors [93].

To identify cancer cells in their early stages, graphene has been combined with TiO_2 and shown sensitivity to cancer cells in several organs [94]. A multilayer plasmonic sensor analyzed human tears for signs of breast cancer using Raman scattering. Another study focused on the early detection and treatment of inflammatory bowel disease (IBD) and osteoarthritis (OA) [95, 96]. Another study suggested that a surface plasmon resonance (SPR) biosensor might detect the RA biomarker miR-15a, with a limit of detection (LOD) of 0.56 fM. Researchers described a 2 nM LOD electrochemical biosensor based on a polyhydrogel film for use in osteoarthritis. To avoid joint injury and articular cartilage deterioration, both biosensors stress the need for early detection and therapy [3].

iii. **Tissue Engineering Usage**

The field of tissue engineering is flexible and uses biological building blocks like cells, scaffolds, and stimulating molecules to create a possible way to fix damaged or aging tissues. When weighed against the high expense of organ and tissue transplantation, this technique is overwhelmingly well-received [97]. When it comes to regenerative medicine and tissue engineering, biosensors have very little use. Biosensors are often used in tissue engineering to track vital signs, including nutrition and oxygen intake as well as metabolite release, to guarantee correct tissue development [98]. Figure 8 clearly shows a general perspective on the use of biosensors to examine the quality of 3D structures. Biosensors are able to be implanted into tissues and organs and may be reasonably shaped to match the cellular environment due to their modest dimensions. It has been shown that a platinum-wire enzymatic biosensor is biocompatible and useful for tracking the condition of tissue implants after they have been implanted into brain tissues [99]. For in vitro applications, electrochemical sensors may be helpful in tracking the redox chemicals produced by implants after biological implantation. In recent times, amperometric devices have been used in vitro and in vivo to test 3D build quality by measuring dopamine, or ascorbic acid, sugar levels, oxygen consumption, and nitric oxide release [100, 101].

iv. **Microfluidic Systems**

The manipulation of sensing via tiny amounts of fluids is made possible by microfluidic devices, which use lithographic microchannels. Lab-on-a-chip (LOC) devices may assist in separating microparticles by simulating the milieu surrounding traditional cell growth [102]. A microfluidic system was added to a biosensor that picks up this signal for use in breast cancer diagnosis. The detection limit of an extracellular in-nature vesicle microRNA biomarker was set at 84 MM [103]. The whole process took half an hour, and the device could detect concentrations as low as 0.5 PM [104]. Scientists were able to distinguish between two influenza A virus subtypes, H5 and H9, by using microfluidic technology in conjunction with quantum dots (QDs). The result was intense fluorescence, with a limit of detection (LOD) of 0.26 for H9 and 0.017 HAU for H5, respectively [105].

v. **Diagnostics**

Since biosensors allow for the quick and precise identification of molecular biomarkers in actual samples, they may completely alter the landscape of real-time microbial diagnostics. Researchers are developing wearable bioelectronic devices that utilize label-free and lab-on-a-chip technologies. To help in wound healing monitoring, one such gadget is a stretchy electrochemical immunological biosensor that can detect tumor necrosis factor (TNF) [69, 106]. Pathogenic bacteria, such as E.

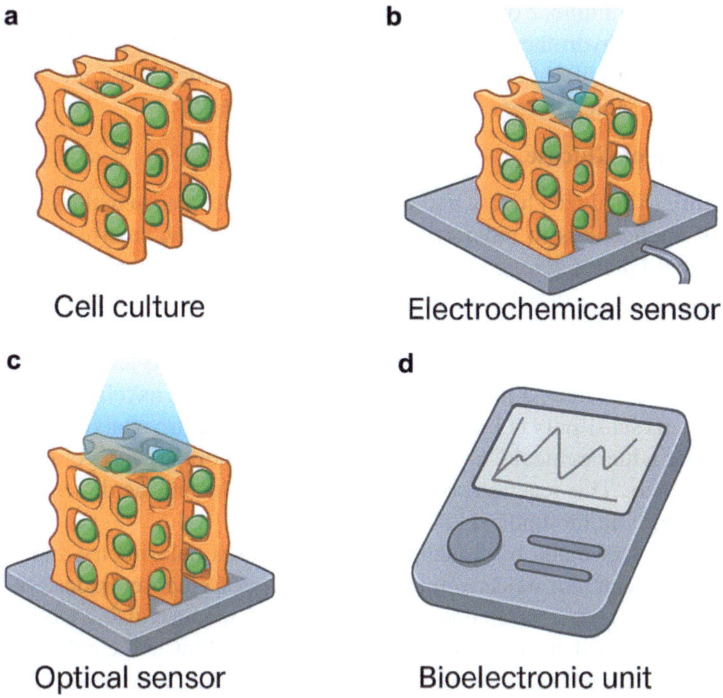

Fig. 8 A general perspective on the use of biosensors to examine the quality of 3D structures, **a** cell culture, **b** electrochemical sensor, **c** optical sensor, and **d** bioelectronic unit

coli, may be detected in drinking water using other biosensors. Glycol nanoparticle-based immunosensors for SPR influenza detection are recent advances. Additionally, in Uganda, researchers have developed a smartphone-based test for IgG antibody detection and investigated the feasibility of point-of-care Ebola virus testing in human survivors. These innovations show how biosensors may be used for microbiological diagnoses in real time [107, 108].

vi. **Porphyrin and Phthalocyanine**

The photophysical properties of porphyrins (Pp) and phthalocyanines (Pc), two different macrocyclic organic compounds, may be useful for biomedical sensors, drug delivery, and fluorescence imaging. Their remarkable light-absorbing capabilities are a result of the highly conjugated configuration of their 18 electrons, which gives them their optical features. The far infrared spectrum is where Pp does the greatest absorption, and there are two kinds of fluorescence caused by electron transitions brought about by relaxation between vibrational modes [109, 110]. Some examples of biological diagnostic applications of the fluorescence property include aflatoxin B1 detection in pharmaceuticals and cancerous ovarian antigen CA 125 detection in human blood samples. A fluorescent biosensor based on carboxyl Pp was made that can pick up a number of 8.38 fg mL^{-1} in a straight line from 100 fg mL^{-1} to 100 ng mL^{-1}. The hydrothermal technique was used to produce a porphyrin-Co_9S_8 nanocomposite for colorimetric sensing of H_2O_2 that achieved a limit of detection (LOD) of 6.803 M in a linear range of 7–100 M [111, 112].

vii. **Detection of Glucose**

The ability to analyze glucose and other biofluid molecules in real time is critical for clinical applications in monitoring diabetes, cardiovascular disease, hypertension, and renal failure, among others. Enzymes used in traditional sensing, such as glucose oxidase, have drawbacks including expensiveness, instability, and lack of repeatability [113, 114]. The innovative electrochemical biosensors suggested by Ahmad et al. for glucose sensing, which are based on CuO nano-leaves, have a linear range of 0.005–5.89 mM, a limit of detection of 12 nM, and excellent stability and repeatability. Its sensitivity is 1467.32 A/(mM cm^2). An electrocatalytic reaction was carried out by a hierarchical Cu nanoflower that was adorned with gold and coated on GO nanofibers. This reaction converted glucose to gluconic acid, which could be measured linearly in biofluids [115].

A biosensor was made by Rashtbari et al. that uses colorimetry and nanolayered manganese-calcium oxide to mimic the activity of glucose oxidase. A human blood sample was tested for glucose and H_2O_2 simultaneously using a colorimetric approach. We obtained the results promptly. Oxygen and $MnCaO_2$ catalyze an oxidation reaction that transforms glucose into gluconic acid; this process also changes the reaction solution's color to red–orange, making it easier to detect glucose by spectrophotometry without the need for tools [116]. The experiment successfully identified glucose within a linearity range of (0–82.3) 10^{-6} M, with a limit of detection (LOD) of 6.12×10^{-6} M.

viii. **Detection of DNA and Protein**

A biosensor is an analytical device that can detect biomarkers such as proteins and DNA. A transducer transforms biological processes into quantifiable signals [74]. Disease states are typically better understood with the use of protein biomarkers. Researchers developed a label-free, sensitive Si nanowire optical biosensor to detect CRP in human serum. Researchers conjugated an electrochemical sensor with gold DNA to detect the presence of pork in consumable foods. The bioconjugation process, which can detect DNA in as little as one hour, can make biosensors more cheaply and quickly [117, 118]. Table 1 lists nanomaterials used for various biosensing applications in DNA and protein detection. Fast DNA detection and cost reduction are two additional benefits of the bioconjugation approach.

This table highlights how various nanomaterials contribute to biosensing by improving sensitivity, selectivity, and real-time detection of DNA and proteins for advanced diagnostics.

Table 1 Nanomaterials used for various biosensing applications in DNA and protein detection

Nanomaterial	Type of biosensor	Target biomolecule	Detection mechanism	Advantages	References
Gold nanoparticles (AuNPs)	Optical, electrochemical	DNA, proteins	Surface plasmon resonance (SPR), colorimetric, electrochemical signal enhancement	High sensitivity, biocompatibility, easy functionalization	[119]
Quantum dots (QDs)	Fluorescence	DNA, proteins	Fluorescent signal emission upon target binding	High photostability, multiplexed detection	[120]
Carbon nanotubes (CNTs)	Electrochemical, field-effect transistors (FETs)	DNA, proteins	Conductivity changes upon target binding	High electrical conductivity, fast response	[121]
Graphene oxide (GO)	Fluorescence, electrochemical	DNA, proteins	Quenching of fluorescence upon target interaction	High surface area, strong fluorescence quenching ability	[122]
Silica nanoparticles	Optical, electrochemical	DNA, proteins	Encapsulation of fluorescent dyes or biomolecules	Biocompatibility, high stability	[123]
Magnetic nanoparticles (MNPs)	Magnetic, electrochemical	DNA, proteins	Target separation and enrichment using external magnetic fields	High specificity, easy sample recovery	[124]
Silver nanoparticles (AgNPs)	Optical, colorimetric	DNA, proteins	Surface plasmon resonance (SPR), aggregation-induced color change	Strong antibacterial properties, enhanced signal sensitivity	[125]
Polymer-based nanoparticles	Optical, electrochemical	DNA, proteins	Molecular imprinting for selective recognition	Customizable, cost-effective	[45]

3.2.1 Biosensors in Disease Detection and Monitoring

Biosensors have revolutionized the detection and tracking of diseases by enabling faster, more accurate, and real-time testing. Medical professionals rely on these technologies to improve patient outcomes and healthcare efficiency via early illness identification, prognosis, and continuous monitoring. Biosensors have a very high sensitivity for detecting infectious diseases, viruses, and poisons [126]. Biosensors, which allowed for quick testing at the point of care, were critical in the detection of COVID-19, HIV, and TB, for instance. As part of chronic illness monitoring, electrochemical biosensors can monitor glucose levels, which helps diabetic patients avoid unnecessary hospitalizations. By measuring cardiac biomarkers, wearable biosensors can allow for the early diagnosis of cardiovascular illnesses and strokes. Nanomaterials-based biosensors provide better cancer diagnostics by finding tumor markers and cancer cells in blood samples. This makes it possible to diagnose cancer in its early stages. Additionally, optical biosensors enhance targeted medication administration and real-time imaging [127].

Furthermore, biosensors aid in the identification of neurological diseases by identifying biomarkers for conditions such as epilepsy, Parkinson's disease, and Alzheimer's, which in turn supports early intervention techniques. Biosensors are transforming into networked, smart diagnostic instruments that provide predictive and tailored healthcare solutions, thanks to breakthroughs in nanotechnology and artificial intelligence. They are essential in contemporary medicine because of the rapid, non-invasive, and cost-effective diagnoses they provide [128].

3.2.2 Biosensors in Modern Diagnostics

Biosensors have brought about a paradigm shift in contemporary diagnostics by enabling the prompt, precise, and real-time identification of illnesses and biological markers. These devices use transducers in conjunction with biological recognition components to transform biological interactions into quantifiable signals [129]. The medical field, environmental scientists, the food industry, and the pharmaceutical industry all make extensive use of biosensors. With biosensors that use electrochemistry, optics, or piezoelectricity, metabolic problems, cancer biomarkers, and infectious diseases can be found early on. The ability to continuously monitor vital indicators like heart rate, glucose levels, and more is a major step forward for customized care made possible by wearable biosensors. Nanotechnology allows for more sensitive biosensors, which in turn lowers detection limits and makes ultra-precise diagnostics possible [130].

Biosensors are important for point-of-care tests and distant diagnostics because they are portable, cheap, and easy to use. Biosensors, enabled by developments in artificial intelligence and wireless technology, are becoming smarter, allowing proactive medical intervention via real-time monitoring of health [131]. Biosensors have several uses in contemporary medicine, as shown in Table 2.

Table 2 Applications of biosensors in modern diagnostics

Application	Description	References
Disease detection	Identifies biomarkers for cancer, diabetes, and infections	[132]
Wearable monitoring	Tracks glucose, heart rate, and vital signs in real-time	[133]
Point-of-care testing	Enables rapid diagnostics in remote or clinical settings	[134]
Drug discovery	Assesses drug efficacy and biomolecular interactions	[135]
Environmental monitoring	Detects toxins and pollutants in air, water, and food	[136]

4 Imaging Agents and Nanotechnology for Nanomaterials

4.1 Nanomaterials in Medical Imaging

Medical imaging is being transformed by nanomaterials, which improve contrast, resolution, and sensitivity. This, in turn, allows for earlier and more accurate illness identification. Fluorescence, magnetic resonance imaging (MRI), computed tomography (CT), and positron emission tomography (PET) are all areas where their exceptional optical, magnetic, and biological characteristics shine. The following are:

- Iron oxide nanoparticles could be used as contrast agents in magnetic resonance imaging (MRI), which could make it much easier to find brain problems, cancers, and inflammatory diseases. Because of their superparamagnetic properties, they improve picture clarity by amplifying signal contrast.
- With CT (Computed Tomography), organs, blood vessels, and tumors can be seen very clearly because gold nanoparticles are very good at absorbing X-rays. They are perfect for targeted imaging due to their biocompatibility and scalability adjustments.
- PET, or positron emission tomography, is one of the most important tools for diagnosing cancer and keeping an eye on treatment. It uses radiolabeled nanoparticles, which are made of nanomaterials like silica or polymers, to give very clear pictures of how cells are using energy.
- Imaging with Fluorescence: Nanomaterials made of carbon and quantum dots allow for the real-time imaging of molecular and cellular structures, which aids in the early detection of neurological diseases and cancer.

Clinicians can increase diagnosis accuracy, pave the way for individualized treatment plans, and boost patient outcomes by combining nanotechnology with medical imaging.

4.2 Functionalized Nanoparticles for Targeted Imaging

Functionalized nanoparticles are transforming medical diagnostics by increasing specificity, sensitivity, and resolution, thereby reshaping targeted imaging. The nanoparticles include ligands, antibodies, and peptides on their surfaces that bind to disease biomarkers. This allows for high-resolution imaging of afflicted tissues with little background noise.

One example is the use of tumor-specific ligands coupled to gold nanoparticles, which increase the contrast of optical and CT scans, making them more useful for cancer imaging. The use of antibody-functionalized iron oxide nanoparticles allows for high-resolution MRI to identify neurological disorders in their early stages. Quantum dots, known for their luminous properties, enable real-time cellular imaging in situations like infections and cardiovascular disorders. Personalized medicine, better treatment planning, and less intrusive diagnostic procedures are all made possible by functionalized nanoparticles, which allow early and accurate illness diagnosis. Their ongoing development will revolutionize non-invasive imaging, leading to more accurate and efficient diagnosis.

4.3 Role in Early Disease Detection

Nanotechnology-powered imaging agents revolutionize early disease diagnosis. These agents allow for the real-time observation of molecular and cellular structures with unprecedented resolution. Magnetic resonance imaging (MRI), computed tomography (CT), and positron emission tomography (PET) scans are examples of conventional imaging methods that often identify illnesses at advanced stages, which limits the effectiveness of therapy. Nanoprobes and contrast agents made from nanomaterials, on the other hand, make sensitivity and specificity much better. This makes it possible to find illnesses early and treat them.

For example, quantum dots and gold nanoparticles enhance cancer diagnosis by focusing on certain tumor biomarkers, allowing for accurate imaging even before symptoms manifest. Neurological diseases, including Alzheimer's and Parkinson's may be better detected using magnetic resonance imaging (MRI) images that include iron oxide nanoparticles. Improving patient outcomes, lowering healthcare costs, and enabling early treatments are all made possible by imaging agents based on nanotechnology that combine focused imaging with molecular recognition and real-time monitoring. Their impact on preventive diagnostics is really groundbreaking and is changing the face of contemporary medicine.

5 Integration of Multifunctional Nanomaterials

5.1 Hybrid Nanomaterials for Simultaneous Detection and Imaging

Hybrid nanomaterials combine many functional characteristics to allow molecular imaging and detection of illnesses all at once. Polymeric architectures, carbon nanoparticles, quantum dots, and metallic nanoparticles are all part of these nanomaterials' arsenal for improving biocompatibility, sensitivity, and multifunctional imaging. For instance, mixtures of gold nanoparticles and quantum dots improve optical and fluorescence imaging. Meanwhile, iron oxide and gold composites allow photothermal nanotechnology imaging agents coof carbon also enhance bioimaging and electrochemical sensing.

There is no need for several diagnostic methods when using hybrid nanomaterials for real-time illness monitoring due to their multifunctionality. The bioconjugation with antibodies or peptides can improve their ability to target and make it easier to find harmful cells or infectious organisms. Nanotechnological materials are paving the way for theranostics, an integrated diagnosis and treatment system that could aid in early illness intervention and personalized medicine.

5.2 Smart Nanoplatforms for Real-Time Monitoring of Nanomaterials

Smart nanoplatforms are a significant advancement in real-time monitoring (Table 3) due to their integration of wireless communication technologies, responsive materials, and nanosensors. These systems allow for the real-time monitoring of biological indicators, environmental pollutants, and illness development, with immediate input allowing for prompt action. For finding tiny molecular changes in the body, smart nanoplatforms use quantum dots, functionalized nanoparticles, and carbon-based nanostructures. Wearable biosensors coated with nanomaterials, for instance, enable diabetic patients to check their glucose levels non-invasively. Implanted nanosensors have similar capabilities, allowing for real-time monitoring of tumor development or medication metabolism.

These platforms are perfect for customized therapy because they are very sensitive, respond quickly, and cause little harm to patients. Furthermore, data analytics enabled by AI improve their capacity to identify early illness signs, which in turn decreases diagnostic mistakes. Smart nanoplatforms are revolutionizing remote healthcare monitoring by integrating nanotechnology biosensors and wireless connections to speed up diagnosis and improve patient care.

Table 3 Smart nanoplatforms are a significant advancement in real-time monitoring

Application	Function	References
Glucose monitoring	Real-time tracking of blood sugar levels via wearables	[137]
Cancer detection	Nanosensors detect tumor biomarkers at early stages	[138]
Cardiac monitoring	Wearable nanosensors track heart rate and abnormalities	[139]
Drug delivery tracking	Monitors drug absorption and metabolism in the body	[140]
Environmental sensing	Detects pollutants and toxins in air and water	[141]

5.3 Future Trends in Integrated Diagnostics for Nanomaterials

Advanced nanotechnology, artificial intelligence, and tailored medicine are propelling the fast evolution of integrated diagnostics. The direction of this discipline is toward flawless illness identification, monitoring, and therapy, thereby improving accuracy, speed, and accessibility. Smart nanomaterials, imaging agents, and biosensors are about to reshape medical treatment. Artificial intelligence (AI)-powered diagnostics is one new trend; in this approach, machine learning algorithms examine data collected in real-time by nanobiosensors to identify illnesses in their early stages. By miniaturizing diagnostic platforms, lab-on-a-chip technologies provide quick, portable, reasonably priced testing. Furthermore, multifunctional nanoprobes will enable simultaneous imaging and treatment, thereby producing theranostics of the next generation.

Cancer, cardiovascular, and infection biomarker surveillance will be continuous using smart wearable devices and wireless health monitoring systems. The novel approaches to organ-on-a-chip modeling and 3D bioprinting promise to revolutionize the way diseases are modeled and drugs are tested. Integrated diagnostics will change as nanotechnology develops toward real-time, non-invasive, customized healthcare solutions, therefore enhancing world health outcomes.

6 Challenges and Ethical Considerations for Nanomaterials

6.1 Biocompatibility and Toxicity Issues

Nanomaterials have the potential to revolutionize diagnostics, yet there are still serious worries about their biocompatibility and toxicity. Depending on their chemical make-up, surface charge, size, and shape, nanomaterials may interact with biological systems in ways that cause cytotoxicity, inflammation, or chronic accumulation in organs. For clinical applications, it is necessary to provide safe biodegradability with a minimum of negative effects. To evaluate their immunological response, toxicity limits, and long-term impacts, comprehensive in vitro and in vivo investigations are necessary. The goal of recent developments in green nanotechnology, biofunctional coatings, and surface changes is to increase biocompatibility so that medical diagnostics using nanomaterials may be done safely and effectively [142].

6.2 Regulatory Challenges in Clinical Applications

Nanomaterials have a complicated make-up that makes it hard for regulators to allow imaging agents, biosensors, and nanoprobes to be used in therapy. Diagnostics based on nanomaterials must pass rigorous safety, effectiveness, and long-term impact evaluations before they can be approved by regulatory agencies, including the FDA, EMA, and ISO. The absence of established protocols for evaluating nanomaterials is a major obstacle. Their biocompatibility and toxicity are affected by their size, surface charge, functionalization, and other physicochemical properties that need to be evaluated in a certain way. Additionally, there is a lack of data from research examining long-term toxicity, which makes risk evaluations questionable [143].

The lack of clarity in nanodiagnostics' categorization is another concern. There is significant overlap between pharmaceutical and medical device regulatory regimes because many nanotechnology-based technologies can simultaneously diagnose and treat. Such complexity often slows down the approval and commercialization processes. Adaptive approval procedures, standardized worldwide rules, and risk assessment frameworks are crucial to overcoming these obstacles. To make sure that using nanotechnology for diagnostics is safe and effective in healthcare, researchers, regulatory bodies, and industry players need to work together more to come up with strict, flexible, and consistent rules [144]. The main regulatory challenges and potential remedies for use in clinical settings, as shown in the Table 4.

Table 4 The main regulatory challenges and potential remedies

Regulatory challenges	Impact	Potential remedies	References
Lack of standardized testing protocols	Unclear safety and toxicity assessments	Develop global testing standards for nanomaterials	[145]
Long-term toxicity concerns	Potential accumulation risks	Conduct extended in vivo and clinical studies	[146]
Classification ambiguity	Delays in regulatory approval	Define clear categories for nano diagnostics	[147]
Multi-functional device regulation issues	Slower approval due to mixed classification	Create hybrid regulatory pathways	[148]
Limited risk assessment frameworks	Uncertainty in long-term safety	Introduce adaptive risk assessment models	[149]
Complex approval processes	Increased time and cost for commercialization	Streamline regulatory pathways	[150]

6.3 *Ethical Concerns in Nanomedicine*

The rapid development of nanomedicine has raised many ethical questions, focusing on patient privacy, accessibility, safety, and long-term consequences. Nanomaterials may build up in organs and cause unexpected side effects, raising significant concerns regarding long-term health risks. It is critical to conduct comprehensive clinical and preclinical studies before broad use. Another concern is ensuring universal access to nanotechnology-powered tests. The healthcare gap between the wealthy and the poor can widen because of high development costs leading to restricted supply. Innovative nanomedicine solutions should be broadly available and inexpensive, according to ethical standards [151].

Smart nanodevices and real-time biosensors both gather private patient data in real-time, which raises security concerns. The rights of patients must be protected by stringent data protection regulations and informed consent procedures. Finally, to ensure that ethical values drive sustainable progress in nanomedicine, it is necessary to address the environmental effect of producing and disposing of nanomaterials [152].

7 International Trade and Upcoming of Nanomaterials

When it comes to basic biological problems in healthcare, medicine, and human existence, nanotechnology offers a brilliant solution. It facilitates the identification of infections, poisons, and pesticides; increases cancer imaging; and improves drug delivery. Research institutions are studying nanosystems, and agencies like the FDA and the National Center for Nanotechnology and Technologies have approved some of them [153]. The biomedical applications sector led the market, with 29.98% of sales in 2020. It is expected that the world market for nanoparticles will grow at a rate of 19.86% per year from 2021 to 2026 and hit USD 57,608 million. The biomedical industry's extensive use of nanomaterials in imaging, targeted medication delivery, surgical robots, nanodiagnosis, cell healing, and nanobiosensors is responsible for this expansion [154].

An excess of novel biomedical applications based on recently identified nanomaterials are becoming possible thanks to the significance of nanotechnology and nanosciences. But there are several obstacles that will slow down the expansion of nanomaterials' application, such as:

(a) making a lot of nanomaterials at a reasonable price;
(b) coming up with manufacturing methods that can be scaled up to meet the costs of going after large markets;
(c) quickly deciding on research priorities to make sure that nanotechnologies and nanomaterials will be safe in the future;
(d) filling in research gaps in risk, toxicity, and safety evaluation for the fields of nanotechnology;

(e) creating an international standard for nanomaterial safety to help people figure out the best way to handle nanomaterial risks [155].

Another major challenge in formulation development will be the repeatability and scaling up of nanomedicines [156]. A new chapter will be written in the annals of medicine in the next decade as nanomaterials make strides toward their long-term viability. Treatment tactics for several illnesses will undergo a revolutionary shift in the coming years due to the synthesis of novel nanomaterials. The future applications of nanomaterial research include biomedicine (bioimaging, biosensors), gardening, environmental safety, and food preparation, to name just a few [157].

8 Forthcoming Perspectives

8.1 *Innovations in Next-Generation Biosensors and Imaging Agents*

Modern biosensors and imaging agents are being developed thanks to breakthroughs in nanotechnology and biotechnology. These innovations will improve the ability to diagnose and monitor diseases, as well as provide individualized medication. To greatly enhance sensitivity, selectivity, and the ability to detect in real-time, modern biosensors use nanomaterials like graphene, quantum dots, carbon nanotubes, and gold nanoparticles. Cancer, viruses, and metabolic problems can now be diagnosed quickly and with few samples thanks to these new technologies [158].

Wearable biosensors are one of the most promising innovations, allowing for continuous health monitoring through non-invasive methods like saliva or sweat analysis. Lab-on-a-chip technologies also integrate many diagnostic purposes in a small system, therefore lowering costs and improving access. Real-time viewing of cellular events, made possible by smart imaging probes, helps diagnose early diseases and guide therapy precisely [159].

Artificial intelligence-powered biosensors can instantly analyze complex biological data, enhancing accuracy and enabling customized diagnoses. The advent of more accessible, quicker, and more accurate diagnostic tools is revolutionizing healthcare, leading to better patient outcomes and better disease management. Biosensors and imaging agents' future resides in their downsizing, automation, and multi-functionality, hence redefining biomedical innovation norms [126].

8.2 *Potential Impact on Personalized Medicine*

Personalized medicine is seeing a revolution because of the integration of imaging agents and modern biosensors. Such technology allows for therapies to be individualized based on unique profiles of genetics, biochemistry, and physiological factors.

Thanks to advances in nanotechnology and artificial intelligence, next-generation biosensors are now able to monitor biomarkers in real time. The development opens up new possibilities for early illness identification, accurate diagnoses, and personalized treatment plans [160]. Imaging agents derived from nanomaterials improve the localization of illnesses like cancer, letting physicians track the disease's development and the efficacy of treatments with greater precision. Drug delivery systems also benefit from these advancements, which help in dosing, reducing adverse effects, and increasing pharmaceutical effectiveness [161].

Medical tools like lab-on-a-chip and wearable biosensors let people keep track of their health, which cuts down on hospital visits and encourages responsible healthcare management. Precision healthcare is rapidly approaching, with personalized medicine spearheading the transition from a standard approach to one that is customized for each individual patient through the utilization of genomes, big data analytics, and smart diagnostics [162].

8.3 Emerging Research and Clinical Applications

Clinical diagnosis and treatment have advanced significantly as a result of creative research on biosensors and imaging agents. The ability of nano-engineered biosensors to identify biomarkers with previously unheard-of sensitivity may allow for the early detection of conditions including diabetes, cancer, and infectious diseases. With innovations like wearable biosensors and lab-on-a-chip, health signs can be tracked in real time without invasive procedures, which lowers the number of clinical tests that need to be done [163]. MRI, PET, and fluorescence imaging are all getting better with the help of nanoparticle-based contrast agents that make it easier to find cancer and other sick cells in specific areas. Functionalized nanoparticles enhance drug distribution, minimizing adverse effects and improving therapeutic efficacy [164].

Biosensors driven by artificial intelligence are revolutionizing data analysis by providing more accurate automated real-time diagnosis. These developments are changing the face of disease management in hospitals and clinics, leading to better patient outcomes and personalized and predictive medicine. Improved biocompatibility, downsizing, and integration are the goals of future research on next-generation medical applications [126].

9 Conclusion

In this summary, recent advancements in imaging agents, biosensors, and nanoprobes are revolutionizing medical diagnostics by improving sensitivity, accuracy, and real-time monitoring. Nanotechnology-driven biosensors, such as graphene, quantum dots, and gold nanoparticles, can rapidly and precisely identify biomarkers for infections, cancer, and diabetes. Wearable and lab-on-a-chip biosensors will reduce the reliance on traditional clinical testing by enabling non-invasive, real-time health monitoring. Additionally, nanoprobes like fluorescence and plasmonic nanoparticles simplify the detection of diseases at the molecular level. This leads to earlier diagnoses and more accurate treatment. Nanoparticle-based contrast agents improve optical, MRI, and PET scans, enabling more precise and detailed viewing of tumors and tissues. Biosensors with AI built in enhance data processing even further, leading to more precise diagnoses and the possibility of tailored treatment plans. A new standard in precision medicine and revolutionary disease management is being established by these innovations, which together move healthcare toward quicker, cheaper, and patient-specific diagnostics.

Nanotechnology possesses the capacity to fundamentally transform the diagnostic landscape by facilitating less invasive, more precise, and expedited disease diagnosis. Emerging advancements within the discipline will concentrate on intelligent biosensors, diagnostics powered by artificial intelligence, and multifunctional nanoprobes capable of monitoring health in real-time. The evolution of wearable nanotechnology devices has the potential to enhance early disease diagnosis and personalized therapy through the continuous monitoring of biomarkers. Magnetic Resonance Imaging (MRI), Positron Emission Tomography (PET), and optical imaging techniques will demonstrate greater precision when utilized with nano-enhanced imaging agents. These innovations will support targeted treatments. Efforts are underway to create more accessible and cost-effective diagnostic solutions, driven by a commitment to biocompatibility and miniaturization in ongoing research. Nanotechnology is significantly influencing the realms of predictive healthcare, personalized medicine, and transformative disease management.

Acknowledgements The authors are grateful to Dr. C. V. Raman University, Bilaspur, Chhattisgarh, India for this work.

Authors' Contributions Milan Hait and Amit Kumar Chaturwedi wrote different sections of the manuscript. Milan Hait and Amit Kumar Chaturwedi edited the manuscript. All authors read and approved the final manuscript.

Ethics Approval Not Applicable.

Competing Interests The authors have no conflicts of interest to declare that are relevant to the content of this chapter.

References

1. Sim, S., Wong, N.K.: Nanotechnology and its use in imaging and drug delivery (Review). Biomed. Rep. **14**(5), 42 (2021). https://doi.org/10.3892/br.2021.1418
2. Adir, O., Poley, M., Chen, G., et al.: Integrating artificial intelligence and nanotechnology for precision cancer medicine. Adv. Mater. **32**(13), e1901989 (2020). https://doi.org/10.1002/adma.201901989
3. Pirzada, M., Altintas, Z.: Nanomaterials for healthcare biosensing applications. Sensors **19**(23), 5311 (2019). https://doi.org/10.3390/s19235311

4. Thwala, L.N., Ndlovu, S.C., Mpofu, K.T., et al.: Nanotechnology-based diagnostics for diseases prevalent in developing countries: current advances in point-of-care tests. Nanomaterials **13**(7), 1247 (2023). https://doi.org/10.3390/nano13071247
5. Combes, G.F., Vučković, A.M., Perić Bakulić, M., et al.: Nanotechnology in tumor biomarker detection: the potential of liganded nanoclusters as nonlinear optical contrast agents for molecular diagnostics of cancer. Cancers **13**(16), 4206 (2021). https://doi.org/10.3390/cancers13164206
6. Prasad, R., Karvekar, O., Powar, P., et al.: A review on nanotechnological aspects in veterinary medicine. Res. J. Life Sci. Bioinform. Pharm. Chem. Sci. **7** (2021). https://doi.org/10.26479/2021.0706.04
7. Dudchenko, N., Pawar, S., Perelshtein, I., et al.: Magnetite nanoparticles: synthesis and applications in optics and nanophotonics. Materials **15**(7), 2601 (2022). https://doi.org/10.3390/ma15072601
8. Jeevanandam, J., Barhoum, A., Chan, Y.S., et al.: Review on nanoparticles and nanostructured materials: history, sources, toxicity and regulations. Beilstein J. Nanotechnol. **9**, 1050–1074 (2018). https://doi.org/10.3762/bjnano.9.98
9. Alapan, Y., et al.: Micro- and nanodevices integrated with biomolecular probes. Biotechnol. Adv. **33**(8), 1727–1743 (2015). https://doi.org/10.1016/j.biotechadv.2015.09.001
10. Hassan, R.Y.A.: Advances in electrochemical nano-biosensors for biomedical and environmental applications: from current work to future perspectives. Sensors **22**(19), 7539 (2022). https://doi.org/10.3390/s22197539
11. Banerjee, A., et al.: Nanotechnology for biosensors: a review (2021). https://doi.org/10.20944/preprints202101.0152.v1
12. Hassan, S.A.H., Almaliki, M.N.S., Hussein, Z.A., et al.: Development of nanotechnology by artificial intelligence: a comprehensive review. J. Nanostruct. **13**(4), 915–932 (2023). https://doi.org/10.22052/JNS.2023.04.002
13. Zhang, X., et al.: Recent advances in nanotechnology applied to biosensors. Sensors **9**(2), 1033–1053 (2009). https://doi.org/10.3390/s90201033
14. Vigneshvar, S., Sudhakumari, C.C., Senthilkumaran, B., et al.: Recent advances in biosensor technology for potential applications—an overview. Front. Bioeng. Biotechnol. **4**, 11 (2016). https://doi.org/10.3389/fbioe.2016.00011
15. Bhatia, D., Paul, S., Acharjee, T., et al.: Biosensors and their widespread impact on human health. Sens. Int. **5**, 100257 (2024). https://doi.org/10.1016/j.sintl.2023.100257
16. Suslu, B., et al.: Understanding the role of sensor optimization in complex systems. Sensors **23**(18), 7819 (2023). https://doi.org/10.3390/s23187819
17. Joudeh, N., Linke, D.: Nanoparticle classification, physicochemical properties, characterization, and applications: a comprehensive review for biologists. J. Nanobiotechnol. **20**, 262 (2022). https://doi.org/10.1186/s12951-022-01477-8
18. Lamba, R., Yukta, Y., Mondal, J., et al.: Carbon dots: synthesis, characterizations, and recent advancements in biomedical, optoelectronics, sensing, and catalysis applications. ACS Appl. Bio Mater. **7**(4), 2086–2127 (2024). https://doi.org/10.1021/acsabm.4c00004
19. Altammar, K.A.: A review on nanoparticles: characteristics, synthesis, applications, and challenges. Front. Microbiol. **14**, 1155622 (2023). https://doi.org/10.3389/fmicb.2023.1155622
20. Yusaf, T., Mahamude, A.S.F., Farhana, K., et al.: A comprehensive review on graphene nanoparticles: preparation, properties, and applications. Sustainability **14**(19), 12336 (2022). https://doi.org/10.3390/su141912336
21. Kolahdouz, M., Xu, B., Nasiri, A.F., et al.: Carbon-related materials: graphene and carbon nanotubes in semiconductor applications and design. Micromachines **13**(8), 1257 (2022). https://doi.org/10.3390/mi13081257
22. Venkataraman, A., Amadi, E.V., Chen, Y., et al.: Carbon nanotube assembly and integration for applications. Nanoscale Res. Lett. **14**, 220 (2019). https://doi.org/10.1186/s11671-019-3046-3

23. Rahamathulla, M., Bhosale, R.R., Osmani, R.A.M., et al.: Carbon nanotubes: current perspectives on diverse applications in targeted drug delivery and therapies. Materials (Basel) **14**(21), 6707 (2021). https://doi.org/10.3390/ma14216707
24. Sacco, L.N., Vollebregt, S.: Overview of engineering carbon nanomaterials such as carbon nanotubes (CNTs), carbon nanofibers (CNFs), graphene and nanodiamonds and other carbon allotropes inside porous anodic alumina (PAA) templates. Nanomaterials **13**(2), 260 (2023). https://doi.org/10.3390/nano13020260
25. Elmaghraby, N.A., Hassaan, M.A., Zien, M.A., et al.: Fabrication of carbon black nanoparticles from green algae and sugarcane bagasse. Sci. Rep. **14**(1), 5542 (2024). https://doi.org/10.1038/s41598-024-56157-4
26. Singh, J., Dutta, T., Kim, K.H., et al.: 'Green' synthesis of metals and their oxide nanoparticles: applications for environmental remediation. J. Nanobiotechnol. **16**, 84 (2018). https://doi.org/10.1186/s12951-018-0408-4
27. Burlec, A.F., Corciova, A., Boev, M., et al.: Current overview of metal nanoparticles' synthesis, characterization, and biomedical applications, with a focus on silver and gold nanoparticles. Pharmaceuticals **16**(10), 1410 (2023). https://doi.org/10.3390/ph16101410
28. Ijaz, I., Gilani, E., Nazir, A., et al.: Detail review on chemical, physical and green synthesis, classification, characterizations and applications of nanoparticles. Green Chem. Lett. Rev. **13**(3), 223–245 (2020). https://doi.org/10.1080/17518253.2020.1802517
29. Pach, A., Szot, A., Fitzner, K., et al.: Opportunities and challenges in the synthesis of noble metal nanoparticles via the chemical route in microreactor systems. Micromachines **15**(9), 1119 (2024). https://doi.org/10.3390/mi15091119
30. Palza, H.: Antimicrobial polymers with metal nanoparticles. Int. J. Mol. Sci. **16**, 2099–2116 (2015). https://doi.org/10.3390/ijms16012099
31. Turner, R.J.: The good, the bad, and the ugly of metals as antimicrobials. Biometals **37**, 545–559 (2024). https://doi.org/10.1007/s10534-023-00565-y
32. Fazio, E., Spadaro, S., Corsaro, C., et al.: Metal-oxide based nanomaterials: synthesis, characterization and their applications in electrical and electrochemical sensors. Sensors **21**(7), 2494 (2021). https://doi.org/10.3390/s21072494
33. Gudkov, S.V., Burmistrov, D.E., Pa, F., et al.: Antibacterial properties of copper oxide nanoparticles (review). Int. J. Mol. Sci. **25**(21), 11563 (2024). https://doi.org/10.3390/ijms252111563
34. Khairy, T., Amin, D.H., Salama, H.M., et al.: Antibacterial activity of green synthesized copper oxide nanoparticles against multidrug-resistant bacteria. Sci. Rep. **14**(1), 25020 (2024). https://doi.org/10.1038/s41598-024-75147-0
35. Di Marco, M., Sadun, C., Port, M., et al.: Physicochemical characterization of ultrasmall superparamagnetic iron oxide particles (USPIO) for biomedical application as MRI contrast agents. Int. J. Nanomed. **2**(4), 609–622 (2007). https://doi.org/10.2147/IJN.S2.4.609
36. Naz, S., Gul, A., Zia, M., et al.: Synthesis, biomedical applications, and toxicity of CuO nanoparticles. Appl. Microbiol. Biotechnol. **107**(4), 1039–1061 (2023). https://doi.org/10.1007/s00253-023-12364-z
37. Rahman, M.: Magnetic resonance imaging and iron-oxide nanoparticles in the era of personalized medicine. Nanotheranostics **7**(4), 424–449 (2023). https://doi.org/10.7150/ntno.86467
38. Adeeyo, A.O., Alabi, M.A., Oyetade, J.A., et al.: Magnetic nanoparticles: advances in synthesis, sensing, and theragnostic applications. Magnetochemistry **11**(2), 9 (2025). https://doi.org/10.3390/magnetochemistry11020009
39. Israel, L.L., Galstyan, A., Holler, E., et al.: Magnetic iron oxide nanoparticles for imaging, targeting and treatment of primary and metastatic tumors of the brain. J. Control. Release **320**, 45–62 (2020). https://doi.org/10.1016/j.jconrel.2020.01.009
40. Mavridi-Printezi, A., Guernelli, M., Menichetti, A., et al.: Bio-applications of multifunctional melanin nanoparticles: from nanomedicine to nanocosmetics. Nanomaterials **10**(11), 2276 (2020). https://doi.org/10.3390/nano10112276

41. Wang, Y.X.: Superparamagnetic iron oxide-based MRI contrast agents: current status of clinical application. Quant. Imaging Med. Surg. **1**(1), 35–40 (2011). https://doi.org/10.3978/j.issn.2223-4292.2011.08.03
42. Al-Harbi, N., Abd-Elrahman, N.K.: Physical methods for preparation of nanomaterials, their characterization and applications: a review. J. Umm Al-Qura Univ. Appl. Sci. (2024). https://doi.org/10.1007/s43994-024-00165-7
43. Iqbal, J., Abbasi, B.A., Ahmad, R., et al.: Biogenic synthesis of green and cost-effective iron nanoparticles and evaluation of their potential biomedical properties. J. Mol. Struct. **1199**, 126979 (2020). https://doi.org/10.1016/j.molstruc.2019.126979
44. Petrovic, S., et al.: Nano formulations in pharmaceutical and biomedical applications: green perspectives. Int. J. Mol. Sci. **25**(11), 5842 (2024). https://doi.org/10.3390/ijms25115842
45. Zielińska, A., Carreiró, F., Oliveira, A.M., et al.: Polymeric nanoparticles: production, characterization. Toxicol. Ecotoxicol. Mol. **25**(16), 3731 (2020). https://doi.org/10.3390/molecules25163731
46. Din, F.U., Aman, W., Ullah, I., et al.: Effective use of nanocarriers as drug delivery systems for the treatment of selected tumors. Int. J. Nanomed. **12**, 7291–7309 (2017). https://doi.org/10.2147/IJN.S146315
47. Jung, H.N., Lee, S.Y., Lee, S., et al.: Lipid nanoparticles for delivery of RNA therapeutics: current status and the role of *in vivo* imaging. Theragnostic **12**(17), 7509–7531 (2022). https://doi.org/10.7150/thno.77259
48. Bayda, S., Adeel, M., Tuccinardi, T., et al.: The history of nanoscience and nanotechnology: from chemical-physical applications to nanomedicine. Molecules **25**(1), 112 (2019). https://doi.org/10.3390/molecules25010112
49. Khan, Y., Sadia, H., Ali Shah, S.Z., et al.: Classification, synthetic, and characterization approaches to nanoparticles, and their applications in various fields of nanotechnology: a review. Catalysts **12**(11), 1386 (2022). https://doi.org/10.3390/catal12111386
50. Fu, L.-M., Hsu, J.-H., Shih, M.-K., et al.: Process optimization of silver nanoparticle synthesis and its application in mercury detection. Micromachines **12**(9), 1123 (2021). https://doi.org/10.3390/mi12091123
51. Malatesta, M.: Transmission electron microscopy as a powerful tool to investigate the interaction of nanoparticles with subcellular structures. Int. J. Mol. Sci. **22**(23), 12789 (2021). https://doi.org/10.3390/ijms222312789
52. Skoglund, S., Hedberg, J., Yunda, E., et al.: Difficulties and flaws in performing accurate determinations of zeta potentials of metal nanoparticles in complex solutions-four case studies. PLoS ONE **12**(7), e0181735 (2017). https://doi.org/10.1371/journal.pone.0181735
53. Pasieczna-Patkowska, S., et al.: Application of Fourier transform infrared (FTIR) spectroscopy in characterization of green synthesized nanoparticles. Molecules **30**(3), 684 (2025). https://doi.org/10.3390/molecules30030684
54. Ali, A., et al.: X-ray diffraction techniques for mineral characterization: a review for engineers of the fundamentals, applications, and research directions. Minerals **12**(2), 205 (2022). https://doi.org/10.3390/min12020205
55. De la Mata, M., Molina, S.I.: STEM tools for semiconductor characterization: beyond high-resolution imaging. Nanomaterials **12**(3), 337 (2022). https://doi.org/10.3390/nano1203033
56. Syduzzaman, Md., Islam Saad, Md.S., Piam, M.F., et al.: Carbon nanotubes: structure, properties and applications in the aerospace industry. Results Mater. **25**, 100654 (2025). https://doi.org/10.1016/j.rinma.2024.100654
57. Sivasubramaniyam, V., Ramasamy, S., Venkatraman, M., et al.: Carbon nanotubes as an alternative to copper wires in electrical machines: a review. Energies **16**(9), 3665 (2023). https://doi.org/10.3390/en16093665
58. Sang, M., Shin, J., Kim, K., et al.: Electronic and thermal properties of graphene and recent advances in graphene based electronics applications. Nanomaterials **9**(3), 374 (2019). https://doi.org/10.3390/nano9030374
59. Elizalde-Herrera, F.J., Flores-Soto, P.A., Mora-Cortes, L.F., et al.: Recent development of graphene-based composites for electronics, energy storage, and biomedical applications: a review. J. Compos. Sci. **8**(11), 481 (2024). https://doi.org/10.3390/jcs8110481

60. Giroux, M., Zahra, Z., Salawu, O.A., et al.: Assessing the environmental effects related to quantum dot structure, function, synthesis and exposure. Environ. Sci. Nano **9**(3), 867–910 (2022). https://doi.org/10.1039/d1en00712b
61. Shabbir, H., Wojnicki, M.: Recent progress of non-cadmium and organic quantum dots for optoelectronic applications with a focus on photodetector devices. Electronics **12**(6), 1327 (2023). https://doi.org/10.3390/electronics12061327
62. Mcoyi, M.P., Mpofu, K.T., Sekhwama, M., et al.: Developments in localized surface plasmon resonance. Plasmonic (2024). https://doi.org/10.1007/s11468-024-02620-x
63. Mohammed, H., Mia, M.F., Wiggins, J., et al.: Nanomaterials for energy storage systems—a review. Molecules **30**(4), 883 (2025). https://doi.org/10.3390/molecules3004088
64. Liu, X., Long, Y.Z., Liao, L., et al.: Large-scale integration of semiconductor nanowires for high-performance flexible electronics. ACS Nano **6**(3), 1888–1900 (2012). https://doi.org/10.1021/nn204848r
65. Mim, J.J., Hasan, M., Chowdhury, M.S., et al.: A comprehensive review on the biomedical frontiers of nanowire applications. Heliyon **10**(8), e29244 (2024). https://doi.org/10.1016/j.heliyon.2024.e29244
66. Abbasi, R., Shineh, G., Mobaraki, M., et al.: Structural parameters of nanoparticles affecting their toxicity for biomedical applications: a review. J. Nanopart. Res. **25**, 43 (2023). https://doi.org/10.1007/s11051-023-05690-w
67. Yusuf, A., Almotairy, A.R.Z., Henidi, H., et al.: Nanoparticles as drug delivery systems: a review of the implication of nanoparticles' physicochemical properties on responses in biological systems. Polymers **15**(7), 1596 (2023). https://doi.org/10.3390/polym15071596
68. Bhalla, N., Jolly, P., Formisano, N., et al.: Introduction to biosensors. Essays Biochem. **60**(1), 1–8 (2016). https://doi.org/10.1042/EBC20150001
69. Kumari, M., Gupta, V., Kumar, N., et al.: Microfluidics-based nanobiosensors for healthcare monitoring. Mol. Biotechnol. **66**, 378–401 (2024). https://doi.org/10.1007/s12033-023-00760-9
70. Naresh, V., Lee, N.: A review on biosensors and recent development of nanostructured materials-enabled biosensors. Sensors **21**(4), 1109 (2021). https://doi.org/10.3390/s21041109
71. Singh, A., Sharma, A., Ahmed, A., et al.: Recent advances in electrochemical biosensors: applications, challenges, and future scope. Biosensors **11**(9), 336 (2021). https://doi.org/10.3390/bios11090336
72. Peltomaa, R., Glahn-Martínez, B., Benito-Peña, E., et al.: Optical biosensors for label-free detection of small molecules. Sensors **18**(12), 4126 (2018). https://doi.org/10.3390/s18124126
73. Pohanka, M.: Overview of piezoelectric biosensors, immunosensors and dna sensors and their applications. Materials **11**(3), 448 (2018). https://doi.org/10.3390/ma11030448
74. Mehrotra, P.: Biosensors and their applications—a review. J. Oral Biol. Craniofac. Res. **6**(2), 153–159 (2016). https://doi.org/10.1016/j.jobcr.2015.12.002
75. Mokhtarzadeh, A., Eivazzadeh-Keihan, R., Pashazadeh, P., et al.: Nanomaterial-based biosensors for detection of pathogenic virus. Trends Anal. Chem. TRAC **97**, 445–457 (2017). https://doi.org/10.1016/j.trac.2017.10.005
76. Ramesh, M., Janani, R., Deepa, C., et al.: Nanotechnology-enabled biosensors: a review of fundamentals, design principles, materials, and applications. Biosensors **13**(1), 40 (2022). https://doi.org/10.3390/bios13010040
77. Lin, P.C., Lin, S., Wang, P.C., et al.: Techniques for physicochemical characterization of nanomaterials. Biotechnol. Adv. **32**(4), 711–726 (2014). https://doi.org/10.1016/j.biotechadv.2013.11.006
78. Malhotra, B.D., Ali, M.A.: Nanomaterials in biosensors: fundamentals and applications. Nanomater. Biosens. 1–74 (2018). https://doi.org/10.1016/B978-0-323-44923-6.00001-7
79. Piroozmand, F., et al.: Emerging biosensors in detection of natural products. Synth. Syst. Biotechnol. **5**(4), 293–303 (2020). https://doi.org/10.1016/j.synbio.2020.08.002
80. Lopez Carrasco, I., Cuniberti, G., Opitz, J., et al.: Evaluation of transducer elements based on different material configurations for aptamer-based electrochemical biosensors. Biosensors **14**(7), 341 (2024). https://doi.org/10.3390/bios14070341

81. Hyde, K.D., Baldrian, P., Chen, Y., et al.: Current trends, limitations and future research in the fungi? Fungal Divers. **125**, 1–71 (2024). https://doi.org/10.1007/s13225-023-00532-5
82. Ghafar-Zadeh, E., et al.: Advances in electronic biosensors. In: Advanced CMOS Biochips. Analog Circuits and Signal Processing. Springer, Dordrecht (2025). https://doi.org/10.1007/978-94-007-0099-4_5
83. Damborský, P., et al.: Optical biosensors. Essays Biochem. **60**(1), 91–100 (2016). https://doi.org/10.1042/EBC20150010
84. Luo, T., Liu, W., Wen, Z., et al.: A high-sensitivity gravimetric biosensor based on s_1 mode lamb wave resonator. Sensors **22**(15), 5912 (2022). https://doi.org/10.3390/s22155912
85. Kawashima, M., Varnava, M., Ozawa, S., et al.: Evaluation of monthly variations in electrometer calibration coefficients using a charge generator for radiation therapy. Radiol. Phys. Technol. **17**, 770–775 (2024). https://doi.org/10.1007/s12194-024-00830-w
86. Chircov, C., Bîrcă, A.C., Grumezescu, A.M., et al.: Biosensors-on-chip: an up-to-date review. Molecules **25**(24), 6013 (2020). https://doi.org/10.3390/molecules25246013
87. Khan, M.Q., Khan, J., Alvi, M.A.H., et al.: Nanomaterial-based sensors for microbe detection: a review. Discov. Nano **19**(1), 120 (2024). https://doi.org/10.1186/s11671-024-04065-x
88. Tovar-Lopez, F.J.: Recent progress in micro- and nanotechnology-enabled sensors for biomedical and environmental challenges. Sensors **23**(12), 5406 (2023). https://doi.org/10.3390/s23125406
89. Uddin, R., Koo, I.: Real-time remote patient monitoring: a review of biosensors integrated with multi-hop IoT systems via cloud connectivity. Appl. Sci. **14**(5), 1876 (2024). https://doi.org/10.3390/app14051876
90. Macdonald, A., Hawkes, L.A., Corrigan, D.K.: Recent advances in biomedical, biosensor and clinical measurement devices for use in humans and the potential application of these technologies for the study of physiology and disease in wild animals. Phil. Trans. R. Soc. B 20200228 (2021). https://doi.org/10.1098/rstb.2020.0228.
91. Song, Y.: Artificial intelligence algorithms in biomedical application. In: 2023 International Conference on Intelligent Supercomputing and BioPharma (ISBP), Zhuhai, China, pp. 42–47 (2023). https://doi.org/10.1109/ISBP57705.2023.10061317.
92. Danielsson, K., et al.: Personalizing pancreatic cancer medicine: what are the challenges? Pers. Med. **10**(1), 45–59 (2013).
93. Miranda-Castro, R., de-Los-Santos-Alvarez, N., Lobo-Castañón, M.J., et al.: PCR-coupled electrochemical sensing of Legionella pneumophila. Biosens. Bioelectron. **24**(8), 2390–2396 (2009).
94. Cruz, S.M.A., Girão, A.F., Gonçalves, G., et al.: Graphene: the missing piece for cancer diagnosis? Sensors **16**(1), 137 (2016). https://doi.org/10.3390/s16010137
95. Hanna, T., et al.: Treatment, return to play, and performance following meniscus surgery. Curr. Rev. Musculoskelet. Med. **15**(3), 157–169 (2022). https://doi.org/10.1007/s12178-022-09754-7
96. Imas, J.J., Ruiz, Z.C., Zubiate, P., et al.: Optical biosensors for the detection of rheumatoid arthritis (RA) biomarkers: a comprehensive review. Sensors **20**(21), 6289 (2020).
97. Dzobo, K., Senthebane, D.A., Rowe, A., et al.: Cancer stem cell hypothesis for therapeutic innovation in clinical oncology? taking the root out, not chopping the leaf. Omics. **20**(12), 681–691 (2016). https://doi.org/10.1089/omi.2016.0152
98. Fedi, A., Vitale, C., Giannoni, P., et al.: Biosensors to monitor cell activity in 3D hydrogel-based tissue models. Sensors **22**(4), 1517 (2022).
99. Remaggi, G., et al.: 3D printing technologies in biosensors production: recent developments. Chemosensors **10**(2), 65 (2022). https://doi.org/10.3390/chemosensors10020065
100. Vadgama, P.: Monitoring with in-vivo electrochemical sensors: navigating the complexities of blood and tissue reactivity. Sensors **20**(11), 3149 (2020). https://doi.org/10.3390/s20113149
101. Doseděl, M., Jirkovský, E., Macáková, K., et al.: Vitamin C-sources, physiological role, kinetics, deficiency, use, toxicity, and determination. Nutrients **13**(2), 615 (2021). https://doi.org/10.3390/nu13020615

102. Damiati, S., Kompella, U.B., Damiati, S.A., et al.: Microfluidic devices for drug delivery systems and drug screening. Genes **9**(2), 103 (2018). https://doi.org/10.3390/genes9020103
103. El Aamri, M., Yammouri, G., Mohammadi, H., et al.: Electrochemical biosensors for detection of microRNA as a cancer biomarker: pros and cons. Biosensors **10**(11), 186 (2020).
104. Erdem, Y., Inal, S., Sivaz, O., et al.: How does working in pandemic units affect the risk of occupational hand eczema in healthcare workers during the coronavirus disease-2019 (COVID-19) pandemic: a comparative analysis with nonpandemic units. Contact Dermatitis **85**(2), 215–224 (2021). https://doi.org/10.1111/cod.13853
105. Ahmad Najib, M., Selvam, K., Khalid, M.F., et al.: Quantum dot-based lateral flow immunoassay as point-of-care testing for infectious diseases: a narrative review of its principle and performance. Diagnostics **12**(9), 2158 (2022).
106. Banakar, M., Hamidi, M., Khurshid, Z., et al.: Electrochemical biosensors for pathogen detection: an updated review. Biosensors **12**(11), 927. https://doi.org/10.3390/bios12110927
107. Broadhurst, M.J., et al.: Diagnosis of Ebola virus disease: past, present, and future. Clin. Microbiol. Rev. **29**(4), 773–793 (2016).
108. Patel, D., Chaturwedi, A.K., Kashyap, N.K. et al.: Quality control and standardization of nanomaterials. In: Thakur A., Kumar A. (eds.) Sustainability, Safety, and Applications of Nanomaterials-Based Corrosion Inhibitors, pp. 297–325. IGI Global, USA (2024). https://doi.org/10.4018/979-8-3693-7640-9.ch012.
109. Teixeira, W., Pallás-Tamarit, Y., et al.: An all-in-one point-of-care testing device for multiplexed detection of respiratory infections. Biosens. Bioelectron. **213**, 114454 (2022). https://doi.org/10.1016/j.bios.2022.114454
110. Berezin, M.Y., Achilefu, S.: Fluorescence lifetime measurements and biological imaging. Chem. Rev. **110**(5), 2641–2684 (2010). https://doi.org/10.1021/cr900343z
111. Attia, M.F., Anton, N., Wallyn, J., et al.: An overview of active and passive targeting strategies to improve the nanocarriers efficiency to tumour sites. J. Pharm. Pharmacol. **71**(8), 1185–1198 (2019).
112. Omer, S., Zakar, R., Zakar, M.Z., Fischer, F.: The influence of social and cultural practices on maternal mortality: a qualitative study from South Punjab, Pakistan. Reprod. Health **18**, 97 (2021). https://doi.org/10.1186/s12978-021-01151-6
113. Johnston, L., Wang, G., Hu, K., et al.: Advances in biosensors for continuous glucose monitoring towards wearables. Front. Bioeng. Biotechnol. **9**, 733810 (2021). https://doi.org/10.3389/fbioe.2021.733810
114. Mohamad Nor, N., et al.: Progress of enzymatic and non-enzymatic electrochemical glucose biosensor based on nanomaterial-modified electrode. Biosensors **12**(12), 1136 (2022). https://doi.org/10.3390/bios12121136
115. Ibupoto, Z.H., et al.: The synthesis of CuO nanoleaves, structural characterization, and their glucose sensing application. Appl. Phys. Lett. **102**(10), 103701 (2013). https://doi.org/10.1063/1.4795135
116. Jeon, H.J., Kim, H.S., Chung, E., et al.: Nanozyme-based colorimetric biosensor with a systemic quantification algorithm for noninvasive glucose monitoring. Theranostics **12**(14), 6308 (2022). https://doi.org/10.7150/thno.72152
117. Hartati, Y.W., Irkham, I., Sumiati, I., et al.: The optimization of a label-free electrochemical DNA biosensor for detection of sus scrofa mtDNA as food adulterations. Biosensors **13**(6), 657 (2023). https://doi.org/10.3390/bios13060657
118. Wu, Q., Zhang, Y., Yang, Q., et al.: Review of electrochemical DNA biosensors for detecting food borne pathogens. Sensors **19**(22), 4916 (2019). https://doi.org/10.3390/s19224916
119. Dykman, L.A., Khlebtsov, N.G.: Gold nanoparticles in biology and medicine: recent advances and prospects. Acta Naturae **3**(2), 34–55 (2011). PMID: 22649683.
120. Bera, D., Qian, L., et al.: Quantum dots and their multimodal applications: a review. Materials **3**(4), 2260–2345 (2010). https://doi.org/10.3390/ma3042260
121. Li, M., Jia, X., Wang, L., et al.: Research on modified carbon nanotubes in wastewater treatment. Catalysts **12**(10), 1103 (2022). https://doi.org/10.3390/catal12101103.

122. Donato, K.Z., Tan, H.L., Marangoni, V.S., et al.: Graphene oxide classification and standardization. Sci. Rep. **13**, 6064 (2023). https://doi.org/10.1038/s41598-023-33350-5
123. Cho, H.S., Noh, M.S., Kim, Y.H., et al.: Recent studies on metal-embedded silica nanoparticles for biological applications. Nanomaterials **14**(3), 268 (2024). https://doi.org/10.3390/nano14 03026
124. Issa, B., Obaidat, I.M., Albiss, B.A., et al.: Magnetic nanoparticles: surface effects and properties related to biomedicine applications. Int. J. Mol. Sci. **14**(11), 21266–21305 (2013). https://doi.org/10.3390/ijms141121266
125. Zhang, X.F., Liu, Z.G., Shen, W., et al.: Silver nanoparticles: synthesis, characterization, properties, applications, and therapeutic approaches. Int. J. Mol. Sci. **17**(9), 1534 (2016). https://doi.org/10.3390/ijms17091534.
126. Wasilewski, T., et al.: AI-assisted detection of biomarkers by sensors and biosensors for early diagnosis and monitoring. Biosensors **14**(7), 356 (2024). https://doi.org/10.3390/bios14 070356
127. El-Sherif, D.M., Abouzid, M., Gaballah, M.S., et al.: New approach in SARS-CoV-2 surveillance using biosensor technology: a review. Environ. Sci. Pollut. Res. **29**, 1677–1695 (2022). https://doi.org/10.1007/s11356-021-17096-z
128. Karaboğa, M.N.S., Sezgintürk, M.K.: Biosensor approaches on the diagnosis of neurodegenerative diseases: sensing the past to the future. J. Pharm. Biomed. Anal. **209**, 114479 (2022). https://doi.org/10.1016/j.jpba.2021.114479
129. Tetyana, P., et al.: Biosensors: design, development and applications. Intech Open (2021). https://doi.org/10.5772/intechopen.97576
130. Williams, A., Aguilar, M.R., Pattiya Arachchillage, K.G.G., et al.: Biosensors for public health and environmental monitoring: the case for sustainable biosensing. ACS Sustain. Chem. Eng. **12**(28), 10296–10312 (2024). https://doi.org/10.1021/acssuschemeng.3c06112
131. Jin, X., Cai, A., Xu, T., et al.: Artificial intelligence biosensors for continuous glucose monitoring. Interdisp. Mater. **2**(2), 290–307 (2023).https://doi.org/10.1002/idm2.12069
132. Demilie, W.B.: Plant disease detection and classification techniques: a comparative study of the performances. J. Big Data **11**, 5 (2024). https://doi.org/10.1186/s40537-023-00863-9
133. Taherdoost, H.: Wearable healthcare and continuous vital sign monitoring with IoT integration. Comput. Mater. Contin. **81**(1), 79 (2024). https://doi.org/10.32604/cmc.2024.054378
134. Price, C.P., Kricka, L.J.: Improving healthcare accessibility through point-of-care technologies. Clin. Chem. **53**(9), 1665–1675 (2007). https://doi.org/10.1373/clinchem.2006.084707
135. Mohs, R.C., Greig, N.H.: Drug discovery and development: role of basic biological research. Alzheimer's Dement.: Transl. Res. Clin. Interv. **3**(4), 651–657 (2017). https://doi.org/10.1016/j.trci.2017.10.005
136. Cunin, T., et al.: Environmental monitoring based on Internet of Things Technology. In: 12th International Conference on Sensing Technology (ICST), pp. 31–34. Limerick, Ireland (2018). https://doi.org/10.1109/ICSensT.2018.8603658.
137. Hina, A., Saadeh, W.: Noninvasive blood glucose monitoring systems using near-infrared technology—a review. Sensors **22**(13), 4855 (2022). https://doi.org/10.3390/s22134855
138. Gromek, P., Senkowska, Z., Płuciennik, E., et al.: Revisiting the standards of cancer detection and therapy alongside their comparison to modern methods. World J. Methodol. **14**(2), 92982 (2024). https://doi.org/10.5662/wjm.v14.i2.92982
139. Zègre-Hemsey, J.K., et al.: Cardiac monitoring in the emergency department. Crit. Care Nurs. Clin. North Am. **28**(3), 331–345 (2016). https://doi.org/10.1016/j.cnc.2016.04.009
140. Han, Z., Liu, G.: CEST MRI trackable nanoparticle drug delivery systems. Biomed. Mater. **16**(2), 024103 (2021). https://doi.org/10.1088/1748-605X/abdd70
141. Borah, S.S., Khanal, A., Sundaravadivel, P., et al.: Emerging technologies for automation in environmental sensing: review. Appl. Sci. **14**(8), 3531 (2024). https://doi.org/10.3390/app140 83531
142. Xuan, L., Ju, Z., Skonieczna, M., et al.: Nanoparticles-induced potential toxicity on human health: applications, toxicity mechanisms, and evaluation models. Med. Comm. **4**(4), e327 (2023). https://doi.org/10.1002/mco2.327

143. Harish, V., Tewari, D., Gaur, M., et al.: Review on nanoparticles and nanostructured materials: bioimaging, biosensing, drug delivery, tissue engineering, antimicrobial, and agro-food applications. Nanomaterials **12**(3), 457 (2022). https://doi.org/10.3390/nano12030457
144. Alharbi, K.K., Al-Sheikh, Y.A.: Role and implications of nanodiagnostics in the changing trends of clinical diagnosis. Saudi J. Biol. Sci. **21**(2), 109–117 (2014). https://doi.org/10.1016/j.sjbs.2013.11.001
145. Kempf, A.: A lack of accountability: teacher perspectives on equity, accuracy, and standardized testing. In: The Pedagogy of Standardized Testing. Palgrave Macmillan, New York (2016). https://doi.org/10.1057/9781137486653_7.
146. Macko, P., et al.: Extrapolating from acute to chronic toxicity in vitro. Toxicol. In Vitro **76**, 105206 (2021). https://doi.org/10.1016/j.tiv.2021.105206
147. Otani, N., Otsubo, Y., Koike, T., et al.: Binary classification with ambiguous training data. Mach. Learn. **109**, 2369–2388 (2020). https://doi.org/10.1007/s10994-020-05915-2
148. Amaral, C., Paiva, M., Rodrigues, A.R., et al.: Global regulatory challenges for medical devices: impact on innovation and market access. Appl. Sci. **14**(20), 9304 (2024). https://doi.org/10.3390/app14209304
149. Krewski, D., Saunders-Hastings, P., Baan, R.A., et al.: Development of an evidence-based risk assessment framework. Altex **39**(4), 667–693 (2022). https://doi.org/10.14573/altex.2004041
150. Bonet, O.S., Sasangohar, F.: Investigating the food and drug administration biotherapeutics review and approval process: narrative review. JMIR Form. Res. **5**(3), e14563 (2021). https://doi.org/10.2196/14563
151. Resnik, D.B., Tinkle, S.S.: Ethical issues in clinical trials involving nanomedicine. Contemp. Clin. Trials **28**(4), 433–441 (2007). https://doi.org/10.1016/j.cct.2006.11.001
152. Junaid, S.B., Imam, A.A., Balogun, A.O., et al.: Recent advancements in emerging technologies for healthcare management systems: a survey. Healthcare **10**(10), 1940 (2022). https://doi.org/10.3390/healthcare10101940
153. Silva, D.F., Melo, A.L.P., Uchôa, A.F.C., et al.: Biomedical approach of nanotechnology and biological risks: a mini-review. Int. J. Mol. Sci. **24**(23), 16719 (2023). https://doi.org/10.3390/ijms242316719
154. Ma, J., He, Y., Li, F., et al.: Segment anything in medical images. Nat. Commun. **15**(1), 654 (2024). https://doi.org/10.1038/s41467-024-44824-z
155. Karnwal, A., Kumar, Sachan, R. S., Devgon, I. et al.: Gold nanoparticles in nanobiotechnology: from synthesis to biosensing applications. ACS Omega **9**(28), 29966–29982 (2024). https://doi.org/10.1021/acsomega.3c10352.
156. Sun, Y., Zhu, G., Zhao, W., et al.: Engineered nanomaterials for improving the nutritional quality of agricultural products: a review. Nanomaterials **12**(23), 4219 (2022). https://doi.org/10.3390/nano12234219
157. Neme, K., Nafady, A., Uddin, S., et al.: Application of nanotechnology in agriculture, postharvest loss reduction and food processing: food security implication and challenges. Heliyon **7**(12), e08539 (2021).https://doi.org/10.1016/j.heliyon.2021.e08539
158. Hemdan, M., Ali, M.A., Doghish, A.S., et al.: Innovations in biosensor technologies for healthcare diagnostics and therapeutic drug monitoring: applications, recent progress, and future research challenges. Sensors **24**(16), 5143 (2024). https://doi.org/10.3390/s24165143
159. Sharma, A., Badea, M., Tiwari, S., et al.: Wearable biosensors: an alternative and practical approach in healthcare and disease monitoring. Molecules **26**(3), 748 (2021). https://doi.org/10.3390/molecules26030748
160. Goetz, L.H., Schork, N.J.: Personalized medicine: motivation, challenges, and progress. Fertil. Steril. **109**(6), 952–963 (2018). https://doi.org/10.1016/j.fertnstert.2018.05.006
161. Chehelgerdi, M., Chehelgerdi, M., Allela, O.Q.B., et al.: Progressing nanotechnology to improve targeted cancer treatment: overcoming hurdles in its clinical implementation. Mol. Cancer **22**(1), 169 (2023). https://doi.org/10.1186/s12943-023-01865-0
162. Vo, D.K., Trinh, K.T.L.: Advances in wearable biosensors for healthcare: current trends, applications, and future perspectives. Biosensors **14**(11), 560 (2024). https://doi.org/10.3390/bios14110560

163. Cheng, H.-P., Yang, T.-H., Wang, J.-C., et al.: Recent trends and innovations in bead-based biosensors for cancer detection. Sensors **24**(9), 2904 (2024). https://doi.org/10.3390/s24092904
164. Baruah, A., Newar, R., Das, S., et al.: Biomedical applications of graphene-based nanomaterials: recent progress, challenges, and prospects in highly sensitive biosensors. Discover. Nano. **19**, 103 (2024). https://doi.org/10.1186/s11671-024-04032-6

Corrosion-Resistant Nanomaterials: Prolonging Implant Longevity and Ensuring Biocompatibility

Manish Baboo Agarwal, Manu Mehrotra, Manish Kumar Panday, Seema Agarwal, and Pankaj Mittal

Abstract The newer corrosion-resistant nanomaterials have proven revolutionary in the biomedical implant field due to their applications while addressing the need for biocompatibility and longevity in implants. Corrosion is a common issue that causes stainless steel, titanium alloys, cobalt-chromium, and other traditional implants to suffer from material degradation and inflammation, among other side effects. The application of corrosion-resistant coatings and the enhancement of lower-level mechanical properties of nanostructured materials, including chemical and biological integration, utilize advancements in nanotechnology. Hydroxyapatite nanocoatings and graphene-doped layers of nanostructured materials can significantly mitigate corrosion by acting as barriers against bodily fluids. Moreover, newly modified nanostructured titanium and magnesium alloys, with zirconium and tantalum additions, have shown increased resistance to oxidation and ion release, resulting in superior performance. Additionally, these modifications enhance biocompatibility, which is vital for implant functionality. The bioactive surfaces of nanomaterials promote cellular adhesion and tissue integration, while nanomaterials with antimicrobial properties help address post-surgical complications. For example, using silver and copper nanoparticles in implant coatings provides antibacterial characteristics that reduce the likelihood of infection in the long term while remaining completely biocompatible. New and innovative research and development primarily focus on creating multifunctional nanomaterials. These materials are unique because they can resist corrosion, self-heal, and deliver drugs to specific sites. Indeed, nanomaterials can ensure the long-term stability of medical implants. Although we have made tremendous strides in this field, some barriers, such as long-term cytotoxicity, large-scale manufacturing,

---

M. B. Agarwal (✉) · M. Mehrotra
Sharda School of Basic Sciences and Research, Sharda University Agra, Keetham, Agra, Uttar Pradesh, India
e-mail: mba.mems@gmail.com

M. K. Panday · P. Mittal
Department of Applied Sciences, Anand Engineering College, Keetham, Agra, Uttar Pradesh, India

S. Agarwal
Department of Mathematics, SRMIST, Ghaziabad, Uttar Pradesh, India

S. N. Saud Al-Humairi et al. (eds.), *Smart Nanomaterials in Biomedical Applications*, Engineering Materials, https://doi.org/10.1007/978-981-96-8023-8_9

and regulatory approvals, still exist. Achieving these goals could involve collaboration among experts in relevant fields such as materials science, nanotechnology, and biomedical engineering. Research into next-generation implants can investigate non-traditional implant materials and novel surfaces to improve the reliability of orthopedic procedures and patient outcomes. Consequently, the clinical applications of corrosion-resistant nanomaterials will reduce patient implant failure rates.

Keywords Corrosion-resistant nanomaterials · Implant longevity · Biocompatibility · Nanocoatings · Osseointegration · Biomedical implants

1 Introduction

In contemporary medicine, the application of biomedical implants is of great importance as it helps restore the functional capabilities and overall quality of life for patients. Common types of implants include orthopedic prosthetics, dental implants, and cardiovascular stents, which utilize metallic materials such as titanium, stainless steel, and cobalt-chromium alloys due to their mechanical strength and durability [1–4]. However, the constant exposure to bodily fluids, pH fluctuations, and biochemical reactions makes the human body a highly corrosive environment for these implants. The corrosion of these materials poses significant health risks to patients, including the release of ions, structural degradation, and potential failure, especially since implants are continuously exposed to bodily fluids [5, 6]. The emergence of nanotechnology makes enhancing the properties of biomaterials, such as corrosion resistance, possible. These advanced materials help address the risks of medical implant degradation by reducing ion leaching, forming protective barriers, and improving mechanical integrity by mitigating centrifugal forces [7, 8]. The ability to engineer nanomaterials while controlling their composition, morphology, and surface chemistry at the nanoscale opens new possibilities in medical applications and underscores the need for the next generation of corrosion-resistant nanomaterials for biomedical implants. While significant advancements have been made in the use of implants, the corrosion of implants is still a problem affecting implant longevity and the patient's safety. Many problems may occur with the corrosion of an implant's material, such as:

i. **Release of Metal ions**: Some of the implanted metal can corrode and be absorbed by bones or muscles, which may cause inflammation, allergic response, or even poisoning in the body [9, 10].
ii. **Mechanical Weakening**: Implant materials that have undergone corrosion lose their mechanical resistance and strength, which could result in the fracture of the implant or its failure [6].
iii. **Formation of Biofilm**: Biofilm can be a central point for growing bacteria. Biofilm growth at the site of infection may increase the chances of infection or complications, such as inflammation of the tissue surrounding the implant [11].

Alongside corrosion, compatibility with the body is another crucial factor, as it is a key element for a successful implant. Biocompatible materials must exhibit very low toxicity, provoke no immunological reaction, and integrate effectively into the body. The challenging aspect lies in the materials that enhance cell adhesion, promote osseointegration, support tissue regeneration, and remain resistant to corrosion [12, 13]. The durability of biomedical implants is determined by how well they withstand wear, mechanical fatigue, and biodegradation. In contrast to modern materials, traditional resources have demonstrated only moderate success due to the need for revisions and replacements over time, which ultimately increases patient discomfort and healthcare costs. There are gaps in solutions that enhance both the chemical stability and biological performance of implants simultaneously [14]. Advancements in conventional biomedical implants have become feasible through the application of nanotechnology, enabling modifications at molecular and atomic levels. With these implants, several methods exist to address the issue of implant degradation, including corrosion-resistant nanomaterials:

- **Self-healing Nanocoating**: Degradation is prevented using "smart" nanomaterials that respond to environmental shifts by creating protective oxide layers or releasing bioactive substances.
- **Nanostructured Alloys and Composites**: Incorporating nanoparticles into metallic matrices improves corrosion resistance by strengthening mechanical properties, refining grain structures, and lessening defect density, as shown in Fig. 1.
- **Nanocoating and Surface Modification**: Nanoscale coatings such as hydroxyapatite, TiO_2, and graphene make smaller implants less corrosion-resistant.
- **Durable Metal**: Implants made of titanium and its alloys are used in bone implants due to the metal's strength and durability, providing the bone implant an extended lifetime.

The incorporation of nanotechnology in biomedical implants represents a new advancement that increases functionality while reducing lifespan concerns and adverse patient complications. Corrosion-resistant nanomaterials facilitate the triad approach to address the issues of biocompatibility, biocorrosion, and mechanical stability, making it a promising frontier in biomedical engineering [15–17]. In brief, the creation and use of these materials will significantly enhance biomedical implants due to their improved performance and extended lifespan. Corrosion-resistant nanomaterials enhance warranty reliability, improve compatibility with bodily fluids, and reduce health risks associated with traditional implant materials. With advancements in nanotechnology research, future medical implants will be sturdier and more adaptable to the unique requirements of individual patients.

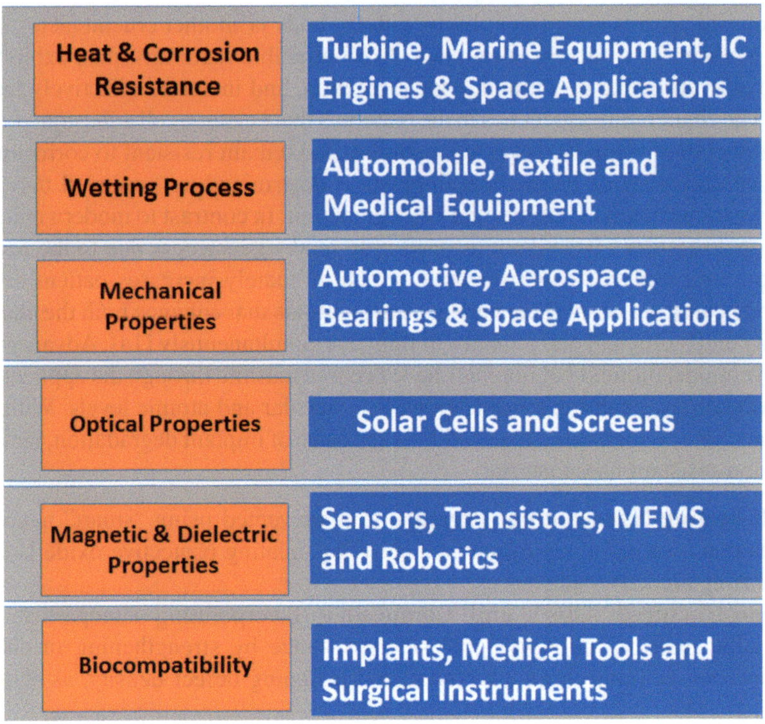

Fig. 1 Application areas of nanostructured coatings

2 Corrosion in Biomedical Implants

Orthopedic prostheses, dental implants, and cardiovascular stents are classified as biomedical implants. These implants are exposed to harsh physiological environments, which can lead to corrosion. Corrosion from mechanical failure and the release of toxic metal ions worsen biocompatibility, a significant concern in biomedical engineering. For optimal health, research aimed at mitigation strategies to reduce corrosion will extend the implant's lifespan while maintaining biocompatibility, thereby enhancing patient health outcomes. Titanium is praised for its exceptional corrosion resistance; however, it is essential to acknowledge that several factors can influence its corrosion behavior, as illustrated in Fig. 2. A detailed examination shows that environmental conditions, material properties, and mechanical stress all play critical roles in determining the corrosion rate of titanium.

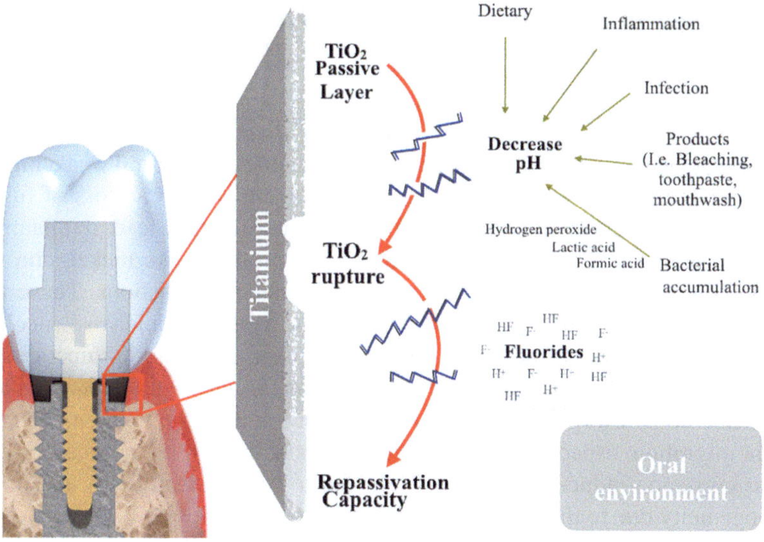

Fig. 2 Factors that affect the corrosion behavior of the titanium [18]

2.1 Mechanisms of Corrosion in Traditional Implant Materials

The primary cause of corrosion in permanent and semi-permanent implants arises from the electrochemical reactions involving water, proteins, oxygen, and ions in bodily fluids. These reactions include, but are not limited to, a crystalline medium covered with steel clamps, cobalt-chromium bolts, and titanium alloy augment:

i. **Pitting Corrosion**: This occurs owing to the breakdown of the passive oxide layer, which results in pits in the implant structure.
ii. **Crevice Corrosion**: It occurs in screw joints and underneath plates with limited diffusion of oxygen where localized acidification accelerates the breakdown of the metal.
iii. **Galvanic Corrosion**: It causes the inferior noble type of metal immersed in the electrolyte to corrode more rapidly when in contact with another metal.
iv. **Fretting Corrosion**: Continuous wear and micro-motion at the implant interfaces cause mechanical corrosion, which breaks the protective layer.
v. **Stress Corrosion Cracking (SCC)**: It is implant failure caused by crack propagation from a combination of mechanical stress and a corrosive environment.
vi. **Bio-corrosion**: It involves the degradation of metals by microbes, primarily found in infected implants.

2.2 *Impacts of Implant Corrosion on Patient Health and Device Functionality*

Corrosion severely affects both the biological environment and the working condition of the device. Mechanical failure of an implant due to structural weakening from corrosion can result in uncontrolled fractures, implant shifts, and loss of functionality, necessitating corrective surgeries. Implant corrosion may lead to inflammation of local tissues due to the presence of damaging products like nickel, chromium, and cobalt ions, as well as being toxic or causing allergic reactions. These factors cause chronic inflammation because the immune system's defensive mechanism prompts capsule tissue formation around the implant, reducing its functional integration. A few metallic particles from corrosion can enhance osteoclast activities and osteoclastic bone resorption of the stabilizing bone around the implant, resulting in implant migration, especially in orthopedic applications. Other systemic problems, such as corrosion-induced metallosis, can lead to severe side effects and potential carcinogenicity due to the by-products of corrosion.

2.3 *Strategies to Reduce Corrosion*

The following process may be considered when discussing the corrosion of biomedical implants and enhancing biocompatibility. The incorporation of titanium, tantalum, and cobalt-chromium alloys increases the strength of the implants. The stability of the passive film is further enhanced by alloying with molybdenum and nitrogen. Protective coatings, such as TiN, hydroxyapatite, or diamond-like carbon (DLC), can be applied to hydroxyapatite to improve protection from corrosive environments [19]. Graphene-based nanomaterials and nanoceramics are known for their higher ionic dissolution and resultant corrosion, which makes them biocompatible. Increasing protective oxide layers through anodizing and passivation enhances resistance against corrosion and damage. The corrosion of biomedical implants remains a significant problem due to the mechanical strength of the implant and safety considerations for the patient. The use of magnesium alloys helps mechanically maintain structural integrity until healing occurs. As a result, the corrosion time period increases. Modifying the surface with chitosan and heparin reduces corrosion and enhances biocompatibility and bioactivity by altering surface properties [20]. The corrosion of biomedical implants remains a significant problem due to the mechanical strength of the implant and safety considerations for the patient. To devise appropriate strategies for mitigation, it is essential to understand how implants corrode and the impact they have through the lifecycle of an implant. The advancements made in the industry of corrosion-resistant nanomaterials are likely to increase the operational lifespan of an implant as long as biocompatibility is considered. Improvements made at the coating, nanotechnology, and new materials levels will ensure that subsequent generations of implants will operate more efficiently, fail less, and, more importantly, yield better patient health results.

3 Nanomaterials for Corrosion Resistance

Biomedical engineering has been able to address some significant issues related to the compatibility, mechanical properties, and corrosion resistance of biomedical implants through the application of nanomaterials. One of the most concerning aspects of implants today is the corrosion resistance an implant possesses, as it deeply relates to the structural integrity of the device and how well it contains potentially dangerous ions that may be released and cause various health complications. The possibility of using hydroxyapatite nanocarriers, graphene-based nanocoatings, and other surface modification resources, along with their implantation, offers hope for improved implant performance [21].

3.1 Nanostructured Coatings and Surface Modifications

These nanostructured coatings and surface modifications are crucial for enhancing the corrosion resistance of metallic implants. Implant materials, including cobalt-chromium alloys, stainless steel, and titanium, demonstrate low corrosion resistance. These alloys are susceptible to corrosion in a harsh physiological environment with high concentrations of body fluids. Nanostructured coatings enhance the implant material's interaction with body fluids, forming a protective barrier against interstitial fluid spaces.

Due to their dense and homogeneous microstructures, zirconium and titanium dioxide coatings exhibit excellent corrosion resistance. Figure 3 shows the different types of coating that can produce a nanostructured coating. These coatings' ability to provide high surface hardness protection, low ion leaching, and improved wear resistance makes them highly valuable. Plasma-sprayed nanoceramic coatings, electrodeposited nanoscale films, and sol–gel-derived nanocoatings have all demonstrated effectiveness in mitigating localized corrosion and have thus been extensively researched [22].

Surface modifications such as anodization, laser surface treatment, and ion implantation add more excellent value to the corrosion resistance of implants. These changes improve the formation of protective oxide layers by controlling surface topography and chemistry at the nanoscale. In addition, nanostructured coatings can be designed to include infection control agents that prevent biofilm formation and help reduce infections and complications associated with implants.

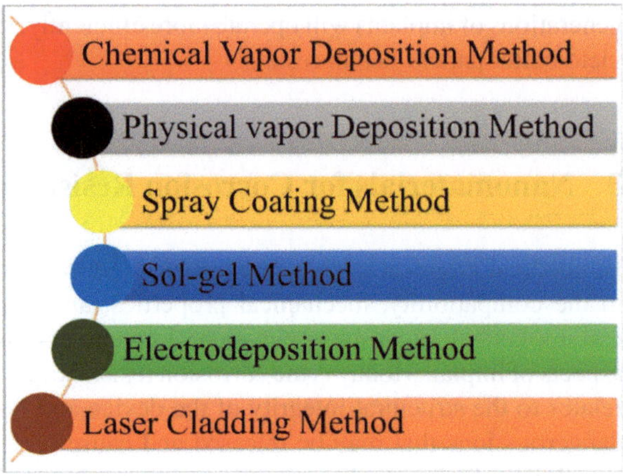

Fig. 3 Different methods to produce nanostructured coatings

3.2 Hydroxyapatite Nanocoatings and Their Role in Implant Protection

The biomedical sector has progressively underscored the significance of employing Hydroxyapatite (HA) nanocoatings owing to their superior biocompatibility osteoconductivity. Given its bone component role, HA is used more frequently as a coating on orthopedic and dental implants to enhance their biological activity and corrosion resistance. Nanostructured HA coatings uniquely improve both corrosion resistance and biological integration simultaneously. These coatings protect the metallic substrate by preventing the permeation of corrosive body fluids while facilitating the adherence and growth of bone cells. Coating with nanohydroxyapatite enhances adhesion, decreases the dissolution rate, and improves mechanical stability. HA nanocoatings 'specific thickness and porosity features have been achieved through plasma spraying, electrophoretic deposition, and sol–gel techniques. Hydroxyapatite (HA) capping with zinc, silver, or silicon dopants enhances the coating's antiseptic properties and corrosion resistance. HA-based nanocoatings improve biocompatibility and reduce implant degradation, thus extending the service life of the implants [23].

3.3 Graphene-Based Nanocoatings for Enhanced Durability

Graphene, a two-dimensional nanomaterial made up of a single layer of carbon atoms, exhibits mechanical, chemical, and electrical strength, making it suitable for use as a corrosion-resistant coating. Graphene-based nanocoatings restrict ion diffusion and other chemical processes that degrade implants due to their barrier properties. The impermeable nature of graphene coatings reduces the occurrence of pitting and crevice corrosion in metallic implants. Graphene oxide (GO) and reduced graphene oxide (rGO) have been extensively utilized to create strong and stable adherent coatings on implant surfaces. Besides minimizing the risk of infection and implant failure, these nanocoatings also serve as effective antimicrobial agents [24, 25].

Some techniques used to coat such surfaces include chemical vapor deposition (CVD), electrophoretic deposition, and spin coating. Graphene-based coatings are more effective when combined with other nanomaterials, such as cerium oxide or polymeric matrices, because they provide multifunctional protection, bioactivity, mechanical enhancement, and heightened wear resistance [26]. The performance and durability of biomedical implants are significantly improved; thus, graphene nanocoatings are corrosion-resistant and exhibit considerable strength and endurance for next-generation biomedical implants. Nanomaterials have enabled implants to achieve outstanding levels of corrosion resistance, biocompatibility, and stability, transforming the landscape of implant coatings. Osseointegration is facilitated by hydroxyapatite nanocoatings, as well as by nanostructured coatings and surface modifications that strengthen protective layers to ensure implant stability. The durability of implants is further augmented by graphene-based nanocoatings, which offer excellent mechanical strength and barrier properties. These nanomaterial advancements incorporated into biomedical implants promise more outstanding durability, fewer complications, and enhanced patient health in dental, orthopedic, and cardiovascular medicine.

4 Advanced Nanostructured Alloys in Implants

Advancements in nanostructured alloys have significantly improved the lifespan of biomedical implants due to increased mechanical strength, biocompatibility, and corrosion resistance. Innovating implants used in orthopedic, dental, and cardiovascular surgery is most important, as material degradation can lead to catastrophic implant failure. Nanostructured alloys with titanium, magnesium, zirconium, and tantalum are promising candidates due to their high biological compatibility and improved corrosion resistance, facilitating durability and functionality inside the human body [27, 28].

4.1 Nanostructured Alloys of Titanium and Magnesium

Biomedical implants require titanium and magnesium alloys due to their unparalleled biocompatibility, mechanical properties, and corrosion resistance. Incorporating nanostructuring methods promises to enhance their capabilities for sustained implants further. Titanium-based nanostructured alloys are also highly beneficial in orthopedic and dental applications due to their remarkable corrosion resistance and bone integration properties. The application of severe plastic deformation (SPD), high-pressure torsion (HPT), and equal channel angular pressing (ECAP) nanostructuring methods results in grain refinement, which increases mechanical properties and reduces the chance of rejection. Adding niobium (Nb) and zirconium (Zr) to titanium enhances biocompatibility and further improves corrosion resistance [29]. The biodegradability of magnesium alloys with a nanostructured design makes them suitable for use as temporary implants, such as orthopedic screws and stents, thus earning them attention in the field. The utmost problem with Mg-based alloys is the high rate of degradation in physiological conditions, which may lead to implant failure. However, it is possible to control the corrosion rate with nanostructure features, ensuring that Mg implants degrade in a manner that is non-obstructive to tissue healing. The addition of some rare earth elements (REEs) such as Yttrium (Y) and Gadolinium (Gd) increases mechanical strength and resistance to corrosion, making them more appropriate for clinical use.

4.2 Zirconium and Tantalum Additions for Enhanced Corrosion Resistance

Zirconium (Zr) and Tantalum (Ta) exhibit remarkable resistance to corrosion and biocompatibility, making them essential alloying materials for biomedical implants. When used with titanium and magnesium-based alloys, these elements enhance the durability and performance of the implants. Zirconium compounds demonstrate excellent biocompatibility and create a protective passive oxide layer that minimizes corrosion in biological environments. Adding Zr to Ti-based alloys improves mechanical performance, wear, and corrosion resistance. These alloys are particularly beneficial in load-bearing applications, such as hip and knee prostheses, where increased strength and fatigue resistance are vital [30, 31]. Tantalum is one of the most corrosion-resistant metals, even in aggressive biological environments. Its stable passive oxide layer protects against degradation and ion leaching. Mechanical implants made from Ti and Mg alloys are stabilized through osseointegration by adding Ta.

Furthermore, tantalum has been structurally engineered for orthopedic use in a porous form to serve as bone ingrowth scaffolds without losing mechanical strength. Zirconium and tantalum additions in nanostructured alloys enhance biocompatibility, mechanical strength, and corrosion resistance. As a result, these modifications improve the durability and reliability of the implants.

4.3 Mechanical and Biological Benefits of the Enhanced Alloys

The implants necessitate nanostructured alloys for modifiable compositions due to their superior biological and mechanical properties. These modified alloys offer numerous benefits over conventional materials. Explicitly, the nanostructuring processes refine plastic deformation, improving alloys' mechanical properties, such as hardness, impact strength, tensile strength, and fatigue strength. As a result, these implants do not break or fail under prolonged mechanical loads. The addition of elements like Zr and Ta, along with nanostructuring, facilitates the formation of protective oxide layers that reduce corrosion rates. This is crucial for implants exposed to bodily fluids, as corrosion contributes to their degradation and failure over time. Nanostructured surfaces promote better cell adhesion, proliferation, and differentiation, improving osseointegration. This is particularly important for orthopedic and dental implants, where direct bone integration is essential for long-term stability.

Regarding magnesium alloys, nanostructuring allows for precise control over the degradation rate, enabling the development of biologically active implants for the duration needed to support tissue regeneration before being safely metabolized by the body. Using antibacterial coatings on the reduced porous nanostructured alloys helps prevent infections after surgery while mitigating adverse immune responses and inflammation associated with joint replacements, thus improving the likelihood of acceptance by the body. Unlike traditional alloys used in biomedical implants, advanced nanostructured alloys address various concerns related to mechanical strength, durability, corrosion resistance, and biocompatibility. Titanium and magnesium-based alloys containing zirconium and tantalum exhibit improved biocompatibility and enhanced wear resistance. With the implementation of nanostructuring techniques, these implants show enhanced mechanical properties and significantly increased survivability, translating to better patient outcomes. Future advancements in nanomaterial research will substantially enhance individuals' overall safety, effectiveness, and quality of life worldwide.

5 Biocompatibility and Cellular Integration

5.1 Importance of Biocompatibility in Implant Success

Incorporated biomedical devices need to be biocompatible, as it is one of the most critical factors that enable the long-term functionality of the device. Biocompatibility refers to the capacity of a material to perform its designated function without invoking any adverse biological reaction in the body. To ensure that the device is functional, biocompatibility guarantees that there is no immune response, inflammation, or hazardous toxic reaction that would lead to either rejection or failure of the implant. Given that implants are set into motion within a dynamic biological system where interaction with bodily fluids, proteins, and cells is inevitable, the device will be exposed to; therefore, they are at risk of provoking fibrosis, infection as well as mechanical failure due to fuel failure pores within the implant due to corrosion. Because of these risks, implants must be expertly designed and fabricated to reduce the probability of toxic and immune reactions and corrosion. There is a great deal of interest in the application of nanomaterials for implants due to their potential to increase biocompatibility by copying the architecture of human tissues and tissue engineering and surface design optimization [32, 33].

5.2 Nanostructures Bioactive Surfaces for Tissues Integration

To promote better integration of both soft and hard tissues with the implant, commonly known as osseointegration in the context of bone implants, several objectives have been outlined within the realm of implant technologies. Surface nanostructures facilitate these advancements due to their bioactivity, which enhances cell adhesion, growth, differentiation, and proliferation. These processes are influenced by nanoscale surfaces that mimic the extracellular matrix, improving cellular interaction with the materials. Bioactive surfaces on nanomaterials can be achieved through various methods. For instance, titanium implants with nanostructured surfaces have been shown to increase adhesion and bone formation by osteoblasts. HA nanoparticles, bioactive glass, and polymeric nanocomposite coatings enhance biointegration by promoting natural bone mineral deposition. Cellular responses can be further optimized by functionalizing nanomaterials with bioactive molecules such as peptides, proteins, or growth factors [34, 35]. For example, the RGD peptide has been demonstrated to promote cell adhesion via integrin receptors when introduced to implant surfaces, thus promptly improving tissue integration with the implant. Moreover, silver or zinc oxide nanoparticle antimicrobial nanocoatings effectively combat infections that may threaten the longevity of the implant.

5.3 Use of Nanocoatings to Stimulate Cell Adhesion

Cellular adhesion is critical for implant success, as it significantly affects tissue integration and stability throughout the device's lifespan. Modifying an implant's surface topography, chemistry, and bioactivity using nanocoatings can greatly enhance cellular adhesion and, consequently, implantation efficacy. One approach to improve adhesion is utilizing nanopatterned surfaces with microscale features facilitating cell adhesion. Studies indicate that osteoblasts and fibroblasts exhibit stronger adhesion and spreading on roughened nanostructured surfaces than smoother ones.

Moreover, nanocoatings created through plasma treatments or hydrophilic self-assembled monolayers enhance protein adsorption and, in turn, improve cell attachment. Another area of focus is bioactive nanocoatings infused with growth factors or signaling molecules intended to promote cell proliferation. For instance, titanium implants coated with calcium phosphate nanoparticles have increased osteointegration due to enhanced ionic release necessary for bone mineralization. Additionally, there is a prevailing assumption that neuronal cell adhesion is improved by graphene nanocoatings, making them suitable for use in neural implants.

Additionally, the use of nanocoatings with antimicrobial properties helps minimize infections related to implants. These nanocoatings reduce bacterial colonization while enhancing tissue adhesion to the implant by utilizing silver, copper, and zinc oxide nanoparticles on the surfaces. This is critical as it improves biocompatibility and extends the longevity of the implant by preventing infection-induced failures. Integrating corrosion-resistant nanomaterials into implants is advancing the field of biomedical engineering due to their enhanced compatibility with biological tissues and integration within cells. The longevity and effectiveness of implants are further improved by bioactive nanostructured surfaces, functionalized nanocoatings, and antimicrobial strategies. The application of nanotechnology allows researchers and medical practitioners to create advanced implants capable of biologically integrating with tissues, evoking minimal immune reactions and succeeding in clinical settings over the long term. The development of this technology will continue to enhance treatment outcomes and create new opportunities for innovative medical implants [36].

6 Antimicrobial Nanomaterials for Implants

Engineering biomedical implants with specific types of antimicrobial nanomaterials represents a new frontier in science aimed at controlling infections without compromising functionality over time. The most significant issue concerning implants is bacterial infection. An infection in implantology can result in the loss of an implant, which is distressing and can lead to other health risks. The use of antimicrobial nanoparticles such as silver (Ag) and copper (Cu) is increasingly seen as a promising alternative to enhance the infection resistance of implants while preserving their physical strength, biocompatibility, and complex interfacial interactions. This section addresses the enhancement of the antibacterial properties of Ag and Cu nanoparticles, their biocompatibility, and the critical balance of these characteristics.

6.1 Role of Silver and Copper Nanoparticles in Infection Prevention

Silver and copper nanoparticles are well-known for their ability to act as potent antimicrobial agents; consequently, they can be effectively utilized to modify implants, their components, and materials for enhanced properties. One of the well-studied applications of silver nanoparticles (AgNPs) is their strong antibacterial effect on a wide range of bacteria, including antibiotic-resistant strains such as MRSA. AgNPs effectively make implant surfaces less conducive to bacterial colonization and biofilm formation, reducing infection risks. The antibacterial action of silver nanoparticles is facilitated by the release of silver ions (Ag^+), which attach to bacterial cell membranes and disrupt essential cell functions, leading to bacterial death [37, 38]. Copper nanoparticles (CuNPs) exhibit broad-spectrum antimicrobial properties similar to silver, as they also produce reactive oxygen species (ROS) that damage bacterial membranes. The low cost of copper and its osteogenic properties make it highly favorable, as it improves bone integration with orthopedic implants [39]. Gradually, the combined use of Ag and Cu nanoparticles in implant materials is gaining popularity due to their contrasting antimicrobial properties, which provide better control. These synergistic profiles enable the sustained release of the two metals and enhance their effectiveness against bacterial infection post-implantation.

6.2 Mechanisms of Antibacterial Action in Nanomaterials

The distinct guarantees supporting the antimicrobial features of silver and copper nanoparticles are sufficient in the following ways:

i. **Destruction of Membrane Cell Identity**: The interaction of the ions with the membrane Ag^+ and Cu^{2+} of the bacteria disrupts the membrane of the bacteria and changes the permeability, leading to a rupture of the cell.
ii. **The creation of Reactive Oxygen Species (ROS)**: The production of ROS detriments silver and copper nanoparticles, leading to oxidative stress by destroying pivotal biomolecules such as proteins, lipids, and DNA in bacterial cells.
iii. **Interruption of Metabolic Function and Enzyme Activity**: The structural–functional silver and copper ions impede the activity of bacterial cell enzymes responsible for respiration and ATP synthesis, halting metabolism and leading to bacteria's death.
iv. **Damage to DNA and Proteins**: Bacterial nucleotides and cellular proteins are substituted with standard Ag^+ and Cu^{2+} ions, inhibiting replication and active cellular division.
v. **Disruption of Biofilms**: Antibiotic and immune system evasion forms protective biofilms around bacteria. Permeable or penetrating nanoparticles disrupts existing biofilms, making bacteria easier to eliminate and treat with immune systems and antimicrobial antibiotics.

6.3 Balancing Biocompatibility with Antimicrobial Properties

The antimicrobial activity of Ag and Cu nanoparticles is undeniable, but they pose a challenge in the context of biocompatibility when used in biomedical implants. Large amounts of silver or copper ions in solution can damage the membranes of mammalian cells, harming resulting inflammatory reactions and the integration of tissues or changes in their consumer. Some processes of balancing biocompatibility and antimicrobial properties include:

i. **Controlled Release and Surface Modification**: This strategy defends the containment of the release of antimicrobial ions from nanocarriers encased in biodegradable coatings to a specific process and level that is therapeutic but not toxic.
ii. **Nanoparticle Size and Concentration Optimization**: The amount of Ag and Cu nanoparticles used is dictated by their size and cytotoxic effects but still guarantees bactericidal activity.
iii. **Composite Coatings**: Biocompatible polymer or ceramic nanoparticle coatings allow for controlled interfaces that prevent excessive metal ion leaching while promoting osseointegration.
iv. **Hybrid Approaches**: The use of biocompatible materials like hydroxyapatite or chitosan with nanomaterials of antimicrobial activity enhances biocompatibility without losing the antimicrobial activity. When designing nanomaterials for implants, especially durable biomedical implants, a judicious balance between protection from infection and the compatibility of host tissue should be achieved. Smart drug release systems and advanced nanotechnology-coated infection-resistant implants are currently being developed based on research focused on meeting these guidelines.

The application of antimicrobial nanomaterials, such as silver and copper nanoparticles, to the surfaces of implants is a novel method aimed at infection control and increasing the longevity of the implants. Silver and copper particles exhibit potent antimicrobial activity as they can destroy bacterial cell membranes, inhibit biofilm formation, and hinder bacterial metabolism. However, the nanoparticles' biocompatibility and release mechanisms must be closely examined in order to be utilized in biomedical technologies. Further research will focus on developing implant coatings with nanomaterials that can optimize the control of antibacterial agents without increasing cytotoxicity, thereby enhancing patient health and the effectiveness of the implants.

7 Multifunctional Nanomaterials for Next-Generation Implants

Multifunctional nanomaterials in biomedical implants are now regarded as one of the most advanced solutions to biocompatibility and implant durability. These novel implants exhibit improved resistance to mechanical stress and corrosion and can self-heal, respond intelligently, and deliver medication. The advancement of such technologies facilitates the development of smart implants that react to the body's variables, reducing the likelihood of complications and enhancing patient outcomes [40].

7.1 Self-Healing and Smart Nanomaterials

Healing bone or tissue damage caused by an implant can occur through self-healing nanomaterials representing new technological advancements. These self-healing materials contain a nanoengineered structure, such as polymeric nanocapsules with healing aids that activate only when mechanically provoked or influenced by various environmental factors. Other relevant structures, such as shape-memory alloys, can also be included [41]. For instance, consider nanocoatings with embedded microcapsules containing bioactive agents that can rupture and release healing aids to seal the implant. Additionally, shape-memory alloys and nanostructured hydrogels can adjust to new temperatures and positions. The healing mechanisms significantly enhance the operational life of an implant, increasing durability while reducing the risk of surgical replacements and revisions. Unlike other materials, smart nanomaterials now provide immediate responsiveness to living conditions. Other pH-sensitive and thermoresponsive nanoparticles facilitate changes within the body. A prime example is titanium implants coated with nanostructured hydrogels, which improve tissue integration while mitigating inflammation. Overall, the addition of self-healing technologies alongside advanced automation will ensure better implants in terms of reliability, adaptability, and overall sturdiness.

7.2 Nanostructured Coatings Therapeutic Agents Delivery

The introduction of nanostructured coatings onto implants has significantly improved drug-targeting delivery systems by allowing the controlled release of therapeutic agents directly at the implantation site. Such coatings contain nanoscale carriers, such as liposomes, polymeric nanoparticles, or mesoporous silica, which can incorporate antimicrobial, anti-inflammatory, or osteoinductive agents [42]. Post-surgical infection prevention is one of the most significant benefits of nanostructured drug-delivery coatings. The controlled release of antimicrobial agents mitigates the risk of biofilm development, one of the leading causes of implant failure. In addition, incorporating growth factors or osteogenic pharmaceuticals in implant coatings is thought to improve osseointegration, enhance bone healing, and reduce rejection time. Nanostructured coatings can further be optimized for modulated stimuli-responsive drug release. Coatings that release anti-inflammatory agents due to acidic surroundings instigated by infection or inflammation are helpful in averting complications before they worsen. The potential for advanced aerosols and improved drug delivery systems from these devices yields minimal systemic side effects while augmenting the functionality of the implant and comfort for the patients.

7.3 Future Advances with Implantable Nanomaterials

Like any technology, implantable nanomaterials will always have room for innovation, focusing on enhancing biocompatibility, integrating nanomaterials, monitoring functions, and addressing specific targeted medical applications to meet commercial market needs. With advancements in technology for producing nanoparticles, the emphasis is shifting toward improving the integration of implants with tissues such as bone and cartilage. The incorporation of biomimetic coatings with nanoscale topographies enhances biological responses for cell adhesion and proliferation. This significantly benefits cell interfaces in biology, improving cellular productivity. Smart nanodevices, such as implants, can now monitor physiological phenomena for healthcare. Changes in pH, infection markers, or particular mechanical stress produce results that specialists can receive in real-time. These instruments assist in better issue detection, thereby enhancing patient treatment and management strategies. With advancements in 3D printing, it is now possible to manufacture patient-specific implants using nanotechnology for scalability. By integrating nanomaterials with 3D-printed scaffolds, implants can be designed with customized strength, porosity, shape, and drug release features tailored for individual treatment plans. The focus is on developing these eco-friendly materials to create nanostructured materials for long-term implantation. Gradually decomposing nanostructures that promote tissue development while dissolving will help ensure patient safety.

8 Challenges and Future Directions

The medical field faces numerous obstacles regarding corrosion-resistant implantable biomedical devices utilizing nanotechnology. These problems include nanomaterials' cytotoxicity and biodegradability, the ethical and legal implications of using them in implants, and the scalability of clinical-grade manufacturing. Addressing these challenges is crucial for ensuring safety, efficacy, and widespread medical adoption.

8.1 Addressing Long-Term Cytotoxicity and Biodegradability Concerns

One of the most significant concerns regarding the application of corrosion-resistant nanomaterials for implants is their long-term cytotoxicity and biodegradability. Issues related to corrosion-resistant nanomaterials tend to overshadow other features of implantable devices. Nevertheless, understanding how these materials interact with biological systems over extended periods remains challenging. Degradable materials capable of supporting some form of life could trigger undesirable immune responses and toxic cellular effects. Biodegradability is also associated with difficulties, as certain types of implants are designed to dissolve safely within the body over long durations. Still, the presence of these materials may lead to adverse biological effects. To address these problems, researchers focus on surface modifications, biocompatible coatings, and hybrid materials that are both safe to degrade and durable. Advanced studies in vitro and in vivo, along with clinical trials, will assess the usability of these materials in the medical field.

8.2 Expanding Industry Scale for Clinical Usage

Using nano corrosion resistance in the clinic poses significant challenges in scaled manufacturing compared to the laboratory setting. Medical implants demand high precision and reproducibility, which can be challenging to achieve at scale. The complex synthesis routes necessary for nanomaterials incur costs, require high purity from starting materials, and demand specialized equipment, which can add to expenses and reduce output. New nano-fabrication techniques such as electrospinning, atomic layer deposition, and 3D printing are being researched to address these challenges. Their scalability and reproducibility greatly surpass current methods, which are crucial for meeting regulatory requirements.

Furthermore, standard manufacturing practices that balance cost and quality will require collaboration among universities, industries, and government agencies. Achieving these objectives would enable a broader application of these materials in medical implants. Automated production and machine learning-powered quality control promise to enhance consistency and efficiency.

8.3 Legal and Ethical Considerations of Nanomaterial Implant Integration

The incorporation of nanomaterials into implants raises several notable ethical and legal issues. Government monitoring agencies, such as the FDA and EMA, must ensure that the materials are adequately tested for safety and efficacy, adding another hurdle. The potential toxicity of nanomaterial compounds the ethical issues, the stability of the device, and the overarching concern of patient safety. Establishing testing frameworks for implants that utilize nanomaterials remains a work in progress due to anticipated biocompatibility assessments not being sufficiently comprehensive. Additionally, concerns arise regarding treatment consent, risk evaluation, and the provision of medical care involving advanced technologies. One way to mitigate potential backlash is to provide patients and other healthcare providers with all information regarding the implants, including advantages and disadvantages. Efforts to develop specific regulatory frameworks that address the needs of nanomaterials in the biomedical engineering field are still essential. Involving scientists in bioethics alongside other stakeholders such as company leaders, public policy experts, medical professionals, and policymakers will be beneficial in creating policies that enhance healthcare delivery, ensure patient safety, and uphold ethical standards. Despite existing shortcomings, including the use of corrosion-resistant nanomaterials, there is undeniable value in the numerous positive perspectives regarding biocompatibility and the longevity of implants. Addressing issues of cytotoxicity and biodegradability, scaling up production for clinical application, and tackling regulatory and ethical challenges are fundamental for successfully integrating these materials into modern medicine. The incorporation of nanomaterials into biomedical engineering has the potential to enhance patient care and outcomes. Achieving these objectives will require collaboration among specialists from various disciplines.

9 Conclusion and Future Direction

Recent advancements in corrosion-resistant nanomaterials have significantly enhanced the durability and biocompatibility of biomedical implants. Implants like hydroxyapatite, titanium dioxide (TiO_2), and graphene-coated devices show less oxidative degradation than those lacking nanostructured coatings. Nanocomposite materials containing zirconium, tantalum, and niobium have also exhibited remarkable resistance to physiological corrosion. Self-healing coatings, surface modifications, and bioactive nanomaterials have contributed to extending the functional lifespan of implants. These developments improve implant performance through enhanced osteointegration, reduced toxic ion wear, and increased mechanical strength. There is considerable concern about how incorporating corrosion-resistant nanomaterials will impact patient health and the durability of implants. Additionally, there is a worrying rise in issues related to implant failures, revisions, and other problems caused by material degradation when using these nanomaterials.

Furthermore, the enhanced corrosion resistance of the implant improves its stability and integration into the surrounding tissues, which reduces inflammatory responses and post-implantation infections. Moreover, the controlled release of therapeutic agents from the nano-coatings can assist in tissue healing and regeneration, bone healing, and ultimately protect the patient. As a result, patients face high healthcare costs due to the implant's extended lifespan, improved quality of life, and lower healthcare costs, which categorize these innovations as remarkable achievements in modern medical technology. The advances in smart nanomaterials have already been proposed to several fields, with the closest area expected to adopt them in the coming years being biomedical implants, primarily for the self-regulating release of therapeutic agents, particularly anti-inflammatory or antibacterial ones. Moreover, the integration of personalized medicine is likely to incorporate nanotechnology into implants, optimizing them for individual patients, which will further enhance the precision of treatments. Additionally, advancements in the 3D printing of nanomaterials will allow for the creation of custom-designed implants in a more compatible and efficient manner. The combination of AI and medicine, along with nanomedicine and regenerative therapies, may facilitate the development of the next generation of corrosion-resistant implants with remarkable performance capabilities, providing enhanced long-term solutions for patients worldwide.

References

1. Al-Shalawi, F.D., Ariff, A.H.M., Jung, D., Ariffin, M.K.A.M., Kim, C.L.S., Brabazon, D., Al-Osaimi, M.O.: Biomaterials as implants in the orthopedic field for regenerative medicine: metal versus synthetic polymers. Polymers **15**(12), 2601 (2023). https://doi.org/10.3390/polym15122601
2. Moghadasi, K., Isa, M.S.M., Ariffin, M.A., Jamil, M.Z.M., Raja, S., Wu, B., Yamani, M., Muhamad, M.R.B., Yusof, F., Jamaludin, M.F., Karim, M.S.B.A., Razak, B.B.A., Yusoff, N.B.: A review on biomedical implant materials and the effect of friction stir based techniques on their mechanical and tribological properties. J. Market. Res. **17**, 1054–1121 (2022). https://doi.org/10.1016/j.jmrt.2022.01.050
3. Marin, E., Lanzutti, A.: Biomedical applications of titanium alloys: a comprehensive review. Materials **17**(1), 114 (2023). https://doi.org/10.3390/ma17010114
4. Gautam, S., Bhatnagar, D., Bansal, D., Batra, H., Goyal, N.: Recent advancements in nanomaterials for biomedical implants. Biomed. Eng. Adv. **3**, 100029 (2022). https://doi.org/10.1016/j.bea.2022.100029
5. Szczęsny, G., Kopec, M., Kowalewski, Z.L.: Toxicity, irritation and allergy of metal implants: historical perspective and Modern solutions. Coatings **15**(3), 361 (2025). https://doi.org/10.3390/coatings15030361
6. Eliaz, N.: Corrosion of metallic biomaterials: a review. Materials **12**(3), 407 (2019). https://doi.org/10.3390/ma12030407
7. Hakim, L.K., Yari, A., Nikparto, N., Mehraban, S.H., Cheperli, S., Asadi, A., Darehdor, A.A., Nezaminia, S., Dortaj, D., Nazari, Y., Dehghan, M., Hojjat, P., Mohajeri, M., Jebelli, M.S.H.: The current applications of nano and biomaterials in drug delivery of dental implant. BMC Oral Health **24**(1) (2024). https://doi.org/10.1186/s12903-024-03911-9
8. Abraham, A.M., Venkatesan, S.: A review on application of biomaterials for medical and dental implants. Proc. Inst. Mech. Eng. Part L J. Mater. Des. Appl. **237**(2), 249–273 (2022). https://doi.org/10.1177/14644207221121981
9. Sansone, V.: The effects on bone cells of metal ions released from orthopaedic implants. A review. Clin. Cases Mineral Bone Metab. (2013). https://doi.org/10.11138/ccmbm/2013.10.1.034
10. Teo, W.Z.W., Schalock, P.C.: Metal hypersensitivity reactions to orthopedic implants. Dermatol. Ther. Ther. **7**(1), 53–64 (2016). https://doi.org/10.1007/s13555-016-0162-1

11. Sharma, S., Mohler, J., Mahajan, S.D., Schwartz, S.A., Bruggemann, L., Aalinkeel, R.: Microbial biofilm: a review on formation, infection, antibiotic resistance, control measures and innovative treatment. Microorganisms **11**(6), 1614 (2023). https://doi.org/10.3390/microorganisms11061614
12. Freese, H., Volas, M., Wood, J., Textor, M.: Titanium and its alloys in biomedical engineering. In: Elsevier eBooks, pp. 9374–9380 (2001). https://doi.org/10.1016/b0-08-043152-6/01695-8
13. Nuss, K.M.R., Von Rechenberg, B.: Biocompatibility issues with modern implants in bone—a review for clinical orthopedics. Open Orthop. J. **2**(1), 66–78 (2008). https://doi.org/10.2174/1874325000802010066
14. Hussain, M., Khan, S.M., Shafiq, M., Abbas, N., Sajjad, U., Hamid, K.: Advances in biodegradable materials: degradation mechanisms, mechanical properties and biocompatibility for orthopedic applications. Heliyon **10**(12), e32713 (2024). https://doi.org/10.1016/j.heliyon.2024.e32713
15. Raymond, D., Weerarathna, I.N., Jalah, J.K., Kumar, P.: Nanoparticles in biomedical implants: pioneering progress in healthcare. AIMS Bioeng. **11**(3), 391–438 (2024). https://doi.org/10.3934/bioeng.2024019
16. Abaszadeh, F., Ashoub, M.H., Khajouie, G., Amiri, M.: Nanotechnology development in surgical applications: recent trends and developments. Eur. J. Med. Res. **28**(1) (2023). https://doi.org/10.1186/s40001-023-01429-4
17. Farooq, S.A., Raina, A., Mohan, S., Singh, R.A., Jayalakshmi, S., Haq, M.I.U.: Nanostructured coatings: review on processing techniques, corrosion behaviour and tribological performance. Nanomaterials **12**(8), 1323 (2022). https://doi.org/10.3390/nano12081323
18. Souza, J.C.M., Apaza-Bedoya, K., Benfatti, C.A.M., Silva, F.S., Henriques, B.: A comprehensive review on the corrosion pathways of titanium dental implants and their biological adverse effects. Metals **10**(9), 1272 (2020). https://doi.org/10.3390/met10091272
19. Saad, K.S.K., Saba, T., Rashid, A.B.: Application of PVD coatings in medical implantology for enhanced performance, biocompatibility and quality of life. Heliyon **10**(16), e35541 (2024). https://doi.org/10.1016/j.heliyon.2024.e35541
20. Mohammadi, F., Golafshan, N., Kharaziha, M., Ashrafi, A.: Chitosan-heparin nanoparticle coating on anodized NiTi for improvement of blood compatibility and biocompatibility. Int. J. Biol. Macromol. **127**, 159–168 (2019). https://doi.org/10.1016/j.ijbiomac.2019.01.026
21. Rafiei, M., Mohammadloo, H.E., Khorasani, M., Kargaran, F., Khonakdar, H.A.: Hydroxyapatite-based coatings on MG and TI-based implants: a detailed examination of various coating methodologies. Heliyon **11**(2), e41813 (2025). https://doi.org/10.1016/j.heliyon.2025.e41813
22. Ahmad, Z., Ahsan, M.: Corrosion studies on the plasma-sprayed nanostructured titanium dioxide coatings. Anti-Corros. Methods Mater. **56**(4), 187–195 (2009). https://doi.org/10.1108/00035590910969310
23. Kowalski, S., Gonciarz, W., Belka, R., Góral, A., Chmiela, M., Lechowicz, Ł, Kaca, W., Żórawski, W.: Plasma-sprayed hydroxyapatite coatings and their biological properties. Coatings **12**(9), 1317 (2022). https://doi.org/10.3390/coatings12091317
24. Ruben, B., Zhang, G., Xin, T., Giorgio, S., Victor, M., Gloria, G., Michele, F., Filippo, P., Shuhui, S., Nadhira, L., Tavares, A.C.: Graphene oxide/reduced graphene oxide films as protective barriers on lead against differential aeration corrosion induced by water drops. Nanoscale Adv. **2**(11), 5412–5420 (2020). https://doi.org/10.1039/d0na00212g
25. Parhizkar, N., Ramezanzadeh, B., Shahrabi, T.: Corrosion protection and adhesion properties of the epoxy coating applied on the steel substrate pre-treated by a sol-gel based silane coating filled with amino and isocyanate silane functionalized graphene oxide nanosheets. Appl. Surf. Sci. **439**, 45–59 (2018). https://doi.org/10.1016/j.apsusc.2017.12.240
26. Ryu, S., Kwon, Y.J., Kim, Y., Lee, J.U.: Corrosion protection coating of three-dimensional metal structure by electrophoretic deposition of graphene oxide. Mater. Chem. Phys. **250**, 123039 (2020). https://doi.org/10.1016/j.matchemphys.2020.123039
27. Mosas, K.K.A., Chandrasekar, A.R., Dasan, A., Pakseresht, A., Galusek, D.: Recent advancements in materials and coatings for biomedical implants. Gels **8**(5), 323 (2022). https://doi.org/10.3390/gels8050323

28. Ashish Daniel, S., Suya Prem Anand, P., Naveen, J., Khan, T., Khahro, S.H.: Advancement in biomedical implant materials—a mini review. Front. Bioeng. Biotechnol. **12** (2024). https://doi.org/10.3389/fbioe.2024.1400918
29. Agarwal, K.M., Tyagi, R., Singhal, A., Bhatia, D.: Effect of ECAP on the mechanical properties of titanium and its alloys for biomedical applications. Mater. Sci. Energy Technol. **3**, 921–927 (2020). https://doi.org/10.1016/j.mset.2020.11.002
30. Su, B., Wang, B., Luo, L., Wang, L., Su, Y., Xu, Y., Wang, F., Han, B., Huang, H., Guo, J., Fu, H.: Effect of zirconium content on the microstructure and corrosion behavior of as-cast Ti–Al–Nb–Zr–Mo alloy. J. Market. Res. **15**, 4896–4913 (2021). https://doi.org/10.1016/j.jmrt.2021.10.102
31. Nikoomanzari, E., Fattah-Alhosseini, A., Karbasi, M., Nourian, A.: A versatile TiO_2/ZrO_2 nanocomposite coating produced on Ti–6Al–4V via plasma electrolytic oxidation process. Surf. Interfaces **32**, 102128 (2022). https://doi.org/10.1016/j.surfin.2022.102128
32. Miramini, S., Fegan, K.L., Green, N.C., Espino, D.M., Zhang, L., Thomas-Seale, L.E.: The status and challenges of replicating the mechanical properties of connective tissues using additive manufacturing. J. Mech. Behav. Biomed. Mater. **103**, 103544 (2019). https://doi.org/10.1016/j.jmbbm.2019.103544
33. Lu, P., Ruan, D., Huang, M., Tian, M., Zhu, K., Gan, Z., Xiao, Z.: Harnessing the potential of hydrogels for advanced therapeutic applications: current achievements and future directions. Sig. Transduct. Target. Ther. **9**(1) (2024). https://doi.org/10.1038/s41392-024-01852-x
34. Liang, J., Lu, X., Zheng, X., Li, Y.R., Geng, X., Sun, K., Cai, H., Jia, Q., Jiang, H.B., Liu, K.: Modification of titanium orthopedic implants with bioactive glass: a systematic review of in vivo and in vitro studies. Front. Bioeng. Biotechnol. **11** (2023). https://doi.org/10.3389/fbioe.2023.1269223
35. Sarian, M.N., Iqbal, N., Sotoudehbagha, P., Razavi, M., Ahmed, Q.U., Sukotjo, C., Hermawan, H.: Potential bioactive coating system for high-performance absorbable magnesium bone implants. Bioactive Mater. **12**, 42–63 (2021). https://doi.org/10.1016/j.bioactmat.2021.10.034
36. Sahoo, J., Sarkhel, S., Mukherjee, N., Jaiswal, A.: Nanomaterial-based antimicrobial coating for biomedical implants: new age solution for biofilm-associated infections. ACS Omega **7**(50), 45962–45980 (2022). https://doi.org/10.1021/acsomega.2c06211
37. Bruna, T., Maldonado-Bravo, F., Jara, P., Caro, N.: Silver nanoparticles and their antibacterial applications. Int. J. Mol. Sci. **22**(13), 7202 (2021). https://doi.org/10.3390/ijms22137202
38. Franci, G., Falanga, A., Galdiero, S., Palomba, L., Rai, M., Morelli, G., Galdiero, M.: Silver nanoparticles as potential antibacterial agents. Molecules **20**(5), 8856–8874 (2015). https://doi.org/10.3390/molecules20058856
39. Ermini, M.L., Voliani, V.: Antimicrobial nano-agents: the copper age. ACS Nano **15**(4), 6008–6029 (2021). https://doi.org/10.1021/acsnano.0c10756
40. Vishwakarma, V., Kaliaraj, G., Mosas, K.A.: Multifunctional coatings on implant materials—a systematic review of the current scenario. Coatings **13**(1), 69 (2022). https://doi.org/10.3390/coatings13010069
41. Brannigan, K., Griffin, M.: An update into the application of nanotechnology in bone healing. Open Orthop. J. **10**(1), 808–823 (2016). https://doi.org/10.2174/1874325001610010808
42. Kurul, F., Turkmen, H., Cetin, A.E., Topkaya, S.N.: Nanomedicine: how nanomaterials are transforming drug delivery, bio-imaging and diagnosis. Next Nanotechnol. **7**, 100129 (2025). https://doi.org/10.1016/j.nxnano.2024.100129

Next Generation Implant and Prostheses: Nanomaterials in Orthopedics, Dentistry, and Cardiovascular Solution

Nida Iqbal, Hareem Hayee, Huma Waheed, Ukasha Jawad, and Saman Iqbal

Abstract Traditional implants and prostheses have a variety of drawbacks, such as inflammation of tissues surrounding the implant, osteolysis, lack of osseointegration, metal allergies, and implant breakage, which eventually lead to long-term damage to patients' health. However, rapid advancements in technology have revolutionized the medical field. Integrating nanotechnology and nanomaterial in implants offers an innovative solution to overcome these drawbacks. Among these technologies, nanocoatings play a considerable role in improving the performance of implants. These coatings are mainly used for surface modification to enhance mechanical strength and biocompatibility and reduce stress shielding. Introducing antimicrobial materials in the implants protects against microbes and prevents cytotoxicity. Nanomaterials have recently been integrated into orthopedic implants and prostheses, enhancing wear resistance and bioactivity to increase osseointegration and strong mechanical strength. Additionally, 3D-printed implants offer material and design personalization to the patients. Furthermore, nanostructure modifications and coatings have been incorporated into dental implants that can improve antimicrobial activity. Nanomaterials enhance their durability and seamless drug delivery in cardiovascular implants like stents and pacemakers. This chapter provides a comprehensive review of the integration of nanotechnology with orthopedic, dental, and cardiovascular implants and prostheses. This chapter highlights the state-of-the-art technological advancements in these implants and how they have revolutionized the medical field.

Keywords Implants · Prostheses · Osseointegration · Antimicrobial · Drug delivery · Stress shielding · 3D printing

---

N. Iqbal (✉) · H. Hayee · H. Waheed · U. Jawad
Department of Biomedical Engineering, University of Engineering and Technology (UET), Lahore, Punjab, Pakistan
e-mail: nidaiqbal@uet.edu.pk

S. Iqbal
Department of Physics, University of the Punjab, Lahore, Pakistan

S. N. Saud Al-Humairi et al. (eds.), *Smart Nanomaterials in Biomedical Applications*, Engineering Materials, https://doi.org/10.1007/978-981-96-8023-8_10

1 Introduction

Nanotechnology rapid progress has led to groundbreaking medical implantation and prosthetics improvements. Traditional implants and prostheses have a variety of drawbacks, such as inflammation of tissues surrounding the implant, osteolysis, lack of osseointegration, metal allergies, and implant breakage that can reduce the life span of implants [1]. Nanomaterial-based medical implants offer unique solutions to problems in orthopedic, dental, and cardiovascular applications by enhancing their functionality and durability.

The introduction of nanotechnology in orthopedic implants and prostheses addresses drawbacks such as implant loosening, poor osseointegration, and inflammation caused by wear debris [2]. Moreover, the current advancement of 3D Printing has revolutionized this field of implants and prostheses. Using the 3D printing technology, a user-specific design can be made, reducing alignment-related problems. In dentistry, the use of nanotechnology has significantly improved the properties of dental implants and various orthodontic devices, such as increasing their biocompatibility and bio-activity [3, 4]. The bioactive materials improve the biological response by increasing the rate of osseointegration and the rate of recovery [3]. The use of nanotechnology in cardiovascular implants addresses the limitations of traditional implants, such as elevated mortality rates and resulting heart failure. The size, extensive surface area, and customizable properties of nanomaterials are ideal for addressing cardiovascular treatment [5].

The chapter investigates how nanotechnology drives next-generation implant and prostheses research, examining the applications, benefits, and technological developments that have altered the medical industry.

2 Nanomaterials in Orthopedic Implants and Prostheses

The worldwide population suffers from orthopedic conditions, including rheumatoid arthritis, osteoarthritis, fractures, injuries, and age-related problems. Such conditions lead to intense pain, which restricts and decreases physical movement and substantially deteriorates life quality [6]. These conditions need proper treatment to restore the natural process of mobility and improve the quality of life. For this purpose, medical devices such as implants and prostheses replace damaged or missing bones and joints [7].

Advancements in medical technology have also improved the functionality of orthopedic implants and prostheses, and surgical interventions have improved the lives of people suffering from different kinds of fractures and musculoskeletal disorders, helping them restore their mobility. However, there are also some issues with surgical intervention, such as the fracture fixation method, which usually involves two steps: in the first phase, an implant is placed for fixation, while in the second surgical process, the implant is removed. Moreover, this method has also shown significant challenges like poor osseointegration, implant loosening, tissue rejection, wear debris, and many other infections.

Nanomaterials are attractive in developing scaffolds that act as short-term assistive structures for tissue engineering processes. To provide seamless growing support, ideal scaffolds must be porous and show high biocompatibility, strength, and controlled degradation abilities. Different structures of nanomaterials, such as nanotubes, demonstrate excellent potential for medical applications because these structures boost functionality, leading to superior bone integration when compared to ordinary materials, which are larger [8]. Moreover, the current advancement of 3D Printing has revolutionized this field of implants and prostheses and made it possible to design user-specific implants that can reduce alignment-related problems.

2.1 Orthopedic Implants

Orthopedic implants are medical devices that play an essential role in today's medical field. It helps restore mobility by replacing damaged parts and improves the quality of life of people suffering from various fracture types, musculoskeletal disorders, and degenerative diseases. Several types of implants, like plates, rods, screws, etc., have been designed to integrate with the natural structure of bone. Over time, this approach presents significant difficulties, including poor tissue integration with bone and implant movement, tissue refusal, material fragment creation, and numerous microbial infections. Advancements in nanotechnology have replaced these implants with nanostructured implants like nanorods, nanotubes, etc., along with nano coating, which has enhanced the osseointegration and life of the implant.

i. **Knee Implants**

Human knee joints are the most enormous synovial joints, which allow a significant degree of mobility. Therefore, proper care and treatment are necessary for fractures or damage [9]. Knee implants are medical devices that are used to replace diseased or damaged knee joints. They are typically designed to alleviate pain and restore the natural function or mobility of the knee. These implants are usually composed of metals, ceramics, and composite components that mimic the natural structure of the knee. Different types of knee implants are discussed below.

ii. **Partial Knee Replacement**

Partial or compartmental knee replacement is a conservative design method for damaged knee implants. In this procedure, a less invasive technique is used, in which only the damaged portion is replaced while the healthy portion of bone and cartilage remains intact so that faster recovery of the knee occurs. The main components that are replaced in a partial procedure usually include the femoral, tibia, and patellar parts of the knee. Advanced nanotechnology has revolutionized this process, which is more suitable for better osseointegration and wear resistance.

iii. **Total Knee Replacement**

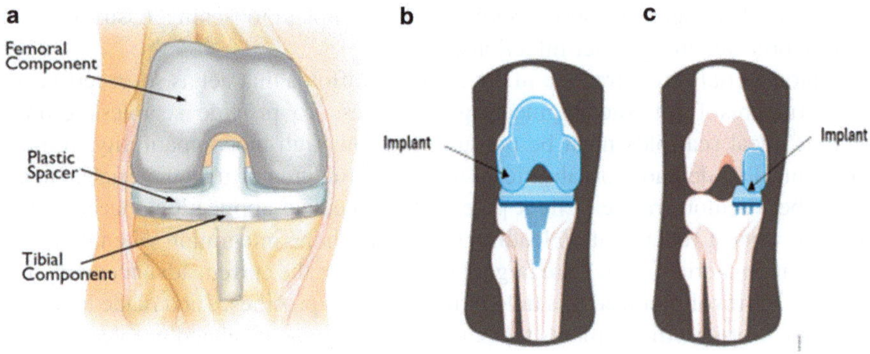

Fig. 1 **a** Knee implant, types of knee implant, **b** permanent knee replacement, **c** partial knee replacement

Total knee replacement is advised for patients with advanced arthritis or when all three compartments of the bones are damaged. The entire knee replacement surgery is more invasive and involves the procedure on the femoral, patella, and tibial components, as shown in Fig. 1. The recent implant technology includes nanomaterials that enhance the durability and lifespan of total knee replacements.

iv. **Hip Implants**

A human hip joint is one of the essential weight-bearing components across the body, which enables the extensive motion required for basic everyday tasks. The significant components of hip joints usually involved in replacement include the acetabulum, liner, femoral head, and femoral stem. The human hip joint sustains significant quality-of-life effects when such conditions as arthritis, fractures, or degeneration damage it. These implants use metal alloys, ceramics, and high-strength materials to create devices that imitate hip joint movements. The primary purpose of a hip implant is to relieve discomfort and improve mobility so patients can resume their daily activities [10]. Below are some different types of hip implants.

v. **Total Hip Replacement (THR)**

Total Hip Replacement (THR) constitutes the most prevalent surgical intervention that removes a patient's femoral head and acetabulum to substitute them with durable surgical materials. New artificial joints comprise strong titanium components, cobalt-chromium alloys, and polyethylene materials [11]. This procedure is defined for patients who have extreme arthritis and joint deterioration because it brings substantial pain relief along with proper functionality of the hip.

vi. **Hip Resurfacing**

The hip resurfacing procedure involves less body invasion than entire hip replacement operations. During this procedure, medical practitioners use a metal surface to cap the remaining femoral head portion. The hip resurfacing procedure preserves natural

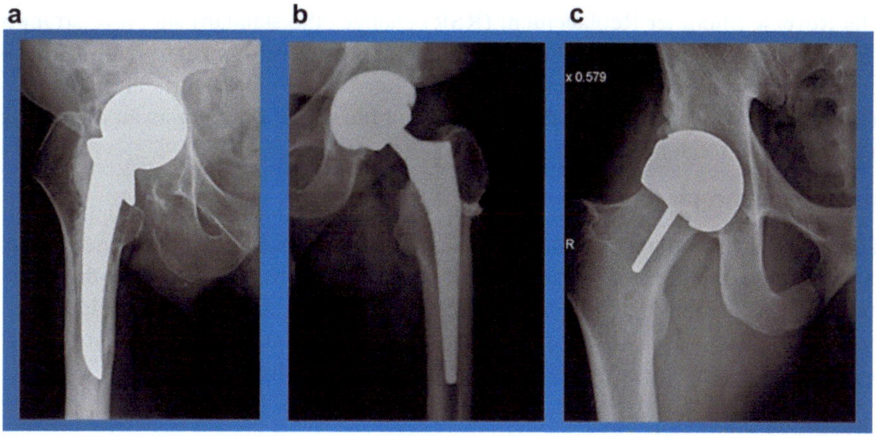

Fig. 2 Types of hip implants **a** partial hip replacement, **b** total hip replacement, **c** resurfacing total hip implant

bone structure, making it suitable for active patients with healthy acetabular bones. The method enables higher levels of mobility and function than total hip replacement, thus enabling patients to move actively.

vii. **Partial Hip Replacement**

Figure 2 shows the Partial Hip Replacement preserves the acetabulum socket by surgically replacing the femoral head. Compared to total hip replacement surgery, this procedure has a minimal impact on patients' original bone structure, which results in better healing.

viii. **Shoulder Implants**

The shoulder joint is one of the most mobile and flexible human joints because it allows significant motion. However, when diseases affect the shoulder joint, a person's ability to carry out daily tasks may suffer. The human body requires shoulder implants to replace damaged joints with pain relief and improved functionality. The implants use titanium, cobalt-chromium alloys, ceramics, and polyethylene as their materials due to their combination of strength, biocompatibility, and durability characteristics [12].

Below are some different types of shoulder implants.

ix. **Total Shoulder Replacement (TSR)**

Total Shoulder Replacement (TSR) is the most frequently performed treatment, and it involves replacing the humeral head (ball) and glenoid (socket) with implant parts. Different materials are used to substitute natural shoulder movement and help pain relief for patients whose joints suffer from severe arthritis or damage [13].

x. **Reverse Shoulder Replacement**

The Reverse Shoulder Replacement (RSR) changes the standard shoulder arrangement by placing the glenoid ball and humeral socket in opposite positions. This implant's design benefits patients who lack functional rotator cuffs and advanced arthritis patients because it enables the deltoid muscle to function as an alternative to the missing rotator cuff [14].

xi. **Hemiarthroplasty**

The surgical procedure of hemiarthroplasty entails substituting the humeral head but maintains the glenoid without replacement. Surgeons usually select this treatment approach when fractures or localized joint damage occur. The procedure gives patients a speedier healing experience than complete shoulder replacement because it maintains a substantial portion of their original bone structure.

xii. **Short Stem Implants**

Short Stem Implants (Fig. 3) provide current surgical options with a shorter stem structure when compared to standard implant designs. The design preserves maximum bone tissue while maintaining patient-specific anatomical accuracy. The implant type grants enhanced surgical revisability, which improves flexibility during the long term. The choice of implant type depends on how much joint damage patients have and the condition of their bones [15].

Stemless Shoulder Implants
Stemless Shoulder Implants (Fig. 4) represent a newer surgical approach without the traditional extended humeral stem component. The implants maintain higher natural bone content, thus making them suitable for people with healthy bone structures. Stemless implants showcase multiple benefits through their minimal invasiveness, lower risk for stress shielding, and more convenient revision surgeries that may be required [14].

Uses of Nanotechnology in Orthopedic Implants

- Different ceramic-based nano biomaterials, such as alumina, zirconia, silica, and calcium phosphates, are used for various types of implant coating. Ceramic materials are used for enhanced mechanical properties, specifically for load-bearing joints, and increased biocompatibility is achieved by using phosphates, like the composition of natural bone [16]. Table 1 shows the main uses of nanotechnology in orthopedic applications.
- Orthopedic implants receive wide-scale bioceramic coatings consisting of calcium phosphate compounds with hydroxyapatite (HA) and octa calcium phosphate (OCP) along with tricalcium phosphates (TCP) for improving biocompatibility and osseointegration and fostering bone regeneration. The coatings duplicate bone tissue while supporting cell population growth and possess additional room for integration with growth factors and antibacterial features. These coatings are applied through plasma spray and electrodepositing methods with fluorinated HA, demonstrating broader antibacterial properties. Implant duration expands while

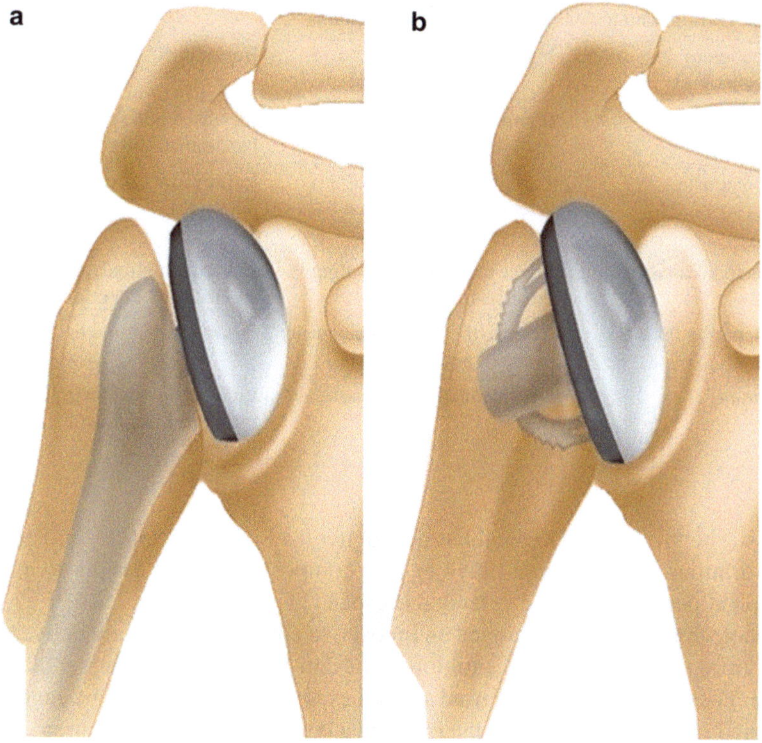

Fig. 3 **a** Short stem shoulder implant, **b** stemless shoulder implant

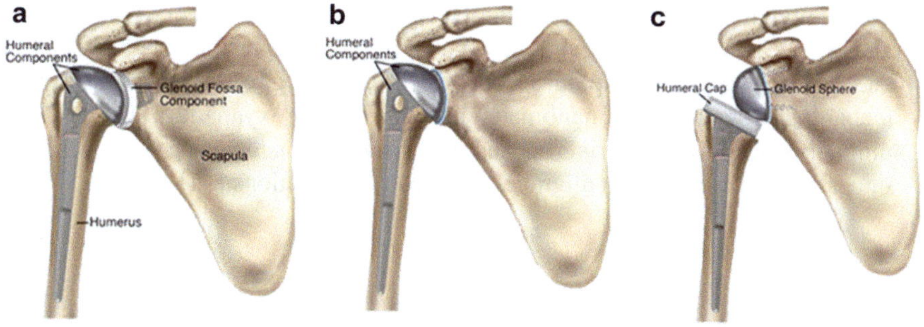

Fig. 4 Types of shoulder implant **a** total shoulder implant, **b** hemiarthroplasty and reverse shoulder replacement

bacterial threats decrease due to these performance-enhancing characteristics [17, 18].

- Incorporating nanomaterials, including carbon nanotubes, graphene, and nano-hydroxyapatite, into implant polymeric parts, boosts their mechanical properties,

Table 1 Uses of nanotechnology in orthopedic implants

Application of nanotechnology	Functions	Benefits
Nanofiber	It enhances tissue integration with implants	Increased bone growth and biocompatibility
Nanocomposites	Carbon nanotubes and graphene are combined for strength	Improved wear resistance and durability
Antibacterial nanocoating	Silver nanoparticle coating is done on implants	Enhanced antibacterial resistance and reduced infection risk
Smart coating for drug delivery	Release the therapeutic agents in the target area	Faster healing process
3D Printing	User-defined implants are made	Enhanced bone integration and minimizes alignment-related issues

elastic behavior, and fatigue tolerance. Better load-bearing capacity and decreased deformities exist in these polymer nanocomposites, which results in enhanced implant performance and extended lifespan [11].

- Nanotechnology has developed an innovative coating that directly delivers therapeutic agents in the form of antibiotics and anti-inflammatory drugs at the implant site. These nanoparticles are used as needles, spheres, cubic bars, and cages. These coated nanoparticles release the drugs in a controlled manner and promote faster healing in high-infection areas. Thus, a high implant success rate can be achieved using smart, innovative coating technology [19].
- The advancement in 3D Printing has revolutionized nanotechnology, which has helped prepare personalized knee implants as per the requirement of the patient's anatomy. The tailored design leads to better implant fittings that enable a natural and efficient operation. Implant success can be enhanced by the use of 3D printing technology together with nanomaterials such as hydroxyapatite (HA) and titanium dioxide (TiO_2) materials because this combination results in improved bone integration and minimizes alignment-related complications [11].

3 Orthopedic Prostheses

3.1 Upper Limb Prostheses

The field of prosthetic device science creates replacement devices for the full or partial arm territory as well as the hand and forearm section for patients who need these body segments reconstructed after surgical amputations or congenital abnormalities. The prosthetic devices restore essential limb functions that enable users to grip, lift, and complete common everyday activities. The recent development of upper limb prostheses now offers patients better command over their devices and enhanced comfort by implementing myoelectric prostheses control systems based on user muscle signals.

3.2 Lower Limb Prostheses

The three common lower limb prostheses are below-the-knee and above-the-knee or trans-femoral devices, which replace different areas below or above the knee. Higher limb prosthetic devices (Fig. 5) enable users to regain the ability to both walk and carry their weight after experiencing the loss of their leg through accidents, disease, or amputation. The current generation of lower limb prosthetics provides greater mobility while being lightweight and allowing users to use them comfortably with enhanced control so they can do their normal activities or participate in activities such as distance walking or running.

Uses of Nanotechnology in Orthopedic Prostheses

Recent technology in orthopedic prostheses has evolved through nanotechnology (See in Fig. 6), making life more comfortable by producing durable and lightweight devices for the proper functionality of limbs. Prosthetic improvement through nanomaterials and nanostructures has achieved notable advances, mainly regarding increased functionality and improved human-body connection [20].

- Nano-fiber scaffolds integrated into limb prostheses enable better attachment of the prosthetic to residual limbs. The scaffold's structure duplicates the extracellular matrix properties to encourage cell development, creating stronger tissue-prostheses attachments. Scientific advances allow the prosthetic device to fit better with patient bodies to enhance comfort alongside better performance capabilities [6].

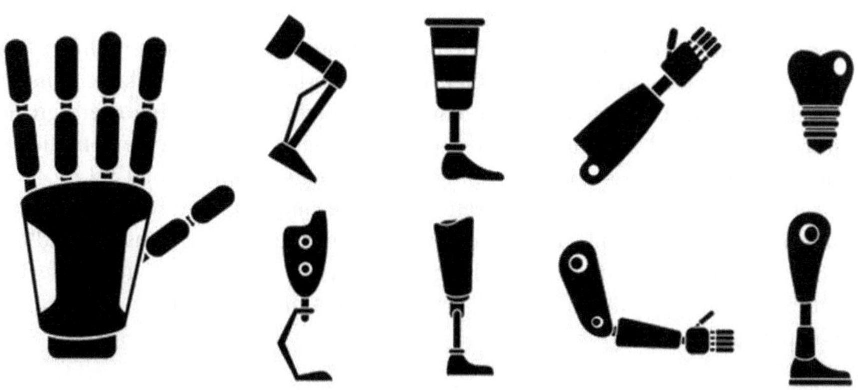

Fig. 5 Upper and lower limb prostheses

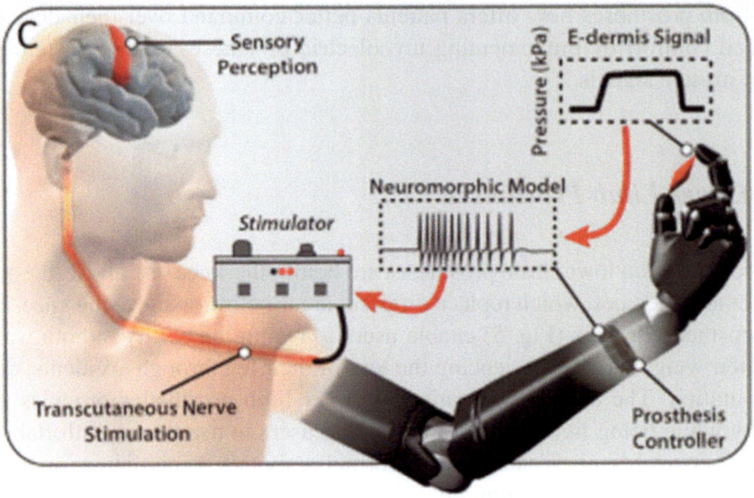

Fig. 6 Upper limb prostheses with sensory feedback

- The strength of prostheses remains unaltered by weight reduction using nanocomposites based on carbon nanotubes and graphene. The material properties of lightweight prosthetics enable better comfort for users and improved mobility, particularly within upper and lower-limb prostheses. Such prosthetic devices can handle everyday stress due to newly incorporated materials that function more efficiently for users, irrespective of the applications. [21]
- Nanotechnology enables the production of smart prostheses through sensors embedded in these devices to offer instant information about pressure levels, motion, and temperature measurements. The feedback system allows users to control their prostheses, which then adapt to their natural movements and create responsive, natural motion. The monitoring system provides for condition checks of prostheses, which help users receive maintenance in time to extend their devices' operational lifespan [22].
- Limb prosthetics depend on carbon nanotube-reinforced materials to achieve flexibility, durability, and strong character. These materials enable prosthetic arms and hand components to endure more damaging conditions, granting users greater movement control. Prosthetic hands need these materials to execute complicated functionalities such as object grasping, holding, and manipulation [22].
- Users can experience pressure sensations, texture, and temperature through touch-sensitive prosthetic limbs developed through nanotechnology. Sensory feedback enhances prosthetic devices, making them easier to operate without feeling artificial; thus, prosthetic devices begin to function like genuine human body components.

Table 2 Uses of nanotechnology in orthopedic prostheses

Application of nanomaterial	Functions	Benefits
Nanofiber scaffold	Scaffold materials are attached to the residual limb	Increased attachment and biocompatibility
Nanocomposites	Carbon nanotubes and graphene are combined for strength	Improved wear resistance, flexibility, and durability
Antibacterial nanocoating	Silver nanoparticle coating is done on prostheses	Enhanced antibacterial resistance and reduced infection risk
Improved osseointegration	Hydroxyapatite coating is done on prostheses	Better bone integration
3D printing	User-specific prostheses are made	Enhanced bone integration and minimizes alignment-related issues

- Combining 3D Printing with nanotechnology has created unique user-defined prosthesis designs, which have helped achieve more precision in prosthetic development. Table 2 shows the use of nanotechnology in orthopedic prostheses. The exact personal dimensions of users drive prosthetic design improvements through 3D printing methods, which simultaneously decrease manufacturing prices and expansion times for affordable access [21].

4 Nanomaterials in Dental Implants and Prostheses

Dental implants are artificial teeth that replace natural human teeth and restore the function of a damaged tooth [23, 24]. However, these implants have certain drawbacks, including inflammation of tissues surrounding the implant, osteolysis, lack of osseointegration, metal allergies, implant breakage due to stress shielding, etc. These drawbacks can lead to temporary to permanent damage to tissues inside the body, making them unsuitable for long-term use [25].

The introduction of nanotechnology in dentistry has significantly improved the properties of dental implants and various orthodontic devices. Integrating nanotechnology and bioactive materials has gained much popularity due to its increasing success rate. The bioactive materials improve the biological response by increasing the rate of osseointegration and the rate of recovery [3]. Nanomaterials and nanostructured surfaces enhance the properties of the implant, providing personalized and regenerative dental care [26]. The use of nanomaterials in dental implants and prostheses has opened the door to various strategies to enhance the integration of implants to the bone and increase the rate of recovery without any infections and inflammations, ensuring the long-term stability of the implants and prosthetic devices inside the body [27].

4.1 Dental Implants

Dental implants are permanent replacements for natural teeth. They look and act exactly like human teeth. These implants are directly screwed into the jawbone.

There are three main types of dental implants.

4.1.1 Endosteal Implants

Endosteal implants are the most commonly used dental implants. They consist of three main parts: a screw, abudment, and crown. They are made up of titanium alloys and zirconia. These implants are used to replace a single tooth and are screwed into the jaw bone [28].

4.1.2 Subperiosteal

Subperiosteal implants are patient-specific and specifically designed for patients with uneven bone height and density. These implants consist of a framework of metallic wires inserted into the jaw bone through the gums. A complete set of teeth is then fixed on the framework [28].

4.1.3 Zygomatic

These implants are specifically for patients who suffer from bone loss in the upper jaw, as shown in Fig. 7. These implants consist of a network of screws which are inserted into the zygomatic bone. They support prostheses like permanent dentures and bridges [29].

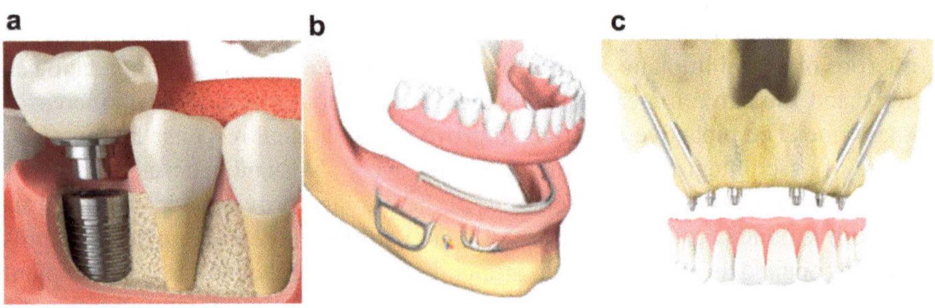

Fig. 7 Visual representation of **a** endosteal implant [30], **b** subperisteal implant [31], and **c** zygomatic implant [32]

Table 3 Nanomaterials in dental implants

Nanomaterials	Function	Benefits
Titanium oxide (TiO_2)	High corrosion resistance and antimicrobial	Offers long-term stability without any inflammations or infections
Hydroxyapatite (HAp)	Enhance bone formation and osseointegration	Faster healing rate, long-term stability
Graphene oxide (GO)	Strong mechanical strength, osseointegration, and antibiotic	Protection against microbes decreases fracture rate

Use of Nanomaterials in Dental Implants

Nano-coating implants is the most reliable and advanced method for increasing their performance and duration inside the body [33]. Coating an implant on a nanoscale enhances its surface properties and increases its surface area, providing more area for osseointegration. The performance and properties of the coated implant depend upon the coating material [23]. Table 3 shows the list of nanomaterials in dental implants.

- Titanium oxide (TiO_2) is a widely used bio-compatible nanomaterial for coating an implant due to its high corrosion resistance, giving the implant a better performance without leaching harmful chemicals into the body [23]. Studies have shown that TiO_2 exhibits strong antimicrobial properties. Under ultraviolet (UV) light, reactive oxygen species (ROS) are generated, which destroy the bacterial cell via oxidative stress and DNA degradation [34].
- Hydroxyapatite (HAp) is a significant component of bone composition. HAp significantly enhances bone formation and osseointegration, which reduces the chances of implant failure and increases the implant's stability [23].
- Graphene oxide (GO) makes an excellent candidate for dental implant coating due to its high surface energy, strong mechanical strength, and osseointegration, giving the implant long-term durability [35]. GO generates ROS to induce oxidative stress and break down the cell membrane of bacterial cells. The hydrophobic nature of GO also exhibits an antibiotic effect by preventing the bacterial cells from clumping together and damaging the cell membrane [36].

4.2 Dental Prostheses

Dental prostheses are personalized devices that can replace a damaged piece of tooth. They are used for both functional and aesthetic purposes. There are three main types of dental prosthetic devices: crowns, bridges, and dentures [37].

4.2.1 Crowns

Dental crowns are caps that are fixed over a chipped, broken, or decaying tooth. Crowns are used to restore the function and look of a regular tooth. This procedure is typically done after a root canal. Dental crowns comprise metals such as gold or silver, ceramics such as zirconia, and resin material [37, 38].

4.2.2 Bridges

Dental bridges are essentially crowns connected to form a bridge. They are used to replace more than one damaged tooth. The damaged teeth should be in the same place to be treated with a dental bridge [37, 38].

4.2.3 Dentures

Dentures are removable sets of custom-made teeth, as shown in Fig. 8. They are designed to be fixed on soft tissues where bridges cannot be placed using a clasp. There are two types: complete and partial [37].

Use of Nanomaterials in Dental Prostheses

Nanomaterials enhance the properties of dental prosthetics in the following ways (Table 4):

- 3D-printed crowns and bridges in nano-fibers show stronger wear resistance and a smoother surface than traditionally manufactured prostheses [39].
- Integrating nanodiamonds in dental prostheses enhances the mechanical strength and reduces the surface roughness, giving a more substantial and smoother finish. Decreased surface roughness reduces the rate of microbial adhesion to the surface. Quaternized nanodiamonds integrated with acrylic resin give noticeable protection against *S. mutans* [40].
- Silver (Ag) integrated prostheses exhibit substantial antimicrobial activities. 20 nm Ag integrated with polymethyl methacrylate (PMMA) reduces the surface

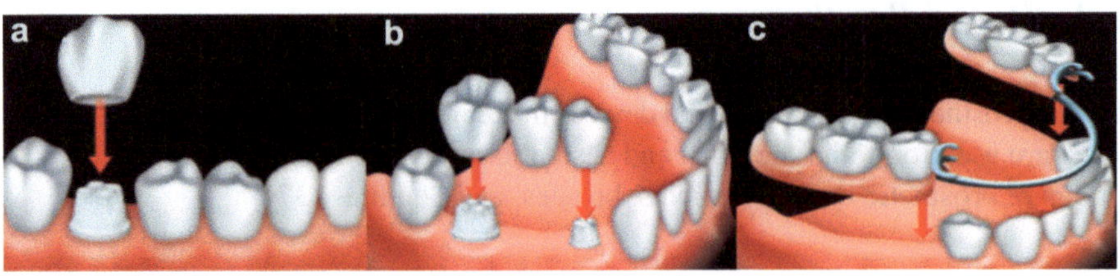

Fig. 8 Visual representation of **a** crown, **b** bridge, and **c** partial dentures [37]

Table 4 Nanomaterials in dental prostheses

Nanomaterial	Function	Benefits
Nanodiamonds	High mechanical strength and antimicrobial	Results in a stronger material without risk of infections
Silver (Ag)	Antimicrobial and low porosity	Protection against microbes and high mechanical strength
Chitosan	Antimicrobial	Protection against inflammations and infections
Copper oxide (CuO)	Antimicrobial	Protection against inflammations and infections

roughness and porosity of the material. It also shows strong anti-fungal properties [40].

- Chitosan is a natural polymer that exhibits intense antimicrobial activity against *C. albicans*, *S. mutans*, *P. aeruginosa*, and *E. faecalis* [40].
- PMMA dentures containing a 2.5–7.5% ratio of copper oxide (CuO) exhibit antimicrobial activity against *candida* and *streptococcus* [40].

4.3 Orthodontic Devices

Orthodontic devices are used to correct the position of the teeth by the jaw. These devices are typically used to correct the bite and for aesthetic purposes. There are two basic types of orthodontic devices.

4.3.1 Orthodontic Brackets

Brackets or braces are devices that use force to pull teeth into place. They come in many different colors, shapes, and sizes. They are typically made of metals, ceramics, and plastics. The traditionally made brackets induce plaque on the teeth' surface, hindering teeth cleaning [41].

4.3.2 Orthodontic Aligners

An acrylic-based clear aligner is a common orthodontic device, as shown in Fig. 9. Aligners are also used to keep teeth in place. They are more comfortable than braces. However, they are a significant site of bacterial adhesion [42].

Use of Nanomaterials in Orthodontic Devices

Nanomaterials enhance the properties of orthodontic devices (Table 5) in the following ways:

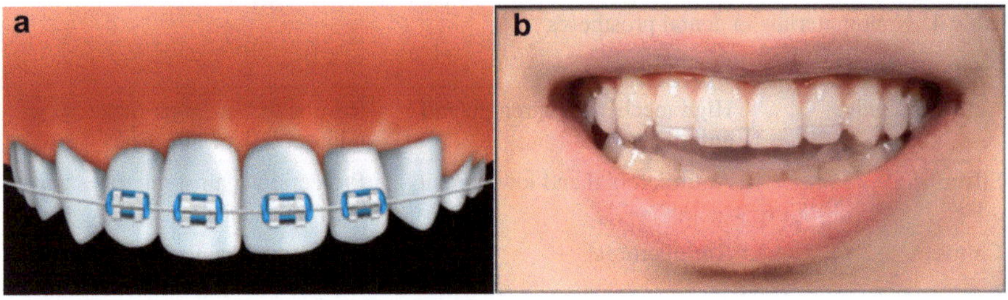

Fig. 9 Visual representation of **a** brackets (braces) and **b** aligners [37]

Table 5 Nanomaterials in orthodontic devices

Nanomaterial	Function	Benefits
TiO_2 and nitrogen	Prevents discoloration and antimicrobial	Maintains natural tooth color and prevents tooth decay
Bioactive glass	Antibacterial	Protection against infections and inflammations
Yttrium fluoride	Antibacterial	Protection against inflammations and infections
Zinc oxide (ZnO)	Increases flexural strength and antimicrobial	Strong structural strength and protection against inflammations and infections

- 50:50 of TiO_2 and Nitrogen (N) to diurethane dimethacrylate resin gives brackets, prevents discoloration, and gives protection against cariogenic bacteria *S. mutans* [40].
- Integrating bioactive glass and acrylic protects against gram-positive bacteria *S. mutans* [40].
- Yttrium fluoride nanomaterial used with acrylic and resin gives antibacterial protection against *S. mutans.* However, increasing the ratio of nanomaterials decreases the antibacterial activity [40].
- The integration of zinc oxide (ZnO) with trans-bond acrylic increases the flexural strength and protects *S. mutans, S. sobrinus,* and *L. acidophilus* [40].

5 Nanomaterial for Cardiovascular Implants

The cardiovascular (CV) system is made up of the heart and blood vessels. The heart's pumping action delivers blood throughout the body via arteries, veins, and capillaries. Nevertheless, issues within the CV system can lead to various diseases, including coronary heart disease, peripheral artery disease, rheumatic heart disease, and cerebrovascular disease. If these conditions go untreated, they can result in myocardial infarction, heart failure, and even death [43]. Cardiovascular diseases are the leading cause of mortality around the world, with approximately 17.7 million fatalities recorded in 2015, as reported by the World Health Organization. By 2030, this number is projected to rise to 23.6 million. The mortality rate from cardiovascular disease is 51% for women, compared to 42% for men. For many years, treatments for cardiovascular diseases have included herbal remedies, dietary changes, and commercially available medications [44]. During the middle of the twentieth century, various interventional therapies for cardiovascular disease have developed. These therapies encompass stent angioplasty, balloon angioplasty, coronary artery bypass grafting, pacemakers, and implantable cardioverter-defibrillators [45]. Nonetheless, the elevated mortality rates and resulting heart failure suggest that these therapies have certain drawbacks. As a result, the treatment of cardiovascular diseases is shifting towards an innovative nanotechnology strategy. Due to their small size, extensive surface area, and customizable properties, nanomaterials are ideal for addressing cardiovascular treatment [5].

5.1 Cardiovascular Stent

A cardiovascular stent is a tiny, mesh-like device. It is employed to retrieve the arteries that have been blocked by plaque, as shown in Fig. 10. The stent is implanted into the body during the angioplasty operation. Stents are typically composed of bare metal, like stainless steel, which offers structural support but might result in complications, including neointimal hyperplasia. By incorporating nanotechnology, stent performance has increased tremendously. Incorporating nanomaterial with the stent has improved drug delivery, bioavailability, and biocompatibility. Nanotechnology has helped reduce thrombosis, inflammation, and delayed endothelialization issues that were prevalent with traditional stents. The Application of nanocoatings in stents exemplifies a strategic approach to advancing cardiovascular implant technology. The following are some advanced nanomaterial-based coating [46].

- Tantalum coating is used on titanium alloy stents to enhance their mechanical characteristics and biocompatibility. Tantalum is an ideal material for prolonged implantation because of its ability to withstand corrosion. Additionally, tantalum prevents restenosis and enhances stent adhesion to surrounding tissue. With improved functionality and fewer side effects, tantalum-coated stents can potentially improve patient outcomes in cardiovascular therapy [5].
- Titanium Oxynitride (TiOxNy) coating has an adjustable ratio of oxygen to nitrogen. Its high nitrogen content lowers the risk of thrombogenicity. It promotes endothelization, boosts cell proliferation, and improves biocompatibility. It is a viable option for a cardiovascular solution because of its excellent corrosion resistance and limited solubility in saline [48].
- GO coating on stents offers several advantages. GO has a strong tensile strength, a large surface area of approximately 2630 m^2/g, and flexibility. GO prevents the

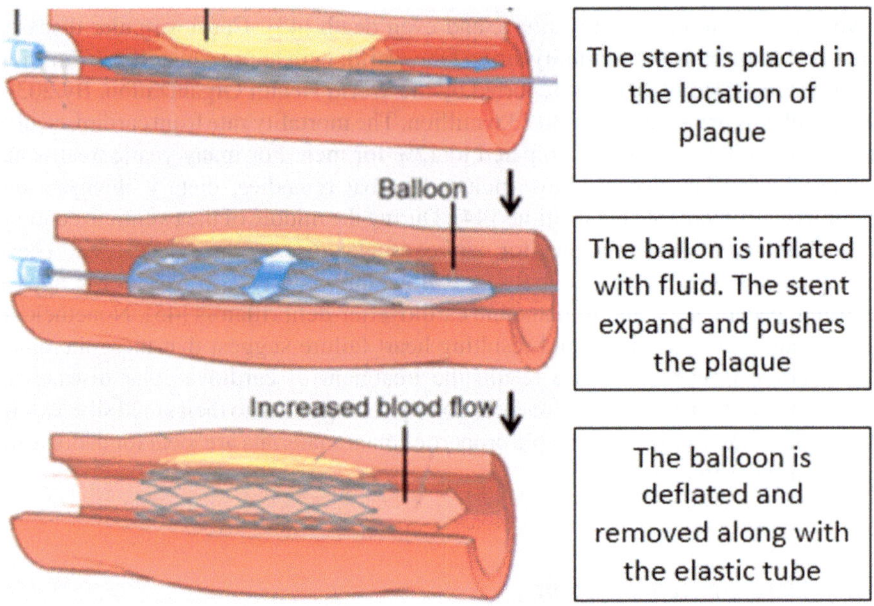

Fig. 10 Restoring blood flow: balloon expansion and stent placement in a blocked artery [47]

body from releasing hazardous compounds and ions from the underlying metal, serving as a protective layer [49]. Table 6 shows the different types of coating that been used for the stent applications.

Table 6 Comparison of various coatings for stent application

Feature	Tantalum coating	Titanium oxynitride coating	Graphene oxide coating	Cationic nano-particle coating
Bio-compatibility	Excellent; enhances stent adhesion	High: boosts cell proliferation and endothelialization	Excellent promotes cell adhesion and proliferation	Variable designed for enhanced cell interaction
Surface properties	Smooth, dense surface; can be textured for better adhesion	Adjustable roughness; tunable nitrogen-to-oxygen ratio	Nanoscale roughness; high surface area for drug delivery	Designed for cell interaction and targeted adsorption
Functionality	Improves stent adhesion; prevents restenosis	Low thrombogenicity; high stability in saline	Anti-thrombogenic; supports drug elution	Enables targeted therapy and magnetic field interactions

- Cationic nanoparticles are superior to polymer-eluting stents because it has a positive charge and rapidly adhere to harmful particles in the body, boosting cellular re-absorption. Cationic nanoparticles are utilized for targeted therapy and have potential in cardiovascular medicine [50].

- Polylactic acid and polyglycolic acid are biodegradable polymers used to create bioresorbable nanostents. These stents show regulated resorption and increased tissue contact due to nanostructuring [51].
- The development of imaging-integrated nano-stents, which improve their diagnostic and real-time monitoring capabilities, heavily relies on nanotechnology. Gold, iron oxide, or quantum dots are examples of nanoparticles used as contrast agents in these stents, which enhance visibility during imaging. It allows physicians to monitor vascular healing, track the position of the stent, and evaluate possible side effects [52].
- Smart nanostructured stents, also known as precision stents, use nanopatterning and surface changes to interact with bodily tissues. Tiny pores and tubes on the surface of the stent are formed by electro-spinning, laser etching, and nanolithography. This provides rapid healing and improved blood flow [53].
- Drug-eluting stent causes inflammation and delayed responses, while the polymer-free nano stent (PFNS) (Fig. 11) does not cause harmful effects.

5.2 *Cardiac Pacemaker*

A cardiac pacemaker regulates the heart rate. It generates electrical activity that is transmitted to the heart walls. Leads and a pulse generator make up the pacemakers. The pulse generator generates the heart's electrical current to stimulate the cardiac muscle. Additionally, the leads carry the electrical current to the heart muscle. Pneumothorax, Venous thrombosis, cardiac perforation, and mechanical lead problems are associated with traditional pacemakers [54]. This limitation of conventional pacemakers (Fig. 12) can be overcome by integrating nanotechnology. Nanotechnology improves pacemaker design by eliminating battery restrictions through energy-harvesting nanogenerators and reducing infection concerns using nanomaterial-based antimicrobial coatings. These advancements increase device longevity, enhance patient safety, and lessen the need for invasive treatments [55].

- Traditional pacemakers need their batteries changed every ten years, as shown in Fig. 13, which requires invasive procedures that increase the risk of infection. To eliminate the need for battery changes, triboelectric and piezoelectric nanogenerators have been developed to produce energy from breathing, heartbeats, and body motion [57]. Table 7 shows the difference between the TENGS and PENGS.

- Infections are a frequent concern following pacemaker placement. To overcome this issue, nanomaterial scoating loading antibiotics have been developed to

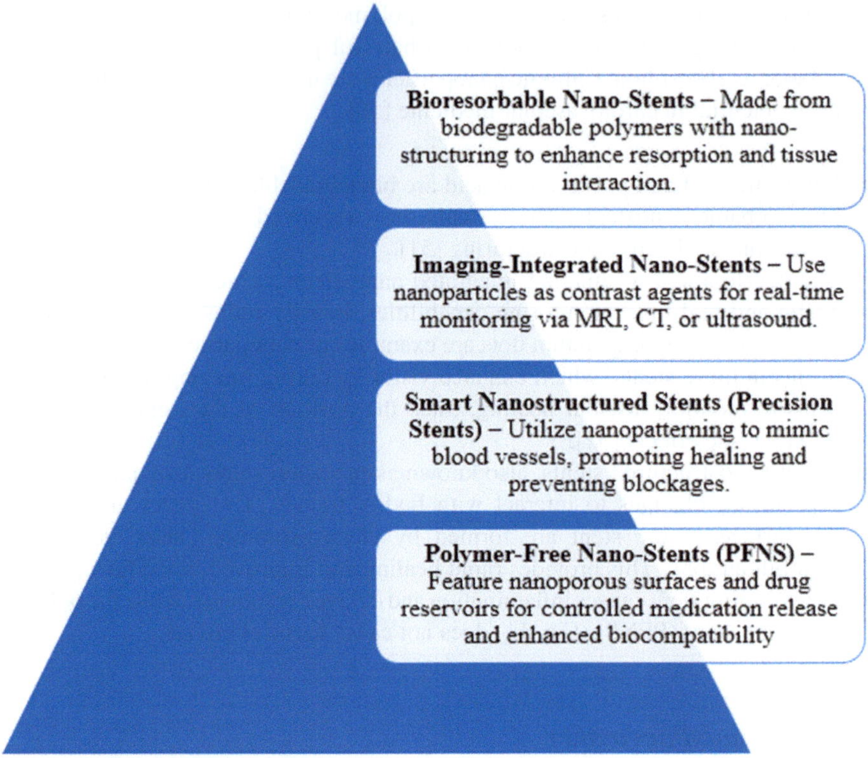

Fig. 11 Classification of nano-stent

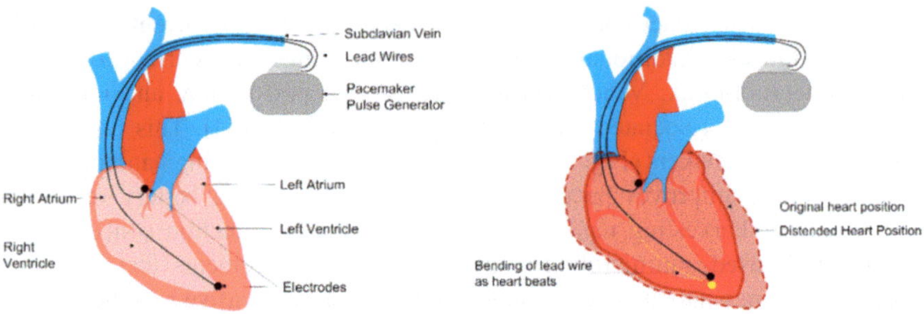

Fig. 12 Schematic showing the implantation of a pacemaker [56]

reduce the risk of infection. L-valine polyester urea film offers localized antibacterial protection by releasing cefazolin. For high-risk individuals, the polypropylene mesh envelope disperses minocycline and rifampicin to prevent infections [55].

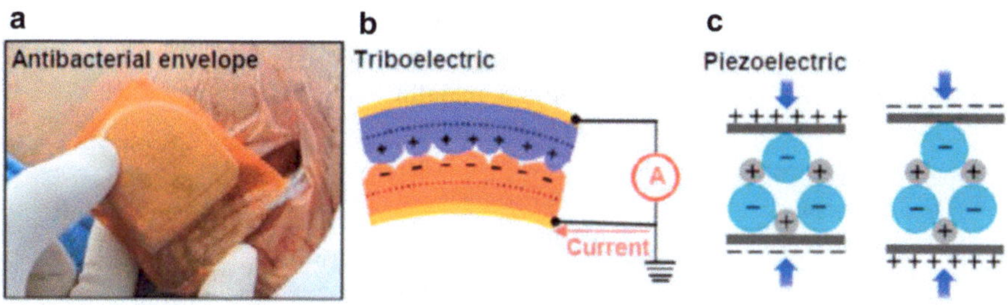

Fig. 13 **a** Antibacterial enveloped pacemaker **b** triboelectric **c** piezoelectric [57]

Table 7 Differences between TEGNS and PENGS

Feature	Triboelectric nanogenerators (TENGs)	Piezoelectric nanogenerators (PENGs)
Working mechanism	Generates electricity through contact electrification and electrostatic induction when two materials come into contact and separate	Generates electricity by converting mechanical stress or deformation into an electric charge using piezoelectric materials
Material composition	Uses dielectric materials like PTFE, PDMS, and nylon combined with electrodes	Uses piezoelectric materials like ZnO, PZT, $BaTiO_3$, and PVDF
Efficiency in pacemakers	Suitable for low-power, self-sustaining sensors in pacemakers	More efficient for direct energy harvesting to power pacemakers
Energy source	Works best with low-frequency body movements (e.g., breathing, limb motion, blood flow)	Works best with continuous mechanical pressure (e.g., heartbeat, muscle contractions)

- Nanowire-based optical pacemakers minimize invasion and improve biocompatibility as compared to electrical pacemakers. The silicon nanowire mesh adheres to cardiac tissue, decreasing mechanical stress and inflammation. There is no electrode degradation or battery replacement issue with it. Thus, it has a high potential in cardiovascular solutions [58].
- Nanotechnology has been crucial in improving optical pacemaking using silicon nanowires. These nanoscale waveguides effectively deliver light energy that activates heart tissue, leading to efficient signal transmission with minimal power loss [58].

5.3 *Cardiovascular Graft*

Cardiovascular grafts repair blocked blood vessels during the treatment of coronary artery disease and then redirect blood flow to improve circulation, as shown in Fig. 14. Three major problems with cardiovascular grafts in conventional usage

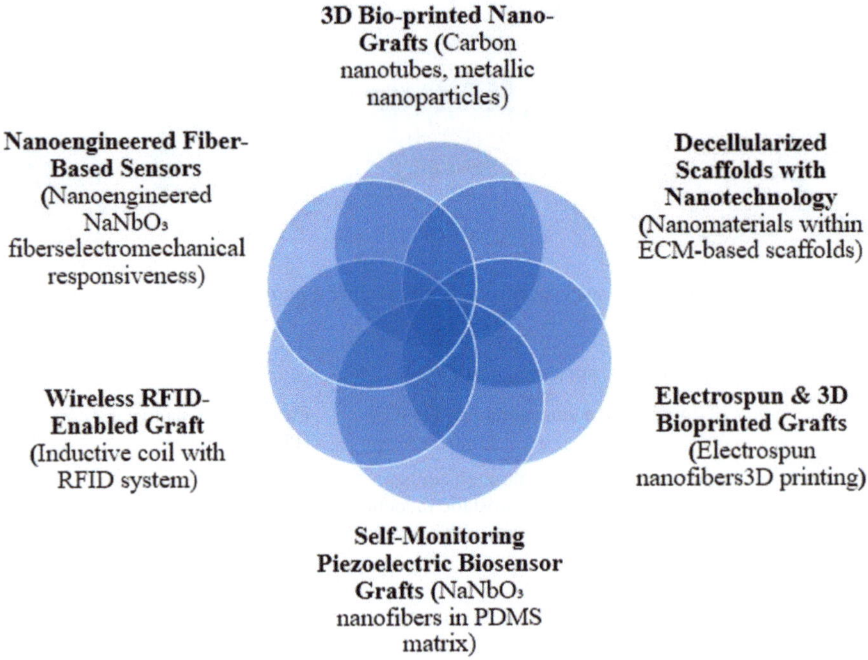

Fig. 14 Emerging technologies in cardiovascular graft

include poor tissue-material compatibility, growing mechanical weakness despite aging, and increased risks of blood clot formation and tissue restenosis. Nanotechnology transforms contemporary cardiovascular grafts using nanoparticles, bioactive coatings, and sophisticated drug delivery systems. The modern cardiac grafting technique produces robust self-healing and bioresorbable vascular products that improve patient healthcare outcomes because of advancements driven by nanotechnology [59].

- Advanced 3D bioprinting methods combine carbon nanotubes and metallic nanoparticles to build vascular grafts with increased electrical response and bioactivity properties. These vascular grafts keep blood vessels open due to their nanostructured characteristics, promoting endothelial cell development [60].
- Electro-spinning and 3D bioprinting techniques have contributed to developing new vascular grafts with improved mechanical qualities, biological functionality, and porosity levels. Nanofiber-based grafts that mimic extracellular matrix features increase cell interaction with graft material and development and specialization capabilities [60].
- Researchers created a self-monitoring piezoelectric biosensor-based vascular graft using a polydimethylsiloxane matrix and sodium niobate ($NaNbO_3$) fibers. The sensor monitors changes in blood pressure and clotting patterns in real time, expediting thrombosis diagnostic processes. This monitoring system allows for the

early detection of thrombosis or hypertension, eliminating the need for intrusive exams [61].

- Wireless Monitoring with Radio Frequency Identification Enabled Grafts employs an inductive coil sensor to detect pressure changes as clots develop, allowing for noninvasive healthcare examinations. The biosensor will monitor wireless using radio frequency identification RFID technology [61].
- Traditional sensors frequently struggle with rigid materials that do not interact effectively with soft tissues. Nanoscale $NaNbO_3$ fibers improve sensor electromechanical responsiveness. The sensor has a low filler percentage (5%), which allows for flexibility while maintaining great sensitivity to variations in blood flow [61].

6 Conclusion

In conclusion, nanomaterials in implants and prostheses improve biocompatibility, osteointegration, stress shielding, etc. Nanotechnology integration in orthopedic, dental, and cardiovascular implants enhances durability, reduces tissue swelling, and reduces material fatigue over time. Engineers can employ nanotechnology to improve the performance and durability of medical equipment, making it appropriate for long-term usage. Combining higher antimicrobial nanoparticles, 3D Printing, and smart coating technology reduces treatment time and protects patients from infections. Healthcare products employing nanotechnology show continued potential in medical implants, resulting in novel healthcare and lifestyle improvements for patients. Medical implants and prostheses will benefit significantly from current nanotechnology research and technological advancements. Future medical developments will provide more treatment options and a higher quality of life.

References

1. Cheng, Y.: Advancements and challenges in nanomaterial-based medical implants. Appl. Comput. Eng. **58**, 60–67 (2024)
2. Pang, A.S.-R., et al.: Nanotechnology-enhanced orthopaedic surgery. J. Nanotheranostics **5**(4), 167–187 (2024)
3. Al-Shalawi, F.D., et al.: Biomaterials as implants in the orthopedic field for regenerative medicine: metal versus synthetic polymers. Polymers **15**(12), 2601 (2023)
4. Marin, E.: History of dental biomaterials: biocompatibility, durability and still open challenges. Heritage Sci. **11**(1), 207 (2023)
5. Hu, Q., et al.: Nanotechnology for cardiovascular diseases. Innovation **3**(2), 100214 (2022)
6. Bandyopadhyay, A., et al.: Improving biocompatibility for next generation of metallic implants. Prog. Mater. Sci. **133**, 101053 (2023)
7. Kulkarni, P.G., et al.: Overcoming challenges and innovations in orthopedic prosthesis design: an interdisciplinary perspective. Biomed. Mater. Devices **2**(1), 58–69 (2024)
8. Liang, W., et al.: Current developments and future perspectives of nanotechnology in orthopedic implants: an updated review. Front. Bioeng. Biotechnol. **12** (2024)

9. Younas, A., et al.: Novel approaches of the nanotechnology-based drug delivery systems for knee joint injuries: a review. Int. J. Pharm. **608**, 121051 (2021)
10. Oladapo, B.I., Zahedi, S.A., Ismail, S.O.: Mechanical performances of hip implant design and fabrication with PEEK composite. Polymer **227**, 123865 (2021)
11. Abaszadeh, F., et al.: Nanotechnology development in surgical applications: recent trends and developments. Eur. J. Med. Res. **28**(1), 537 (2023)
12. Amirtharaj Mosas, K.K., et al.: Recent advancements in materials and coatings for biomedical implants. Gels **8**(5), 323 (2022)
13. Leafblad, N., Asghar, E., Tashjian, R.Z.: Innovations in shoulder arthroplasty. J. Clin. Med. **11**(10), 2799 (2022)
14. Sanchez-Sotelo, J.: Current concepts in humeral component design for anatomic and reverse shoulder arthroplasty. J. Clin. Med. **10**(21), 5151 (2021)
15. Yasmin, H.: Design and research of biomaterials. Coatings **12**(11), 1684 (2022)
16. Liang, W., et al.: Prospective applications of bioactive materials in orthopedic therapies: a review. Heliyon **10**(16), e36152 (2024)
17. Charles Oladipupo, F.: Advancements in hydroxyapatite nanocomposites for orthopedic implants (2025)
18. Shang, J., et al.: Recent developments in nanomaterials for upgrading treatment of orthopedics diseases. Front. Bioeng. Biotechnol. **11** (2023)
19. Jeyaraman, M., et al.: Silver nanoparticle technology in orthopaedic infections. World J. Orthop. **14**(9), 662–668 (2023)
20. Chen, L., et al.: Translation of nanotechnology-based implants for orthopedic applications: current barriers and future perspective. Front. Bioeng. Biotechnol. **11** (2023)
21. Ornaghi, H.L., Monticeli, F.M., Agnol, L.D.: A review on polymers for biomedical applications on hard and soft tissues and prosthetic limbs. Polymers **15**(19), 4034 (2023)
22. Vinchurkar, K., et al.: Advances in sustainable biomaterials: characterizations, and applications in medicine. Discov. Polym. **2**(1), 2 (2025)
23. Bokobza, L.: On the use of nanoparticles in dental implants. Materials **17**(13), 3191 (2024)
24. Fiorillo, L., Meto, A., Cicciù, M.: Bioengineering applied to oral implantology, a new protocol: "digital guided surgery." Prosthesis **5**(1), 234–250 (2023)
25. Sun, Z., et al.: The overview of antimicrobial peptide-coated implants against oral bacterial infections. Aggregate **4**(3), e309 (2023)
26. Soe, Z.C., et al.: Application of nanoparticles as surface modifiers of dental implants for revascularization/regeneration of bone. BMC Oral Health **24**(1), 1175 (2024)
27. Sreenivasalu, P.K.P., et al.: Nanomaterials in dentistry: current applications and future scope. Nanomaterials **12**(10), 1676 (2022)
28. Gulener, F., et al.: Rehabilitation of edentulous patient using endosteal and subperiosteal implants: a case report. J. Craniomaxillofac. Res. 27–30 (2023)
29. Cardoso, F.R.G.R., Grillo, R.: Maxillary rehabilitation after zygomatic implant sequelae using custom subperiosteal implants: a case study. J. Stomatology Oral Maxillofac. Surg. **126**(5), 102154 (2025)
30. smiles, S.: How dental implants replace missing teeth—serene smiles. Serene Smiles (2020)
31. Miso Web, D.: Federal Way Dental Excellence. Federalwaydentalexcellence.com (2025)
32. NoFrills, D.: NoFrills Dental (2024)
33. Nicolae, C.-L., et al.: An overview of nanotechnology in dental medicine. J. Compos. Sci. **8**(9), 352 (2024)
34. Serov, D.A., et al.: Review of antimicrobial properties of titanium dioxide nanoparticles. Int. J. Mol. Sci. **25**(19), 10519 (2024)
35. Kwak, J.M., et al.: Graphene oxide as a biocompatible and osteoinductive agent to promote implant osseointegration in a rabbit tibia model. Adv. Mater. Interfaces **9**(28), 2201116 (2022)
36. Daulbayev, C., et al.: Effect of graphene oxide/hydroxyapatite nanocomposite on osteogenic differentiation and antimicrobial activity. Surf. Interfaces **28**, 101683 (2022)
37. Canadian Dental Association. Cda-adc.ca (2025)

38. Sailer, I., et al.: Prosthetic failures in dental implant therapy. Periodontology 2000 **88**(1), 130–144 (2022)
39. Alzahrani, S.J., et al.: Mechanical properties of three-dimensional printed provisional resin materials for crown and fixed dental prosthesis: a systematic review. Bioengineering **10**(6), 663 (2023)
40. Yudaev, P., et al.: Polymeric dental nanomaterials: antimicrobial action. Polymers **14**(5), 864 (2022)
41. Toz Ertop, M., et al.: Evaluation of the demineralization development around different types of orthodontic brackets. Materials **16**(3), 984 (2023)
42. Charavet, C., et al.: Cleaning and disinfection protocols for clear orthodontic aligners: a systematic review. Healthcare **10**(2), 340 (2022)
43. Olvera Lopez, E., Ballard, B.D., Jan, A.: Cardiovascular Disease. In: StatPearls. StatPearls Publishing. Copyright © 2025, StatPearls Publishing LLC.: Treasure Island (FL) (2025)
44. Bachheti, R.K., et al.: [Retracted] Prevention and treatment of cardiovascular diseases with plant phytochemicals: a review. Evid. Based Complement. Altern. Med. **2022**(1), 5741198 (2022)
45. Netala, V.R., et al.: A comprehensive review of cardiovascular disease management: cardiac biomarkers, imaging modalities, pharmacotherapy, surgical interventions, and herbal remedies. Cells **13**(17) (2024)
46. Udrişte, A.S., et al.: Coatings for cardiovascular stents—an up-to-date review. Int. J. Mol. Sci. **25**(2), 1078 (2024)
47. Shah, P., Chandra, S.: Review on emergence of nanomaterial coatings in bio-engineered cardiovascular stents. J. Drug Deliv. Sci. Technol. **70**, 103224 (2022)
48. Cherian, A.M., et al.: Surface engineering at the nanoscale: a way forward to improve coronary stent efficacy. APL Bioeng. **5**(2) (2021)
49. Coronel-Meneses, D., et al.: Strategies for surface coatings of implantable cardiac medical devices. Front. Bioeng. Biotechnol. **11**, 1173260 (2023)
50. Kuznetsov, K.A., et al.: Vascular stents coated with electrospun drug-eluting material: functioning in rabbit iliac artery. Polymers **12**(8), 1741 (2020)
51. Islam, P., et al.: Nanotechnology in development of next generation of stent and related medical devices: current and future aspects. Wiley Interdisc. Rev. Nanomed. Nanobiotechnol. **16**(2), e1941 (2024)
52. Sikkander, A.M., et al.: A review of advances in the development of bioresorbable nano stents: part (II). J. Chem. Rev. **6**(3), 304–330 (2024)
53. Hassan, S., Ali, M.N., Ghafoor, B.: Evolutionary perspective of drug eluting stents: from thick polymer to polymer free approach. J. Cardiothorac. Surg. **17**(1), 65 (2022)
54. Lak, H.M., Goyal, A.: Pacemaker types and selection. In: StatPearls [internet]. StatPearls Publishing (2022)
55. Tetali, S., et al.: Intelligent biomaterials for cardiovascular applications. Curr. Opin. Biomed. Eng. **28**, 100474 (2023)
56. Hu, C., Behdinan, K., Moradi-Dastjerdi, R.: PVDF energy harvester for prolonging the battery life of cardiac pacemakers. Actuators. MDPI (2022)
57. Zhao, L., et al.: Triboelectric nanogenerators and piezoelectric nanogenerators for preventing and treating heart diseases. BMEMat **1**(2), e12020 (2023)
58. Song, S., et al.: Recent advances in 1D nanomaterial-based bioelectronics for healthcare applications. Adv. NanoBiomed. Res. **2**(3), 2100111 (2022)
59. Aydin, S., et al.: The cardiovascular system and the biochemistry of grafts used in heart surgery. Springerplus **2**, 1–7 (2013)
60. Matějková, J., Kaňoková, D., Matějka, R.: Current status of bioprinting using polymer hydrogels for the production of vascular grafts. Gels **11**(1), 4 (2024)
61. D'Ambrogio, G., et al.: Piezoelectric biosensor for smart cardiovascular grafts based on $NaNbO_3$ fibers/PDMS structured composite. Mater. Des. **223**, 111195 (2022)

Smart Nanomaterials in Tissue Engineering: Scaffolds, Bioprinting, and Stem Cell Integration

Trang Thi Minh Nguyen, Xiangji Jin, Qiwen Zheng, and Tae-Hoo Yi

Abstract Smart nanomaterials have emerged as a pivotal innovation in tissue engineering, offering enhanced structural and functional properties for regenerative medicine. This chapter explores the role of these advanced materials in scaffolds, bioprinting, and stem cell integration, emphasizing their impact on exosome-based therapies. Smart nanomaterials facilitate controlled exosome delivery, improving stability, targeting, and therapeutic efficacy in skin regeneration, cartilage repair, and bone healing. The discussion covers nanomaterial-enhanced scaffolds, bio-inks, and bioprinting technologies that enable precise tissue modeling. Additionally, a systematic review of in vitro, in vivo, and clinical studies highlights recent advancements and challenges in translating nanomaterial-based exosome therapies to clinical applications. The chapter concludes with future perspectives on nanomaterial functionalization, personalized regenerative strategies, and regulatory considerations, underscoring the transformative potential of smart nanomaterials in next-generation tissue engineering. Key findings from this chapter highlight that nanomaterial-integrated scaffolds reinforce mechanical strength and bioactivity and guide stem cell differentiation, accelerating tissue repair and regeneration. Moreover, the incorporation of nanoparticles into bioinks enables high-precision bioprinting, fostering the development of biomimetic architectures tailored for personalized regenerative medicine. Despite these advancements, challenges related to large-scale production, regulatory compliance, and long-term biocompatibility persist. Addressing these limitations through interdisciplinary collaboration and technological refinement will be essential to harnessing the full potential of smart nanomaterials in next-generation regenerative medicine.

Keywords Bioprinting · Exosomes · Nanomaterials · Regenerative medicine · Scaffolds · Stem cells · Tissue engineering

T. T. M. Nguyen · Q. Zheng · T.-H. Yi (✉)
Graduate School of Biotechnology, Kyung Hee University, Yongin-Si 17104, Republic of Korea
e-mail: drhoo@khu.ac.kr

X. Jin
Department of Dermatology, School of Medicine, Kyung Hee University, Seoul, Republic of Korea

S. N. Saud Al-Humairi et al. (eds.), *Smart Nanomaterials in Biomedical Applications*, Engineering Materials, https://doi.org/10.1007/978-981-96-8023-8_11

1 Introduction

As a multidisciplinary domain bridging biology, engineering, and medicine, tissue engineering has emerged as a transformative solution for addressing challenges in tissue repair and organ regeneration [1, 2]. The field has evolved significantly since its inception in the late twentieth century, transitioning from basic tissue grafting to sophisticated bioengineered constructs that closely mimic native tissues [1, 3].

The development of tissue engineering can be categorized into four distinct stages (Fig. 1). The initial phase focused on *early grafting techniques and cryopreservation methods,* which laid the groundwork for modern regenerative strategies. In the nineteenth century, pioneers like Reverdin and Wolfe developed the first successful skin grafting techniques, addressing the challenge of skin repair for burn victims and those with severe injuries. Early refrigeration methods, such as Wentscher's 1903 technique and Carrel's 1911 refinement, demonstrated that skin grafts could survive extended periods, making transplants more feasible. In 1944, Webster's successful refrigeration of skin grafts expanded clinical applications, while [4] cryopreservation breakthrough enabled long-term tissue storage, addressing the challenge of preserving biological materials for future use. These developments laid the foundation for modern regenerative medicine, influencing today's tissue repair and preservation strategies. This phase encompasses the foundational practices of tissue repair, such as skin grafting and the development of methods for cell and tissue preservation. These early efforts established the groundwork for future advancements [4–6].

The second phase, development of synthetic substitutes and early tissue engineering foundations, introduced biocompatible synthetic materials such as polylactic acid (PLA) and collagen, enhancing scaffold development by improving structural integrity and biomimicry of the extracellular matrix (ECM), enabling cell adhesion, growth, and differentiation [7–11].

With the advent of advanced fabrication techniques such as 3D bioprinting and electrospinning, the third phase of *Advances in organ engineering and bioprinting* enabled precise control over tissue architecture, facilitating breakthroughs in organ engineering. This period witnessed the successful development of bioartificial tissues, including engineered bladders and tracheostomies. Emerging bioprinting technologies and scaffold fabrication techniques have further expanded the potential of engineered tissues by enabling precise spatial control of cellular arrangements, fostering tissue vascularization, and enhancing the integration of functional biomaterials [12–15]. The fourth and current phase is marked by the integration of smart nanomaterials, which impart novel functionalities such as enhanced mechanical strength, electrical conductivity, and targeted bioactivity. These innovations have accelerated commercialization efforts, making engineered tissues more viable for clinical applications [16–19].

Smart nanomaterials represent a pivotal advancement in tissue engineering, offering unique properties that enhance scaffold design, stem cell behavior, and bioprinting technologies. These materials are characterized by their responsiveness to environmental stimuli, such as pH, temperature, and light, which enables precise

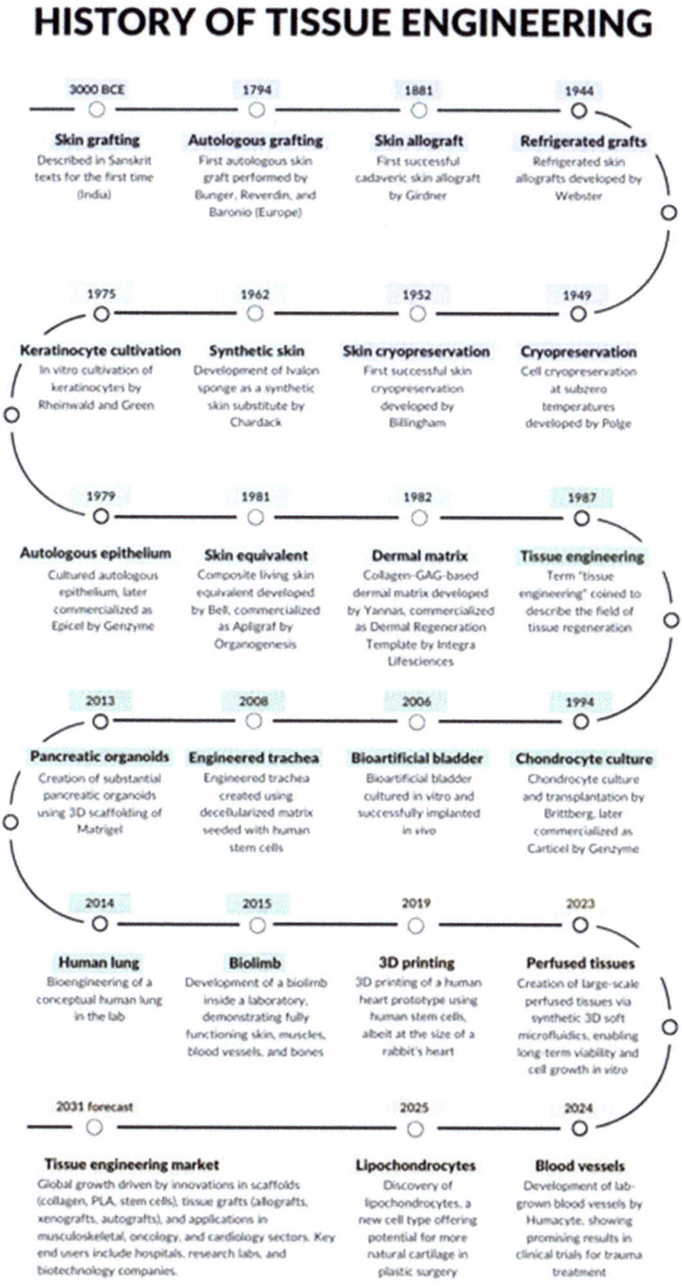

Fig. 1 Stages in the evolution of tissue engineering. The timeline illustrates the four key stages of tissue engineering history: (1) early tissue grafting and cryopreservation techniques (purple), (2) development of synthetic substitutes and early tissue engineering foundations (blue), (3) Advances in organ engineering and bioprinting (green), and (4) Advanced tissue engineering and market growth (yellow)

control over their behavior in biological systems. This responsiveness allows smart nanomaterials to dynamically interact with their surroundings, making them highly versatile for regenerative applications [20, 21].

One of the defining features of smart nanomaterials is their ability to mimic the extracellular matrix (ECM) at the nanoscale. This capability enables the creation of scaffolds that closely resemble the natural cellular environment, promoting cell adhesion, proliferation, and differentiation. For instance, stimuli-responsive hydrogels, which change their physical properties in response to external cues, have been successfully used to enhance cartilage regeneration. These hydrogels not only replicate the structural characteristics of the ECM but also facilitate minimally invasive applications through injectable formulations [22].

Incorporating nanomaterials into scaffolds significantly improves their mechanical and biological properties. For example, conductive hydrogels have been developed to mimic the electrical microenvironment of electroactive tissues, such as myocardium and neural tissues. These hydrogels enable better cell communication by transmitting electrical signals, promoting functional integration and tissue regeneration. Advanced fabrication methods, including in situ forming of injectable hydrogels and 3D bioprinting, have further expanded the utility of these materials in constructing complex tissue architectures [23].

Stem cell technologies, 3D bioprinting, and scaffolds have the potential to significantly advance clinical practices, particularly in the field of regenerative medicine [2, 24]. Pluripotent and mesenchymal stem cells are increasingly utilized for tissue regeneration, with applications ranging from skin wound healing to cartilage repair and spinal cord injury recovery. 3D bioprinting facilitates the precise fabrication of complex tissue structures, allowing for the creation of scaffolds that support stem cell growth and direct differentiation [21, 25]. These scaffolds, often constructed from biocompatible materials such as hydrogels or synthetic polymers, provide structural guidance and mimic the extracellular matrix, promoting tissue regeneration. In clinical settings, this triad of technologies is being investigated for applications in skin grafts, organ regeneration, and bone healing [23]. For instance, 3D-printed scaffolds seeded with stem cells are being explored to repair osteoarthritis cartilage defects, while stem cell-based scaffolds improve skin regeneration for burn patients. Although challenges such as vascularization, immune rejection, and scalability remain, addressing these obstacles could result in significant advancements in personalized medicine, offering the possibility of custom-engineered tissues and organs that better integrate with the patient's biological systems, enhancing healing outcomes and reducing dependence on donor organs [19].

Nanomaterials also play a critical role in enhancing stem cell behavior. By providing nanoscale topographical cues and delivering localized biochemical signals, these materials guide stem cell differentiation into specific lineages. For instance, rare earth nanomaterials have demonstrated osteogenic and angiogenic properties, making them highly effective in bone tissue engineering. Additionally, nanomaterials such as graphene oxide have been employed to optimize scaffold architectures, supporting cell viability and proliferation while sustaining the formation of functional tissues [21].

In the context of bioprinting, smart nanomaterials have revolutionized the development of bioinks [26, 27]. Functionalized bioinks containing nanomaterials offer enhanced mechanical strength, rapid gelation, and spontaneous crosslinking capabilities. For example, neural stem cell-laden scaffolds fabricated using bioinks with nanomaterials have shown high cell viability and facilitated neuronal differentiation. These scaffolds not only improve the cell-material interaction but also promote the regeneration of neural networks, as demonstrated in spinal cord injury repair models [28].

Additionally, nanomaterials have been instrumental in the development of bioadhesives for minimally invasive procedures. These bio-glues, which respond to near-infrared (NIR) light, solidify to create strong tissue adhesion in both wet and dry conditions. This technology has been particularly effective in wound healing and hemostatic applications. Recent studies highlight the potential of metallic nanomaterials, such as gold and silver nanoparticles, which exhibit antimicrobial effects while promoting tissue repair and regeneration. These multifunctional materials not only optimize scaffold performance but also support infection control and accelerated healing [29–31].

Despite their numerous advantages, challenges remain in the clinical translation of smart nanomaterials. Issues such as biodegradability, long-term biocompatibility, and scalability need to be addressed to realize their potential. However, with ongoing advancements in material science and nanotechnology, smart nanomaterials are poised to become an integral component of next-generation tissue engineering and regenerative medicine.

This chapter explores the integration of smart nanomaterials into tissue engineering, focusing on their role in scaffold development, bioprinting technologies, and stem cell-based regenerative strategies. It highlights how nanomaterials enhance scaffold bioactivity, structural integrity, and biomimicry, facilitating controlled cell adhesion, proliferation, and differentiation. The discussion extends to exosome-based therapies, emphasizing their potential in targeted delivery, intercellular communication, and regenerative signaling.

Additionally, the chapter examines recent advancements in functionalized bioinks, nanocomposite scaffolds, and biofabrication techniques that enable precise tissue modeling. A systematic review of in vitro and in vivo studies provides insights into emerging trends, challenges, and future opportunities in nanomaterial-assisted tissue engineering. By addressing the latest innovations in scaffold design, exosome functionalization, and smart biomaterials, this chapter aims to provide a comprehensive understanding of the evolving landscape of regenerative medicine.

2 Smart Nanomaterials in Scaffolds, Bioprinting, and Stem Cells for Tissue Engineering

The extracellular matrix (ECM) is pivotal in providing structural and functional support to tissues and organs. By offering three-dimensional (3D) structural integrity, the ECM regulates essential cellular functions such as proliferation, differentiation, and migration. Mimicking the ECM is a cornerstone of tissue engineering and regenerative medicine, facilitating the development of organoids in vitro and promoting tissue regeneration in vivo. Scaffolds designed to replicate the ECM's properties have become indispensable tools in these fields, providing the necessary cues for cellular behavior and tissue formation [32, 33].

To replicate the ECM effectively, scaffolds must integrate structural and biochemical cues. Decellularization techniques have emerged as a key strategy to remove cellular components from tissues while preserving the ECM's structural and biochemical properties. These techniques employ physical methods (e.g., freeze–thaw cycles), chemical agents (e.g., detergents), and enzymatic treatments to retain critical ECM components such as collagen, fibronectin, and glycosaminoglycans. Tissue-specific decellularized ECMs, such as cardiac or liver dECM, offer tailored microenvironments that closely mimic native tissue, enhancing the potential for successful tissue regeneration [33–35].

Recent advancements in scaffold design have focused on incorporating smart nanomaterials to mimic the intricate architecture and functional properties of the ECM. These nanomaterials provide scaffolds with enhanced mechanical stability, bioactivity, and the ability to deliver bioactive molecules in a controlled manner [36]. Nanoparticle-functionalized scaffolds, such as those using gold or silver nanoparticles, offer antimicrobial properties and enhance wound healing by reducing infection risk [37]. Additionally, carbon-based nanomaterials like graphene and carbon nanotubes improve electrical conductivity, making them suitable for neural tissue engineering applications. By integrating nanostructures into scaffolds, researchers have developed platforms capable of closely replicating the native ECM's nanotopography and biochemical composition, thereby improving cellular adhesion, proliferation, and differentiation.

Applications of ECM-mimicking scaffolds span diverse fields. For instance, cardiac dECM scaffolds have shown promise in heart tissue engineering, where recellularization with cardiomyocytes or stem cells is a key step in creating functional cardiac tissue. Similarly, in cancer research, ECM scaffolds replicate the tumor microenvironment, enabling the development of 3D cancer models to study tumor behavior and drug responses. Moreover, scaffolds embedded with magnetic nanoparticles allow for remote control of cellular responses, improving targeted tissue regeneration. In chronic wound healing, scaffolds derived from adipose-derived stem cells (ASCs) have demonstrated efficacy in supporting fibroblast and endothelial cell adhesion, promoting tissue repair [38].

Advancements in scaffold design have integrated nanotopographical and biochemical cues to enhance functionality. For example, electrospun fibers infused with dECM improve mechanical stability and hydrophilicity, supporting cell attachment and growth. Hybrid scaffolds combining decellularized ECM with synthetic materials, such as silk fibroin or polycaprolactone, offer enhanced biocompatibility and functionality for applications in liver and cardiac tissue engineering.

Building on these advancements, the recent evolution in scaffold development has focused on optimizing both natural and synthetic materials to achieve enhanced clinical performance. The use of decellularized ECM scaffolds derived from cartilage has shown promise in cartilage tissue engineering, with magnetized ECM scaffolds demonstrating improved knee joint function recovery in preclinical trials [39]. The creation of human adipose tissue-derived extracellular matrix sheets, enhanced with crosslinked hyaluronic acid, shows promise for wound healing. However, scalability, variability in tissue sourcing, crosslinking standardization, and high processing costs may limit clinical use, particularly in resource-limited settings [40]. Similarly, electrospun nanofiber scaffolds made from microcrystalline cellulose and chitosan exhibit excellent mechanical properties and bactericidal activity, yet face challenges in large-scale production, fiber alignment, and long-term stability. Potential immune responses and the need for sustained functionality in dynamic wound environments hinder their broader clinical adoption [41]. These examples underscore the ongoing efforts to refine scaffold technologies while addressing the challenges inherent in their clinical applications.

Nanomaterials have further revolutionized scaffold design by enabling dynamic interactions with cells. Smart nanomaterials can be engineered to respond to external stimuli, such as changes in pH, temperature, or mechanical stress, allowing scaffolds to adapt to the biological environment. For example, thermos-responsive nanomaterials, such as poly(N-isopropylacrylamide) (PNIPAM)-based scaffolds, exhibit temperature-dependent swelling behavior, making them ideal for controlled drug release and tissue expansion. Additionally, these materials facilitate the controlled release of growth factors, cytokines, and other bioactive molecules, enhancing their therapeutic potential [36].

The integration of smart nanomaterials in scaffold design marks a significant advancement in tissue engineering and regenerative medicine. By leveraging the unique properties of these materials, researchers can create scaffolds that replicate the ECM's architecture and actively support cellular processes essential for tissue repair and regeneration. As the field progresses, nanomaterial synthesis, characterization, and functionalization innovations are expected to expand the capabilities of ECM-mimicking scaffolds further.

2.1 Functionalized Scaffolds with Smart Nanomaterials

The integration of smart nanomaterials into scaffolds has revolutionized tissue engineering by imparting advanced functionalities such as stimuli responsiveness, controlled release, and enhanced cellular interactions. These properties enable scaffolds to mimic the dynamic behavior of native tissues, facilitating tissue regeneration and repair.

Electrospinning, a versatile nanotechnology, is pivotal in fabricating polymeric nanofibers with nanoscale structures. Electrospun fiber materials (EFMs) are widely used in biomedical applications, including tissue engineering, wound dressing, and drug delivery, due to their high surface-to-volume ratio, porosity, and ease of fabrication. Nanofiber-based scaffolds functionalized with metallic nanoparticles, such as silver or zinc oxide, provide antibacterial properties, while graphene oxide-enhanced fibers improve mechanical strength and bioactivity. For example, electrospun PLA/OMMT/resveratrol (RSV) nanofibers combine antioxidant and osteogenic properties, with RSV promoting osteogenic differentiation and organophilic montmorillonite (OMMT) providing mechanical reinforcement. These scaffolds exhibit significant antioxidant activity (83.75% in DPPH assay) and support cell adhesion, proliferation, and osteogenic differentiation, making them promising for bone tissue regeneration [42].

Nanogels, another class of smart nanomaterials, are gaining attention for their ability to deliver growth factors and bioactive molecules in a controlled manner. Nanogels can modify their architecture, texture, and mechanical properties when incorporated into scaffolds, regulating cellular behavior. Nanogel-integrated scaffolds have demonstrated enhanced bioavailability of growth factors such as VEGF and BMP-2, ensuring sustained release for improved tissue regeneration. For instance, nanogels have been effectively used in skin and bone tissue regeneration, promoting cell adhesion and differentiation. This synergy enhances the functionality of scaffolds, facilitating the healing of chronic wounds and critical-size bone fractures [43].

3D printing of nanocomposite scaffolds has further revolutionized regenerative medicine by enabling the precise deposition of nanomaterials such as hydroxyapatite, graphene, and bioactive glass nanoparticles. By utilizing bioinks composed of natural and synthetic polymers, 3D printing allows the creation of scaffolds that mimic the mechanical and biological properties of native tissues. These scaffolds can be loaded with growth factors, cells, or drugs to enhance tissue regeneration. The ability to print scaffolds with varying levels of complexity and functionality opens up new avenues for the development of patient-specific implants for bone and cartilage repair [44].

The Sol–Gel process is versatile for producing nanostructured scaffolds by transitioning a liquid precursor (sol) into a solid gel phase. This technique allows for the incorporation of bioactive molecules, such as growth factors or drugs, into the scaffold during its formation. Sol–gel-derived scaffolds incorporating silica-based nanomaterials have demonstrated superior osteoconductivity and mechanical reinforcement, making them ideal for bone tissue engineering applications. The sol–gel process is beneficial for fabricating ceramic-based scaffolds, such as hydroxyapatite (HA) and tricalcium phosphate (TCP), which are commonly used for bone tissue engineering due to their osteoconductive properties. These scaffolds can be tailored to control degradation rates and promote cell differentiation [45].

Phase Separation is commonly used to create porous structures in scaffolds by exploiting the immiscibility of two or more polymers or solvents. This process can generate scaffolds with high porosity and interconnectivity, essential for tissue regeneration. Phase separation is particularly effective for producing polymer-based scaffolds with controlled pore sizes, enhancing cell infiltration and tissue formation. By incorporating nanoparticles such as carbon nanotubes or hydroxyapatite during phase separation, scaffolds exhibit improved mechanical properties and bioactivity, further enhancing their regenerative potential. The method allows for the creation of scaffolds that mimic the extracellular matrix (ECM), promoting cellular adhesion and growth [46].

Electrospun Nanocomposites combine the high surface area and porosity of electrospun nanofibers with the functional properties of nanomaterials such as nanoparticles, nanotubes, or nanorods. These composite scaffolds can be tailored to provide additional functionalities, such as antibacterial properties, mechanical reinforcement, or controlled drug release. For example, PLA-based electrospun nanocomposites reinforced with hydroxyapatite nanoparticles improve osteogenesis while providing mechanical support for bone tissue regeneration. The combination of nanomaterials with electrospun fibers allows for the creation of scaffolds with unique properties that can be fine-tuned for specific tissue engineering applications [47].

Freeze-drying, or lyophilization, is a method used to create highly porous scaffolds by removing water from a frozen material under a vacuum. This process preserves the scaffold's structure, resulting in a porous, lightweight material with high surface area and interconnectivity. Freeze-dried scaffolds are often used to fabricate biopolymer-based scaffolds for bone, cartilage, and skin regeneration. The method is particularly advantageous for creating scaffolds with controlled porosity and mechanical properties, essential for supporting cell growth and tissue formation.

Melt electrospinning writing (MEW) is a novel technique that combines electrospinning with melt extrusion to fabricate 3D scaffolds with precise control over fiber alignment and orientation. This method uses a molten polymer electrospun onto a substrate to form 3D scaffolds with defined geometries. MEW allows for the fabrication of nanofiber scaffolds incorporating bioactive nanomaterials, enhancing cellular interactions, and accelerating tissue regeneration. MEW provides for the creation of scaffolds with high mechanical strength and tailored microstructures, making it ideal for applications in bone and cartilage tissue engineering. MEW scaffolds can promote tissue regeneration and repair by incorporating bioactive molecules or cells into the melt [48].

DNA origami-based scaffolds are also emerging as a novel approach for the precise spatial arrangement of biomolecules, enabling advanced functionalities such as single-molecule analysis and interaction with external stimuli. DNA-based nanoscaffolds have demonstrated potential in controlled drug delivery and gene therapy, offering new perspectives in regenerative medicine. These scaffolds provide insights into cellular processes and radiation-induced effects on DNA, contributing to the development of precision therapies [49].

Functionalized scaffolds with smart nanomaterials, such as those utilizing electrospinning, 3D printing, sol–gel processes, phase separation, electrospun nanocomposites, freeze-drying, and melt electrospinning writing, represent significant advancements in regenerative medicine. These scaffolds leverage properties like controlled release, stimuli responsiveness, and multifunctionality, creating dynamic environments that closely mimic native tissues and enhance the efficacy of tissue engineering strategies.

2.2 *Bioprinting Using Nanomaterials*

Three-dimensional (3D) bioprinting has transformed tissue engineering by enabling the precise fabrication of biomimetic scaffolds that replicate native tissue architecture. This additive manufacturing approach allows for controlled deposition of biomaterials, cells, and bioactive molecules, making it highly promising for applications in regenerative medicine. However, a significant challenge in bioprinting is the development of bioinks with suitable printability, mechanical stability, and biological functionality. Nanomaterials have been incorporated into bio-inks to enhance their rheological properties, structural integrity, and bioactivity [26, 28] to address these limitations.

Nanomaterial-enhanced bioinks include metallic, ceramic, polymeric, and carbon-based nanomaterials, each offering distinct advantages for scaffold fabrication. Metallic nanoparticles such as gold, silver, and zinc oxide provide antimicrobial effects and electrical conductivity, making them valuable for neural and cardiac tissue applications. Ceramic-based nanoparticles, including hydroxyapatite and bioactive glass, are widely used in bone and cartilage regeneration due to their osteoinductive properties. Polymeric nanomaterials, such as nanocellulose and peptide-based nanostructures, enhance bioink viscosity, improving printability and structural fidelity. Carbon-based nanomaterials, including graphene and carbon nanotubes, contribute to mechanical reinforcement and facilitate cell signaling, particularly in musculoskeletal and neural tissue engineering [27].

The integration of nanomaterials has also refined different bioprinting modalities. Inkjet bioprinting (Fig. 2), which involves controlled droplet deposition, benefits from nanomaterials that optimize fluid behavior and maintain cell viability. Extrusion bioprinting, the most commonly used approach, relies on nanocomposites to enhance shear-thinning properties and structural stability. Laser-assisted bioprinting leverages nanoparticle-enriched bioinks to improve cell adhesion and differentiation. Stereolithography-based bioprinting, which uses light-based crosslinking, employs nanomaterials to modulate degradation rates and fine-tune mechanical properties.

An emerging technique, electrohydrodynamic (EHD) bioprinting, has gained attention for its ability to fabricate high-resolution fibrous scaffolds at micro- and nanoscale precision. By incorporating conductive nanomaterials, EHD bioprinting allows for the biofabrication of functional tissues with tunable electrical properties,

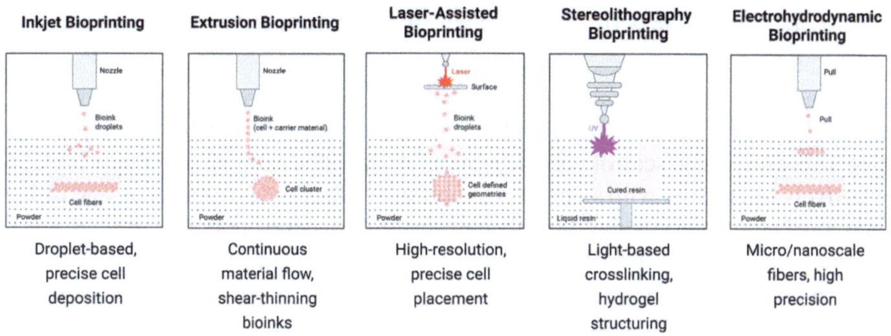

Fig. 2 Schematic representation of different 3D bioprinting modalities used in tissue engineering. Inkjet bioprinting relies on droplet-based deposition for precise cell placement. Extrusion bioprinting enables continuous material flow, supporting shear-thinning bioinks for large-scale scaffold fabrication. Laser-assisted bioprinting offers high-resolution cell deposition using laser-induced transfer. Stereolithography bioprinting utilizes light-based crosslinking to fabricate hydrogel-based structures with fine detail. Electrohydrodynamic (EHD) bioprinting enables the creation of micro- and nanoscale fibrous scaffolds, mimicking native extracellular matrix structures for advanced biofabrication

making it particularly useful for neural, cardiac, and musculoskeletal applications [50, 51].

Applications of nanomaterial-assisted bioprinting span diverse fields of tissue engineering. In bone and cartilage repair, nanocomposite bio-inks reinforced with hydroxyapatite and bioceramic nanoparticles have demonstrated enhanced osteogenic and chondrogenic potential. Vascular tissue engineering has benefited from the use of graphene-based bioinks that promote endothelial cell organization and angiogenesis. Neural regeneration has been supported by carbon nanotube-integrated bioinks, which improve scaffold conductivity and guide neuronal differentiation [50, 51].

While nanomaterials have significantly advanced bioprinting technologies, several challenges remain, including ensuring uniform nanoparticle dispersion, optimizing bioink compositions, and addressing long-term biocompatibility concerns. Future research should focus on multi-functional bioinks that integrate dynamic properties such as controlled drug release, bioresponsiveness, and adaptable mechanical behavior. Additionally, the application of artificial intelligence in bioprinting workflows may enhance scaffold precision and reproducibility, further driving the clinical translation of nanomaterial-assisted bioprinting in tissue engineering.

2.3 Characteristics of Stem Cells in Regeneration

The integration of nanomaterials into scaffolds has emerged as a transformative strategy in tissue engineering, enabling enhanced stem cell delivery and functionality. These advanced scaffolds go beyond mimicking the extracellular matrix (ECM) by actively interacting with stem cells to optimize adhesion, proliferation, and differentiation. Nanomaterials significantly elevate the therapeutic potential of scaffolds in regenerative medicine through their nanoscale features, functionalization potential, and dynamic properties.

i. **Nanomaterials for scaffold-stem cell integration**

Nanomaterials provide scaffolds with unique physicochemical properties that closely replicate the microenvironment of the native ECM. Their nanoscale topography offers structural cues that influence cell behavior at the molecular level, promoting adhesion and cytoskeletal reorganization. For example, nanofibrous scaffolds created via electrospinning closely mimic the fibrous architecture of the ECM, enhancing stem cell attachment and enabling lineage-specific differentiation. The stiffness and elasticity of nanocomposite scaffolds can also be tuned to match target tissues, further supporting cellular integration [21].

Moreover, nanomaterials such as carbon nanotubes and graphene derivatives enhance electrical conductivity within scaffolds, making them particularly beneficial for neural tissue engineering. These conductive scaffolds have been shown to promote neural stem cell differentiation and functional integration by simulating the electrical stimuli in native neural tissues. Similarly, hydroxyapatite and other ceramic nanoparticles provide ionic cues that facilitate osteogenic differentiation, emphasizing the versatility of nanomaterials in tissue-specific applications [52].

ii. **Functionalized scaffolds for enhanced regeneration**

Functionalizing scaffolds with bioactive molecules represents a critical advancement in stem cell delivery. Nanomaterials carry growth factors, cytokines, and other therapeutic agents, enabling their controlled and localized release. For instance, polymeric nanoparticles embedded within scaffolds can release vascular endothelial growth factor (VEGF) or bone morphogenetic protein-2 (BMP-2) in a sustained manner, thereby promoting angiogenesis and osteogenesis at the injury site.

Additionally, nanogels and liposomal nanocarriers incorporated into scaffolds enhance their regenerative potential. These carriers allow spatiotemporal control over drug release, reducing systemic side effects and increasing therapeutic efficacy. Antibacterial scaffolds combining ciprofloxacin-loaded nanoparticles with BMP-2 have demonstrated dual functionality: eradicating infection and promoting bone tissue regeneration. Such multifunctional scaffolds are particularly valuable in complex regenerative contexts, such as critical-sized bone defects [43].

iii. **Mechanisms of interaction**

The interactions between nanomaterials and stem cells involve a synergistic combination of topographical, chemical, and mechanical cues. Nanoscale features, such as nanopillars, grooves, and aligned nanofibers, provide physical guidance for cell alignment, elongation, and differentiation. Aligned nanofibrous scaffolds, for example, encourage neural stem cells to differentiate into mature neurons, while randomly oriented fibers promote embryoid body formation.

Chemical functionalization of nanomaterials further augments scaffold performance by introducing bioactive moieties that enhance cell adhesion and signaling. For example, RGD (Arg-Gly-Asp) peptide-functionalized scaffolds improve integrin-mediated adhesion, creating a microenvironment conducive to cellular growth and differentiation. In addition, nanomaterials modulate mechanotransduction pathways by altering cell morphology and cytoskeletal dynamics, influencing key signaling cascades such as Wnt/β-catenin and integrin-mediated pathways [53, 54].

3 Systematic Review of Exosomes in Tissue Engineering

3.1 Overview of Recent in Vitro, in Vivo, and Clinical Studies up to December 2024

Exosome and stem cell research in tissue engineering has gained significant momentum over the past decade, with an increasing number of in vitro, in vivo, and clinical studies evaluating their regenerative potential. In vitro studies have demonstrated the ability of exosomes derived from mesenchymal stem cells (MSCs) to regulate inflammation, promote angiogenesis, and enhance cellular proliferation and differentiation. These effects have been observed in multiple tissue types, including bone, cartilage, skin, and neural tissues [21].

In vivo studies have further validated these findings, showing that exosome-based therapies can accelerate wound healing, improve bone regeneration, and restore tissue function following injury. Recent animal studies indicate that exosomes can modulate immune responses, facilitating a pro-regenerative environment. Clinical trials have started to explore translating these findings into human applications, with promising results in regenerative therapies for nerve injuries, cardiovascular diseases, and musculoskeletal disorders. However, challenges remain in standardizing exosome isolation and ensuring batch-to-batch consistency for therapeutic applications (Table 1).

3.2 Trends in Exosome-Based Therapies for Tissue Engineering

Emerging trends in exosome-based therapies highlight their potential as a cell-free alternative to stem cell transplantation. Exosomes have demonstrated superior safety profiles compared to live cell therapy, reducing the risks associated with

Table 1 Summary of recent studies on exosome-based tissue engineering and regeneration

Year	Title	Findings
2025	Exosome loaded 3D printed magnetic PLA constructs: a candidate for bone tissue engineering [57]	3D-printed scaffolds + exosomes: • Magnetic PLA + osteosarcoma-derived Exo enhance RUNX2, COL1A1, OPN, OSC expression → boosts hMSC differentiation and bone mineralization • Growth-factor-free osteogenesis with high cytocompatibility
2025	PI3K/Akt pathway-mediated enhancement of bone and vascular regeneration by gelatin/hyaluronic acid/exosome composite scaffold in bone tissue engineering [58]	Gel/HA/Exo complex • Exo (TGF-β) activates PI3K/Akt → promotes osteoblast proliferation, differentiation and fracture healing • Biocompatible, controlled degradation, strong clinical potential
2024	Exosome-loaded hyaluronic acid hydrogel composite with oxygen-producing 3D printed polylactic acid scaffolds for bone tissue repair and regeneration [59]	O_2-supplying 3D scaffold + Exo • CaO_2-PLA scaffold provides O_2 and HAP → enhances angiogenesis and osteogenesis • BMSC-Exo in HA hydrogel modulates inflammation, BMSC proliferation and bone formation • Improved hypoxia relief, effective bone repair
2024	Bone tissue regeneration by 58S bioactive glass scaffolds containing exosome: an in vivo study [60]	EXOs in bone regeneration • EXOs from human endometrial MSCs + 58S bioactive glass scaffolds show potential for bone defect repair • Limited studies, mechanisms remain unclear
2024	COTiR: Molecular Communication Model for Synthetic Exosome-Based Tissue Regeneration [61]	EXOs for COVID-19 tissue damage • MSC-EXOs alleviate inflammation-induced tissue damage • CoTiR model (DMS-based) improves EXO targeting efficiency vs. random propagation • 99.96% collagen reduction without inflammatory cytokines
2024	A new method regulates bone fracture tissue exosome lncRNA-mRNA to promote mesenchymal stem cell proliferation and migration [62]	Post-injury adaptation (PIA) and bone healing via EXOs • PIA enhances BMSC proliferation/ migration via fracture-site tissue EXOs • Upregulated Akt1, Actb, Uba52 in EXOs; Kif11 in BMSCs • High-throughput sequencing: 12,211 target genes identified

(continued)

Table 1 (continued)

Year	Title	Findings
2023	Bone-Targeting Exosome Mimetics Engineered by Bioorthogonal Surface Functionalization for Bone Tissue Engineering [63]	Bone-targeting exosome mimetics (EMs) • Engineered EMs with hydroxyapatite-binding moieties improve bone targeting • SAG-loaded EMs activate hedgehog signaling, enhancing osteogenesis • In vivo studies: EM/SAG scaffolds improve bone regeneration
2023	Assessment of Aggregated and Exosome-Associated α-Synuclein in Brain Tissue and Cerebrospinal Fluid Using Specific Immunoassays [64]	α-Synuclein EXOs in Parkinson's disease (PD) • New ELISA detects fibrillar and oligomeric α-synuclein in CSF and brain EXOs • Aggregated α-synuclein correlates with PD onset • CSF analysis: low abundance of pathogenic α-synuclein species
2022	Effectiveness of exosome mediated miR-126 and miR-146a delivery on cardiac tissue regeneration [65]	Exosome-based therapy for myocardial infarction (MI) • miR-126 and miR-146a-loaded EXOs promote angiogenesis and cardiac regeneration • Improved cell migration, VEGF expression, CD31, Connexin 43 • Alginate hydrogel-encapsulated EXOs reduce infarct size, fibrosis and enhance vascular integrity
2021	Cell-free exosome-laden scaffolds for tissue repair [66]	EXOs in regenerative medicine • EXOs facilitate cell communication, angiogenesis, immune response, and tissue repair • Advantages over stem cell therapy: avoids immune rejection and ethical concerns • Potential applications: molecular/cellular repair strategies in tissue regeneration
2021	Impact of repeat dosing and mesh exposure chronicity on exosome-induced vaginal tissue regeneration in a porcine mesh exposure model [67]	Vaginal wound healing (porcine model) • Weekly exosome dosing increased epithelial proliferation (1.8×) and capillary density (1.5×) versus single dose • Subacute versus acute exposure: similar regeneration, supporting delayed treatment efficacy • Clinical relevance: potential therapy for pelvic floor disorders

(continued)

Table 1 (continued)

Year	Title	Findings
2021	Mesenchymal stromal exosome–functionalized scaffolds induce innate and adaptive immunomodulatory responses toward tissue repair [68]	Exosome-Laden scaffolds for skin repair • Exosome-immobilized scaffolds recruit immune cells, enhancing macrophage and regulatory T-cell responses • T-helper activation in remote organs suggests systemic immunomodulation • Wound healing and immune-based tissue engineering
2021	Exosome-Induced Vaginal Tissue Regeneration in a Porcine Mesh Exposure Model [69]	Vaginal mesh exposure repair (porcine model) • Exosome injections improved surgical closure and standalone therapy for vaginal mesh exposure • Combination with sutures enhanced healing
2020	Exosome-like vesicles derived from Hertwig's epithelial root sheath cells promote the regeneration of dentin-pulp tissue [70]	Dental pulp-dentin regeneration • HERS-derived exosome-like vesicles (ELVs-H1) induced DPC migration, proliferation, and odontogenic differentiation • Activated Wnt/β-catenin signaling and promoted vascular and neural differentiation • Potential use: tooth regeneration via biomimetic exosome therapy
2020	Exosome loaded alginate hydrogel promotes tissue regeneration in full-thickness skin wounds: an in vivo study [71]	Wound healing with exosome-loaded hydrogel • ADSC-derived exosomes in alginate hydrogel improved wound closure, collagen synthesis, and vessel formation • Biodegradable, biocompatible scaffold provided sustained exosome release • Potential use: cell-free exosome therapy for chronic wound healing

*3D: Three−dimensional; 58S bioactive glass: Silica−based bioactive glass used in bone regeneration; ADSC: Adipose−derived stem cell; Akt1: AKT serine/threonine kinase 1; Actb: Beta−actin; α−Synuclein: Alpha−synuclein, a protein linked to Parkinson's disease; BMSC: Bone marrow−derived mesenchymal stem cell; CaO_2: Calcium peroxide; CD31: Platelet endothelial cell adhesion molecule−1; COL1A1: Collagen type I alpha 1 chain; CoTiR: Controlled targeted release; Connexin 43: Gap junction protein alpha 1; CSF: Cerebrospinal fluid; DMS: Directed molecular strategy; DPC: Dental pulp cell; ELISA: Enzyme−linked immunosorbent assay; EMs: Exosome mimetics; Exo: Exosomes; HA: Hydroxyapatite; HAP: Hydroxyapatite; hMSC: Human mesenchymal stem cell; HERS: Hertwig's epithelial root sheath; Kif11: Kinesin family member 11; MI: Myocardial infarction; MSC: Mesenchymal stem cell; OPN: Osteopontin; OSC: Osteocalcin; PD: Parkinson's disease; PI3K/Akt: Phosphoinositide 3−kinase/protein kinase B signaling pathway; PIA: Post−injury adaptation; PLA: Polylactic acid; RUNX2: Runt−related transcription factor 2; SAG: Smoothened agonist; TGF−β: Transforming growth factor beta; Uba52: Ubiquitin A−52 residue ribosomal protein fusion product 1; VEGF: Vascular endothelial growth factor immune rejection and tumorigenesis. Recent studies emphasize the use of engineered exosomes with targeted delivery capabilities, enhancing their therapeutic efficacy. Modifications such as exosome surface functionalization and cargo loading with specific growth factors or miRNAs have further improved their regenerative potential.

Exosome-based therapies are being actively investigated in orthopedic applications, including bone and cartilage repair, where they have been shown to enhance osteogenesis and chondrogenesis. Additionally, exosome-derived therapeutics are being explored for soft tissue regeneration, particularly in skin and wound healing. Their ability to facilitate intercellular communication and modulate the immune response has positioned exosomes as a promising tool for tissue engineering. Exosomes derived from MSCs harvested from oral tissues are also being explored for craniofacial tissue engineering, where they promote bone tissue repair and fracture healing. Studies suggest that exosome-mediated osteoblastic differentiation involves pathways such as Wnt and PI3K/Akt, providing insights into their regenerative mechanisms [55, 56].

3.3 Challenges in Clinical Translation and Regulatory Considerations

Despite the promising preclinical and early clinical data, several challenges hinder the widespread adoption of exosome-based therapies. One major challenge is the lack of standardized exosome isolation, characterization, and quality control protocols. Variability in the source of stem cells, culture conditions, and purification techniques can lead to inconsistencies in therapeutic outcomes (Fig. 3).

Regulatory agencies, including the FDA and EMA, are working on frameworks to establish quality control guidelines for exosome-based products. The classification of exosomes as either biologics or drug-like entities remains debatable, impacting their regulatory approval pathways. Furthermore, the large-scale production of exosomes with reproducible potency remains an obstacle, necessitating advancements in biomanufacturing techniques.

Additionally, long-term safety and efficacy concerns need to be addressed through well-designed clinical trials. The immunogenicity and biodistribution of exosome-based therapies require thorough investigation to ensure their suitability for human applications. As the field progresses, collaboration between researchers, clinicians, and regulatory bodies will be essential to overcoming these hurdles and translating exosome-based therapies into routine clinical practice.

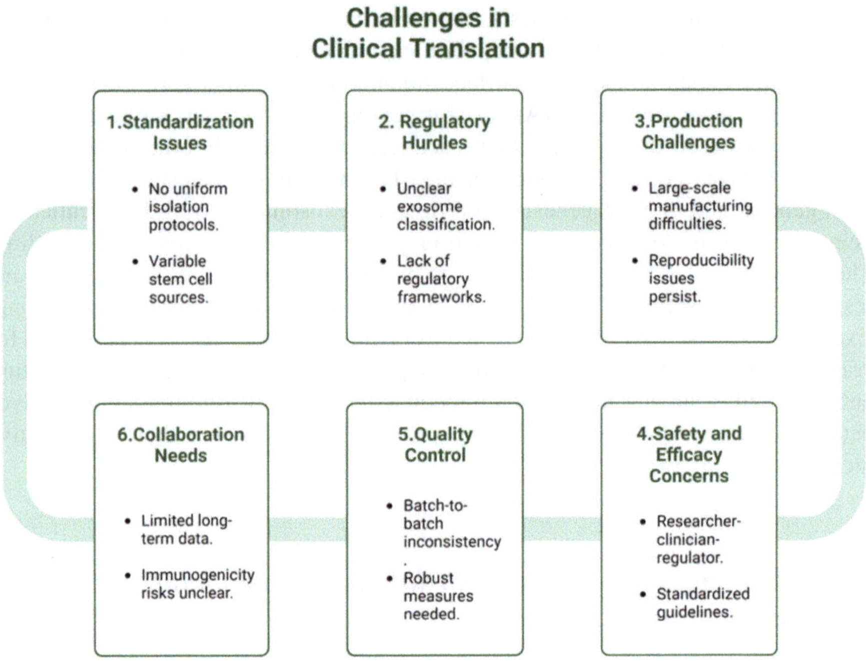

Fig. 3 Challenges in clinical translation of exosome-based therapies

3.4 Application of Exosomes in Tissue Engineering

Exosome-based strategies have demonstrated significant potential in various tissue engineering applications, contributing to enhanced regeneration across multiple tissue types (Fig. 4). These vesicles facilitate intercellular communication, regulate inflammatory responses, and promote cellular proliferation and differentiation, making them a promising tool for regenerative medicine. Figure 5 shows the main signaling pathways in exosome-mediated tissue regeneration.

Recent studies have explored exosome-based therapies in bone, cartilage, skin, and neural tissue regeneration, highlighting their ability to modulate key signaling pathways such as Wnt and PI3K/Akt. Additionally, exosomes derived from mesenchymal stem cells (MSCs), particularly those harvested from oral tissues, have shown potential in craniofacial tissue engineering by promoting osteoblastic differentiation and fracture healing. Table 1 summarizes recent studies investigating the role of exosomes in tissue engineering and regeneration. It outlines key findings from in vitro, in vivo, and clinical studies, detailing the tissue types targeted, the exosome sources, and their regenerative effects.

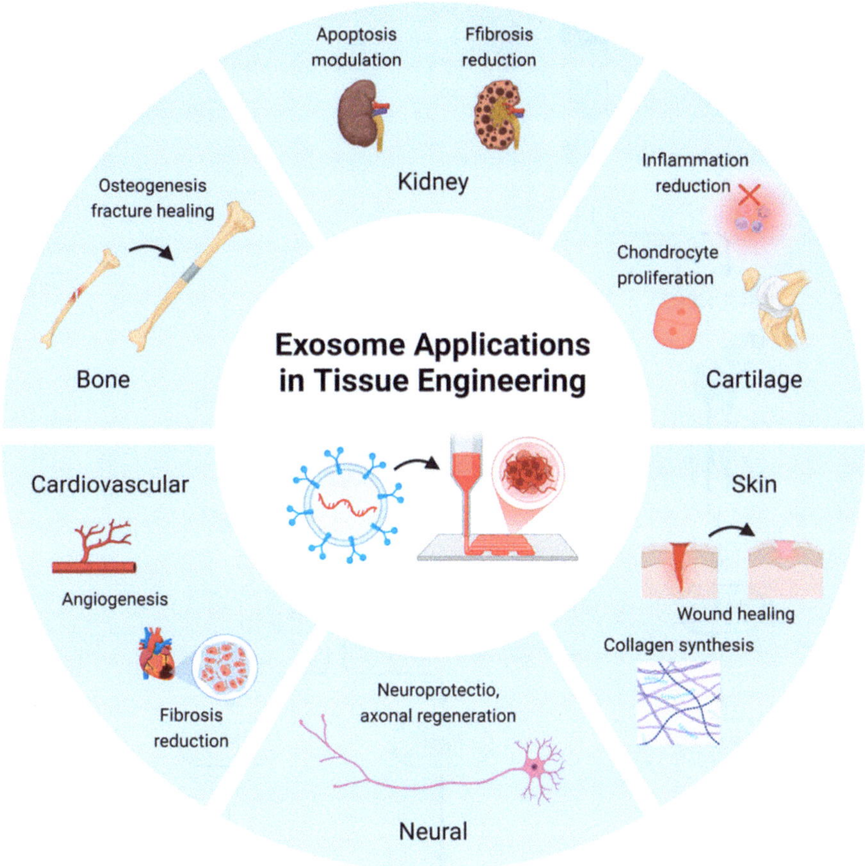

Fig. 4 Applications of exosomes in tissue engineering

i. **Exosomes for Central Nervous System Reparation**

Exosomes have shown potential in central nervous system (CNS) repair by modulating oxidative stress and inflammation. They can cross the blood–brain barrier (BBB) through endothelial interactions, enabling targeted delivery to neurons and glial cells. Studies using exosome-loaded therapies have demonstrated neuroprotective effects in ischemic stroke and traumatic brain injury models. Astrocyte-secreted exosomes carrying LAMB2 have been found to enhance oligodendrocyte precursor cell proliferation, while exosomes from bone marrow MSCs (BMSCs) contribute to axonal regeneration by promoting myelin debris clearance [72].

ii. **Exosomes for Heart Regeneration**

Cardiac tissue has limited regenerative capacity, making heart disease a major challenge in regenerative medicine. MSC-derived exosomes have been shown to reduce fibrosis, enhance angiogenesis, and improve myocardial repair post-infarction.

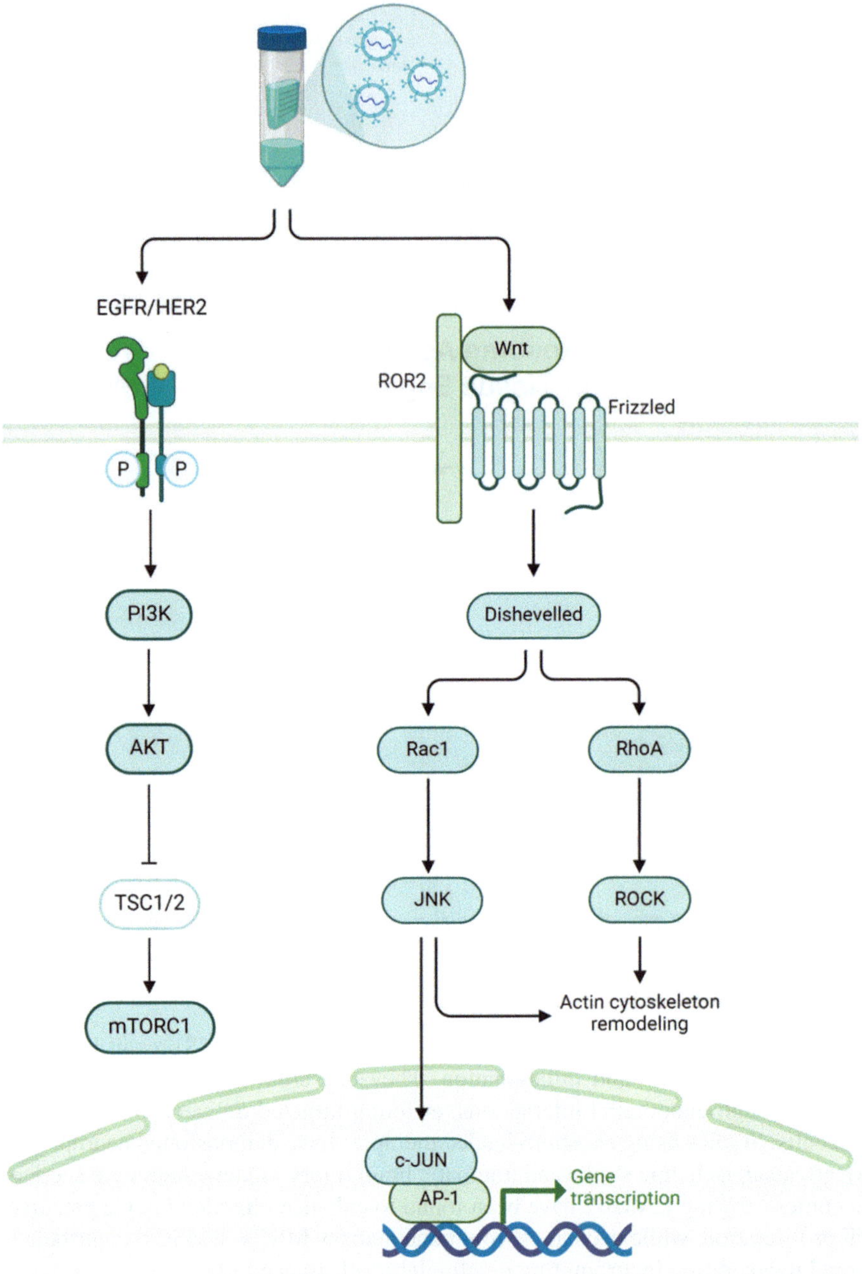

Fig. 5 Key signaling pathways in exosome-mediated tissue regeneration

Exosomes carrying miR-146a and miR-210 have been identified as key regulators in cardiomyocyte survival and recovery. Additionally, cardiac patches incorporating exosomes have improved cardiac function by enhancing myocardial retention and reducing inflammation [73].

iii. **Exosomes for Kidney Regeneration**

Stem cell-derived exosomes have demonstrated therapeutic effects in acute and chronic kidney injury by modulating apoptosis, inflammation, and fibrosis. Preclinical studies have shown that MSC-derived exosomes can restore renal function and reduce fibrosis in chronic kidney disease models. Exosome-based therapies hold promise as novel interventions for kidney repair, leveraging their regenerative and immunomodulatory properties.

iv. **Exosomes for Bone Regeneration**

Exosomes play a crucial role in bone repair by mediating osteogenic differentiation and matrix mineralization. In CD9−/− mice, exosome treatment accelerated fracture healing, highlighting their therapeutic potential. Mechanistically, exosome cargo regulates key signaling pathways such as PI3K/Akt and Wnt, driving osteoblast activation and bone formation. Additionally, exosome-functionalized biomaterials, including titanium nanotube implants, have shown enhanced osteoconductivity and bone integration [59].

v. **Exosomes for Cartilage Regeneration**

Cartilage regeneration is challenging due to the avascular nature of cartilage tissue. Exosome-based therapies have demonstrated efficacy in osteoarthritis models by reducing inflammation and promoting chondrocyte proliferation. Exosomes carrying miR-100-5p modulate mTOR/autophagy pathways, maintaining cartilage homeostasis and preventing degradation. Studies indicate that MSC-derived exosomes can improve extracellular matrix synthesis and suppress inflammatory mediators, making them promising candidates for cartilage repair [74].

These advancements illustrate the potential of exosome-based therapies in regenerative medicine. As research unravels their mechanisms and applications, the future of exosome-based therapeutics holds promise for transforming tissue engineering and clinical medicine.

4 Future Perspectives

The landscape of smart nanomaterials in tissue engineering is undergoing rapid transformation, marked by notable advancements in scaffold fabrication, bioprinting methodologies, and stem cell applications. Future investigations will emphasize the creation of highly biomimetic, multifunctional nanomaterials that emulate the native extracellular matrix, thereby optimizing cellular adhesion, proliferation, and differentiation. The incorporation of stimuli-responsive nanostructures will facilitate the dynamic modulation of the cellular microenvironment, enhancing tissue regeneration with greater precision.

In bioprinting, the convergence of artificial intelligence and machine learning is set to refine the design and fabrication of complex tissue architectures with superior spatial and temporal control. Real-time monitoring of printed constructs through cutting-edge imaging and biosensing modalities will improve reproducibility and structural fidelity. Additionally, bioinks enriched with functionalized nanoparticles and bioactive factors will be instrumental in fostering optimal cell interactions and tissue maturation.

Stem cell-based tissue engineering stands to gain significantly from nanotechnological advances, particularly in the engineering of tailored microenvironments using nanostructured scaffolds. The deployment of exosome-loaded nanomaterials and nano-engineered extracellular vesicles will augment intercellular communication, thereby enhancing regenerative outcomes. Furthermore, the integration of CRISPR-mediated gene editing with nanomaterials holds the potential to usher in a new era of personalized regenerative therapies. CRISPR technology allows for targeted genetic modifications that improve stem cell differentiation, accelerate tissue regeneration, and correct genetic abnormalities. Meanwhile, AI-driven models optimize scaffold designs and predict cellular interactions within engineered environments. By integrating these advanced tools, researchers can develop more adaptive and personalized regenerative therapies, ultimately improving treatment outcomes and expanding the possibilities of tissue engineering [75, 76].

The global biomaterials market for tissue engineering is poised for substantial growth, with a projected compound annual growth rate (CAGR) of 4.67% between 2021 and 2028. Market valuation is expected to rise from USD 1344.32 million in 2020 to USD 1936.78 million by 2028 (CAGR). This trajectory is fueled by the increasing demand for advanced biomaterials, the rising incidence of chronic conditions necessitating tissue regeneration, and continuous innovations in scaffold fabrication and nanomaterials. Leading industry players are actively engaged in enhancing biomaterial efficacy and biocompatibility, further driving market expansion [77, 78].

Despite these promising advancements, challenges persist in large-scale manufacturing, standardization of production methodologies, and the long-term biocompatibility assessment of these materials. Regulatory frameworks must adapt to accommodate the clinical translation of these next-generation smart nanomaterials, ensuring their safety and efficacy for medical applications. In the foreseeable future, interdisciplinary collaboration among materials scientists, bioengineers, clinicians, and regulatory authorities will be paramount in addressing these challenges and harnessing the full potential of smart nanomaterials in regenerative medicine. The intersection of nanotechnology, biotechnology, and computational modeling is expected to catalyze transformative breakthroughs in tissue engineering, with profound implications for personalized medicine and organ transplantation.

5 Conclusion

The integration of smart nanomaterials in tissue engineering represents a transformative advancement, enhancing scaffold design, bioprinting precision, and stem cell-based therapies. These materials mimic the extracellular matrix at the nanoscale, improving mechanical properties, bioactivity, and cellular interactions. Their role in modulating exosome stability and targeted delivery has further expanded their applications in regenerative medicine, particularly in skin regeneration, bone repair, and vascular tissue engineering.

The systematic review of in vitro, in vivo, and clinical studies underscores the therapeutic potential of nanomaterial-enhanced exosome therapies. Yet, challenges remain in large-scale manufacturing, standardization, and long-term biocompatibility. Addressing these barriers requires interdisciplinary collaboration to optimize scaffold functionalization, bioprinting methodologies, and regulatory frameworks.

Future research should focus on refining nanomaterial properties for personalized regenerative medicine, leveraging artificial intelligence for scaffold design, and integrating CRISPR-based gene editing for enhanced therapeutic outcomes. As these technologies advance, smart nanomaterials will continue to drive innovation in tissue engineering, bridging the gap between laboratory research and clinical applications.

Glossary of Terms

Terms	Definition
3D Bioprinting	This is a type of 3D printing used in tissue engineering where layers of cells and materials are printed to create structures that mimic natural tissues. It allows for precise placement of cells and bioactive molecules
Angiogenesis	The formation of new blood vessels. In tissue engineering, promoting angiogenesis is important for ensuring that tissues have the necessary blood supply to survive and grow
Bioactive Glass	A type of glass that can interact with biological tissues and promote healing. It is often used in scaffolds for bone regeneration
Bioinks	Special materials used in 3D bioprinting that contain living cells and other biological components. These inks are used to print tissue-like structures for regenerative medicine
Cell Differentiation	The process by which a less specialized cell becomes a more specialized type of cell. In tissue engineering, scaffolds often promote the differentiation of stem cells into the specific cell types needed for tissue regeneration
Controlled Release	The ability to release bioactive molecules, like drugs or growth factors, from a scaffold in a controlled manner over time. This helps improve the effectiveness of tissue regeneration by ensuring that cells receive the right signals at the right time
Decellularization	This is a process used to remove all the cells from a tissue while leaving behind the ECM. It helps create a "blank slate" that can be repopulated with new cells for tissue regeneration

(continued)

(continued)

Terms	Definition
DNA Origami-Based Scaffolds	Scaffolds made from DNA, which can be folded into specific shapes to create complex structures. These scaffolds can be used to deliver drugs or study cellular interactions at the molecular level
Electrospinning	A process where a charged polymer (a type of plastic) is stretched into fibers that are used to make scaffolds. These fibers can mimic the natural structure of tissues, which helps cells grow and differentiate properly
Extracellular Matrix (ECM)	The ECM is a network of proteins and other molecules that surround and support cells in tissues. It provides structural support and helps cells grow, divide, and move
Freeze-Drying (Lyophilization)	A process used to remove water from materials to create lightweight and porous structures. Freeze-dried scaffolds are often used in tissue engineering because they have a high surface area and can support cell growth
Graphene and Carbon Nanotubes	Graphene is a single layer of carbon atoms arranged in a hexagonal pattern, and carbon nanotubes are hollow tubes made from carbon. Both are highly conductive and strong, and they are used in scaffolds for applications like nerve or muscle regeneration
Hydroxyapatite	A naturally occurring mineral form of calcium apatite. It is a key component of bone and is often added to scaffolds for bone tissue engineering
Melt Electrospinning Writing (MEW)	A technique where a molten polymer is electrospun into fibers to create 3D scaffolds. This method allows for precise control over the structure and alignment of the fibers, which is important for creating strong and functional tissues
Nanocomposites	Materials made by combining nanomaterials (such as nanoparticles) with other materials to improve their properties. In scaffolds, nanocomposites can enhance strength, flexibility, and bioactivity
Nanogels	Nanogels are tiny, gel-like materials that can change their structure in response to stimuli, such as temperature. They are used in scaffolds to deliver growth factors or drugs to cells in a controlled way
Nanoparticles	Small size particles (smaller than 100 nm) made from materials like metals, carbon, or ceramics. When added to scaffolds, they can provide special features like improved strength, electrical conductivity, or the ability to kill bacteria
Osteogenesis	The process by which new bone is formed. Scaffolds that support osteogenesis help in healing bone fractures or creating new bone tissue
Phase Separation	A process used to create porous structures in scaffolds. By using materials that do not mix well together, it is possible to create a scaffold with tiny holes that allow cells to grow and tissues to form
Scaffolds	Scaffolds are materials designed to mimic the ECM and provide a supportive structure for cells to grow and form tissues. They are used in tissue engineering to help repair or replace damaged tissues

(continued)

(continued)

Terms	Definition
Smart Nanomaterials	These are materials at the nanoscale (extremely small) that can respond to changes in their environment, such as temperature, pH, or light. They are used to improve the performance of scaffolds in tissue engineering by making them more adaptable and functional
Stem Cells	Special cells that have the ability to develop into different types of cells in the body. They are often used in regenerative medicine to repair or replace damaged tissues
Stimuli-Responsive Materials	Materials that can change their properties in response to external stimuli like changes in temperature, pH, or light. These materials are used in scaffolds to make them more dynamic and adaptable to the biological environment

References

1. Berthiaume, F., Maguire, T.J., Yarmush, M.L.: Tissue engineering and regenerative medicine: history, progress, and challenges. Annu. Rev. Chem. Biomol. Eng. **2**, 403–430 (2011)
2. Zhu, W., Ma, X., Gou, M., Mei, D., Zhang, K., Chen, S.: 3D printing of functional biomaterials for tissue engineering. Curr. Opin. Biotechnol. **40**, 103–112 (2016)
3. Caplan, A.I.: Tissue engineering: then, now, and the future. Tissue Eng. Part A **25**(7–8), 515–517 (2019)
4. Polge, C., Smith, A.U., Parkes, A.S.: Revival of spermatozoa after vitrification and dehydration at low temperatures. Nature [Internet] **164**(4172), 666–666 (1949). https://www.nature.com/articles/164666a0
5. Herman, A.R.: The history of skin grafts. J. Drugs Dermatol. JDD. **1**(3), 298–301 (2002)
6. Webster, J.P.: Refrigerated skin grafts. Ann. Surg. **120**(4), 431–448 (1944)
7. Bell, E., Ehrlich, H.P., Buttle, D.J., Nakatsuji, T.: Living tissue formed in vitro and accepted as skin-equivalent tissue of full thickness. Science **211**(4486), 1052–1054 (1981)
8. Chardack, W.M., Bruseske, D.A., Santomauro, A.P., Fazekas, G.: Experimental studies on synthetic substitutes for skin and their use in the treatment of burns. Ann. Surg. LWW **155**(1), 127–139 (1962)
9. Green, H., Kehinde, O., Thomas, J.: Growth of cultured human epidermal cells into multiple epithelia suitable for grafting. Proc. Natl. Acad. Sci. U. S. A. **76**(11), 5665–5668 (1979)
10. Rheinwald, J.G., Green, H.: Serial cultivation of strains of human epidermal keratinocytes: the formation of keratinizing colonies from single cells. Cell **6**(3), 331–343 (1975)
11. Yannas, I.V., Burke, J.F., Orgill, D.P., Skrabut, E.M.: Wound tissue can utilize a polymeric template to synthesize a functional extension of skin. Science **215**(4529), 174–176 (1982)
12. Atala, A., Bauer, S.B., Soker, S., Yoo, J.J., Retik, A.B.: Tissue-engineered autologous bladders for patients needing cystoplasty. Lancet Lond. Engl. **367**(9518), 1241–1246 (2006)
13. Macchiarini, P., Jungebluth, P., Go, T., Asnaghi, M.A., Rees, L.E., Cogan, T.A., et al.: Clinical transplantation of a tissue-engineered airway. Lancet Lond. Engl. **372**(9655), 2023–2030 (2008)
14. Vacanti, J.P., Morse, M.A., Saltzman, W.M., Domb, A.J., Perez-Atayde, A., Langer, R.: Selective cell transplantation using bioabsorbable artificial polymers as matrices. J. Pediatr. Surg. **23**(1 Pt 2), 3–9 (1988)
15. Viola, J., Lal, B., Grad, O.: The emergence of tissue engineering as a research field [Internet] (2003). http://www.nsf.gov/pubs/2004/nsf0450/exec_summ.pdf

16. Grebenyuk, S., Abdel Fattah, A.R., Kumar, M., Toprakhisar, B., Rustandi, G., Vananroye, A., et al.: Large-scale perfused tissues via synthetic 3D soft microfluidics. Nat. Commun. [Internet] **14**(1), 193 (2023). https://www.nature.com/articles/s41467-022-35619-1
17. Lee, A., Hudson, A.R., Shiwarski, D.J., Tashman, J.W., Hinton, T.J., Yerneni, S., et al.: 3D bioprinting of collagen to rebuild components of the human heart. Science **365**(6452), 482–487 (2019)
18. Ostrovidov, S., Salehi, S., Costantini, M., Suthiwanich, K., Ebrahimi, M., Sadeghian, R.B., et al.: 3D bioprinting in skeletal muscle tissue engineering. Small Weinh. Bergstr. Ger. **15**(24), e1805530 (2019)
19. US Food and Drug Administration.: FDA Approves First Acellular Tissue Engineered Vessel to Treat Vascular Trauma in Extremities [Internet] (2024). https://www.fda.gov/news-events/press-announcements/fda-approves-first-acellular-tissue-engineered-vessel-treat-vascular-trauma-extremities
20. Ahadian, S., Khademhosseini, A.: Smart scaffolds in tissue regeneration. Regen. Biomater. [Internet] **5**(3), 125–128 (2018). https://academic.oup.com/rb/article/5/3/125/4975526
21. Zhao, C., Tan, A., Pastorin, G., Ho, H.K.: Nanomaterial scaffolds for stem cell proliferation and differentiation in tissue engineering. Biotechnol. Adv. [Internet] **31**(5), 654–668 (2013). https://linkinghub.elsevier.com/retrieve/pii/S0734975012001401
22. Smith, I.O., Liu, X.H., Smith, L.A., Ma, P.X.: Nanostructured polymer scaffolds for tissue engineering and regenerative medicine. WIREs Nanomed. Nanobiotechnol. [Internet] **1**(2), 226–236 (2009). https://wires.onlinelibrary.wiley.com/doi/10.1002/wnan.26
23. Lim, H.L., Hwang, Y., Kar, M., Varghese, S.: Smart hydrogels as functional biomimetic systems. Biomater. Sci. [Internet] **2**(5), 603–618 (2014). https://xlink.rsc.org/?DOI=C3BM60288E
24. Bacakova, L., Zarubova, J., Travnickova, M., Musilkova, J., Pajorova, J., Slepicka, P., et al.: Stem cells: their source, potency and use in regenerative therapies with focus on adipose-derived stem cells—a review. Biotechnol. Adv. [Internet] **36**(4), 1111–1126 (2018). https://linkinghub.elsevier.com/retrieve/pii/S0734975018300624
25. Yoon, J., Han, H., Jang, J.: Nanomaterials-incorporated hydrogels for 3D bioprinting technology. Nano Converg. [Internet] **10**(1), 52 (2023). https://nanoconvergencejournal.springeropen.com/articles/10.1186/s40580-023-00402-5
26. Cai, Y., Chang, S.Y., Gan, S.W., Ma, S., Lu, W.F., Yen, C.-C.: Nanocomposite bioinks for 3D bioprinting. Acta Biomater. [Internet] **151**, 45–69 (2022). https://linkinghub.elsevier.com/retrieve/pii/S1742706122004901
27. Maan, Z., Masri, N.Z., Willerth, S.M.: Smart bioinks for the printing of human tissue models. Biomolecules [Internet] **12**(1), 141 (2022). https://www.mdpi.com/2218-273X/12/1/141
28. Chandra, D.K., Reis, R.L., Kundu, S.C., Kumar, A., Mahapatra, C.: Nanomaterials-based hybrid bioink platforms in advancing 3D bioprinting technologies for regenerative medicine. ACS Biomater. Sci. Eng. [Internet] **10**(7), 4145–4174 (2024). https://pubs.acs.org/doi/10.1021/acsbiomaterials.4c00166
29. Naskar, A., Kim, K.: Recent advances in nanomaterial-based wound-healing therapeutics. Pharmaceutics [Internet] **12**(6), 499 (2020). https://www.mdpi.com/1999-4923/12/6/499
30. Shin, K., Choi, J.W., Ko, G., Baik, S., Kim, D., Park, O.K., et al.: Multifunctional nanoparticles as a tissue adhesive and an injectable marker for image-guided procedures. Nat. Commun. [Internet] **8**(1), 15807 (2017). https://www.nature.com/articles/ncomms15807
31. Zhang, H., Zhang, J., Peng, X., Li, Z., Bai, W., Wang, T., et al.: Smart internal bio-glues. Adv. Sci. [Internet] **9**(27), 2203587 (2022). https://onlinelibrary.wiley.com/doi/10.1002/advs.202203587
32. Cajkova, J., Trebunova, M., Bacenkova, D.: Extracellular matrix decellularization approaches for 3D tissue printing. Acta Tecnol. [Internet] **10**(03), 81–86 (2024). https://www.actatecnologia.eu/issues/2024/III_2024_01_Cajkova_Trebunova_Bacenkova.pdf
33. Ebrahimi Sadrabadi, A., Baei, P., Hosseini, S., Baghaban Eslaminejad, M.: Decellularized extracellular matrix as a potent natural biomaterial for regenerative medicine. In: Turksen, K. (ed.) Cell Biology and Translational Medicine, vol. 13, pp. 27–43 [Internet]. Springer International Publishing, Cham (2020). https://link.springer.com/10.1007/5584_2020_504

34. Brown, M., Li, J., Moraes, C., Tabrizian, M., Li-Jessen, N.Y.K.: Decellularized extracellular matrix: new promising and challenging biomaterials for regenerative medicine. Biomaterials [Internet] **289**, 121786 (2022). https://linkinghub.elsevier.com/retrieve/pii/S0142961222004264
35. Gilpin, A., Yang, Y.: Decellularization strategies for regenerative medicine: from processing techniques to applications. BioMed Res. Int. [Internet] **2017**, 1–13 (2017). https://www.hindawi.com/journals/bmri/2017/9831534/
36. Omidian, H., Gill, E.J.: Nanofibrous scaffolds in biomedicine. J. Compos. Sci. [Internet] **8**(7), 269 (2024). https://www.mdpi.com/2504-477X/8/7/269
37. Wei, G., Ma, P.X.: Nanostructured biomaterials for regeneration. Adv. Funct. Mater. [Internet]. **18**(22), 3568–3582 (2008). https://onlinelibrary.wiley.com/doi/10.1002/adfm.200800662
38. Dvir, T., Timko, B.P., Kohane, D.S., Langer, R.: Nanotechnological strategies for engineering complex tissues. Nat. Nanotechnol. [Internet] **6**(1), 13–22 (2011). https://www.nature.com/articles/nnano.2010.246
39. Huang, H., Li, J., Wang, C., Xing, L., Cao, H., Wang, C., et al.: Using decellularized magnetic microrobots to deliver functional cells for cartilage regeneration. Small [Internet] **20**(11), 2304088 (2024). https://onlinelibrary.wiley.com/doi/10.1002/smll.202304088
40. Kim, D.Y., Ko, E., Ryu, Y.H., Lee, S.J., Jun, Y.J.: Hyaluronic acid based adipose tissue-derived extracellular matrix scaffold in wound healing: histological and immunohistochemical study. Tissue Eng. Regen. Med. **21**(6), 829–842 (2024)
41. El-Ghoul, Y., Altuwayjiri, A.S., Alharbi, G.A.: Synthesis and characterization of new electrospun medical scaffold-based modified cellulose nanofiber and bioactive natural propolis for potential wound dressing applications. RSC Adv. [Internet] **14**(36), 26183–26197 (2024). https://xlink.rsc.org/?DOI=D4RA04231J
42. Xing, J., Zhang, M., Liu, X., Wang, C., Xu, N., Xing, D.: Multi-material electrospinning: from methods to biomedical applications. Mater. Today Bio. **21**, 100710 (2023)
43. Barati, D., Shariati, S.R.P., Moeinzadeh, S., Melero-Martin, J.M., Khademhosseini, A., Jabbari, E.: Spatiotemporal release of BMP-2 and VEGF enhances osteogenic and vasculogenic differentiation of human mesenchymal stem cells and endothelial colony-forming cells co-encapsulated in a patterned hydrogel. J. Control. Release Off. J. Control. Release Soc. **223**, 126–136 (2016)
44. Li, T., Zhang, X., Ma, L., Qi, X., Wang, H., Zhou, Q., et al.: 3D printing of stiff, tough, and ROS-scavenging nanocomposite hydrogel scaffold for in situ corneal repair. Acta Biomater. **15**(192), 189–205 (2025)
45. Montheil, T., Maumus, M., Valot, L., Lebrun, A., Martinez, J., Amblard, M., et al.: Inorganic sol-gel polymerization for hydrogel bioprinting. ACS Omega **5**(6), 2640–2647 (2020)
46. Banani, S.F., Rice, A.M., Peeples, W.B., Lin, Y., Jain, S., Parker, R., et al.: Compositional control of phase-separated cellular bodies. Cell **166**(3), 651–663 (2016)
47. Lasenko, I., Sanchaniya, J.V., Kanukuntla, S.P., Ladani, Y., Viluma-Gudmona, A., Kononova, O., et al.: The mechanical properties of nanocomposites reinforced with PA6 electrospun nanofibers. Polymers **15**(3), 673 (2023)
48. Wunner, F.M., Wille, M.-L., Noonan, T.G., Bas, O., Dalton, P.D., De-Juan-Pardo, E.M., et al.: Melt electrospinning writing of highly ordered large volume scaffold architectures. Adv. Mater. Deerfield Beach Fla. **30**(20), e1706570 (2018)
49. Kong, G., Xiong, M., Liu, L., Hu, L., Meng, H.-M., Ke, G., et al.: DNA origami-based protein networks: from basic construction to emerging applications. Chem. Soc. Rev. **50**(3), 1846–1873 (2021)
50. Lin, Q., Si, Y., Zhou, F., Hao, W., Zhang, P., Jiang, P., et al.: Advances in polysaccharides for probiotic delivery: properties, methods, and applications. Carbohydr. Polym. **1**(323), 121414 (2024)
51. Qi, S., Craig, D.: Recent developments in micro- and nanofabrication techniques for the preparation of amorphous pharmaceutical dosage forms. Adv. Drug Deliv. Rev. **1**(100), 67–84 (2016)

52. Kerativitayanan, P., Carrow, J.K., Gaharwar, A.K.: Nanomaterials for engineering stem cell responses. Adv. Healthc. Mater. [Internet] **4**(11), 1600–1627 (2015). https://onlinelibrary.wiley.com/doi/10.1002/adhm.201500272
53. Contreras, J.L., Wu, H., Smyth, C.A., Eckstein, C.P., Young, C.J., Seki, T., et al.: Double genetic modification of adenovirus fiber with RGD polylysine motifs significantly enhances gene transfer to isolated human pancreatic islets. Transplantation **76**(1), 252–261 (2003)
54. Gentile, P., Garcovich, S.: Advances in regenerative stem cell therapy in androgenic alopecia and hair loss: Wnt pathway, growth-factor, and mesenchymal stem cell signaling impact analysis on cell growth and hair follicle development. Cells [Internet] **8**(5), 466 (2019). https://www.mdpi.com/2073-4409/8/5/466
55. Hade, M.D., Suire, C.N., Suo, Z.: Mesenchymal stem cell-derived exosomes: applications in regenerative medicine. Cells [Internet] **10**(8), 1959 (2021). https://www.mdpi.com/2073-4409/10/8/1959
56. Huber, J., Griffin, M.F., Longaker, M.T., Quarto, N.: Exosomes: a tool for bone tissue engineering. Tissue Eng. Part B Rev. [Internet] **28**(1), 101–113 (2022). https://www.liebertpub.com/doi/10.1089/ten.teb.2020.0246
57. Ksouri, R., Odabas, S., Yar Sağlam, A.S.: Exosome loaded 3D printed magnetic PLA constructs: a candidate for bone tissue engineering. Prog. Addit. Manuf. [Internet] **10**(1), 247–260 (2025). https://link.springer.com/10.1007/s40964-024-00619-8
58. Wan, L., Liu, F., Wang, A., He, Y., Pan, J., Liu, Y., et al.: PI3K/Akt pathway-mediated enhancement of bone and vascular regeneration by gelatin/hyaluronic acid/exosome composite scaffold in bone tissue engineering. Biomater. Adv. [Internet] **166**, 214064 (2025). https://linkinghub.elsevier.com/retrieve/pii/S2772950824003078
59. Zhang, Y., Fang, M., Zhu, J., Li, T., Li, N., Su, B., et al.: Exosome-loaded hyaluronic acid hydrogel composite with oxygen-producing 3D printed polylactic acid scaffolds for bone tissue repair and regeneration. Int. J. Biol. Macromol. [Internet] **274**, 132970 (2024). https://linkinghub.elsevier.com/retrieve/pii/S0141813024037759
60. Ranjbar, F.E., Ranjbar, A.E., Malekshahi, Z.V., Taghdiri-Nooshabadi, Z., Faradonbeh, D.R., Youseflee, P., et al.: Bone tissue regeneration by 58S bioactive glass scaffolds containing exosome: an in vivo study. Cell Tissue Bank. [Internet] **25**(1), 389–400 (2024). https://link.springer.com/10.1007/s10561-023-10120-1
61. Pal, S., Misra, S., Mallik, R.K.: COTiR: molecular communication model for synthetic exosome-based tissue regeneration. IEEE Trans. NanoBiosci. [Internet] **23**(1), 202–209 (2024). https://ieeexplore.ieee.org/document/10210307/
62. Wang, D., Diao, S., Zhou, X., Zhou, J., Liu, Y.: A new method regulates bone fracture tissue exosome lncRNA-mRNA to promote mesenchymal stem cell proliferation and migration. Injury [Internet] **55**(2), 111210 (2024). https://linkinghub.elsevier.com/retrieve/pii/S0020138323009270
63. Lee, C.-S., Fan, J., Hwang, H.S., Kim, S., Chen, C., Kang, M., et al.: Bone-targeting exosome mimetics engineered by bioorthogonal surface functionalization for bone tissue engineering. Nano Lett. [Internet] **23**(4), 1202–1210 (2023). https://pubs.acs.org/doi/10.1021/acs.nanolett.2c04159
64. Anagnostou, D., Sfakianaki, G., Melachroinou, K., Soutos, M., Constantinides, V., Vaikath, N., et al.: Assessment of aggregated and exosome-associated α-synuclein in brain tissue and cerebrospinal fluid using specific immunoassays. Diagnostics [Internet] **13**(13), 2192 (2023). https://www.mdpi.com/2075-4418/13/13/2192
65. Shafei, S., Khanmohammadi, M., Ghanbari, H., Nooshabadi, V.T., Tafti, S.H.A., Rabbani, S., et al.: Effectiveness of exosome mediated miR-126 and miR-146a delivery on cardiac tissue regeneration. Cell Tissue Res. [Internet] **390**(1), 71–92 (2022). https://link.springer.com/10.1007/s00441-022-03663-4
66. Huang, J., Xiong, J., Yang, L., Zhang, J., Sun, S., Liang, Y.: Cell-free exosome-laden scaffolds for tissue repair. Nanoscale [Internet] **13**(19), 8740–8750 (2021). https://xlink.rsc.org/?DOI=D1NR01314A

67. Kisby, C.K., Shadrin, I.Y., Peng, L.T., Stalboerger, P.G., Trabuco, E.C., Behfar, A., et al.: Impact of repeat dosing and mesh exposure chronicity on exosome-induced vaginal tissue regeneration in a porcine mesh exposure model. Female Pelvic Med. Reconstr. Surg. [Internet] **27**(3), 195–201 (2021a). https://journals.lww.com/10.1097/SPV.0000000000001017
68. Su, N., Hao, Y., Wang, F., Hou, W., Chen, H., Luo, Y.: Mesenchymal stromal exosome–functionalized scaffolds induce innate and adaptive immunomodulatory responses toward tissue repair. Sci. Adv. [Internet] **7**(20), eabf7207 (2021). https://www.science.org/doi/10.1126/sciadv.abf7207
69. Kisby, C.K., Shadrin, I.Y., Rolland, T.J., Stalboerger, P.G., Zhou, B., Trabuco, E.C., et al.: Exosome-induced vaginal tissue regeneration in a porcine mesh exposure model. Female Pelvic Med. Reconstr. Surg. [Internet] **27**(10), 609–615 (2021b). https://journals.lww.com/10.1097/SPV.0000000000001005
70. Zhang, S., Yang, Y., Jia, S., Chen, H., Duan, Y., Li, X., et al.: Exosome-like vesicles derived from Hertwig's epithelial root sheath cells promote the regeneration of dentin-pulp tissue. Theranostics [Internet] **10**(13), 5914–5931 (2020). http://www.thno.org/v10p5914.htm
71. Shafei, S., Khanmohammadi, M., Heidari, R., Ghanbari, H., Taghdiri Nooshabadi, V., Farzamfar, S., et al.: Exosome loaded alginate hydrogel promotes tissue regeneration in full-thickness skin wounds: an in vivo study. J. Biomed. Mater. Res. A [Internet] **108**(3), 545–556 (2020). https://onlinelibrary.wiley.com/doi/10.1002/jbm.a.36835
72. Zakeri, Z., Heiderzadeh, M., Kocaarslan, A., Metin, E., Hosseini Karimi, S.N., Saghati, S., et al.: Exosomes encapsulated in hydrogels for effective central nervous system drug delivery. Biomater. Sci. **12**(10), 2561–2578 (2024)
73. Li, J., Sun, S., Zhu, D., Mei, X., Lyu, Y., Huang, K., et al.: Inhalable stem cell exosomes promote heart repair after myocardial infarction. Circulation **150**(9), 710–723 (2024)
74. Wu, J., Kuang, L., Chen, C., Yang, J., Zeng, W.-N., Li, T., et al.: MiR-100-5p-abundant exosomes derived from infrapatellar fat pad MSCs protect articular cartilage and ameliorate gait abnormalities via inhibition of mTOR in osteoarthritis. Biomaterials **206**, 87–100 (2019)
75. Jia, Y., Li, L., Lin, Y.-H., Gopal, P., Shen, S., Zhou, K., et al.: In vivo CRISPR screening identifies BAZ2 chromatin remodelers as druggable regulators of mammalian liver regeneration. Cell Stem Cell **29**(3), 372-385.e8 (2022)
76. Ringel, T., Frey, N., Ringnalda, F., Janjuha, S., Cherkaoui, S., Butz, S., et al.: Genome-scale CRISPR screening in human intestinal organoids identifies drivers of TGF-β resistance. Cell Stem Cell **26**(3), 431-440.e8 (2020)
77. CAGR. Global Biomaterials for Tissue Engineering Market—Industry Trends and Forecast to 2028 [Internet]. https://www.databridgemarketresearch.com/reports/global-biomaterials-for-tissue-engineering-market
78. Revealed: the fatty cells that are the 'bubble wrap' of the body [Internet]. Nature (2025). https://www.nature.com/articles/d41586-025-00012-7

Bridging the Gap: Regulatory Pathway and Clinical Translation of Nanomaterials in Medicines

Neha Raghuwanshi, Vinita Isad, R. K. Shanmathi, and Chaitanya Kale

Abstract The applications of nanomaterials in medicine are extremely promising, particularly in the fields of drug transport, therapeutic interventions, and diagnostics. But moving these cutting-edge nanomaterials from lab studies to clinical settings requires negotiating intricate regulatory processes and guaranteeing their usefulness and safety for human usage. This chapter aims to provide an in-depth analysis of the regulatory challenges and concerns surrounding the clinical translation of nanomaterials. The main elements influencing the creation of nanomedicines are covered, such as the requirements for thorough preclinical and clinical testing, material characteristics, and scale-up procedures. The chapter also discusses the regulatory frameworks set up by agencies, including the Food and Drug Administration (FDA), European Medicines Agency (EMA), and the International Council for Harmonization of Technical Requirements for Pharmaceuticals for Human Use (ICH), with an emphasis on their rules for approving pharmaceutical items and medical devices made of nanomaterials. It also discusses the significance of long-term safety profiles, biocompatibility, and nanotoxicology in guaranteeing that nanomaterials fulfill the requirements for clinical use. The conclusion of the chapter highlights future trends, focusing on the role of personalized medicine and the potential of nanomaterial-based therapies in treating complex diseases such as infections, cancer, and neurological disorders. To successfully integrate nanomaterials into clinical practiceseeks to bridge the gap between scientific study and real-world medicinal applications by offering a thorough analysis of the regulatory and clinical factors.

N. Raghuwanshi (✉)
Amity Institute of Pharmacy, Amity University Maharashtra, Mumbai, Maharashtra 410206, India
e-mail: raghuwanshi41@gmail.com

School of Pharmacy, RK University, Tramba, Rajkot, Gujarat, India

V. Isad · R. K. Shanmathi
Bombay College of Pharmacy, Kalina, Santacruz (E), Mumbai, Maharashtra 400098, India

R. K. Shanmathi · C. Kale
Poona College of Pharmacy, Bhartiya Vidyapeeth (Deemed to Be University), Pune, India

C. Kale
Industrial Pharmacy, Poona College of Pharmacy, Pune 411038, India

S. N. Saud Al-Humairi et al. (eds.), *Smart Nanomaterials in Biomedical Applications*, Engineering Materials, https://doi.org/10.1007/978-981-96-8023-8_12

Keywords Nanomaterials · Regulatory pathways · Clinical translation · Nanomedicines · Biocompatibility · Personalized medicine

1 Introduction

Drug delivery methods, diagnostic instruments, and therapy modalities have all undergone radical changes due to the application of nanomaterials in medicine. These materials present unprecedented possibilities for precision medicine and targeted therapies because of their distinctive physicochemical properties, including high surface area, nanoscale size, and adjustable features. However, the absence of established regulatory frameworks that can effectively address their new characteristics and potential risks remains the primary reason why the clinical translation of nanomaterials is so challenging.

The chapter begins by providing a thorough insight into nanomaterials in medicine, covering their various classifications, inherent characteristics, and current applications across a range of therapeutic and diagnostic fields. Next, we examine the importance of robust regulatory channels for the development and marketing of nanomedicines. For these products to be safe, effective, and consistent in clinical settings, regulatory oversight is essential. The chapter emphasizes the need for a flexible and dynamic regulatory framework to manage the complexities of nanomaterials, enabling their effective clinical implementation while protecting public health.

1.1 Overview of Nanomaterials in Medicine

A new technological era is being ushered in by nanotechnology, an interdisciplinary field that connects engineering, medicine, physics, chemistry, and biology via biology. The research and innovation of nanomaterials were made possible by the "magic bullet" idea. Nowadays, nanomaterials with sizes between 1 and 100 nm are employed in many medical applications, such as medication administration, imaging, biosensors, and biomarkers. The creation, identification, and characterisation of several nanomaterials for a variety of biomedical applications are demonstrated in many studies. Materials with nanoscale sizes have unique physicochemical (mechanical, chemical, magnetic, electrical, and optical) and biological properties [1]. Nanomaterials' unique mechanical, optical, and electrical properties have led to their widespread use in pharmacology and medicine. Nanomaterials are used in novel therapies, sick tissue imaging, and biological molecule identification [2].

In contrast to standard materials, nanoparticles possess distinct and unique features, as shown in Fig. 1. Numerous routes, including the digestive tract, skin, respiratory system, and drug injections, can allow nanomaterials to enter the body. Once inside, they can travel to organs and cause biological consequences such as

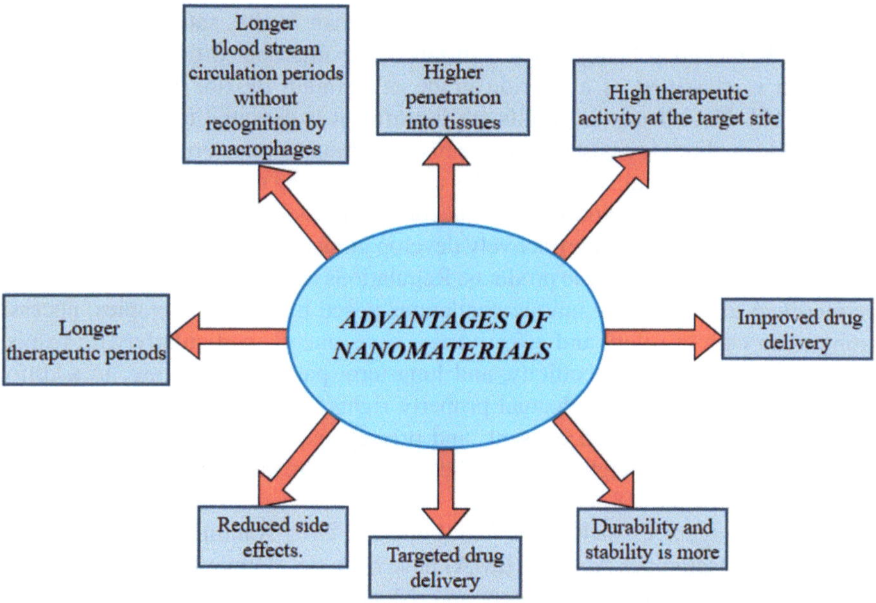

Fig. 1 Advantages of nanomaterials in comparison to the micro-sized particles [10]

oxidative stress, cellular death, inflammatory reactions, and DNA damage. Significant progress has been made in targeted, selective, and patient-centered medication therapy, and there are excellent opportunities when considering materials and medications developed from nanotechnology [3]. Because of their unique qualities, nanoparticles are utilized to improve drug delivery efficiency [4–9].

1.2 Importance of Regulatory Pathways in Nanomedicine

Although there is a great need to develop these nanomedicines, industry and regulatory bodies face significant obstacles due to their unique characteristics [11–13]. The failure of these nanomedicines in the latter stages of clinical trials may have been caused by a widespread absence of specific procedures to describe them at the biological, physiological, and physicochemical levels. As a result, there is a chance to offer more precise direction for developing those novel nanomedicines, as their regulatory environment has been progressively challenged by critical difficulties [14].

The combination of increased reactivity and the ratio of surface area to volume, which quantum effects could cause, makes characterizing nanoparticles much more challenging and complicates drawing simple conclusions regarding their potential toxicity [15]. Regulators worldwide are also preparing for the possible integration of nanotechnology into novel therapeutic uses. Currently, the law does not require testing of nanoparticles for their impacts on human health, safety, and the environment. Rather than being a new substance, the micronized titanium dioxide in sunscreen is classified as a specific grade of titanium dioxide that was initially studied [16]. Not all nanomaterials are harmful; it is essential to filter those with undesirable qualities to assist businesses in developing nanomaterials with the lowest possible risk [17]. Regulatory authorities are emphasizing the necessity for standardized characterization methodologies to accurately measure the safety and efficacy of nanomedicine products as they actively develop guidelines and standards to evaluate the hazards and safety of these products. Regulations concerning gene vaccines must be modified to address the unique challenges posed by genetic therapies, necessitating extra care in safety and effectiveness assessments, particularly with respect to vector design, target specificity, and long-term patient monitoring. In addition to the complex issue of intellectual property rights, ethical considerations such as patient autonomy, informed consent, and privacy require thorough consideration. These factors must be weighed against the need to ensure these life-saving treatments are widely accessible [18, 19].

Regulatory agencies such as the U.S. Food and Drug Administration (FDA) and the European Medicines Agency (EMA) have been developing guidelines for evaluating nanomedicine products' safety and risk assessments. These guidelines stress the need for appropriate preclinical and clinical research to assess the safety and efficacy of nanomaterials, as well as their characterisation, toxicity, and environmental impact [20].

2 Historical Background and Development of Nanomaterials in Medicine

New therapeutic, diagnostic, and medication delivery methods have been made possible by the development of nanotechnology in biomedical applications, drastically changing modern medicine. Early ideas for altering matter at the nanoscale date back to the middle of the twentieth century, when nanotechnology in medicine first emerged. Advances in molecular biology, materials science, and nanofabrication techniques over the years have made it easier to create nanomaterials with special qualities that allow for more precise and targeted medicinal interventions. The successful use of nanocarriers for targeted cancer therapy, the creation of nanosensors for early illness diagnosis, and the manufacturing of nanoparticles for drug delivery systems are essential turning points in nanomedicine research. The substantial advancement in this sector is exemplified by noteworthy accomplishments like the approval of liposomal formulations (like Doxil®) and the application of gold nanoparticles for imaging and therapy. This chapter follows these developments, emphasizing how a paradigm change in biomedical innovation has occurred because of the gradual integration of nanomaterials in clinical applications brought about by advances in nanotechnology.

2.1 Evolution of Nanotechnology in Biomedical Applications

Beginning with investigating colloidal gold and silver nanoparticles for their optical characteristics in the early twentieth century, nanotechnology has remarkably evolved in biomedical applications [21, 22]. Material science made tremendous strides in the 1990s, creating carbon nanomaterials, metallic nanoparticles, polymeric nanoparticles, and quantum dots, transforming drug delivery and biosensing [23, 24]. Figure 2 shows the evolution of nanotechnology.

By the early 2000s, advances in imaging and diagnostics were made possible by nanotechnology, including biosensors for early illness detection and imaging agents based on nanoparticles [25]. FDA-approved treatments like Doxil and Abraxane, in addition to the creation of stimuli-responsive nanoparticles for precise drug administration, are the results of advancements in nanomedicine. Nanotechnology improves tissue engineering in regenerative medicine by creating nanostructured scaffolds that resemble the extracellular matrix and encourage cell growth and differentiation. While tackling issues like nanoparticle toxicity and scalability, recent developments are propelling breakthroughs in personalized medicine, immunotherapy, nanorobotics, and CRISPR delivery systems. These developments demonstrate nanotechnology's potential to revolutionize regenerative medicine, therapeutics, and diagnostics, providing a bright future for healthcare.

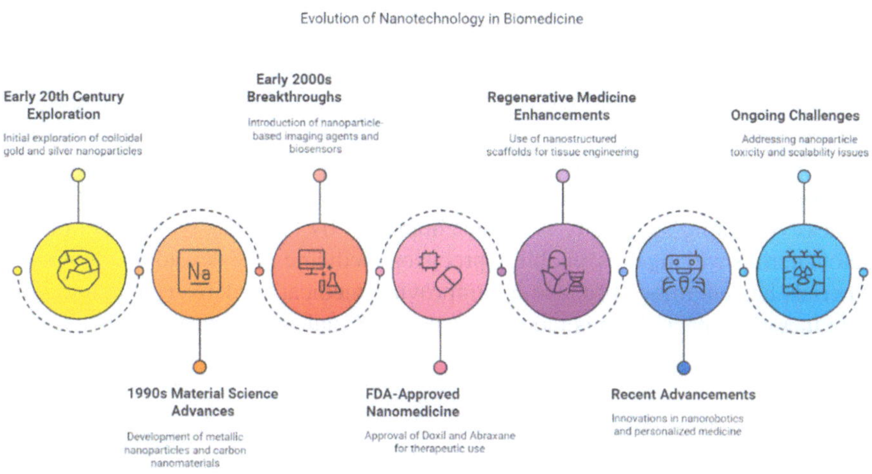

Fig. 2 Evolution of nanotechnologies

2.2 Key Milestones in Nanomedicine Research

Since Gris-PRO®, the first medication based on nanocrystals, was introduced in 1975, nanomedicine has undergone substantial change. This was the first time nanotechnology was used in medicine, opening the door for creative medicinal breakthroughs. The first PEGylated liposomal medication, Doxil®, was approved in 1992, improving drug circulation and efficacy. AmBisome®, which was approved in 1990, introduced nanotechnology for the treatment of antifungals.

Additional advances in metal-based nanomedicines (CosmoFer 1995) and albumin-bound nanoparticles (Abraxane® 1997) occurred in the late 1990s and early 2000s, as shown in Fig. 3. The growth of nanomedicine applications is demonstrated by two-drug liposome treatments (Vyxeos® 2017) and polymeric micelle paclitaxel (Genexol® PM, 2007). Nanomedicine's place in contemporary medicine was established by the approval of inhalation nanomedicine (Arikayce®, 2018) and mRNA COVID-19 vaccines (BNT162b2, mRNA-1273, 2021). These developments shape the future of nanotechnology in therapeutic applications, highlighting its regulatory and translational challenges.

3 Nanomaterials in Drug Delivery Systems

Nanomaterials have completely transformed drug delivery systems because they provide focused therapy, increased bioavailability, and better control over drug release. Their tiny size and large surface area allow for effective drug encapsulation, which enhances solubility and prevents degradation. When delivering therapeutic drugs directly to the site of action, nanoparticles, liposomes, dendrimers, and nanoshells are frequently utilized to reduce adverse effects and maximize effectiveness. Targeting specificity is further improved by the capacity to functionalize nanomaterials with ligands, enabling precision medicine. The development of nanomaterials has enormous promise for improving medication delivery methods for a few illnesses, especially in personalized medicine and cancer treatment.

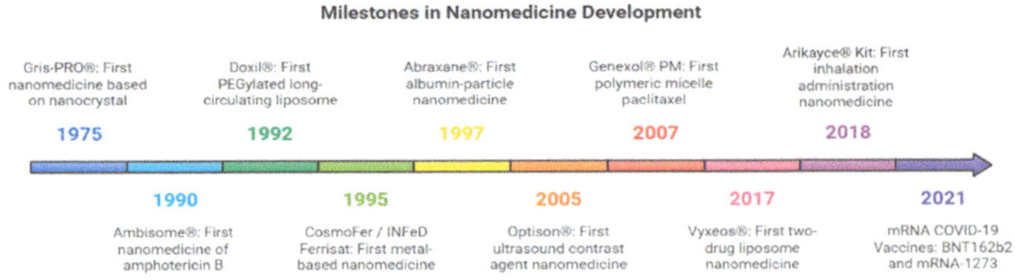

Fig. 3 Milestones in nanomedicine development

3.1 Nanomaterials in Drug Delivery

Nanomaterials in drug delivery (Fig. 4) are categorized based on composition and functional properties:

(a) **Lipid-Based Nanomaterials**: Liposomes, solid lipid nanoparticles, and nanostructured lipid carriers enhance bioavailability and stability via amphiphilic architectures.
(b) **Inorganic Nanomaterials**: Gold nanoparticles, mesoporous silica, and iron oxide nanoparticles provide high drug-loading capacity, controlled release, and theranostic applications.
(c) **Polymeric Nanomaterials**: Dendrimers and polymeric micelles facilitate tunable drug encapsulation and sustained release via molecular self-assembly.
(d) **Carbon-Based Nanomaterials**: Carbon nanotubes and graphene derivatives exhibit high surface area, π-π interactions, and functionalization potential for targeted delivery.
(e) **Biomimetic Nanomaterials**: Engineered to mimic biological systems, these enhance immune evasion, prolong circulation, and improve biocompatibility.

3.1.1 Lipid-Based Nanomaterials

Lipid-based nanomaterials are versatile drug delivery systems known for their biocompatibility, controlled release, and ability to enhance drug solubility. Key types include liposomes, which encapsulate drugs in phospholipid bilayers; solid lipid nanoparticles (SLNs), offering improved stability and controlled release; and nanostructured lipid carriers (NLCs), which enhance drug loading and reduce crystallinity. These nanocarriers are widely used in cancer therapy, vaccines, and gene delivery [26].

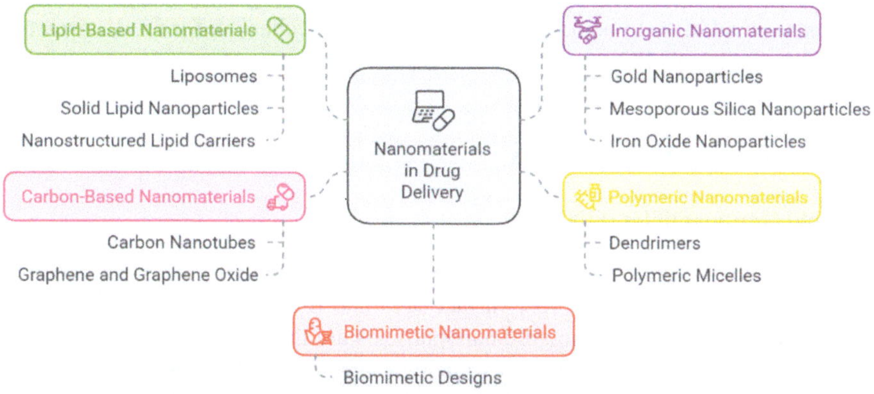

Fig. 4 Types of nanomaterials in drug delivery

i. **Liposomes**

Drug solubility, pharmacokinetics, toxic side effects, and in vitro and in vivo anticancer efficacy have all been investigated in relation to liposomes [liposome encapsulation by liposomes, the pH, osmotic gradient, liposome composition, and ambiance all affect drug release [27, 28]. Liposomes have been employed as drug carriers due to their capacity to target drugs, prevent drug degradation, and reduce side effects. However, the primary drawbacks of liposomes are their poor storage durability, low encapsulation effectiveness, and rapid drug leakage in biological fluids. A phospholipid's positive or negative charge determines the liposome's surface charge. Liposomes reduce toxicity and stop drug breakdown by encapsulating a medication after it has been administered to the body.

ii. **Solid Lipid Nanoparticles (SLNs)**

In the 1990s, the phrases "solid lipid nanoparticles" (SLN) [29] and "nanostructured lipid carriers" (NLC) seemed to fit together perfectly. It combines the advantages of parenteral emulsions, which are lipid-based and composed of non-toxic and biodegradable lipid components, with NPs, primarily polymeric and metallic at the time [30]. Since lipid nanoparticles are made of a solid, they were promoted as a protective substitute for other nanosystems. Matrix that is more reliable (and undoubtedly less costly) than earlier phospholipid-based liposomes and permits controlled drug release [31].

iii. **Nanostructured Lipid Carriers (NLCS)**

These efficient and accurate drug delivery systems, also known as second-generation lipid nanocarriers, have attracted much attention globally due to their many applications and advantages. NLCs—lipid-based formulations with a solid matrix at room temperature offer advantages like enhanced stability, low toxicity, extended shelf life, increased drug loading capacity, and biocompatibility compared to other conventional lipid-based nanocarriers like solid lipid nanoparticles and nanoemulsions. NLCs have recently been used for brain targeting, gene therapy, and cancer treatment, in addition to being an essential delivery mechanism for food, cosmetics, and pharmaceuticals [32].

3.1.2 Inorganic Nanomaterials

i. **Gold Nanoparticles (GNPs)**

GNPs consist of a gold core surrounded by an outer coating of organic ligands for protection [33]. Due to their vast surface area and remarkable optical, plasmonic, and magnetic capabilities, metallic nanoparticles are the safest and least hazardous agents [34, 35]. They are the perfect vehicle for medication delivery because of their ease of modification and the potential to be loaded with pharmacological therapies.

ii. **Mesoporous Silica Nanoparticles (MSNS)**

Numerous scientific and industrial fields, including molecular biology, electrochemistry [36], analytical chemistry, heterogeneous catalysis, and more, have used them since their discovery in the early 1990s. These nanostructured materials were first used in a pharmaceutical environment in 2001, when Vallet-Regi reported using MCM-41 silica in the ibuprofen release process [37]. The dynamic development of research on the use of mesoporous particles as carriers in drug delivery systems was initiated by this seminal work. The US Food and Drug Administration, or FDA, has approved silica as "Generally Recognized as Safe" (GRAS), meaning it can be used as a food ingredient and in cosmetics [38]. Following administration, mesoporous silica nanoparticles degrade in three stages, according to studies conducted in a simulated bodily fluid. This degradation is advantageous for the drug release process. Furthermore, tests of mice's in vivo absorption, distribution, and excretion after receiving MSN via different routes show that the drug is tissue biocompatible for both intravenous and oral injection.

iii. **Iron Oxide Nanoparticles (IONPs)**

The abundant varieties of iron oxide nanoparticles (IONPs) have gained popularity in various applications due to their exceptional qualities, making it easier to advance high technology based on nanoscale materials. According to their distinct phases, the IONPs can be classified as magnetite, hematite, maghemite, or wustite [39]. However, because of their particular characteristics and physicochemical qualities, γ-Fe_2O_3, α-Fe_2O_3, and Fe_3O_4 are the most employed IONP forms [40]. Interestingly, IONPs at the nanoscale provide materials unique magnetic, electrical, and optical characteristics, which oversee the quantum size phenomena [41–43]. It has been observed that IONPs' superparamagnetic behavior improves with decreasing size, whereas their ferromagnetic behavior deteriorates [44]. Interestingly, IONPs' (10–20 nm) size indicates that Fe_3O_4 and γ-Fe_2O_3 have high superparamagnetic characteristics, leading to notable performance in various applications. Interestingly, IONPs can be employed for a range of pharmacological applications [45], including biomedicine [46], biotechnology [47], and agriculture [48] because of their potential biocompatibility, high stability, biodegradability, nontoxicity, and low preparation cost.

3.1.3 Polymeric Nanomaterials

Polymeric nanomaterials are advanced drug delivery systems for controlled release, high drug-loading capacity, and targeted therapy. They include dendrimers, highly branched macromolecules enabling precise drug conjugation, polymeric micelles, and self-assembling amphiphilic structures enhancing solubility and bioavailability. These nanocarriers improve pharmacokinetics and reduce systemic toxicity and are widely used in cancer therapy, gene delivery, and regenerative medicine for efficient and sustained therapeutic effects [49, 50].

i. **Dendrimers**

Dendrimers are drug-delivery polymeric structures with a precisely determined size [49]. Because of the high percentage of chemical groups on their surface relative to their molecular volume, they have a lot of potential for targeted medication administration. They are the perfect candidate for drug delivery systems because of their spherical form, nanosize, monodispersity, lipophilic nature, hyper-branching points, and capacity to pass through cell walls [51]. Their employment as drug delivery agents is facilitated by their capacity to entrap and/or conjugate with hydrophobic or hydrophilic groups of high molecular weight. Bioactive compounds and antibodies can be attached to the terminal chemical groups to make them soluble, miscible, and reactive. For instance, as the first platinum-based drug, cisplatin, also known as cis-diamine platinum dichloride, aids with the anti-neoplastic application of cisplatin DNA-adducts in the treatment of malignancies of the testicles, lungs, head and neck, ovaries, colorectal, and bladder. However, its use is limited by low absorption and high efflux from cisplatin-resistant cell types. To create pH and temperature-sensitive MoS_2 and magnetic nanoparticle dendrimers, it was found that when stimulated by near-infrared laser light, cisplatin was released at 86% and 92%, respectively [52].

ii. **Polymeric Micelles**

In aqueous solutions, amphiphilic block copolymers containing hydrophilic (A) and hydrophobic (B) blocks self-assemble to create polymeric micelles, which are nanoscale drug delivery devices that usually range in size from 10 to 100 nm. These micelles have a two-phase structure that is typically spherical and is defined by a hydrophilic outer shell and a hydrophobic inner core. The core–shell architecture of PMs is the basis for some of their applications. It protects the hydrophobic portion and, consequently, the therapeutic load, from biological invasion. At the same time, the hydrophilic shell lessens protein adsorption on the micelle surface, which can result in systemic circulation clearance [53]. Micelles are typically spherical. However, depending on the copolymers used, the self-assembled system can come from polymer vesicles, polymersomes, or worm-like or cylindrical micelle morphologies. The hydrophilic–lipophilic balance (HLB) of the copolymers and the solvent used to prepare the micelles impact the structure of the produced PM [50, 54].

3.1.4 Carbon-Based Nanomaterials

The most common geometries for carbon nanotubes (CNTs) are tubes or cylinders. Their diameters, lengths, number of layers, and chirality can all vary. Their special blend of stiffness, strength, and elasticity gave them a lot of appealing qualities. Due to their different structural topologies, CNTs can be classified as single, double, multi-walled, or functionalized. CNTs are synthesized using various techniques, including high-pressure carbon monoxide disproportionation, chemical vapor deposition, arc discharge, and laser ablation. Graphene and graphene oxide: Graphene comprises a single carbon layer with partially filled sp2 orbits that are both above and below the plane. It has a vast surface area, great mechanical and elastic strength, optical clarity, and excellent thermal conductivity. It is frequently used as a semiconductor. The interaction between graphene and electrons results in the formation of new quasi-particles. This class's two most crucial elements are quantum dots and graphene nanoribbons. These particles demonstrate Ballistic transport, allowing electrons to go up to micrometers without scattering. The electrical conductivity of graphene is good. Its solubility in water is lower. Many alterations have been made to create chemically altered graphene, graphene oxide, and layered graphene oxide. In sol–gel chemistry, graphene oxide and its derivatives are widely used and have been proven appropriate for creating biocompatible nanocomposites [55].

3.2 Challenges in Nanomaterials Development for Medicines

The development of nanomaterials for medicines faces several challenges, which are as follows:

i. **Unconventional Pharmacokinetics**

Due to their small size, nanoparticles can exhibit unusual behavior in the body, leading to unpredictable absorption, distribution, metabolism, and excretion (ADME). This may result in unexpected toxicity or reduced efficacy.

ii. **Blood–Brain Barrier Permeability**

Smaller nanoparticles' increased permeability and motility can compromise the blood–brain barrier, potentially disrupting normal brain function.

iii. **Scalability and Stability Concerns**

Ensuring the stability of nanomedicines and nanomaterials during large-scale production is a significant challenge. Variations in production processes can affect the final product's quality, safety, and efficacy.

iv. **Environmental Impact**

The potential environmental impact of nanomaterials is a growing concern. Releasing nanoparticles into the environment can have unforeseen consequences on ecosystems and human health.

v. **Regulatory Framework**

The lack of a unified international regulatory framework for nanomaterials and nanomedicines creates uncertainty and inconsistencies in their development, testing, and approval.

vi. **Accumulation in Organs**

Nanoparticles can accumulate in organs, potentially leading to toxicity and other adverse effects.

vii. **Genotoxicity Concerns**

The potential genotoxicity of nanoparticles is a concern, as they may interact with DNA and cause mutations.

viii. **Limitations of in Vitro Testing**

Conventional in vitro testing may not accurately reflect the complex interactions between nanoparticles and the human body.

ix. **Variability in Nanomaterial Definitions and Classifications**

Different countries and organizations have varying definitions and classifications for nanomaterials and nanomedicines, which can create confusion and inconsistencies.

x. **Characterization Challenges**

Variations in nanomorphology can lead to characterization challenges, making it difficult to ensure the quality and consistency of nanomaterials. To overcome these challenges, researchers and regulatory agencies must work together to develop standardized testing protocols, regulatory frameworks, and characterization methods. Additionally, continued research into the behavior and effects of nanomaterials in the body and environment is crucial for ensuring their safe and effective development.

4 Regulatory Framework for Nanomedicine

The regulatory framework for nanomedicine guarantees the quality, safety, and effectiveness of treatments based on nanotechnology. Organizations such as the FDA (United States), EMA (Europe), and ICH establish guidelines for characterization, toxicity assessment, and clinical evaluation. Substantial obstacles include standardizing nanoparticle characteristics, biodistribution research, and long-term safety monitoring. Regulatory processes change to accommodate new nanomedicines, necessitating thorough preclinical and clinical research to guarantee patient safety and therapeutic efficacy.

4.1 Global Regulatory Agencies and Guidelines

International regulatory bodies, such as the FDA (United States), EMA (Europe), and WHO, set guidelines for the safety, effectiveness, and quality of pharmaceuticals, including nanomedicines. These organizations impose strict preclinical, clinical, and post-market surveillance to guarantee patient safety. Nanomaterial characterisation, toxicity, and pharmacokinetics are covered by regulatory frameworks, which ensure adherence to changing global standards for personalized medicine, novel medicines, and sophisticated drug delivery systems.

i. **European Medicines Agency**

The 28 Member States of the European Union and the European Economic Area have a network of national regulatory bodies for both human and veterinary medications, which forms the foundation of the European Union's regulatory framework for pharmaceuticals. In order to protect public health, a single set of laws (1) and regulatory science requirements (2) oversee the supervision of pharmaceuticals intended for human use in the Member States. The following is the definition of a nanomaterial suggested by the European Commission (5).

Any natural, accidental, or artificial material that contains particles in an unbound state, aggregation, or agglomerate, and where at least 50% of the particles in the number size distribution have one or more exterior dimensions in the 1–100 nm range is referred to as a nanomaterial. Nanomedicines are typically approved in all EU member states through the Centralized Marketing Authorization (MA) process. Innovative medications like nanotechnology-based ones are especially well-suited for this concentrated channel. The following vital committees supervise the assessment and authorization of nanomedicines:

(a) The main body responsible for evaluating the quality, safety, and effectiveness of nanomedicines is the Committee for Medicinal Products for Human Use (CHMP). It provides scientific advice on regulatory clearances and conducts benefit-risk analyses.
(b) The Committee for Advanced Therapies (CAT) is an expert in assessing advanced therapy medicinal products (ATMPs), such as gene therapy, cell-based treatments, and delivery methods based on nanoparticles. This committee offers scientific guidance on new formulations of nanomedicine.
(c) The Pharmacovigilance Risk Assessment Committee (PRAC) monitors the safety of nanomedicines once they are put on the market by identifying and reducing any possible side effects that might surface.

The European Pharmacopoeia (Ph. Eur.) sets guidelines for nanoparticle characterization to guarantee the quality and uniformity of nanomedicines throughout the EU market.

2020: Pharmaceutical Quality of Nasal and Inhalation Products. This recommendation is especially pertinent to nanomedicines delivered via nasal or inhalation methods. It offers thorough suggestions for maintaining drug delivery repeatability, improving aerosol characteristics, and guaranteeing the stability of nanoparticle suspensions. The recommendation emphasizes the necessity of using specific analytical techniques to assess particle size distribution, surface changes, and the possibility of systemic absorption because nanoparticles employed in inhalable formulations must adhere to strict quality criteria [56, 57]

ii. **U.S. Food and Drug Administration (FDA)**

The Food and Drug Administration (FDA) of the United States oversees the regulation of nanomedicine in the country and ensures that strict safety, efficacy, and quality criteria are met. Existing pharmaceutical laws serve as the main regulatory framework for nanomedicine in the United States, with supplementary recommendations addressing the unique difficulties presented by medications based on nanotechnology. This chapter thoroughly examines the FDA's role, guidance guidelines, and regulatory procedures for nanomedicine. In the United States, the FDA is crucial in assessing and approving nanomedicines using a risk–benefit, scientific, and evidence-based methodology. Despite lacking a distinct regulatory category, the FDA assesses nanomedicines in the context of current pharmaceutical product frameworks, considering their hazards and nanoscale characteristics. The FDA has issued guidance documents and regulatory policies outlining key nanomedicine development and evaluation considerations. These documents help ensure nanomedicines' safety, quality, and efficacy while addressing their unique challenges. Nanomedicines usually adhere to one of the FDA's defined regulatory routes for drug and biologic approvals in the United States. These consist of:

(a) The novel Drug Application (NDA) procedure is required for nanomedicine-based small-molecule drugs to meet stringent safety and efficacy standards before being put on the market.
(b) The Biologics License Application (BLA) Pathway is used for nanomedicines that fall under the biologic product category, including gene treatments, monoclonal antibodies, and vaccinations based on nanoparticles.
(c) Generic nanomedicines that exhibit bioequivalence to an approved nanomedicine under the NDA process are submitted via the Abbreviated New Drug Application (ANDA) pathway.
(d) Clinical trials in nanomedicine require submitting an Investigational New Drug (IND) application, which calls for thorough documentation of the drug's pharmacokinetics, safety, and expected therapeutic benefits.

The FDA's primary centers for evaluating and approving nanomedicines are as follows:

(a) The Center for Drug Evaluation and Research (CDER) oversees the assessment of small-molecule nanomedicines, including chemotherapeutic medications made with nanoparticles.
(b) The Center for Biologics Evaluation and Research (CBER) is in charge of biological nanomedicines, which include formulations of nanovaccines, gene therapy, and cell therapy.
(c) The Center for Devices and Radiological Health (CDRH) regulates medical equipment based on nanotechnology, including diagnostic instruments, contrast agents, and nanoscale imaging agents.
(d) The National Center for Toxicological Research (NCTR) carries out studies to evaluate the long-term safety and possible toxicity of nanomaterials in biological systems.

To preserve quality and reproducibility, physicochemical characterization involves exact control over nanoparticle size, shape, surface charge, and aggregation behavior. Understanding how nanomedicines spread throughout the body, get past biological barriers like the blood–brain barrier, and build up in specific tissues is known as pharmacokinetics and biodistribution.

Toxicology Studies: Evaluating the cytotoxicity, genotoxicity, immunogenicity, and long-term biocompatibility of nanomaterials. Manufacturing Quality and GMP Compliance: Ensuring that GMP is adhered to, especially for intricate nanomedicine formulations that require strict quality control procedures. Post-Market Safety Monitoring: The FDA mandates ongoing post-market surveillance (PMS) and risk management techniques to detect any long-term impacts of nanomedicines. A multi-agency program called the National Nanotechnology Initiative (NNI) funds nanotechnology research and development in the United States. It works with the FDA to:

- create standardized techniques for characterizing nanomedicine.
- identify optimal procedures for nanomedicine preclinical and clinical assessment.
- encourage public–private collaborations to hasten the clinical application of nanomedicine research.
- encourage risk assessment research to determine the long-term safety of products containing nanomedicine.

The NNI is crucial in building the regulatory environment for nanomedicines by working with research institutes and regulatory bodies to ensure that they fulfill strict safety and effectiveness requirements before being administered to patients [58].

iii. **Japan**

The Pharmaceuticals and Medical Devices Agency (PMDA) and the Ministry of Health, Labor, and Welfare (MHLW) oversee medicine regulation in Japan. The Japanese regulatory organizations have not developed a definition or regulations unique to nanomedicines. Guidelines for the creation of liposome medicinal products were released in 2016. The Pharmaceutical Affairs Law framework governs ordinary medical products and regulates nanomedicines individually. It should be mentioned that data on nanomedicine is being compiled and examined by reviewers and regulators. The EMA and the MHLW/PMDA have also worked together to publish reflection papers, particularly on creating drug products based on nanotechnology loaded with nucleic acids (siRNA) and block-copolymer micelle medical goods [59].

iv. **Others**

Although there isn't much regulation in this sector in Asia, countries like China, India, and Thailand are already attempting to set up regulatory and governance structures to deal with the growing issues in the field of nanotechnology. The Department of Science and Technology and the Government of India formed a panel to govern nanotechnology and prepare recommendations to assist policymakers in establishing a roadmap for nanomedicine regulation. The implementation of this three-tiered governance structure has occurred. This addresses the risks associated with nanomedicine while ensuring the ongoing advancement of technology [59].

4.2 Standards and Quality Assurance in Nanomedicine Development

A crucial component of the preclinical evaluation is the examination of the pharmaceutical product's quality, which includes physicochemical characterization, stability, purity, and formulation sterility. The relevant recommendations are provided by the International Conference on Harmonization of Technical Requirements for Pharmaceutical Registration for Human Use (ICH). It may be particularly difficult to characterize the physicochemical properties of drugs enabled by nanotechnology, which comprise several attributes like size distribution, shape, surface coating and charge, hydrophobicity, chemical composition, dispersion stability, and others. Furthermore, it is crucial to evaluate their properties in both their pristine and pertinent biological conditions because the presence of organic molecules may change them and result in formation of a protein corona, dissolution, aggregation, or sedimentation [60]. Many methods for characterizing particles can produce unreliable findings due to the extremely small size and special characteristics of nanoparticles (NPs), especially when proteins are present, or the media is complex [61].

The order in which the nanomaterial is characterized is also important characterization techniques that make use of high radiation or high voltage to analyze the samples, like electron microscopy, zeta potential, ICP, atomic emission spectrometry, mass spectrometry, and FIB, should be performed after the sample has been analyzed using nondestructive characterization techniques in order to prevent altering the nanomaterial's physicochemical properties [62]. Since FDA approval (or something similar) is one of the explicit translational goals of nanomedicine, there are specific qualities that nanomedicines must possess to be eligible for approval. The production process must adhere to good manufacturing practice (GMP) and good laboratory practice (GLP), which means that it must be both reproducible and, at the very least, theoretically scalable. These design concepts are pertinent to researchers studying nanomedicine [63]. The field should gather a vast amount of analyzable data as nanomedicine takes a stricter approach to documentation (i.e., standards). This would then enable machine learning and artificial intelligence to see patterns in the data and forecast which basic nano-bio interactions might be crucial for enabling translation, which can then inform research on the mechanisms of action of nanomedicine [64, 65].

5 Pre-Clinical Development of Nanomaterials

The usage and application of materials in the nanometer range is the basis of the broad definition of nanotechnology, as shown in Fig. 5. This extremely small range offers numerous advantage66s for a variety of applications, such as medicine delivery, electronics, sunscreen, cosmetics, energy storage, and more.6. When these particles are in their nanoscale form, their size frequently confers special and desirable qualities, such as chemical, physical, and biological properties that may be advantageous over their bigger counterparts. Given that many biologically important components, including water, proteins, carbohydrates, enzymes, hemoglobin, antibodies, and receptors [66].

Currently, a variety of nanostructured materials, such as magnetic nanoparticles (NPs), quantum dots, plasmonic NPs, SiO_2, carbonic nanomaterials, polymeric NPs, and chalcogenides metals NPs, have been successfully studied for potential uses in the treatment of leukemia cancer due to the rapid advancement of bio-nanotechnology [67].

AI methods have been extensively used to assess risk and evaluate the safety of preclinical nanomaterials. By carefully developing nanomaterials, adverse effects of nanomaterials on living things—including people, animals, plants, algae, and the environment—present at various food chain levels can be avoided. Premarketing (preclinical) evaluations are therefore required to prevent the use of hazardous nanomaterials (nanomedicines) in consumer goods. The primary factor restricting the use of nanomaterials in consumer goods, particularly in medicine, is still higher

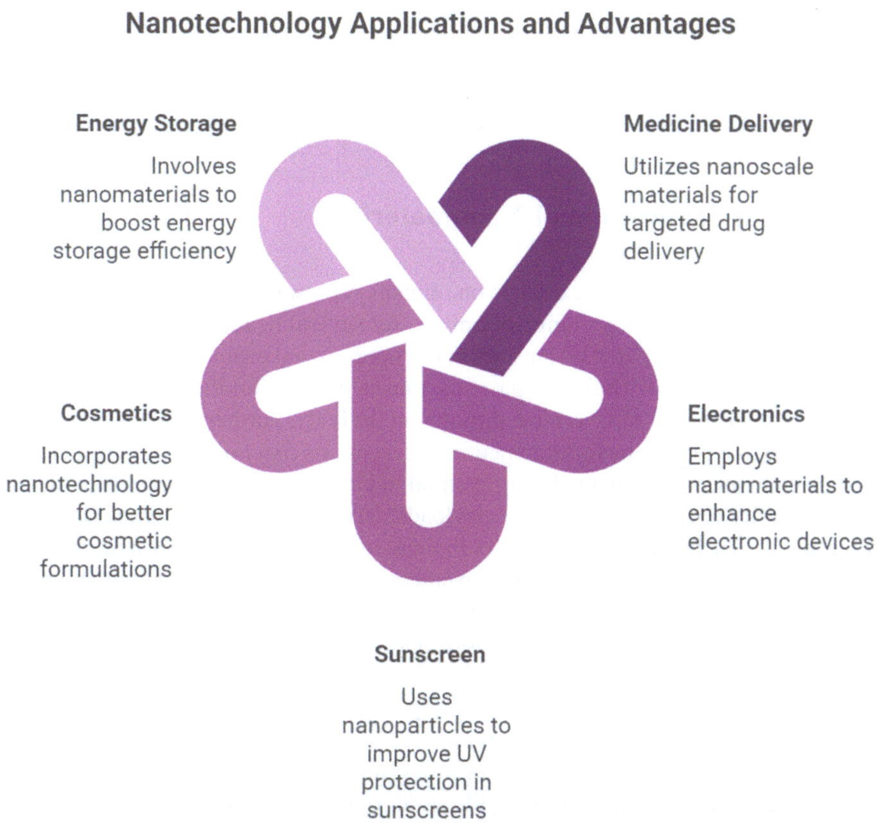

Fig. 5 Nanotechnology applications

nanotoxicity in ambient matrices and during clinical studies. One of the main obstacles to the commercialization of innovative nano-enabled goods is nanomaterial safety. Unexpected harm contributes to unnecessary morbidity and mortality and is a major cause of attrition and concern regarding the spread of nanomaterials into consumer goods, such as medications and nanotherapeutics. Unexpected results from exposure to nanoparticles are known as adverse nanomaterial responses and may vary depending on the physicochemical combinations. Before nanomaterials are included into everyday items, a supplementary system for reporting their safety must exist. To make sure that the safety information for nanoparticles (nanomedicine) is current, the fate of nanomaterials should be tracked through adverse effect (AE) reports and adverse outcome routes once the nano-enabled product is put on the market. This is known as nanomaterial vigilance (NV). By evaluating the adverse effects of nanomaterials on the entire biological system rather than just specific targets, enzymes, metabolites/transcription factors, system toxicology explains how nanomaterials function from the standpoint of system biology. This method reveals the unidentified nanomaterial AE that arises from the complex interaction of different toxicological end points (targets and routes). Since system nanotoxicology (safety and risk assessment) mainly offers the possibilit of off targeting induced bad effects (from a clinical point of view), its applications to adverse nanomaterial (nanomedicine) actions differ from those in drug discovery [67].

6 Clinical Translation of Nanomaterials

Advanced nanoparticle clinical trials on humans represent a major advancement in medical research, offering a range of options for several healthcare domains. These nanoscale-engineered particles have special properties that enable accurate medicinal treatments (see Fig. 6). Their compact size makes it easier for them to enter cellular and subcellular spaces and circulate throughout the body. Furthermore, nanoparticles provide flexible platforms for therapeutic interventions that range from regenerative medicine to cancer treatment. Their potential in customized medicine is highlighted by their ability to deliver therapeutic payloads, modify cellular activity, and interact at the molecular level [68, 69]. The combination of nanotechnology and medicine holds promise for revolutionary improvements in patient care, bringing in an era of precision healthcare, as human clinical trials continue to reveal the varticle uses [70]. In the context of medicine, nanotechnology generally refers to the atomic-level design, production, anvariouszation of molecules for a variety of applications, such as guiding cellular activities, preventing enzymatic degradation, and managing triggered release. The state-of-the-art nanomaterials for tissue regeneration, diagnosis, and treatment are highlighted in this summary review. These nanomaterials fall into three primary categories: lipid-based, polymeric, and inorganic. Quantum dots (QDs), carbon nanotubes (CNTs), and inorganic nanoparticles (INPs) have all been used to create a broad range of novel diagnostic nanomaterials. The two broad categories of therapeutic nanomaterials are soft and hard. Soft nanomaterials are mostly made up of liposomes and polymers, whereas hard nanomaterials are typically defined as metallic, ceramic, and oxide nanoparticles. One important and growing area of nanomedicine is the use of natural and synthetic polymers for medical applications in imaging, diagnostics, and drug delivery, with many polymer nanodrugs already licensed by the FDA and even more in clinical research [70].

Polymeric NPs are perhaps the most easily generated and widely applicable product in the entire industry since they can vary over the entire nanoscale from a single polymer chain to large aggregates, depending on the therapeutic target [71]. Generally speaking, polymer nanodrugs can be divided into two categories: (1) polymer-drug conjugates that enhance drug solubility, biocompatibility, and half-life and circulation time in vivo, and (2) slowly dissolving polymer forms for controlled-release applications. This is the basic, first-generation polymeric NP that was first developed in the 1960s [70]. By allowing the preferential delivery of drugs to the tumor, nanotechnology holds great promise for improving selectivity in targeting neoplastic cells due to the enhanced permeability and retention (EPR) effect caused by the leaky vasculature of tumors and inflammatory tissue [70].

Phase I: To assess a new biomedical intervention's safety (e.g., to identify side effects and determine a safe dosage range), it is first tried in a small group of patients.

Phase II: To ascertain effectiveness and further assess its safety, a larger number of human volunteers (hundreds) are used in the study of the behavioral or biological intervention.

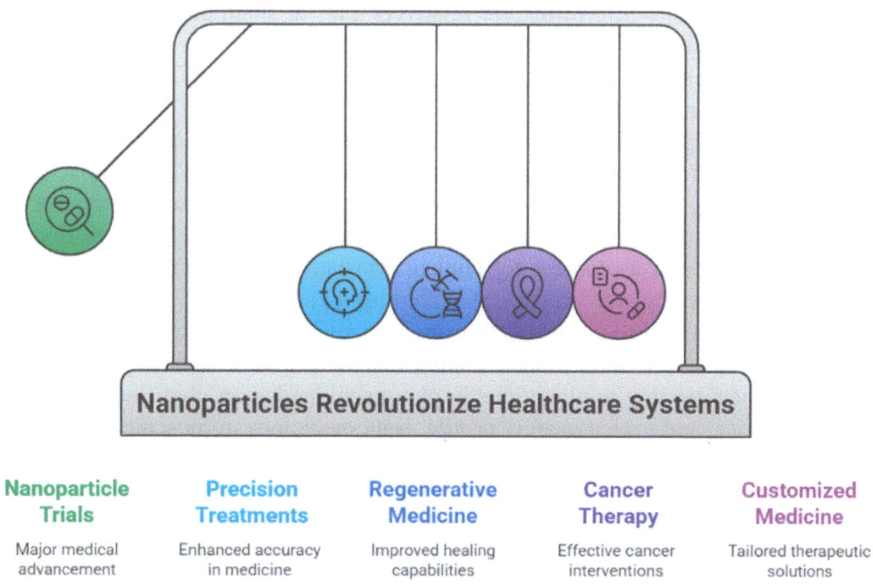

Fig. 6 Nanoparticles revolutionize healthcare systems

Phase III studies: During this stage, the behavioral intervention's or the biomedical intervention's effectiveness is thoroughly examined in larger groups of human subjects (from several hundred to several thousand) by comparing it to other conventional or experimental interventions, keeping an eye out for any negative side effects, and gathering information that will enable the intervention's safe use.

Phase IV studies: Often referred to as post-marketing surveillance trials, these are carried out following a medicine or device's approval for sale to consumers. Pharmaceutical companies have several goals during this phase, including evaluating the drug's efficacy and long-term effects on the patient's quality of life, as well as comparing it to other medications currently on the market. Additionally, the drug's cost-effectiveness is assessed in relation to conventional and novel treatments. Depending on the findings of phase IV research, medication or technology may be removed from the market or subject to limited use [70].

7 Regulatory Challenges in Translating Nanomedicines

There are difficulties in producing and scaling up nanomedicines successfully in controlled environments [72–75]. Numerous processes are frequently used in the nano formulation process, including homogenization, sonication, centrifugation, solvent evaporation, extrusion, size reduction, lyophilization, sterilization, and others, are frequently used in the nano formulation process. Reproducibility and formulation/process optimization are relatively easy at small scales but achieving a robust manufacturing process and minimizing batch-to-batch variability at large scales is a problem. This significantly restricts the application of nanomedicine in clinical and/or commercial settings [76]. Critical physicochemical features like size, crystallinity, drug-loading/-release profile, surface charge, and chemistry can all be significantly impacted by minute changes in formulation and/or manufacturing method, which can eventually impact therapeutic results. The lack of appropriate and trustworthy screening platforms that can assess toxicity and efficacy with good in vivo correlation is a problem for clinical translation of nanomedicine [73, 74]. This is especially true for preclinical screening methods and models that have a significant connection to clinical data from humans. Slight variations in the manufacturing process can cause modest changes in the physicochemical properties of nanomedicines, leading to alterations in pharmacokinetics, in vivo clearance, drug release kinetics, and toxicity/safety profile. Nanomedicines' minuscule size has the potential to significantly impact medication Adsorption, Distribution, Metabolism, and Excretion (ADME). Product interactions with the biological system are characterized by both in vitro and in vivo studies; however, the results of contemporary toxicity tests in preclinical models are not always applicable to humans. The potential of nanomedicine products in clinical practice has been hindered by the absence of clear regulatory criteria for preclinical development and biophysiological assessment [73, 74], as shown in Fig. 7.

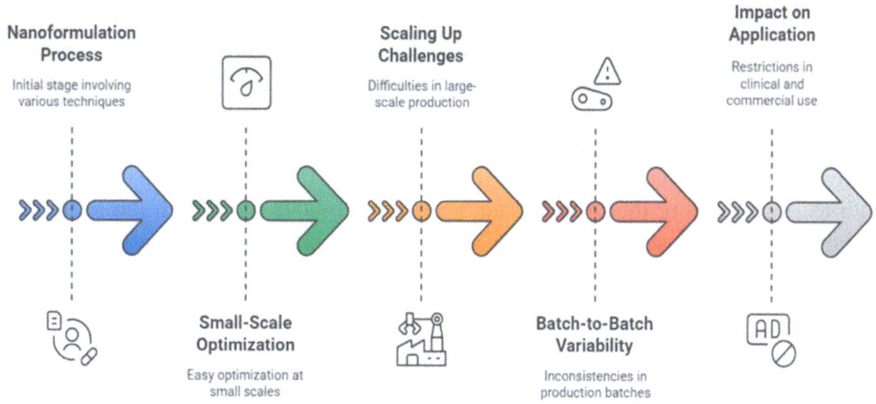

Fig. 7 Challenges in nanomedicine production

The complexity of the product and the frequent adaptation of small-molecule drug evaluation procedures are the leading causes of regulatory pathway barriers for the translation of nanomedicines [77]. Therefore, it is imperative to create a thorough set of assays that address nanomedicines' pharmacology, toxicity, and characterization. Additionally, whether the product satisfies the requirements of a scientifically recognized definition of what constitutes a nanomedicine is a critical challenge for nanomedicine translation. The FDA, EMA, and other medical regulatory bodies recommend that nanomedicine products be assessed and approved case-by-case because the sector is complicated and changing quickly. In order to promote innovation, the best strategy will be adaptable, iterative, scientific, team-oriented, and product-focused [78].

The introduction of nanotechnology-based treatments and pharmaceuticals is set to revolutionize medicine. Nanomedicine R&D involves around 200 firms worldwide, with the majority being small-to-medium-sized businesses or startups as opposed to big pharmaceutical companies [71, 79]. Nanomedicine's revolutionary promise is accompanied by possible risks due to its lack of testing. This new technology raises various ethical issues. Legal issues, privacy, and intellectual property (IP). Can a nanomedicine trial participant's informed consent be genuine if the researchers running the study also provide it do not grasp the full range or nature of the dangers involved [80]. Nanomedicine raises serious ethical concerns that require open dialogue between researchers and the public. To effectively address privacy concerns, biotechnology businesses and healthcare providers must use a bottom-up approach rather than relying solely on top-down legislation from authorities. The ailing IP system is a prime example of legislating without industry participation, as inefficient laws have been upheld despite concerns [81]. Policymakers and businesses must establish a conversation to safeguard nanomedicine breakthroughs while allowing for future advancement. Legislators, businesses, and the general public must work together to effectively regulate nanomedicine and develop a regulatory agency to handle the technology's possible risks.

Although nanotechnology and nanomedicine threaten individual privacy, it is crucial to recognize the value of privacy and consider the historical and legal contexts. Accounting for costs is critical in evaluating benefits and minimizing expenditures. To address privacy concerns, it's essential to include all relevant stakeholders in a discourse [82]. As nanomedicine advances and new fabrication methods emerge, innovators and discoverers may want to safeguard and claim ownership of their innovations. Recognizing their intellectual property provides them with the protection and ownership they seek. Intellectual property owners can seek a court remedy for infringements and, in certain situations, a temporary monopoly on the use and sale of their inventions, which the government protects. The four intellectual property (IP) categories are trade secrets, trademarks, copyrights, and patents. Each kind offers distinct safeguards for a particular IP and varies in duration [83]. Trade secrets, patents, copyrights, and trademarks are all considered intellectual property (IP) forms. Each kind differs regarding protected subject matter, degree of protection, and disclosure requirements. Generally speaking, disclosure laws are closely related to security, including patents and trade secrets [84].

The USPTO grants patents to inventors of "new and useful processes, machines, manufactures, or compositions of matter, or any new and useful improvements" [84]. Patents are classified into design, utility, and plant [84]. Patents are classified into three types: design patents, utility patents, and plant patents. Design patents protect the appearance of an innovation, utility patents preserve its functionality, and plant patents cover asexually generated plant species [84]. For various reasons, patenting nanomedicines and treatments based on nanotechnology is difficult at the USPTO. The USPTO's concept of nanotechnology is debatable and ambiguous [85]. The FDA and the USPTO use the definition of nanotechnology proposed by the National Nanotechnology Initiative (NNI) [85]. The National Nanotechnology Initiative defines nanotechnology as "Understanding and controlling matter at dimensions of one to one hundred nanometers, where unique phenomena enable novel applications" [86].

Despite their challenges, patents continue to be critical for nanomedicine development. Most applicants seek a patent to obtain a government-granted monopoly on the sale and use of their product. Obtaining a patent can help recover R&D costs for nanomedicine development, while failing may lead to market failure due to generic competition [87]. Patents establish startups' reputation with investors and typically precede funding [88, 89]. Companies increasingly invest in acquiring and protecting patents, which is understandable given their significance [89]. In addition to gaining a monopoly on sales and usage, patent applicants actively pursue nanomedicine to control significant portions of the developing field [88]. The holder of a patent has the sole right to license. A company or individual can charge others for using a fundamental nanotechnology component if they have a patent. A company can create a licensing toll booth for the entire industry with the proper patent [88].

8 Ethical Considerations and Societal Impact

The treatment or enhancement argument is frequently central to nanomedicine-related issues, including poor risk assessments in R&D, distributional disparities, and normative concerns in application and ultimate use [71, 90]. Before therapies based on nanotechnology are put on the market, the R&D stage includes premarket research, laboratory testing, and testing on humans and animals. All regulatory requirements for medical items prior to approval are covered at this stage. The primary ethical concern with nanomedicine R&D is insufficient clinician and subject risk assessment, as shown in Fig. 8. Nanomedicines pose unknown dangers to participants, researchers, and manufacturers due to their recent development [91]. This hinders informed consent, a crucial aspect of ethical clinical research [90].

Examining ethical challenges throughout the life cycle of nanotechnology-based treatments through the prism of lay ethics provides a more comprehensive and accessible understanding. One issue with the institutional study of ethics is the use of jargon and obscure concepts that are difficult for the public and scholars from other fields to understand. Ethicists may deny it, yet the public widely understands lay ethics. A society's history and social experiences shape this kind of ethics, which is frequently explained without using ideas like deontology or the categorical imperative. Technology narratives often reflect society's ambivalence towards emerging technologies, as seen with nuclear and biotechnology [91]. As the area of nanomedicine develops, it will surely achieve previously unimaginable things, necessitating a reevaluation of risk, hope, harm, and reward to account for these circumstances [94]. However, a rigidly utilitarian paradigm is insufficient to address the ethical concerns surrounding nanomedicine; instead, the moral framework must consider the values of justice, autonomy, and personal integrity [95].

As nanotechnology advances, social scientists are increasingly interested in analyzing public perceptions of its benefits and risks [96–112]. Nanotechnology, which is still in its exploratory stage and primarily used in academic labs, provides

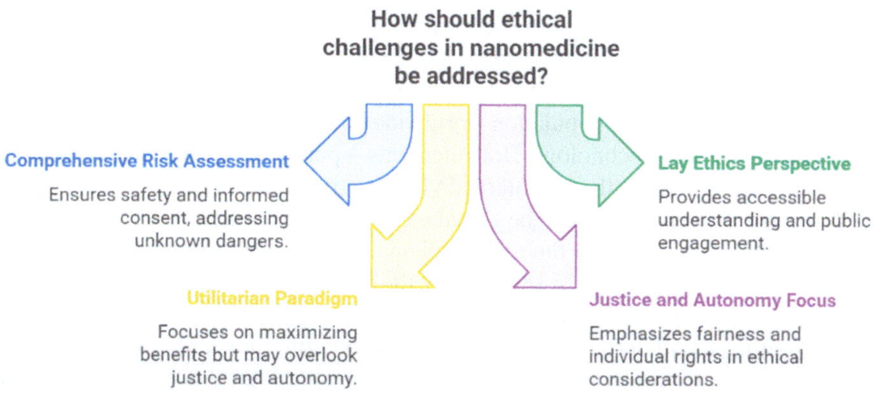

Fig. 8 Ethical challenges in nanomedicines

a unique opportunity to learn about public perception before it becomes a part of everyday life. This also prevents any potential disasters [113]. Conventional risk perception literature suggests that nonfamiliarity and psychometric characteristics (e.g., fear, no observability, novelty, and involuntary exposure) might cause significant public worry [114, 115]. However, the perception model does not describe public attitudes towards nanotechnology. Research indicates that despite poor public awareness of nanotechnology and the intangibility of its products, there is a widespread positive perception of it in the countries questioned. The study found that perceptions of nanotechnology are influenced by factors other than familiarity and psychometric characteristics [108].

9 Future Directions and Innovations in Nanomedicine

The medical use of nanotechnology and drug delivery systems based on nanoparticles for the treatment of cancer over the previous 20 years is described by nanomedicine research. In essence, nanomedicine is a byproduct of nanotechnology, a more recent scientific field. The multidisciplinary scientific field of nanotechnology is revolutionizing the way that illnesses are identified in the human body and damage is treated. Nanomedicine is used in particular medical applications for illness prevention, diagnosis, and therapy. In order to change medicines and enhance current treatment methods, the emerging field of nanomedicine combines active pharmacological agents with nanotechnology. These nanomedicines can improve the delivery of the integrated medicinal components to particular tissues and organs at a predetermined rate by overcoming the biological barriers in the human body. More specifically, it has been noted that nanomedicines alter the way therapeutic compounds are absorbed by cells and tissues, improving the compounds' biodistribution to target areas in vivo. In nanomedicine, pharmaceutical nanostructures of the same size are created by manipulating active biomolecules and their formulations to have a specific positive effect on humans. The dynamic of the nanomedicine is shown in Fig. 9.

These nanomedicines will improve the treatment of diabetes, cancer, Alzheimer's, Parkinson's, and cardiovascular disorders since they provide a great way to diagnose illnesses early and accurately. Only the approval of regulatory bodies and the sincere acceptance of the general population worldwide could truly gauge the factual potential of this innovative technology. Both elements—public acceptance and regulatory approval—are crucial to the commercial viability of nanomedicines [116].

A recent study estimates that the global nanomedicine market is over $150 billion. Furthermore, it predicts that the nanomedicine market, particularly related to anticancer products, will expand by over 15% in the coming years. It is also predicted that in the future, nanomedicine, especially related to central nervous systems and anticancer products, will dominate the market by accounting for nearly 40% of revenue. For the successful implementation of these nanotechnologies, it is imperative that basic research be conducted to address these challenges. Instead of imposing applications for some currently popular materials, the future of nanomedicine will depend

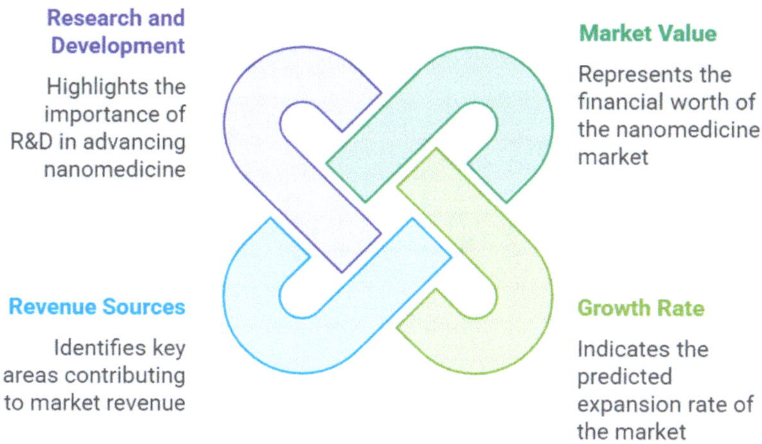

Fig. 9 Dynamics of the nanomedicine market

on the rational development of various materials for creating nanotechnology and advanced tools based on an effective and comprehensive understanding of biology processes.

10 Role of Artificial Intelligence in Nanomedicine Development

Numerous nanomedicine platforms have been investigated for use in imaging and medication administration. These include carbon-based platforms, silicon, lipids, polymers, metals, and natural compounds. These nanomedicine-functionalized treatments and imaging agents have been used to treat a variety of cancers, heart conditions, regenerative medicine, and other conditions. The ability to overcome efflux-mediated drug resistance, better drug delivery localization through different targeting moieties, and higher magnetic resonance imaging contrast efficiency are some advantages that have been shown with nanotechnology-enhanced drugs [90].

10.1 AI Techniques in NPs Design

AI methodologies, such as ML, neural networks, and optimization algorithms, aid in designing NPs with ideal characteristics, including size, shape, surface charge, and drug release profiles. Different AI approaches (Fig. 10) are employed to represent LNPs in various dimensions by applying DL methods like convolutional neural networks (CNNs), recurrent neural networks (RNNs), and transformers to forecast

Fig. 10 AI Application in nanoparticle formulation

their characteristics. ML was used to guide the design of LNPs for delivering mRNA. By employing a curated dataset of 622 published LNPs, the capability of the model to predict the transfection efficiency of novel LNPs was demonstrated accurately; the multilayer perceptron model attained a classification accuracy of 97% on the test set; this method considerably improves the screening efficiency by employing computational methods to prioritize LNP candidates for experimental testing, thus speeding up the development of effective mRNA delivery systems [91].

10.2 ML Models for Predicting NP Properties

The evaluation of LNPs presents significant challenges. Densely qualitative results due to the manual analysis of conservative, traditional electron microscopy often yield highly qualitative results for sample populations. Five methods including dynamic light scattering offer average readouts across an aggregated particle population, which can be limiting. ATEM structural discovery (ATEM) has led the way in employing AI to overcome the challenges associated with cryogenic electron microscopy (cryo-EM). By applying AI, ATEM has effectively resolved the issue of low signal–noise ratios in cryo-EM data. Some Researchers had Explored an evolutionary computational platform aimed at the automatic discovery of nanocarriers specifically designed for cancer treatment. The researchers developed EVONANO, an open-source platform that integrates multiscale and modular simulations with advanced ML techniques. This allows for the assessment of the properties of both nanocarriers and treatments within biologically pertinent tissue models, which are fabricated employing a virtual tumor generated in silico. EVONANO features a simulator capable of developing tumors, extracting representative scenarios, and modelling the transport of NPs through these scenarios to forecast their distribution. They showed in this first such study that ML can be used to recognize NP configurations capable of efficiently targeting and eradicating evident portions of tumour tissue while maintaining minimal dosages. The outcome exhibited that NPs with high diffusion rates (smaller radius) and low binding affinities (higher KD values) are most effective for drug delivery. Some study predicted the delivery of NPs to tumors by employing ML methodologies. This research presented a novel DL neural network model grounded in PBPK principles that precisely predicts the efficacy of NP delivery to tumour sites in mice. The effectiveness of different NPs in delivering drugs to dissimilar tumour types was precisely forecasted by a deep neural network model, surpassing the performance of alternative ML approaches such as random forest, support vector machine, linear regression, and bagged model methods. The revised coefficients of determination (R2) in the complete training set were 0.93, 0.78, 0.79, and 0.77 for the maximum delivery efficiency (DEmax), delivery efficiency after 24 h (DE24), after 168 h (DE168), and at the final sampling time (DETlast). The R2 values for the test dataset were 0.71, 0.47, 0.34, and 0.64, respectively. This research showed that the type of cancer considerably influenced the accuracy of the deep neural network model in predicting the effectiveness of tumours delivery across different measures, with improvements ranging from 20 to 30% [91].

11 Conclusion

In conclusion, successfully translating nanomaterials from the laboratory to clinical practice remains a complex and multifaceted challenge. Integrating innovative nanotechnologies into medicine has the potential to revolutionize treatments across various therapeutic areas, but overcoming regulatory hurdles is crucial for their widespread application.

11.1 Bridging the Regulatory Gap for Effective Clinical Translation

A major obstacle to the clinical application of nanomedicines is the absence of well-defined, uniform regulatory structures in many international jurisdictions. It is necessary to create thorough regulations that can be modified to accommodate the special characteristics of nanomaterials to close this regulatory gap. Refinement of safety assessments, risk assessment procedures, and clinical trial designs customized for nanomedicines must be the focus of efforts. Facilitating communication and cooperation among regulatory agencies like the FDA, EMA, and WHO will also guarantee more unified regulatory processes, which will promote a more effective road to market for potential nanomedicines.

11.2 Collaborative Efforts for Sustainable Nanomedicine Development

Researchers, physicians, regulatory bodies, and industry stakeholders must work together to develop nanomedicine, which is by its very nature an interdisciplinary field. Overcoming the field's technical, moral, and legal obstacles will need these cooperative efforts. Sharing best practices, resources, and information will be encouraged by establishing communication between sectors and participating in public–private partnerships. Additionally, ensuring that regulatory bodies have the skills and resources necessary to properly evaluate nanomaterials will contribute to the establishment of a long-lasting framework that facilitates the ongoing research and clinical application of nanomedicines, ultimately improving healthcare solutions for the benefit of patients everywhere. Overcoming these regulatory obstacles with cooperation, creativity, and a dedication to guaranteeing patient safety and effectiveness in novel therapeutic techniques will be crucial to the future of nanomedicine.

References

1. Moreira, A.F.: Nanomaterials in drug delivery applications. Nanomaterials **12**(20), 3565 (2022). https://doi.org/10.3390/nano12203565
2. Damodharan, J.: Nanomaterials in medicine—an overview. Mater. Today Proc. **37**, 383–385 (2021). https://doi.org/10.1016/j.matpr.2020.05.380
3. Teleanu, D.M., Negut, I., Grumezescu, V., Grumezescu, A.M., Teleanu, R.I.: Nanomaterials for drug delivery to the central nervous system. Nanomaterials **9**(3), 371 (2019). https://doi.org/10.3390/nano9030371
4. Zhang, G., Zeng, X., Li, P.: Nanomaterials in cancer-therapy drug delivery system
5. Park, J., You, X., Jang, Y., et al.: ZnO nanorod matrix based electrochemical immunosensors for sensitivity enhanced detection of Legionella pneumophila. Sens. Actuators B Chem. **200**, 173–180 (2014). https://doi.org/10.1016/j.snb.2014.03.001.[crossreference]

6. Kim, D.K., Chang, J.H., Kang, Y.J.: Efficient internalization of peptide-conjugated SPIONs in dendritic cells for tumor targeting. J. Nanosci. Nanotechnol. **12**(7), 5191–5198 (2012). https://doi.org/10.1166/jnn.2012.6379[crossreference]
7. Li, L., Huang, X., Liu, T., et al.: Overcoming multidrug resistance with mesoporous silica nanorods as nanocarrier of doxorubicin. J. Nanosci. Nanotechnol. **12**(6), 4458–4466 (2012). https://doi.org/10.1166/jnn.2012.6198.[crossreference]
8. Liu, P., Wang, H., Wang, Q., et al.: CRGD Conjugated mPEG-PLGA-PLL nanoparticles for SGC-7901 gastric cancer cells-targeted delivery of fluorouracil. J. Nanosci. Nanotechnol. **12**(6), 4467–4471 (2012). https://doi.org/10.1166/jnn.2012.6213.[crossreference]
9. Kato, S., Kimura, M., Kageyama, K., Tanaka, H., Miwa, N.: Enhanced radiosensitization by liposome-encapsulated pimonidazole for anticancer effects on human melanoma cells. J. Nanosci. Nanotechnol. **12**(6), 4472–4477 (2012). https://doi.org/10.1166/jnn.2012.6180[crossreference]
10. Esim, O., Kurbanoglu, S., Savaser, A., Ozkan, S.A., Ozkan, Y.: Chapter 9—Nanomaterials for drug delivery systems. ScienceDirect (2019). https://www.sciencedirect.com/science/article/abs/pii/B9780128161449000092
11. Sainz, V., Conniot, J., Matos, A.I., et al.: Regulatory aspects on nanomedicines. Biochem. Biophys. Res. Commun. **468**(3), 504–510 (2015). https://doi.org/10.1016/j.bbrc.2015.08.023
12. Wagner, V., Dullaart, A., Bock, A.K., Zweck, A.: The emerging nanomedicine landscape. Nat. Biotechnol. **24**(10), 1211–1217 (2006). https://doi.org/10.1038/nbt1006-1211
13. Gaspar, R.S.: Therapeutic products: Regulating drugs and medical devices (2010). https://www.researchgate.net/publication/287551711_Therapeutic_products_Regulating_drugs_and_medical_devices
14. Gaspar, R., Florindo, H.F., Silva, L.C., et al.: Regulatory aspects of oncologicals: nanosystems main challenges, pp. 425–452 (2014). https://doi.org/10.1007/978-3-319-08084-0_15
15. Environmental and safety impacts of nanotechnology: what research is needed? Hearing before the committee on science house of representatives one hundred ninth congress first session (2005). Accessed Feb 5, 2025. https://www.govinfo.gov/content/pkg/CHRG-109hhrg24464/pdf/CHRG-109hhrg24464.pdf
16. Size Matters! ETC Group. Published April 30, 2003. Accessed Feb 5, 2025. https://www.etcgroup.org/content/size-matters
17. Chan, V.S.W.: Nanomedicine: an unresolved regulatory issue. Regul. Toxicol. Pharmacol. **46**(3), 218–224 (2006). https://doi.org/10.1016/j.yrtph.2006.04.009
18. Sorrentino, C., Ciummo, S.L., Fieni, C., Carlo, E.D.: Nanomedicine for cancer patient-centered care. MedComm **5**(11) (2024). https://doi.org/10.1002/mco2.767
19. Souto, E.B., Blanco-Llamero, C., Krambeck, K., et al.: Regulatory insights into nanomedicine and gene vaccine innovation: safety assessment, challenges, and regulatory perspectives. Acta Biomater. (2024). https://doi.org/10.1016/j.actbio.2024.04.010
20. Halamoda-Kenzaoui, B., Holzwarth, U., Roebben, G., Bogni, A., Bremer-Hoffmann, S.: Mapping of the available standards against the regulatory needs for nanomedicines. Wiley Interdisc. Rev. Nanomed. Nanobiotechnol. **11**(1), e1531 (2018). https://doi.org/10.1002/wnan.1531
21. Faraday, M.X.: The Bakerian lecture—experimental relations of gold (and other metals) to light. Philos. Trans. R. Soc. Lond. **147**, 145–181 (1857). https://doi.org/10.1098/rstl.1857.0011
22. Link, S., El-Sayed, M.A.: Spectral properties and relaxation dynamics of surface plasmon electronic oscillations in gold and silver nanodots and nanorods. J. Phys. Chem. B **103**(40), 8410–8426 (1999). https://doi.org/10.1021/jp9917648
23. Alivisatos, A.P.: Perspectives on the physical chemistry of semiconductor nanocrystals. J. Phys. Chem. **100**(31), 13226–13239 (1996). https://doi.org/10.1021/jp9535506
24. Nie, S.: Probing single molecules and single nanoparticles by surface-enhanced raman scattering. Science **275**(5303), 1102–1106 (1997). https://doi.org/10.1126/science.275.5303.1102
25. Weissleder, R.: A clearer vision for in vivo imaging. Nat. Biotechnol. **19**(4), 316–317 (2001). https://doi.org/10.1038/86684

26. Esim, O., Kurbanoglu, S., Savaser, A., Ozkan, S.A., Ozkan, Y.: Nanomaterials for Drug Delivery Systems, pp. 273–301 (2019). https://doi.org/10.1016/b978-0-12-816144-9.00009-2
27. Gabizon, A.A., Shmeeda, H., Zalipsky, S.: Pros and cons of the liposome platform in cancer drug targeting. J. Liposome Res. **16**(3), 175–183 (2006). https://doi.org/10.1080/08982100600848769
28. dos Santos Giuberti, C., de Oliveira Reis, E.C., Ribeiro Rocha, T.G., et al.: Study of the pilot production process of long-circulating and pH-sensitive liposomes containing cisplatin. J. Liposome Res. **21**(1), 60–69 (2010). https://doi.org/10.3109/08982101003754377
29. Bhardwaj, A.K., Pandit, A.K., Rehalia, A., Singh, V., Sharma, R.: A review on nanomaterials for drug delivery systems and application of carbon based nanomaterials. ES Mater. Manuf. (2023). https://doi.org/10.30919/esmm5f824
30. Sahoo, S.K., Labhasetwar, V.: Nanotech approaches to drug delivery and imaging. Drug Discov. Today **8**(24), 1112–1120 (2003). https://doi.org/10.1016/s1359-6446(03)02903-9
31. Gabizon, A., Goren, D., Cohen, R., Barenholz, Y.: Development of liposomal anthracyclines: from basics to clinical applications. In: 8th International Symposium on recent Advances in Drug Delivery Systems, Salt Lake City, UT, USA (1997). J. Controll. Release **53**(1–3), 275–279 (1998). https://doi.org/10.1016/s0168-3659(97)00261-7
32. Garg, J., Pathania, K., Sah, S.P., Pawar, S.V.: Nanostructured lipid carriers: a promising drug carrier for targeting brain tumours. Future J. Pharm. Sci. **8**(1) (2022). https://doi.org/10.1186/s43094-022-00414-8
33. Yafout, M., Ousaid, A., Khayati, Y., El Otmani, I.S.: Gold nanoparticles as a drug delivery system for standard chemotherapeutics: a new lead for targeted pharmacological cancer treatments. Sci. African **11**, e00685 (2021). https://doi.org/10.1016/j.sciaf.2020.e00685
34. Lukianova-Hleb, E.Y., Wagner, D.S., Brenner, M.K., Lapotko, D.O.: Cell-specific transmembrane injection of molecular cargo with gold nanoparticle-generated transient plasmonic nanobubbles. Biomaterials **33**(21), 5441–5450 (2012). https://doi.org/10.1016/j.biomaterials.2012.03.0770
35. Eustis, S., El-Sayed, M.A.: Why gold nanoparticles are more precious than pretty gold: noble metal surface plasmon resonance and its enhancement of the radiative and nonradiative properties of nanocrystals of different shapes. Chem. Soc. Rev. **35**(3), 209–217 (2006). https://doi.org/10.1039/b514191e
36. Trzeciak, K., Chotera-Ouda, A., Bak-Sypien, I.I., Potrzebowski, M.J.: Mesoporous silica particles as drug delivery systems—the state of the art in loading methods and the recent progress in analytical techniques for monitoring these processes. Pharmaceutics **13**(7), 950 (2021). https://doi.org/10.3390/pharmaceutics13070950
37. Vallet-Regi, M., Rámila, A., del Real, R.P., Pérez-Pariente, J.: A new property of MCM-41: drug delivery system. Chem. Mater. **13**(2), 308–311 (2001). https://doi.org/10.1021/cm0011559
38. Garcia-Bennett, A.E.: Synthesis, toxicology and potential of ordered mesoporous materials in nanomedicine. Nanomedicine (Lond.) **6**(5), 867–877 (2011). https://doi.org/10.2217/nnm.11.82
39. Attia, N.F., El-Monaem, E.M.A., El-Aqapa, H.G., et al.: Iron oxide nanoparticles and their pharmaceutical applications. Appl. Surf. Sci. Adv. **11**, 100284 (2022). https://doi.org/10.1016/j.apsadv.2022.100284
40. Roca, A.G., Gutiérrez, L., Gavilán, H., Fortes Brollo, M.E., Veintemillas-Verdaguer, S., de Puerto Morales, M.: Design strategies for shape-controlled magnetic iron oxide nanoparticles. Adv. Drug Deliv. Rev. **138**, 68–104 (2019). https://doi.org/10.1016/j.addr.2018.12.008
41. Trindade, T., Thomas, P.J.: Defining and Using Very Small Crystals. Elsevier eBooks, pp. 343–369 (2013). https://doi.org/10.1016/b978-0-08-097774-4.00416-2
42. Tringides, M.C., Jałochowski, M., Bauer, E.: Quantum size effects in metallic nanostructures. Phys. Today **60**(4), 50–54 (2007). https://doi.org/10.1063/1.2731973
43. Li, Q., Kartikowati, C.W., Horie, S., Ogi, T., Iwaki, T., Okuyama, K.: Correlation between particle size/domain structure and magnetic properties of highly crystalline Fe_3O_4 nanoparticles. Sci. Rep. **7**(1) (2017). https://doi.org/10.1038/s41598-017-09897-5

44. Sun, Y., Gray, S.K., Peng, S.: Surface chemistry: a non-negligible parameter in determining optical properties of small colloidal metal nanoparticles. Phys. Chem. Chem. Phys. **13**(25), 11814 (2011). https://doi.org/10.1039/c1cp20265k
45. Yusefi, M., Shameli, K., Ali, R.R., Pang, S.W., Teow, S.Y.: Evaluating anticancer activity of plant-mediated synthesized iron oxide nanoparticles using punica granatum fruit peel extract. J. Mol. Struct. **1204**, 127539 (2020). https://doi.org/10.1016/j.molstruc.2019.127539
46. Europe PMC. Europepmc.org. Published 2016. Accessed February 5, 2025. https://europepmc.org/article/MED/32244817
47. Magdanz, V., Gebauer, J., Sharan, P., Eltoukhy, S., Voigt, D., Simmchen, J.: Sperm-particle interactions and their prospects for charge mapping. Adv. Biosyst. **3**(9), e1900061 (2019). https://doi.org/10.1002/adbi.201900061
48. Rui, M., Ma, C., Hao, Y., et al.: Iron oxide nanoparticles as a potential iron fertilizer for peanut (Arachis hypogaea). Front. Plant Sci. **7** (2016). https://doi.org/10.3389/fpls.2016.00815
49. Mishra, M., Prasad, K., Ramakrishn, S., Jha, A.K.: Nanomaterials in drug delivery—promises and limitations. Nano Med. Mater. **3**(1), 38 (2023). https://doi.org/10.59400/nmm.v3i1.38
50. Chavoshy, F., Makhmalzade, B.: Polymeric micelles as cutaneous drug delivery system in normal skin and dermatological disorders. J. Adv. Pharm. Technol. Res. **9**(1), 2 (2018). https://doi.org/10.4103/japtr.japtr_314_17
51. Kannan, R.M., Nance, E., Kannan, S., Tomalia, D.A.: Emerging concepts in dendrimer-based nanomedicine: from design principles to clinical applications. J. Intern. Med. **276**(6), 579–617 (2014). https://doi.org/10.1111/joim.12280
52. Khodabakhshi, M.J., Panahi, H.A., Konoz, E., Feizbakhsh, A., Kimiagar, S.: Synthesis of pH and thermo-sensitive dendrimers based on MoS_2 and magnetic nanoparticles for cisplatin drug delivery system by the near-infrared laser. Polym. Adv. Technol. **32**(4), 1626–1635 (2020). https://doi.org/10.1002/pat.5199
53. Parra, A., Jarak, I., Santos, A., Veiga, F., Figueiras, A.: Polymeric micelles: a promising pathway for dermal drug delivery. Materials **14**(23), 7278 (2021). https://doi.org/10.3390/ma14237278
54. Cho, H.K., Cho, J.H., Jeong, S.H., Cho, D.C., Yeum, J.H., Cheong, I.W.: Polymeric vehicles for topical delivery and related analytical methods. Arch. Pharm. Res. **37**(4), 423–434 (2014). https://doi.org/10.1007/s12272-014-0342-4
55. Debnath, S.K., Srivastava, R.: Drug Delivery with Carbon-Based Nanomaterials as Versatile-Nanocarriers: Progress and Prospects
56. Pita, R., Ehmann, F., Papaluca, M.: Nanomedicines in the EU—regulatory overview. AAPS J. **18**(6), 1576–1582 (2016). https://doi.org/10.1208/s12248-016-9967-1
57. Nanotechnology-Based Medicinal Products for Human Use EU-in Horizon Scanning Report Table of Content. Accessed February 5, 2025. https://www.ema.europa.eu/en/documents/report/nanotechnology-based-medicinal-products-human-use-eu-horizon-scanning-report_en.pdf
58. .Emily, M., Ioanna, N., Scott, B., Beat, F.: Reflections on FDA draft guidance for products containing nanomaterials: is the abbreviated new drug application (ANDA) a suitable pathway for nanomedicines? AAPS J. **20**(5) (2018). https://doi.org/10.1208/s12248-018-0255-0
59. Foulkes, R., Man, E., Thind, J., Yeung, S., Joy, A., Hoskins, C.: The regulation of nanomaterials and nanomedicines for clinical application: current and future perspectives. Biomater. Sci. **8**(17), 4653–4664 (2020). https://doi.org/10.1039/d0bm00558d
60. Halamoda-Kenzaoui, B., Ceridono, M., Colpo, P., et al.: Dispersion behaviour of silica nanoparticles in biological media and its influence on cellular uptake. PLOS ONE **10**(10), e0141593–e0141593 (2015). https://doi.org/10.1371/journal.pone.0141593
61. .Halamoda-Kenzaoui, B., Holzwarth, U., Roebben, G., Bogni, A., Bremer-Hoffmann, S.: Mapping of the available standards against the regulatory needs for nanomedicines. Wiley Interdiscip. Rev. Nanomed. Nanobiotechnol. **11**(1), e1531 (2018). https://doi.org/10.1002/wnan.1531
62. Salvador Morales, C., Khorasani, A.A., Weaver, J.: Closing the gap: accelerating the translational process in nanomedicine by proposing standardized characterization techniques. Int. J. Nanomed. 5729 (2014). https://doi.org/10.2147/ijn.s72479

63. Richardson, J.J., Caruso, F.: Nanomedicine toward 2040. Nano Lett. **20**(3), 1481–1482 (2020). https://doi.org/10.1021/acs.nanolett.0c00620
64. Brown, K.A., Brittman, S., Maccaferri, N., Jariwala, D., Celano, U.: Machine learning in nanoscience: big data at small scales. Nano Lett. **20**(1), 2–10 (2019). https://doi.org/10.1021/acs.nanolett.9b04090
65. Wagner, V., Hüsing, B., Gaisser, S.: Nanomedicine: Drivers for Development and Possible Impacts. http://nanotechradar.com/sites/default/files/isect7_nanomedicine_drivers_development.pdf
66. Ansari, A.A., Parchur, A.K., Thorat, N.D., Chen, G.: New advances in pre-clinical diagnostic imaging perspectives of functionalized upconversion nanoparticle-based nanomedicine. Coord. Chem. Rev. 213971 (2021). https://doi.org/10.1016/j.ccr.2021.213971
67. Biomater. Sci. **9**(5), 1598 (2021). https://doi.org/10.1039/d0bm01672a
68. Pandey, A., Singh, B.K., Gayathiri, E., et al.: Nanoparticles in biomedical and clinical research: a current perspective and future implications. In: Nanomaterials for Biomedical and Bioengineering Applications, pp. 415–457. Springer (2024)
69. Amjad, M.S., Fatima, R., Qureshi, H., Benish, Z.: The application of nanomaterials in biological and biomedical processes: advances and perspectives. In: Nanomaterial for Biomedicine and Bioengineering Applications, pp. 1–25 (2024). https://doi.org/10.1007/978-981-97-0221-3_1
70. Greish, K.: Enhanced permeability and retention (EPR) effect for anticancer nanomedicine drug targeting. Methods Mol. Biol. **624**, 25–37 (2010)
71. Atalla, K., Chaudhary, A., Eshaghian-Wilner, M.M., et al.: Ethical, privacy, and intellectual property issues in nanomedicine. In: Handbook of Clinical Nanomedicine: Law, Business, Regulation, Safety, and Risk, vol. 20 (2016). https://doi.org/10.1002/9781118993620.ch20
72. Agrahari, V., Hiremath, P.: Challenges associated and approaches for successful translation of nanomedicines into commercial products. Nanomedicine **12**(8), 819–823 (2017). https://doi.org/10.2217/nnm-2017-0039
73. Ragelle, H., Danhier, F., Preat, V., Langer, R., Anderson, D.G.: Nanoparticle-based drug delivery systems: a commercial and regulatory outlook as the field matures. Exp. Opin Drug Deliv. (2016). https://doi.org/10.1080/17425247.2016.1244187
74. Kaur, I.P., Kakkar, V., Deol, P.K., et al.: Issues and concerns in nanotech product development and its commercialization. J. Control. Release **193**, 51–62 (2014)
75. Muthu, M.S., Wilson, B.: Challenges posed by the scale-up of nanomedicines. Nanomedicine (Lond.) **7**(3), 307–309 (2012)
76. Desai, N.: Challenges in development of nanoparticle-based therapeutics. AAPS J. **14**(2), 282–295 (2012)
77. Sainz, V., Conniot, J., Matos, A.I., et al.: Regulatory aspects on nanomedicines. Biochem. Biophys. Res. Commun. **468**(3), 504–510 (2015)
78. Hamburg, M.A.: Science and regulation: FDA's approach to regulation of products of nanotechnology. Science **336**(6079), 299–300 (2012)
79. Wagner, V., Dullaart, A., Bock, A.K., Zweck, A.: The emerging nanomedicine landscape. Nat. Biotechnol. **24**(10), 1211–1217 (2006)
80. Wolf, S., Jones, C.: Designing oversight for nanomedicine research in human subjects: systematic analysis of exceptional oversight for emerging technologies. J. Nanopart. Res. **13**(4), 1449–1465 (2011)
81. Bawa, R.: Will the nanomedicine "patent land grab" thwart commercialization? Nanomed. Nanotechnol. Biol. Med. **1**(4), 346–350 (2005)
82. Oberdörster, G., Oberdörster, E., Oberdörster, J.: Nanotoxicology: an emerging discipline evolving from the study of ultrafine particles. Environ. Health Perspect. **113**(7), 823–839 (2005)
83. Emanuel, E., Wendler, D., Grady, C.: What makes clinical research ethical? JAMA **283**(20), 2701–2711 (2000)
84. Renn, O., Roco, M.: White paper on nanotechnology risk governance (2006). Accessed Jan 8, 2016. http://www.irgc.org/IMG/pdf/IRGC_white_paper_2_PDF_final_version-2.pdf
85. Bacchini, F.: Is nanotechnology giving rise to new ethical problems? NanoEthics **7**(2), 107–119 (2013)

86. Bottini, M., Rosato, N., Gloria, F., et al.: Public optimism towards nanomedicine. Int. J. Nanomed. **6**, 3473–3485 (2011). https://doi.org/10.2147/IJN.S26340
87. Singh, R., Nalwa, H.S.: Medical applications of nanoparticles in biological imaging, cell labeling, antimicrobial agents, and anticancer nanodrugs. J. Biomed. Nanotechnol. **7**, 489–503 (2011)
88. Bainbridge, W.S.: Public attitudes toward nanotechnology. J. Nanopart. Res. **4**, 561–570 (2002)
89. Mandal, S.D., Mandal, S., Pathak, Y.V., Patel, J.K.: Current challenges and future directions in nanomedicine. In: Patel, J.K., Pathak, Y.V. (eds.) Emerging Technologies for Nanoparticle Manufacturing. Springer (2021). https://doi.org/10.1007/978-3-030-50703-9_27
90. Ho, D., Wang, P., Kee, T.: Artificial intelligence in nanomedicine. Nanoscale Horizons **4**(2), 365–377 (2019). https://doi.org/10.1039/C8NH00233A
91. Kapoor, D.U., Sharma, J.B., Gandhi, S.M., Prajapati, B.G., Thanawuth, K., Limmatvapirat, S., Sriamornsak, P.: AI-driven design and optimization of nanoparticle-based drug delivery systems. Sci. Eng. Health Stud. 24010003–24010003 (2024). https://doi.org/10.69598/sehs.18.24010003
92. Palazzolo, S., Bayda, S., Hadla, M., et al.: The clinical translation of organic nanomaterials for cancer therapy: a focus on polymeric nanoparticles, micelles, liposomes, and exosomes. Curr. Med. Chem. **24**(10), 3755–3776 (2017). https://doi.org/10.2174/0929867324666170830113755

Pioneering the Future of Nanotechnology: Trends, Challenges, and Ethical Considerations in Biomedical Science

Milan Hait, Benvikram Barman, Nand Kumar Kashyap, Arvind Kumar Bhardwaj, Sarita Diwakar, and Amit Kumar Chaturwedi

Abstract Nanotechnology has the potential to revolutionize healthcare by expanding the boundaries of biological study. Nanotechnology has profoundly affected imaging, stem cell therapy, tailored medication administration, and early disease diagnosis. Some of the distinctive features of nanomaterials include improved contact with live organisms, a regulated medication release mechanism, and a more reactive surface. The book chapter covers recent developments and advancements, such as the use of nanoscale scaffolds for tissue engineering, nano biosensors for real-time monitoring, and the creation of multifunctional nanodevices. However, significant challenges remain, such as long-term biocompatibility, regulatory hurdles, cost-effectiveness, and scalability. These challenges include long-term biocompatibility, regulatory hurdles, cost-effectiveness, and scalability. We talk about some of the ethical issues surrounding nanomedical devices, including data privacy and equitable access to this innovative technology. We employ an interdisciplinary approach that prioritizes collaboration among researchers, doctors, and lawmakers to address these ethical challenges. As the chapter comes to a close, it stresses how important ethical guidelines and sustainable development is for getting the most out of nanotechnologies in biological research for the good of society.

Keywords Nanotechnology · Biomedical innovations · Ethical considerations · Drug delivery · Regenerative medicine · Real-time monitoring

Abbreviations

MIT Massachusetts institute of technology

M. Hait · N. K. Kashyap · A. K. Bhardwaj · S. Diwakar · A. K. Chaturwedi (✉)
Department of Chemistry, Dr. C. V. Raman University, Kota, Bilaspur, Chhattisgarh 495113, India
e-mail: amitchaturwedi1@gmial.com

B. Barman
Department of Chemistry, Govt Naveen College, Thelkadih, K.C.G., Chhattisgarh 491444, India

S. N. Saud Al-Humairi et al. (eds.), *Smart Nanomaterials in Biomedical Applications*, Engineering Materials, https://doi.org/10.1007/978-981-96-8023-8_13

EG	Ethylene glycol
QDs	Quantum dots
Nm	Nanometers
MRI	Magnetic resonance imaging
LCA	Lifecycle assessment
AI	Artificial intelligence
EPR	Enhanced permeability and retention
ECM	Extracellular matrix
3D	Three dimensional
mRNA	Massinger ribonucleic acid
DNA	Deoxyribonucleic acid

1 Introduction

Nanotechnology is the study of tiny particles that can change things at the atomic and molecular levels. It is an area that has become very important to modern scientific discovery. The application of nanotechnology in biomedical research has led to revolutionary breakthroughs that have transformed the field of illness management, treatment, and diagnosis [1]. In the biological sciences, nanotechnology has rapidly proliferated. Dental professionals have employed nanotechnology to develop restorative materials with varying degrees of success. The ways in which nanotechnology has been applied to modify dental restorations to create successful, long-lasting restorations, as well as the possibility that it could limit their longevity. It also discusses how nanotechnology can help with a number of challenging dental issues, including oral biofilm and cancers. To maintain and attain optimal dental health, the most recent innovations in nano dentistry and diagnostic, preventative, and therapeutic approaches connected to oral health are required [2]. In order to effectively manage the breakthrough pain associated with cancer, nanotechnology has also been applied in cancer treatment. A nano-delivery-trans buccal device was developed in order to rapidly and efficiently provide opioid analgesia at a steady and controlled diffusion into the target tissue. Consequently, it prevents patients from receiving needle injections and lowers the risk of overdosing. It also avoids the spontaneous and enzymatic degradation of the medication that occurs with oral administration [3]. Drugs can be delivered directly to tumors using nanoparticles, and because of their small size, the drug's surface area is increased, improving its solubility and absorption. Drug toxicity to tissues that are not the intended target can be decreased using nanoparticles. Conventional therapies including as radiotherapy, chemotherapy, and surgery can be made more successful with the use of nanomaterials. In uncommon or complicated samples, nanomaterials can aid in the identification of cancer cells.

In addition to breaking up kidney stones and clearing plaque from arteries, nanorobots may eventually replace traditional endoscopes in diagnosing intestinal disorders [4]. Making a decision may be aided by knowing the ethical considerations at play, especially for employers, employees, investors, and health authorities. Since preventing illness in employees is the aim of occupational safety and health, the ethically problematic circumstances that most impact employees have been identified. Options for decision makers can be developed as the ethical issues are identified and addressed. These issues include objectively assessing risks and hazards, autonomy, justice, privacy, nonmaleficence (not harm), and fostering respect for persons [5].

2 Overview of Nanotechnology in Biomedical Science

Nanotechnology will make it feasible to detect genetic mutations, tumours, and all disease states early because it is a portable, small instrument that can process and analyse data almost instantly. The size of samples and biosensors used in in vitro diagnostic testing will continue to grow, and formulations of nanoparticles for iron oxides and specialty polymers will improve their imaging capabilities by employing smaller and more potent dosages of the diagnostic substances. Like biotechnology, nanomedicine has sparked significant debate, particularly in relation to safety and privacy. It is impossible to ignore the possibility of abuse or unexpected effects in the usage of antibodies and antigens due to their high affinity for particular targets. Establishing strong regulatory frameworks is essential as the sector develops to guarantee that advancements in nanomedicine are morally and safely sound, striking a balance between the advantages of state-of-the-art technology and the requirement for public safety and confidence. Immunoassays are appropriate applications since they offer high sensitivities and a link. Regenerative immune sensors are an interesting new topic that allows for semi-continuous monitoring and statistical rigour through repetition. The clinic needs to determine the potential impact of nanomedicine because it is a relatively young topic in cancer [6, 7].

2.1 Tissue Engineering and Cell Treatment

Nanomedicine may also be used in regenerative medicine, which includes cell therapy and tissue engineering. Nanotechnology's structural "scaffolding" is essential for creating unique functional structures that mimic natural tissues and enable the introduction and proliferation of live cells. The developing tissues benefit from the cells' and nanoparticles' biocompatibility with the sample source. The resulting tissue's cell creation, growth, and repair processes can subsequently be managed by medical professionals [8–11].

2.2 Therapy Development

The most tremendous impact of nanomedicine on drug delivery and treatment is expected to occur in therapy. Nanoparticles increase drug efficacy and decrease adverse effects by focusing drugs on the disease's root cause. They also present new opportunities to regulate therapeutic releases. Advances in equipment and pharmaceutical manufacture have considerably helped nanomedicine, which has a wide range of applications, including drug delivery, diagnostics, and nanodrugs. A research sector is being formed to develop nanotechnology-based nanoparticles and therapeutics [12].

2.3 Drug Delivery

The first elements of drug delivery systems based on nanotechnology are adsorbent polymer matrices and nanoparticles that carry one or more therapeutic medications that have the ability to bind or disperse. In recent years, tremendous progress has been made in the production of nanodrugs that use imaging, treatments, and diagnostics. Increasing the bioavailability of targeted tissue delivery, extending the half-life of injectable drugs, and administering pharmaceuticals orally are the primary objectives of nano-drug systems. Nanomedicines' pharmacological effects are significantly improved at lower dosages, and there is a lower chance of adverse effects and health risks [13, 14].

2.4 Disease Detection

Nanomedicine is an interesting application for disease prevention, detection, therapy, and monitoring, ranging from biocompatible tissue implantation to bioengineered nanoparticles that target and destroy cells to nanosized implanted biosensors. The molecular level may be able to take advantage of the unique properties of nanoparticles, even if conventional medicine has already embraced basic nanomedicine applications like tailored drugs and medical equipment for more ambitious goals like multipart nanomedical devices. Because more surface area is available for chemical attachment on this minuscule size, handling molecules and altering particle behaviour is simpler. Furthermore, nanoparticles can enter biological cells due to their small size [15–18].

2.5 Detection of Body Oxygen

Nanosensors can measure the body's levels of carbon dioxide, oxygen, and dangerous substances. A person's diet and eating habits might reveal food sensitivity and cancerous digestive organs. Repeated chemotherapy and radiation during treatment pose a major concern for damaging healthy cells. New nanomedical methods are being employed to treat skin cancer because they enable effective medication and other possible therapy delivery to particular tumour locations and target cells with fewer detrimental side effects. Nanomedicine is utilising Nano Flares, a novel technique for identifying cancer cells in the blood. Particles called Nano Flares are made to attach to particular genetic targets of cancer cells and, upon finding that target, emit light [19, 20].

2.6 Improve the Efficiency of Medicine

A captivating domain of study that has the potential to aid pharmaceutical companies in enhancing the efficacy of existing medications while eliminating specific chronic side effects is drug delivery utilizing advanced materials and nanoparticles. This field encompasses the assembly of transport agents; pharmaceutical compounds and imaging technologies are devised to oversee the process and target the affected tissues concurrently. In other words, the distinct functional requirements must be delineated into necessary physicochemical properties and correlated with biological behavior [21–24].

2.7 Management of Illness

With bacterial resistance to antibiotics on the rise globally, nanotechnology offers an even more potent tool for managing and preventing disease. Antimicrobial substances like nano-silver can increase the effectiveness of bacterial membranes at the nanoscale. When incorporated into conventional materials, these characteristics can reduce the requirement for medicines while safeguarding the patient against infection. Treatment for patients could be revolutionized by nanotechnology. This technology is still a long way from becoming ubiquitous in life science research and development and other areas of healthcare. Given the rising cost of healthcare and life science research and development, the economic climate in which nanotechnology is at the vanguard of healthcare treatment may suddenly become too challenging and risky to justify expenditures. Usually, administering oral painkillers would be simpler and provide adequate relief for the patient [25, 26].

2.8 Fight Against Cancer

Nanotechnology is considered one of the most promising developments in drug delivery in the current battle against cancer. Usually not produced by pharmaceutical companies, nanoparticles have the potential to deliver conventional cancer medications to tumors with fewer side effects and enable the targeted destruction of cancer cells in unconventional treatments. More cytotoxic medicines can be administered at the intended site using nanoparticle-permitted chemotherapeutic delivery, and off-target toxicity and efficacy are reduced [27].

2.9 Helpful for Cardiovascular Disease

In the medical field, nanomedicine uses nanotechnological skills to prevent and treat serious illnesses, including heart disease and cardiac illness. Recent developments in nanotechnology have enabled physicians to better comprehend the goals of a living creature's action using nanoscale materials, such as biocompatible nanoparticles and medical nanobots [28].

2.10 Nanobots

The development of nanobots is the most important breakthrough in nanomedicine. Nanobots can replace whole intracellular components and restore damaged cells. Additionally, they can be duplicated to fix a genetic flaw or swap out a DNA molecule to eradicate an illness. In the medical field, nanobots might offer lucrative medical services like clearing clogged arteries or completely replacing an organ. Although nanocarriers may deliver exact dosage concentrations to specific places, their tiny size necessitates a paradoxical trade-off between volume and output. Because artificial intelligence enables coordinated activity and communication, nanobots will be able to complete tasks quickly and methodically. Individual nanobots could be programmed, searched, and moved in different directions. To more effectively monitor the presence, diagnosis, or targeting of damaged tissues, a single nanobot may link with friends and coworkers, each with their own set of rules, and request that they band their coatings, loads, or sizes together [29].

2.11 Better Preventive Procedure

The basic goal of nanomedicine is to make medical treatment a completely preventative process. Early identification of malaise can help repair it before it gets to the point where therapy is required. By removing wrinkles, extra fat, and cellulite, strengthening muscles and bones, restoring hair, and restoring vision, nanomedicine seeks to enhance human life. To prevent transmission, nanoparticles found in cleaning products and personal protective equipment can be used. The immune system may be regulated by nanomedicine to improve defenses against infection [30].

2.12 Radiation Therapy

Nanosensors can detect heart attacks and strokes before the patient does; tumour-attracted radiation nanoparticles can concentrate radiation to increase treatment effectiveness while preserving healthy tissue; nanotechnology particles can continuously monitor blood to find endothelial cells that exist prior to an attack; and nanotechnologies have also promised to stop attacking a patient's body when resetting the immune system. Nanotechnology is used for purposes other than cancer treatment. Many diseases, including cancer, may be detected and prevented via nanomedicine. There are already hundreds of products in clinical trials for nanomedicine that treat all of the main inflammatory, neurological, musculoskeletal, and cardiovascular conditions. It facilitates technology applications across the healthcare industry and is helpful for biomaterials, diagnostics, and medical imaging. Nanomedicine is an essential tool for personalized, targeted, and regenerative medicine. Real advancements in healthcare are required to make the next level of innovative medications, treatments, and implausibility accessible to physicians and patients. It is viewed as crucial to improving and more affordable health care and offers vital new tools for tackling the main issue facing the elderly. Universal access to drugs and therapies is crucial [31–35]. Despite these advancements, challenges such as biocompatibility, toxicity, scalability, and ethical considerations remain [36–38]. Ongoing research aims to address these issues, ensuring that nanotechnology drives innovation in biomedical science, ultimately improving patient outcomes and advancing global healthcare systems [39–41].

3 Historical Milestones and Advancements

American physicist Richard Feynman, who won the Nobel Prize, first proposed the concept of nanotechnology in 1959. At the annual conference of the American Physical Society, Feynman presented a discussion titled "There's Plenty of Room at the Bottom" at the California Institute of Technology. Figure 1 shows the history of nanotechnology. Fifteen years later, Japanese scientist Norio Taniguchi coined the word "nanotechnology" in 1974 and defined it as mainly involving the processing of material separation, consolidation, and deformation by a single atom or molecule [42].

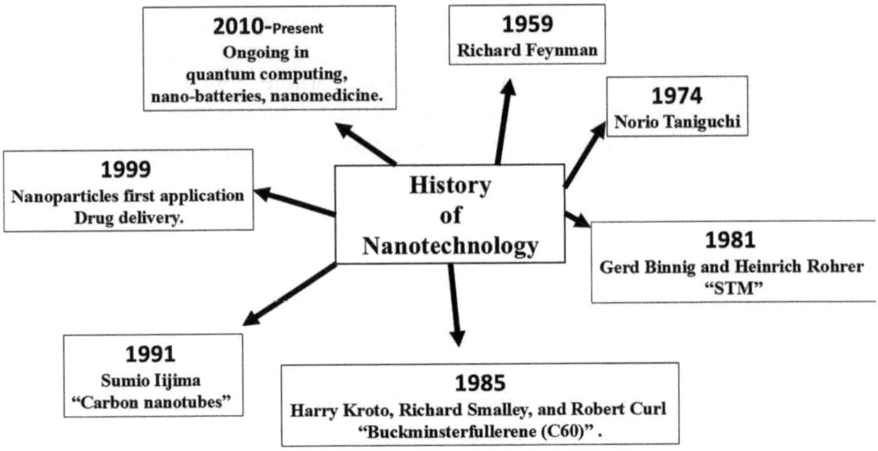

Fig. 1 History of the nanotechnology

In 1981—Atomic-scale imaging is made possible by the invention of the Scanning Tunnelling Microscope (STM) by Gerd Binnig and Heinrich Rohrer. And in 1985—Buckminsterfullerene (C60), a novel type of carbon with special characteristics, is discovered by Harry Kroto, Richard Smalley, and Robert Curl [43].

In 1991—The discovery of carbon nanotubes by Sumio Iijima results in advances in nanomaterials. And in 1999—Drug delivery by nanoparticles is the first application of nanomedicine to appear.

In 2010s–Present—ongoing developments in quantum computing, energy storage (nano-batteries), and nanomedicine [44].

4 Types of Nanoparticles

Clinical trials have studied and approved the use of several nanomaterials we will go over a few of the most typical kinds of nanoparticles show in Table 1 [45].

4.1 *Micelles*

Micelles are incredibly efficient at delivering wide range of medications especially those that are poorly in water. Their core shell structure improves drug solubility and bioavailability by allowing hydrophobic molecules to be encapsulated with their inner core while retaining a hydrophilic outside layer. Micelles are essential to contemporary drug delivery, especially for hydrophobic medication, as they improve targeted distribution, bioavailability, and therapeutic efficacy while lowering side effects. The

Table 1 Carbon nanotube difference between Single-walled carbon nanotubes and multi-walled carbon nanotubes based their important feature

S. n	Feature	Single-walled carbon nanotubes	Multi-walled carbon nanotubes	References
1	Structure	Singel layer of graphene rolled into a tube	Multiple concentric graphene layers	Lombardo et al. [46]
2	Diameter	0.4–2 nm	2–100 nm	
3	Tensile strength	Very high	Higher than single-walled carbon nanotubes due to multiple layer	
4	Flexibility	Very flexible	Less flexible	Tiriveedhi et al. [37]
5	Electrical conductivity	Metallic or semiconducting	General metallic	
6	Reactivity	More reactive	Less reactive	
7	Thermal conductivity	High conductivity	Less conductivity	
8	Production cost	More expansive	Less expansive	
9	Application	Sensors, biomedical, electronics	Composites, shielding, structure reinforcement,	Palmerston Mendes et al. [47]

following are some important medication classes that micelles can efficiently deliver [41, 48–50].

(a) ***Chemotherapy drugs***: hydrophobic anticancer medications are frequently delivered via micelles, which enhanced their solubility, stability, and targeted distribution while lowering systemic toxicity among the examples are, for targeted cancer treatment, doxorubicin is utilized in micelle formulation like Doxils$^{(R)}$.

Cisplatin: Encapsulated in micelles to reduce the renal toxicity and improve tumor penetration.

(b) ***Hydrophobic imaging agent***: To improve imaging methods, micelles can contain contrast agents like:

- A fluorescent imaging chemical called indocyanine (ICG) is used to visualize tumorvisualizesI contrast agents—such as gadolinium-based compounds, which can be integrated into micelles to increase biodistribution and imaging efficiency.

(c) ***Anti-inflammatory medications***: Micelles enhance the solubility and bioavailability of hydrophobic including:

- Cumin is a naturally occurring anti-inflammatory and anti-cancer substance that is poorly soluble in water.

- A corticosteroid called dexamethasone is utilized in formulations based on micelles to improve medication distribution and lessen systemic adverse effects.

(d) ***The delivery of genes and nucleic acid***: Delivery of genes and nucleic acid IThe d is possible using micelles for example, plasmid DNA which is encapsulated in micelles for gene therapy applications, and siRNA (small interfering RNA) which is used in gene silencing the, rapies for cancer and genetic disorders [51, 52].

4.2 Liposomes

Liposomes are spherical vesicles made of lipid bilayer that enhance the stability, bioavailability, and targeted delivery of hydrophilic and hydrophobic medications, various liposome generations have been created over time to improve therapeutic effectiveness and reduce adverse effect. Every liposome generation has been developed of address the shortcoming to the one before it, enhasncing drug delivery effectiveness, targeting capability. The ongoing development of liposome technology has enormous potential to improve the effectiveness of several medical substances [53–56]. These includes:

(a) ***Conventional Liposomes***

The first type of liposomal drug carriers are conventional liposomes, which are mostly made of cholesterol and phospholipids. These liposomes use the enhanced permeability and retention (EPR) effect to passively accumulate in tissues with leaky vasculature, like tumors.

Advantages:

- Enhance the stability and solubility of drug
- Lessen the toxicity of medication that are encapsulated ex. Doxorubicin's.
- Offer regular drug release.

Limitations:

- Quick removal from circulation as a result of opsonization and absorption by the mononuclear phagocyte system (MPS), mostly in the spleen and liver.
- Short circulation during lowers medication efficacy in systemic treatments.

Example:

- **DOXIL®**—Used for cancer therapy.
- ***Stealth Liposomes***

The second generation of liposomes, known as stealth liposomes, are made to avoid quick immune clearance. Polyethylene glycol (PEG) coats them, creating a hydrophilic barrier that decreases opsonization and increases circulation time.

Advantages:

- Longer blood circulation duration, which promotes tumor growth.
- Reduced detection by the immune system, limiting premature clearance.
- Enhanced therapeutic efficacy and drug bioavailability.

Limitations:

- Potential immune reactions to PEG (anti-PEG antibodies).
- High manufacturing cost.

Example:

- **Caelyx®**—A liposomal formulation of doxorubicin is treating for ovarian cancer, myeloma, multiple and Kaposi's sarcoma.

4.3 Carbon Nanotubes (CNTs)

Carbon atoms set up in a hexagonal lattice form the cylindrical nanostructures known as carbon nanotubes (CNTs) [57–59].

4.4 Metallic Nanoparticle's

Metals like gold, silver, platinum and iron make up metallic nanoparticles their distinct optical, electrical and antibacterial characteristics render them valuable for use in electronic, medicinal, catalytic applications and Metallic nanoparticle is modified based of different purpose that is show in below Tables 2 and 3 [60–64].

5 Fundamental Principles of Nanotechnology: Real-World Applications and Considerations

Materials in the nanoscale, which is generally between 1 and 100 nm, have distinctive chemical, physical, and biological characteristics; this is the region where nanotechnology functions. The ability to manipulate molecules and atoms to produce new structures is made possible by its basic concepts, which have their origins in surface science and quantum physics. The main ideas are that traits change with size, that materials have a lot of surface area compared to volume, and that quantum effects make materials more reactive, strong, and able to reflect light. Nanotechnology enables precision engineering of materials for specific applications, from medicine to

Table 2 Metallic nanoparticle modification based on different purpose

S. n	Modification type	Purpose	References
1	Silica coating	Enhances stability, prevents aggregation, and provides a biocompatible shell Ex. Mesoporous silica coating	Kumar et al. [65]
2	Liposome encapsulation	Improves solubility, biocompatibility, targeted drug delivery. Ex. Lipid bilayers, phospholipid coating	Gupta et al. [66]
3	Polymer coating	Improve biocompatibility, reduces toxicity, and targeted drug delery. Ex. Polyethylene glycol (PEG), chitsan, polyvinyl alcohol (PVA)	Mody et al. [67]
4	Ligand functionalization	Enables specific targeting to cells, tissues, or biomolecule by attaching targeting ligands. Ex. Antibodies, peptides, folate, aptamers	Thakkar et al. [68]
5	Magnetic functionalization	Allows external control via magnetic field for targeted therapy or imaging. Ex. Iron oxide core with polymer or silica coatings	Blackman [69]

energy. By harnessing these principles, scientists create innovative solutions that transcend traditional limitations, offering transformative potential across diverse fields, including electronics, healthcare, and environmental sustainability [76, 77].

5.1 Quantum Effects: The Rules Change at the Nanoscale

Quantum mechanics is a special science that helps us understands how small things behave. When we look at really small materials, we seen that they can cat in surprising ways with light, electricity and magnetism. This is due to the ways in which tiny particles called electrons move around [78]. Light shines quantum dots, which are small semiconducting particles. As a result, they aid in better diagnosis, clearer visualization, and very accurate disease detection, including cancer.

5.2 The Power of Small: Understanding Nanotechnology

Nanotechnology means working with things much smaller than what we see, such as one to a hundred nanometers. When materials are so small, they can be much stronger, conduct electricity better, and when they are large, react differently than

Table 3 Types of nanoparticles and compression between their efficacy, safety, applications

S. n	Nanoparticle type	Efficacy	Safety	Applications	References
1	**Metallic nanoparticles** [Au, Ag, FeO]	Catalytic, high antimicrobial, and imaging efficacy. The gold (Au) nanoparticles (NPs) enhancing drug delivery and imaging	Generally biocompatible Gold (Au) NPs, but Silver (Ag) NPs can be cytotoxic at high doses	Used in drug delivery, biosensors, antimicrobial coatings, and medical imaging	Fadeel and Garcia-Bennett [70]
2	**Ceramic nanoparticles** [SiO_2, ZrO_2, TiO_2]	Ultraviolet (UV) resistance, High stability, and enhanced mechanical properties	Biocompatible but concerns about inhalation toxicity (e.g., TiO_2 NPs)	Used in coatings, sunscreens, medicinal delivery, and catalysis	Wolfram et al. [71]
3	**Carbon-based nanoparticles** (Graphene, Fullerenes, Carbon Nanotubes)	Electrical, excellent mechanical, and thermal properties using in electronics and nanomedicine (NM)	Worries about toxicity brought on by oxidative stress and buildup in tissues	Used in sensors, medication delivery, energy storage, and structural materials	Mohajerani et al. [72]
4	**Lipid nanoparticles**	Gene delivery and highly efficient in drug especially for mRNA vaccines (e.g., COVID-19 vaccines)	Though harmless and biodegradable, there is a chance of immunological sensitivities in some situations	Most utilized in gene therapy vaccine formulations and medication delivery	Yang et al. [73]
5	**Quantum dots** (CdSe, CdTe, Perovskite)	Strong optical characteristics make it extremely effective for bio-labeling and fluorescence imaging	Heavy metal content (cadmium based QDs) could be hazards	Used in solar cells, quantum computing, and medical imaging	Giroux et al. [74]
6	**Polymeric nanoparticles** (PLGA, Chitosan)	High biocompatibility, controlled drug release, and adaptability in drug encapsulation	Minimally harmful, biocompatible, and biodegradable	Formulations for controlled release, drug delivery and tissue engineering	Kamaly et al. [75]

similar materials [79]. Think of a super light airplane that can fly just as well but needs less fuel and emits less smoke. That's what carbon nanotubes (CNTs) do. Companies like Boeing use them to make airplane parts stronger and help them use less fuel.

5.3 Self-Assembly: Nature's Blueprint

Independent configurations are processes in which molecules are arranged in structures without any external help, and this is one of the most interesting properties of nanotechnology. As example of the DNA Origami for Biomedical Applications. To reduce the negative side effects of conventional chemotherapy, researchers create small, precise nanostructures by creating DNA channels that can deliver the drug directly to cancer cells.

5.4 More Surface, More Action

Because their surface area is far greater than their volume, nanomaterials are very reactive. Technologies pertaining to drug delivery, catalysis, and filtration can greatly benefit from this concept. Take the COVID-19 mRNA vaccines, for example. They turn nanotechnology into a lifesaving finding by using lipid nanoparticles to protect the fragile mRNA and enable its efficient distribution to human cells [80].

5.5 Manipulating the Atomic World

Scientists can use sophisticated instruments such as scanning tunnelling microscopes (STM) and atomic force microscopes (AFM) to move individual atoms and molecules. IBM has demonstrated the feasibility of atomic-level data storage, paving the way for ultra-dense storage devices that could soon revolutionize computing.

6 Environmental Benefits and Risks

Nanomaterials' actions in living organisms are conditional on their dimensions, geometries, surface charges, and functionalization. These variables, which are important in the biomedical field, affect their biodistribution, cellular interactions, and biocompatibility. Despite their potential, the unique properties of nanomaterials pose challenges such as toxicity, environmental persistence, and unpredictable interactions. Understanding these behaviors is essential to designing safe, effective, and sustainable nanotechnology solutions [81]. Nanotechnology in transforming a number of disciplines, including environmental science and healthcare with its groundbreaking solutions. As we continue to develop and use these little materials, it is important to balance their benefits against safety issues. Despite all the nanotechnology's benefits, there are risks to human health and the environment that must be appropriately considered [62].

Environmental benefits: Cleaning up environmental pollution more effective by using nanoparticles like titanium dioxide, breakdown of harmful pollutants in water and soil.

Environmental Risk: some nanoparticle such as siler and carbon-based chemicals, have the ability to accumulate in the environment and disrupt aquatic and soil ecosystems and its long-term effect.

7 Role of Nanotechnology in Modern Biomedical Applications

Nanotechnology has transformed renewing drug through the routine of nanostructured frameworks in muscle production. These scaffolds mimic natural extracellular matrices, promoting cell growth and tissue repair in areas such as bone, skin, and cartilage regeneration. Despite its promise, nanotechnology in biomedical applications faces challenges, including potential toxicity, scalability, and regulatory hurdles. However, ongoing research continues to refine nanomaterials, paving the way for the day when precision and customized treatment become a reality. Table 4 summarizes how nanotechnology integrates into various aspects of modern biomedical applications, revolutionizing healthcare with enhanced precision, efficiency, and effectiveness [82, 83]. Personalized Healthcare Advancements with Netechnology enabling medicines to be customized to each patient's distinct genetic composition, disease profile, and bioand logical reactions, nanotechnology is transforming personalized medicine [84]. Here are a few innovative instances of how nanotechnology is influencing healthcare in the future:

7.1 Nanoscale Biosensors for Early Disease Detection

Imagine having a small sensor that can identify illnesses in your body before any symptoms show up. Researchers are creating nano sensors that can detect biomarkers for conditions including cancer, Alzheimer's, and cardiovascular illnesses in blood, saliva, or urine. Gold nanoparticles, for instance, are utilized in early prostate cancer detection tests, allowing for faster and more efficient treatment [90].

Table 4 Role of nanotechnology in modern biomedical applications

S. n	Application area	Nanotechnology role	Impact	Example	References
1	Targeted drug delivery	Use of nanoparticles for precise drug targeting	Reduces side effects and improves efficacy of drug	Liposomes have to be used for delivering pharmaceutical agents	Bharathala and Sharma [85]
2	Diagnostics and monitoring	Development of nano-sensors for real-time detection	Enables early diagnosis and disease tracking	Early disease detections, drug delivery for cancer treatments	Labhasetwar and Leslie-Peleck [86]
3	Tissue engineering	Nanoengineered scaffolds for tissue regeneration	Promotes faster and more efficient healing	Hafnium oxide and gold-based nanoparticle used enhance x-ray therapy	Emilian Leucuta [84]
4	Imaging and bioimaging	Quantum dots and nanoparticles for contrast agents	Enhances clarity in imaging techniques	Cancer detection and drug delivery	Bharathala and Sharma [85]
5	Cancer treatment	Nanocarriers for delivering anticancer agents	Focused treatment, minimizing healthy cell damage	Targeted drug delivery and photodynamic therapy	Labhasetwar and Leslie-Pelecky [87]
6	Infectious disease control	Nano-coatings and nanostructures for antimicrobial use	Prevents infections and enhances sterilization	Develop mRNA vaccine against SARS COV-19	Pasika et al. [88]
7	Biosensors	Integration of nanomaterials in wearable devices	Real-time health monitoring and personalized medicine	Gold nano particle, carbon nanotube, quantum dot	Ramos et al. [89]

7.2 Smart Drug Delivery Systems

Drugs can be delivered precisely where they are needed using nanoparticle engineering, which will increase efficacy and decrease negative effects. Lipid nanoparticles (LNPs), for instance, have been created by researchers to deliver chemotherapy medications straight to cancer cells while avoiding healthy tissue. This method reduces the severe side effects of conventional chemotherapy while simultaneously improving treatment results [91].

7.3 Personalized Cancer Therapy

Treatments can be customized by using nanotechnology to analyze the unique tumor characteristics of each patient. To make therapy more successful and less intrusive than traditional treatments, researchers are developing nanoparticles that may carry siRNA (small interfering RNA) to quiet cancer-causing genes in specific patients [92].

7.4 Regenerative Medicine and Tissue Engineering

Regenerative medicine is incorporating nanomaterials to fix damaged organs and tissues. Researchers are creating nanofiber scaffolds that may aid in the regeneration of nerve cells in cases of spinal cord injuries or encourage the formation of new bone in fracture patients. By imitating the body's natural structures, these nanoscale materials promote quicker and more efficient healing [93].

7.5 Fighting Antibiotic-Resistant Infections

Researchers are looking at using nanoparticles to fight bacteria that are resistant to drugs. Because of their potent antibacterial qualities, silver and copper nanoparticles, for example, are being used in targeted antibiotic treatments, medical device coatings, and wound dressings to treat infections that are no longer treatable with conventional antibiotics [94].

7.6 Wearable Nanotech for Continuous Health Monitoring

Real-time vital sign monitoring and disease early detection are possible with wearable nanotechnology, such as smart patches and nano-implants. For diabetics, for instance, researchers are creating nano-based glucose monitoring devices that provide continuous, non-invasive blood sugar monitoring without the need for uncomfortable finger pricks [95].

7.7 Nanomedicine and Targeted Drug Delivery

Nanomedicine, a revolutionary discipline at the intersection of nanotechnology and medicine, has greatly improved the idea of tailored medication delivery. Targeting specific cells or tissues with nanoscale materials and technology makes nanomedicine more effective and less harmful to the body as a whole [96, 97]. Lipid nanoparticles, dendrimers, polymeric nanoparticles, and liposomes are some of the nanocarriers that are used in systems that deliver drugs precisely. By encasing them, these carriers prevent the degradation of therapeutic substances and guarantee their regulated release at the site of action. When targeting ligands like aptamers, peptides, antibodies are turned into nanocarriers, they bind to receptors that are overexpressed on sick cells, like cancer cells, making the treatment more precise [98]. An important benefit of nanomedicine is that it can cross biological barriers, which is a real problem when trying to deliver drugs to the brain or other parts of the nervous system that are central. Nanoparticles can traverse these barriers, enabling treatment for previously inaccessible conditions like brain tumors and neurodegenerative diseases [99]. Nanoparticles can target cancer either actively or passively. Passive targeting uses the EPR effect, where leaky vasculature causes nanoparticles to aggregate in tumor tissues. Active targeting entails modifying surfaces in such a way that they can detect and bind to markers unique to tumors. Nanomedicine also enables the co-delivery of multiple drugs, offering combination therapies that can enhance therapeutic outcomes. For instance, chemotherapy drugs and gene therapy agents can be delivered simultaneously, addressing drug resistance and improving efficacy. Nanotoxicity, obtaining regulatory permission, and producing on a wide scale are still obstacles, despite its potential. Still, nanomedicine has tremendous promise for enhancing healthcare delivery via the introduction of novel, individualized approaches to therapy that are less invasive and more effective [100].

7.8 Nano-Sensors for Diagnostics and Monitoring Nanoengineering in Tissue Regeneration

New diagnostic and real-time monitoring instruments have arisen nano sensors. These take advantage of nanotechnology's capacity to change materials on a molecular and atomic level. These sensors are designed to detect specific biological changes with unparalleled sensitivity and precision, making them invaluable in biomedical applications [101]. Diagnostic nano sensors are particularly impactful in identifying diseases at their earliest stages. Nano sensors allow for the early detection of illnesses, including cancer, and diabetes, by detecting biomarkers at very low concentrations. In colorimetric tests, gold nanoparticles show that certain biomolecules are present. Quantum dot-based sensors, on the other hand, are more sensitive and can find more than one biomarker at the same time.

Nano sensors can measure things like glucose levels in diabetics and cardiac biomarkers in patient who have heart problems in real time so that health can be monitored all the time. By combining implantable nano sensors with wearable technology, we can improve individualized healthcare via the seamless and non-invasive monitoring of health parameters. Nano sensors also find applications in environmental

[102] monitoring, detecting harmful substances [103, 104], toxins, or pathogens in water, air, and food. This capability supports proactive measures to ensure safety and public health. The exceptional sensitivity of nano sensors arises from their high surface-to-volume ratio, enabling efficient interaction with target analytes. Furthermore, advancements in signal transduction mechanisms, such as [105], electrochemical, and plasmonic responses, amplify detection capabilities. Nano sensors have the potential to revolutionize monitoring and diagnostics, propelling advances in proactive healthcare and personalized medicine despite obstacles such as cost, scalability, and their integration into clinical environments [106–109].

7.9 Nanoengineering in Tissue Regeneration

Nanoengineering, which involves designing and manipulating materials at the nanoscale, has revolutionized tissue regeneration by enabling precise control over cellular behavior, scaffold fabrication, and biomaterial functionality. By mimicking the intricate nanostructure of native tissues, nanoengineered platforms enhance cellular interactions, promoting efficient tissue repair and regeneration. Nanostructured Support Structures Nanoengineered scaffolds are made up of hydrogels, nanofibers, and nanoparticles. They are made to look like the extracellular matrix (ECM) in nature so cells can stick to them, grow, and change. We use scaffolds made of biocompatible polymers or nanomaterials like graphene or carbon nanotubes to mimic the biochemical and mechanical characteristics of real tissues. Medications and growth factors were delivered. Scaffolds embedded with nanoparticles enable the precise delivery of growth factors and medicines where they are required. Localizing therapies, like those that help blood vessels grow, or bones heal in hurt areas, makes them more effective and less likely to cause side effects [110]. Cellular Interaction Nanoscale topography, including ridges and pores, facilitates cell adhesion and alignment. These features guide stem cell differentiation into specific cell types, such as osteoblasts for bone regeneration or neurons for nerve repair. Future Prospects While nanoengineering holds immense potential for tissue regeneration, challenges such as scaling production, ensuring long-term biocompatibility, and managing costs require attention. Research is ongoing to integrate 3D printing and nanotechnology to fabricate complex, patient-specific tissue constructs [111]. Table 5 provide further information on the topic of nanoengineering in tissue regeneration. Figure 2 shows an example of a nanoengineered scaffold that successfully supports tissue healing and regeneration. We intend this scaffold to resemble the extracellular matrix.

Table 5 Applications of nanoengineering in tissue regeneration

S. n	Application	Nanomaterial used	Outcome	References
1	Bone regeneration	Hydroxyapatite nanoparticles	Enhanced mineralization	Hasan et al. [112]
2	Neural tissue repair	Carbon nanotubes	Neuronal growth and connectivity	Fabbro et al. [113]
3	Wound healing	Silver nanoparticles	Antibacterial properties, faster healing	Gunasekaran et al. [114]

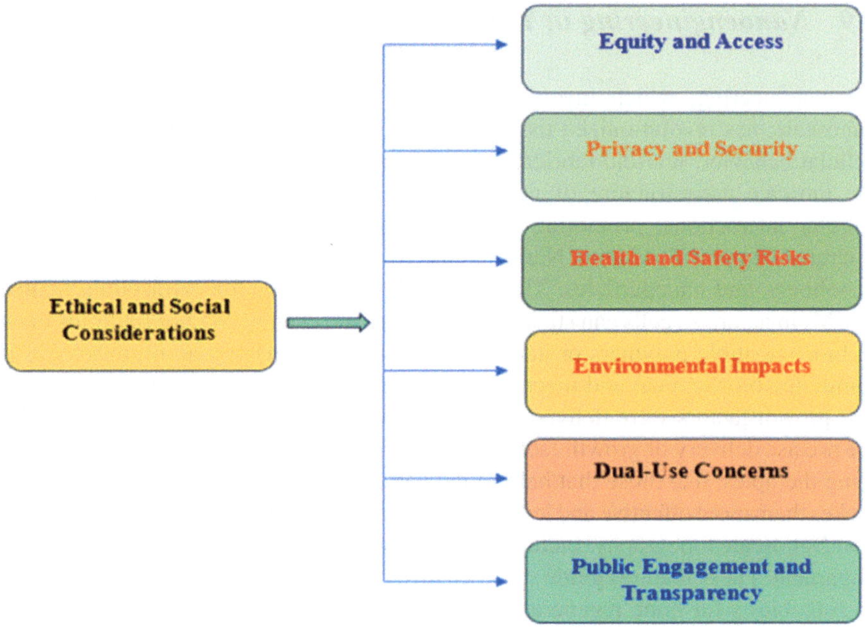

Fig. 2 Ethical and social considerations

8 Challenges in Nanotechnology Applications

Nanotechnology is still transforming environmental research, industry, and medicine in spite of obstacles. We can optimize the advantages of nanotechnology while reducing hazards by implementing strategic solutions including cost-effective production methods, green synthesis processes, and changing safety laws. Nanotechnology holds promise for an innovative, safe, and accessible future across industries as efforts continue (Table 6) [112, 115–125].

Table 6 Key challenges in nanotechnology applications

S. n	Challenge	Description	Potential solutions	References
1	Toxicity and Environmental Safety	Health and ecological risks posed by nanoparticles	Rigorous safety evaluations and green synthesis methods	Köhler [127]
2	High Costs	Expensive research, development, and scaling processes	Economical synthesis and shared infrastructure	Jaldurgam et al. [128]
3	Regulatory Gaps	Lack of unified global standards	Establishing comprehensive international frameworks	Morley et al. [129]
4	Long-Term Effects	Unknown impacts on health and environment	Conducting long-term studies	Anonymous [130]
5	Manufacturing Precision	Difficulty in maintaining quality at scale	Developing advanced fabrication and control technologies	Li [131]

i. ***Toxicity and Environmental Safety***

The health of people and the environment may be at risk from nanoparticles because of their small size and strong reactivity [126].

Solution: To guarantee that there are few health hazards, researchers are creating safer nanoparticles by employing biocompatible coatings and sophisticated toxicity testing.

ii. ***High Research and Development Costs***

Many sectors, particularly those in underdeveloped nations, cannot afford the substantial capital required to create nanotechnology solutions [118].

Solution: To lower the cost of nanotechnology research, governments and business institutions, such as the National Science Foundation (NSF), are funding public–private partnerships.

8.1 Scaling up from Laboratory to Commercial Production

One major obstacle is moving nanotechnology from research labs to large-scale industrial applications.

Solution: To enhance large-scale production, firms like as IBM and Intel are investing in nanomanufacturing breakthroughs, such as precision fabrication and self-assembly processes.

iii. ***Regulatory and Ethical Concerns***

Public trust and safety enforcement are compromised by the absence of uniform international laws for the use of nanomaterials.

Solution: To standardize safety procedures for nanotechnology, the International Council on Nanotechnology (ICON) is aiming to create global regulatory frameworks.

iv. ***Limited Understanding of Long-Term Effects***

The long-term effects of nanomaterials on biological systems and ecosystems require further investigation.

Solution: Researching long-term consequences is the focus of international projects like the Nanotechnology Environmental and Health Implications (NEHI) working group.

9 Ethical and Social Considerations

As nanotechnology advances, its integration into diverse fields raises important ethical and social considerations. These issues are crucial to ensure that nanotechnology serves society responsibly, equitably, and sustainably, without causing unintended harm or exacerbating existing inequalities. Figure 3 show the ethical and social considerations. As nanotechnology develops, significant ethical and societal issues are brought up by its application in several industries [132]. These concerns are essential to guaranteeing that nanotechnology benefits society in a responsible, equitable, and sustainable manner without inadvertently harming others or escalating already-existing disparities [133].

(a) ***Equity and Access***

The fair distribution of the advantages of nanotechnology is a significant ethical issue. There is a chance that its sophisticated uses will disproportionately help affluent people and industrialized countries, leaving impoverished areas and marginalized communities behind.

Case Study: In order to supply clean drinking water in areas such as sub-Saharan Africa, researchers created nano-enabled water filtering devices in 2015. However, widespread adoption is still difficult because of high pricing and inadequate infrastructure. Initiatives from UNICEF and other international collaborations are attempting to close this gap [134].

(b) ***Privacy and Security***

Devices that use nanotechnology, such tracking systems and nano sensors, raise privacy issues. The capacity to gather enormous volumes of data, frequently covertly, may result in improper use of surveillance or unapproved data exchange.

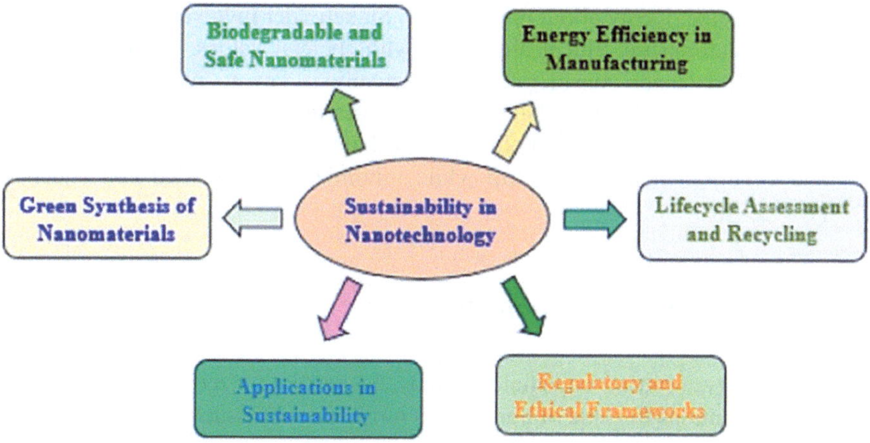

Fig. 3 Sustainability in nanotechnology

Case Study: Nano-biosensors used for continuous glucose monitoring in diabetic patients raised concerns in 2020. These sensors gathered comprehensive personal health data while enhancing patient health. Better encryption standards and more business rules were the results of ethical worries about data security.

(c) ***Health and Safety Risks***

Nanoparticles and nanomaterials can have unanticipated interactions with living systems and the natural world because of their minuscule size and unique properties. It is crucial to protect ecosystems, consumers, and employees [135].

Case Study: Concerns regarding lung toxicity among factory workers were raised when carbon nanotubes were introduced into the manufacturing process. Research organizations responded by creating safer production procedures and enacting protective laws to reduce exposure, guaranteeing a balance between worker safety and innovation.

(d) ***Environmental Impacts***: While nanotechnology offers solutions to environmental problems, its development can also pose risks. The improper disposal of nanomaterials might lead to ecological contamination. Ethical practices must prioritize sustainability, promoting green synthesis and minimizing harmful waste.
(e) ***Dual-Use Concerns***: Certain nanotechnological advancements, such as nanoscale weapons or surveillance tools, can have dual-use applications, raising ethical concerns about misuse in warfare or espionage. International agreements and monitoring mechanisms are required to prevent harmful applications.
(f) ***Public Engagement and Transparency***: The societal impact of nanotechnology depends significantly on public perception and trust. It is critical to educate the public on the potential advantages and hazards of nanotechnology and to

communicate openly about these issues. Stakeholder involvement in decision-making can ensure ethical alignment with societal values. Addressing these ethical and social considerations proactively is key to realizing nanotechnology's potential for global good. By fostering inclusive, sustainable, and responsible innovation, the field can contribute meaningfully to human welfare and environmental stewardship while minimizing unintended consequences [136–138].

10 Sustainability in Nanotechnology

Nanotechnology, a cutting-edge field of study, has revolutionized numerous markets and offered solutions to some of humanity's most pressing issues. To ensure its uses don't lead to social or environmental damage, a strong emphasis on sustainability is essential, especially given its rapid expansion. Achieving sustainability in nanotechnology, as shown in Fig. 3, involves adopting eco-friendly practices across the entire lifecycle of nanomaterials, from synthesis to application and eventual disposal [139, 140].

(a) **Green Synthesis of Nanomaterials**: Sustainable nanotechnology begins with the synthesis of nanomaterials using green chemistry principles. This includes the use of non-toxic solvents, renewable resources, and energy-efficient methods such as microwave or ultrasonic-assisted synthesis. Biological synthesis, employing plant extracts or microorganisms, also provides a sustainable alternative to conventional chemical methods [141, 142].
(b) **Biodegradable and Safe Nanomaterials**: Designing nanomaterials that are biodegradable, or recyclable reduces their long-term environmental impact. These materials can break down into harmless components, minimizing ecological damage. Additionally, focusing on non-toxic materials ensures safety for human health and ecosystems.
(c) **Energy Efficiency in Manufacturing**: Nanotechnology processes can be optimized to consume less energy, reducing their carbon footprint. Techniques like low-temperature synthesis and the integration of renewable energy sources in production facilities contribute to sustainable practices.
(d) **Applications in Sustainability**: Nanotechnology plays a significant role in advancing sustainability across various sectors. Nanomaterials are used in water purification, air filtration, and renewable energy technologies like solar cells and fuel cells. In agriculture, nanotechnology aids in efficient pesticide delivery and nutrient management, reducing resource wastage.
(e) **Lifecycle Assessment and Recycling**: A comprehensive lifecycle assessment (LCA) evaluates the environmental impact of nanomaterials from creation to disposal. Recycling and repurposing nanomaterials further mitigate waste and reduce resource consumption.

Table 7 Strategies for sustainability in nanotechnology

S. n	Category	Sustainable practices	Impact	References
1	Eco-friendly nanomaterial design	Development of biodegradable and non-toxic nanomaterials	Reduces environmental pollution and toxicity risks	Ray [149]
2	Green synthesis techniques	Use of plant extracts and microorganisms for nanoparticle synthesis	Minimizes hazardous chemical use and energy consumption	Bahrulolum et al. [150]
3	Energy-efficient processes	Adoption of low-energy manufacturing methods	Reduces carbon footprint in production	Yurtay [151]
4	Recyclability and reusability	Design of nanomaterials for multiple-use applications	Promotes circular economy and waste reduction	Lingaitiene and Burinskiene [152]
5	Waste management	Safe disposal and recycling of nanotechnology waste	Prevents contamination and ensures environmental safety	Schwab et al. [153]
6	Sustainable applications	Focus on renewable energy, water purification, and eco-friendly products	Aligns nanotechnology with global sustainability goals	Khatoon and Velidandi [154]

(f) **Regulatory and Ethical Frameworks**: Adopting worldwide norms and laws guarantees the ethical development and implementation of nanotechnology. Achieving a balance between nanotechnology progress and social values requires openness and participation from the public.

By integrating sustainability into every phase of nanotechnology, researchers and industries can harness its transformative potential while safeguarding the environment and fostering a balanced relationship between innovation and ecological stewardship [143–148]. Table 7 highlights how sustainable practices in nanotechnology can address environmental concerns, promote eco-friendly innovation, and support long-term ecological balance.

11 Future Perspectives

Nanotechnology is being significantly impacted by artificial intelligence (AI) and machine learning (ML), particularly in the areas of medical diagnosis and treatment. Scientists and medical professionals are creating more accurate, effective, and individualized methods of disease detection and treatment by fusing the power of artificial intelligence with nanoscale materials and technologies [155].

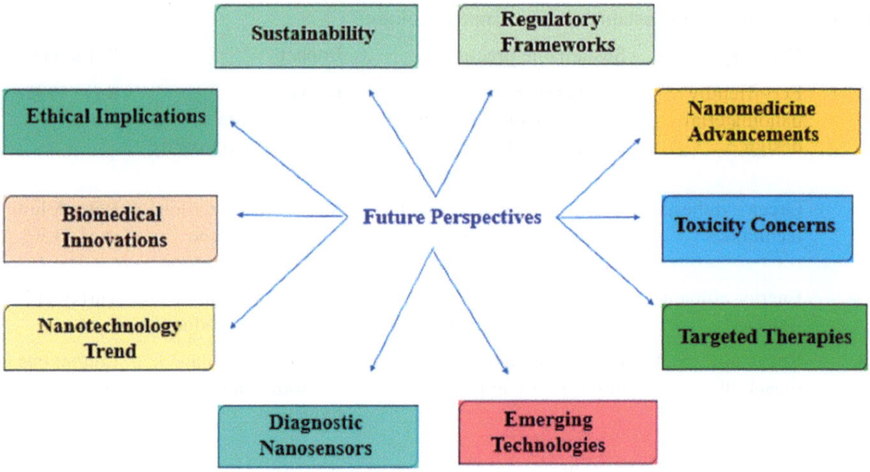

Fig. 4 Future perspectives key points for pioneering the future of nanotechnology

Nanotechnology's role in addressing global health inequities is also significant. Portable, low-cost nanodevices for diagnostics and treatment could bring advanced healthcare solutions to underserved regions, democratizing access and reducing disparities [156, 157]. However, future progress will depend on overcoming persistent challenges. Efforts must focus on improving scalability, reducing production costs, and enhancing the biocompatibility of nanomaterials [158]. Ethical considerations will remain at the forefront, necessitating clear regulatory frameworks and global cooperation to ensure equitable use, safety, and sustainability. As we pioneer these advancements, the future of nanotechnology in biomedical science holds transformative potential, promising not just to enhance human health but also to redefine the way we approach healthcare at every level [159–165]. Thoughtful innovation, guided by ethics and sustainability, will shape this exciting trajectory. Future perspectives key points for pioneering the future of nanotechnology are given in Fig. 4.

12 Conclusion

In the summary, In the field of biomedicine, nanotechnology is in the front, providing the opportunity to revolutionize patient care in areas such as diagnosis and therapy. The potential uses of nanotechnology are limitless and profound, ranging from highly specific systems for drug delivery that transform chemotherapy to tiny sensors that allow for the monitoring of diseases in real time. Nevertheless, strenuous maneuvering is required to overcome the obstacles posed by these developments. Problems with biocompatibility, the inability to scale production, and red tape show how important it is for scientists, politicians, and ethicists to work together. The ethical climate of nanotechnology is also critical. It is critical that we address public concerns around privacy, safety, and equal access while we lead the way in developing new applications. To promote innovation while ensuring that nanotechnology deployment is in line with society's ideals, regulatory structures must change. Additionally, we must prioritize sustainability to ensure eco-friendly operations, given the environmental impact of nanomaterial manufacturing and disposal. As we go forward in biomedical nanotechnology, we can expect to see solutions for healthcare that are more precise than ever before thanks to developments in areas such as AI (artificial intelligence), ML (machine learning), and customized medicine. These events show how important it is for countries to work together to solve problems and get the most out of their resources. This makes it clear that nanotechnology will bring people together to solve common health problems. The future of nanotechnologies in biomedical research will be defined by the integration of technological developments, ethical concerns, and sustainability efforts. Responsibility tapping into its revolutionary potential may lead to advancements that benefit society and the environment while simultaneously improving healthcare. To fully fulfill its potential, nanotechnology must be carefully managed. It is an exciting and promising field.

13 Competing Interests

The authors have no conflicts of interest to declare that are relevant to the content of this chapter.

14 Ethics Approval

Not Applicable.

Acknowledgements The authors are grateful to Dr. C. V. Raman University, Kota, Bilaspur, Chhattisgarh, India and Govt Naveen College, Thelkadih, K.C.G., Chhattisgarh, India for this work collaborative work.

Authors' Contributions Milan Hait, Amit Kumar Chaturwedi and Benvikram Barman wrote different sections of the manuscript. Milan Hait, Benvikram Barman and Amit Kumar Chaturwedi edited the manuscript. All authors read and approved the final manuscript.

References

1. Saleh, T.A.: Detection: from electrochemistry to spectroscopy with chromatographic techniques, recent trends with nanotechnology. Detection **2**(04), 27 (2015). https://doi.org/10.4236/detection.2014.24005

2. Abou Neel, E.A., Bozec, L., Perez, R.A., et al.: Nanotechnology in dentistry: prevention, diagnosis, and therapy. Int. J. Nanomed. **10**, 6371–6394 (2015). https://doi.org/10.2147/IJN.S86033
3. Sprintz, M., et al.: Applied nanotechnology for the management of breakthrough cancer pain. Minerva Anestesiol. **71**(7–8), 419–423 (2005)
4. Chan, B.P., Leong, K.W.: Scaffolding in tissue engineering: general approaches and tissue-specific considerations. Eur. Spine J. **17**(4), 467–479 (2008). https://doi.org/10.1007/s00586-008-0745-3
5. Wu, Y.N., Yang, L.X., Shi, X.Y., et al.: The selective growth inhibition of oral cancer by iron core-gold shell nanoparticles through mitochondria-mediated autophagy. Biomaterials **32**(20), 4565–4573 (2011). https://doi.org/10.1016/j.biomaterials.2011.03.006
6. Yarlagadda, T., Sharma, S., Yarlagadda, P.K., et al.: Recent developments in the field of nanotechnology for the development of medical implants. Proc. Manuf. **30**, 544–551 (2019). https://doi.org/10.1016/j.promfg.2019.02.077
7. Shailaja, R., Sugunthan, S.: Concept of nanotechnology in Siddha medical literature. World J. Pharm. Res. **5**(10), 276–284 (2016). www.wjpr.net
8. Yao, J., Zheng, F., Yao, C., et al.: Rational design of nanomedicine for photothermal chemodynamic bimodal cancer therapy. Wiley Interdiscip. Rev. Nanomed. Nanobiotechnol. **13**(3), e1682 (2021)
9. Maheshwari, P.V., Gupta, N.V.: Advances of nanotechnology in healthcare. Int. J. Pharm. Tech. Res. **4**, 1221–1227 (2012)
10. Jurj, A., Braicu, C., Pop, L.A., et al.: The new era of nanotechnology, an alternative to change cancer treatment. Drug Des. Devel. Ther. **11**, 2871 (2017)
11. Landau, J., Groscurth, C.R., Wright, V., et al.: Visualizing nanotechnology: the impact of visual images on lay American audience associations with nanotechnology. Public Underst. Sci. **18**(3), 325–337 (2009)
12. Wang, Y., Gao, D., Liu, Y., et al.: Immunogenic-cell-killing and immunosuppression-inhibiting nanomedicine. Bioact. Mater. **6**(6), 1513–1527 (2021)
13. Li, T., Liang, W., Xiao, X., et al.: Nanotechnology an alternative with promising prospects and advantages for the treatment of cardiovascular diseases. Int. J. Nanomed. **13**, 7349 (2018)
14. Aithal, P.S., Aithal, S.: Ideal technology concept & its realization opportunity using nanotechnology. Int. J. Appl. Innov. Eng. Manage. **4**(2), 153–164 (2015)
15. Hobson, D.W.: Commercialization of nanotechnology. Wiley Interdiscip. Rev. Nanomed. Nanobiotechnol. **1**(2), 189–202 (2009)
16. Doroudian, M., Loughlin, M., Poynton, R.F., et al.: Nano-technology-based therapeutics for lung disease. Thorax **74**(10), 965–976 (2019)
17. Sarmah, D., Banerjee, M., Datta, A., et al.: Nanotechnology in the diagnosis and treatment of stroke. Drug Discov. Today **26**(2), 585–592 (2021)
18. Grumezescu, A.M.: Essential oils and nanotechnology for combating microbial biofilms. Curr. Organ. Chem. **17**(2), 90–96 (2013)
19. Patil, M., et al.: Future impact of nanotechnology on medicine and dentistry. J. Ind. Soc. Periodontol. **12**(2), 34 (2008)
20. Guptav, J.: Nanotechnology applications in medicine and dentistry. J. Invest. Clin. Dent. **2**(2), 81–88 (2011)
21. Al Kahtani, R.N.: The implications and applications of nanotechnology in dentistry: a review. Saudi Dent. J. **30**(2), 107–116 (2018)
22. Emerich, D.F., Thanos, C.G.: Nanotechnology and medicine. Exp. Opin. Biol. Ther. **3**(4), 655–663 (2003)
23. Solanke, I.M.F.A., et al.: Nanotechnology and its application in dentistry. Ann. Med. Health Sci. Res. **4**(3), 171–177 (2014)
24. Ball, A.S., et al.: Introduction into nanotechnology and microbiology. In: Gurtler, V., Ball, A.S., Soni, S. (eds.) Methods in Microbiology, vol. 46. Academic Press, Waltham (2019)
25. Bozzato, E., et al.: Nanomedicine: a useful tool against glioma stem cells. Cancers **13**(1), 9 (2021)

26. Honda, M., Asai, T., Oku, N., et al.: Liposomes and nanotechnology in drug development: focus on ocular targets. Int. J. Nanomed. **8**, 495 (2013)
27. Horton, M.A., Khan, A.: Medical nanotechnology in the UK: a perspective from the London Centre for Nanotechnology. Nanomed. Nanotechnol. Biol. Med. **2**(1), 42–48 (2006)
28. Buxton, D.B.: Current status of nanotechnology approaches for cardiovascular disease: a personal perspective. Wiley Interdiscip. Rev. Nanomed. Nanobiotechnol. **1**(2), 149–155 (2009)
29. Kroto, H.W., Heath, J.R., O'Brien, S.C., et al.: C60: Buckminsterfullerene. Nature **318**(6042), 162–163 (1985). https://doi.org/10.1038/318162a0
30. Drexler, E.: Engines of creation: the coming era of nanotechnology. Anchor (1987).
31. Patel, B., Darji, P., Fnu, P.I.J., et al.: A comprehensive review and insight into the latest advancements in nanotechnology. Bios. Biot. Res. Asia **21**(3), 985–1000 (2024). https://doi.org/10.13005/bbra/3279
32. Banjare, M.K., Barman, B., Behera, K., et al.: Molecular interaction between three novel amino acid based deep eutectic solvents with surface active ionic liquid: a comparative study. Heliyon **10**, 15 (2024). https://doi.org/10.1016/j.heliyon.2024.e35598
33. Patel, D., Chaturwedi, A.K., Kashyap, N.K., et al.: Chapter 12 Quality control and standardization of nano materials, vol. 1, pp. 297–325. IGI Global (2024). https://doi.org/10.4018/979-8-3693-7640-9.ch012
34. Katsuki, S., Matoba, T., Koga, J.I., et al.: Anti-inflammatory nanomedicine for cardiovascular disease. Front. Cardiovasc. Med. **4**(87) (2017). https://doi.org/10.3389/fcvm.2017.00087
35. Barman, B., Banjare, M.K.: Synthesis and characterization of biologically active five amino acids based deep eutectic solvents. Alochana J. **13**, 1 (2024). 20.14118.AJ.2024.V13I1.2161
36. Morgan, M.T., Carnahan, M.A., Finkelstein, S., et al.: Dendritic supramolecular assemblies for drug delivery. Chem. Commun. (Camb.) **34**, 4309–4311 (2005). https://doi.org/10.1039/b502411k
37. Tiriveedhi, V., Kitchens, K.M., Nevels, K.J., et al.: (2011) Kinetic analysis of the interaction between poly(amidoamine) dendrimers and model lipid membranes. Biochim. Biophys. Acta **1808**(1), 209–218 (2011). https://doi.org/10.1016/j.bbamem.2010.08.017
38. Palmerston Mendes, L., Pan, J., Torchilin, V.P., et al.: Dendrimers as nanocarriers for nucleic acid and drug delivery in cancer therapy. Molecules **22**(9), 1401 (2017). https://doi.org/10.3390/molecules22091401
39. Kukowska-Latallo, J.F., Bielinska, A.U., Johnson, J., et al.: Efficient transfer of genetic material into mammalian cells using Starburst polyamidoamine dendrimers. Proc. Natl. Acad. Sci. U.S.A. **93**(10), 4897–4902 (1996). https://doi.org/10.1073/pnas.93.10.4897
40. Banjare, M.K., Barman, B.: Effect of biologically active amino acids based deep eutectic solvents on sodium dodecyl sulfate: a comparative spectroscopic study. Spectrochim. Acta A Mol. Biomol. Spectrosc. **308**, 123700 (2024). https://doi.org/10.1016/j.saa.2023.123700
41. Svenson, S., Tomalia, D.A.: Dendrimers in biomedical applications–reflections on the field. Adv. Drug Deliv. Rev. **57**(15), 2106–2129 (2005). https://doi.org/10.1016/j.addr.2005.09.018
42. Tawkif, S.: Detection: from electrochemistry to spectroscopy with chromatographic techniques, Recent Trends with Nanotechnology. Detection **2**(4), 27–32 (2014)
43. Bayda, S., Adeel, M., Tuccinardi, T., et al.: The history of nanoscience and nanotechnology: from chemical-physical applications to nanomedicine. Molecules **25**(1), 112 (2020). https://doi.org/10.3390/molecules25010112
44. Bayda, S., Adeel, M., Tuccinardi, T., et al.: The history of nanoscience and nanotechnology: from chemical–physical applications to nanomedicine. Molecules **25**(1), 112 (2019)
45. Katsuki, S., Matoba, T., Koga, J.I., et al.: Anti-inflammatory nanomedicine for cardiovascular disease. Front. Cardiovasc. Med. **4**, 87 (2017). https://doi.org/10.3389/fcvm.2017.00087
46. Lombardo, D., Calandra, P., Caccamo, M.T., et al.: Colloidal stability of liposomes. AIMS Mater. Sci. **6**, 2 (2019). https://doi.org/10.3934/matersci.2019.2.200
47. Palmerston Mendes, L., et al.: Dendrimers as nanocarriers for nucleic acid and drug delivery in cancer therapy. Molecules **22**(9), 1401 (2017). https://doi.org/10.3390/molecules22091401

48. Kukowska-Latallo, J.F., Bielinska, A.U., et al.: Efficient transfer of genetic material into mammalian cells using Starburst polyamidoamine dendrimers. Proc. Natl. Acad. Sci. U.S.A. **93**(10), 4897–4902 (1996). https://doi.org/10.1073/pnas.93.10.4897
49. Figueiras, A., Domingues, C., Jarak, I., et al.: New advances in biomedical application of polymeric micelles. Pharmaceutics **14**(8), 1700 (2022)
50. Bakshi, M.S.: Colloidal micelles of block copolymers as nanoreactors, templates for gold nanoparticles, and vehicles for biomedical applications. Adv. Coll. Interface. Sci. **213**, 1–20 (2014)
51. Nune, S.K., Chanda, N., Shukla, R., et al.: Green nanotechnology from tea: phytochemicals in tea as building blocks for production of biocompatible gold nanoparticles. J. Mater. Chem. **19**(19), 2912–2920 (2009). https://doi.org/10.1039/B822015H
52. Kim, E.M., Jeong, H.J.: Liposomes: biomedical applications. Chonnam Med. J. **57**(1), 27 (2021)
53. Panahi, Y., Farshbaf, M., Mohammadhosseini, M., et al.: Recent advances on liposomal nanoparticles: synthesis, characterization and biomedical applications. Artif. Cells Nanomed. Biotechnol. **45**(4), 788–799 (2017)
54. Gehrke, I., et al.: Innovations in nanotechnology for water treatment. Nanotechnol. Sci. Appl. 1–17 (2015). https://doi.org/10.1021/ja0486059
55. Dai, H.: Carbon nanotubes: opportunities and challenges. Surf. Sci. **500**(1–3), 218–241 (2002)
56. Popov, V.N.: Carbon nanotubes: properties and application. Mater. Sci. Eng. R. Rep. **43**(3), 61–102 (2004)
57. Anzar, N., Hasan, R., Tyagi, M., et al.: Carbon nanotube—a review on synthesis, properties and plethora of applications in the field of biomedical science. Sens. Int. **1**, 100003 (2020). https://doi.org/10.1016/j.sintl.2020.100003
58. Alarifi, I.M.: Investigation the conductivity of carbon fiber composites focusing on measurement techniques under dynamic and static loads. J. Mat. Res. Tech. **8**(5), 4863–4893 (2019). https://doi.org/10.1016/j.jmrt.2019.08.019
59. Shahapurkar, K., et al.: Comprehensive review on polymer composites as electromagnetic interference shielding materials. Polym. Polym. Compos. 30 (2022). https://doi.org/10.1177/09673911221102127
60. Pantarotto, D., Briand, J.P., Prato, M., et al.: Translocation of bioactive peptides across cell membranes by carbon nanotubes. Chem. Comm. **1**, 16–17 (2004). https://doi.org/10.1039/B311254C
61. Acharya, S., Sahoo, S.K.: PLGA nanoparticles containing various anticancer agents and tumour delivery by EPR effect. Adv. Drug Deliv. Rev. **63**(3), 170–183 (2011). https://doi.org/10.1016/j.addr.2010.10.008
62. Probst, C.E., Zrazhevskiy, P., Bagalkot, V., et al.: Quantum dots as a platform for nanoparticle drug delivery vehicle design. Adv. Drug Deliv. Rev. **65**(5), 703–718 (2013). https://doi.org/10.1016/j.addr.2012.09.036
63. Medina-Sánchez, M., Miserere, S., Morales-Narvaez, E., et al.: On-chip magneto-immunoassay for Alzheimer's biomarker electrochemical detection by using quantum dots as labels. Biosens. Bioelectron. **54**, 279–284 (2014). https://doi.org/10.1016/j.bios.2013.10.069
64. Lombardo, D., Calandra, P., Pasqua, L., et al.: Self-assembly of organic nanomaterials and biomaterials: the bottom-up approach for functional nanostructures formation and advanced applications. Materials **13**(5), 1048 (2020). https://doi.org/10.3390/ma13051048
65. Kumar, H., Venkatesh, N., Bhowmik, H., et al.: Metallic nanoparticle: a review. Biomed. J. Sci. Tech. Res. **4**(2), 3765–3775 (2018)
66. Gupta, P., Tamrakar, G., Biswas, S., et al.: Quantum dots as a fluorescent sensor for detecting heavy metal ions in aqueous environment: an overview. ES Chem. Sustain. **1**(1359), 1–12 (2024)
67. Mody, V.V., Siwale, R., Singh, A., et al.: Introduction to metallic nanoparticles. J. Pharm. Bioallied Sci. **2**(4), 282–289 (2010)
68. Thakkar, K.N., et al.: Biological synthesis of metallic nanoparticles. Nanomed. Nanotechnol. Biol. Med. **6**(2), 257–262 (2010)

69. Blackman, J.: Metallic Nanoparticles, vol. 5. Elsevier. (2008)
70. Fadeel, B., Garcia-Bennett, A.E.: Better safe than sorry: understanding the toxicological properties of inorganic nanoparticles manufactured for biomedical applications. Adv. Drug Deliv. Rev. **62**(3), 362–374 (2010)
71. Wolfram, J., Zhu, M., Yang, Y., Shen, J., et al.: Safety of nanoparticles in medicine. Curr. Drug Targets **16**(14), 1671–1681 (2015)
72. Mohajerani, A., Burnett, L., Smith, J.V., et al.: Nanoparticles in construction materials and other applications, and implications of nanoparticle use. Materials **12**(19), 3052 (2019)
73. Yang, L., Gong, L., Wang, P., et al.: Recent advances in lipid nanoparticles for delivery of mRNA. Pharmaceutics **14**(12), 2682 (2022). https://doi.org/10.3390/pharmaceutics14122682
74. Giroux, M., Zahra, Z., Salawu, O.A., et al.: Assessing the environmental effects related to quantum dot structure, function. Synth. Expos. Environ. Sci. Nano **9**(3), 867–910 (2022). https://doi.org/10.1039/d1en00712b
75. Kamaly, N., Yameen, B., Wu, J., et al.: Degradable controlled-release polymers and polymeric nanoparticles: mechanisms of controlling drug release. Chem. Rev. **116**(4), 2602–2663 (2016). https://doi.org/10.1021/acs.chemrev.5b00346
76. Batra, S.: Core Principles and Practices of Nanotechnology. Educohack Press (2025)
77. Mc, I.: Common nano-materials and their use in real world applications. Sci. Prog. **95**(1), 1–22 (2012)
78. Ratner, M.A., Ratner, D.: Nanotechnology: A Gentle Introduction to the Next Big Idea. Prentice Hall Professional (2003)
79. Bhushan, B.: Introduction to nanotechnology. In: Springer Handbook of Nanotechnology, pp. 1–19. Springer, Berlin, Heidelberg (2017)
80. Barman, B., Banjare, M.K.: Deep understanding of corrosion inhibition mechanism based on first-principle calculations. In: Computational Modelling and Simulations for Designing of Corrosion Inhibitors, pp. 55–78. Elsevier (2023). https://doi.org/10.1016/B978-0-323-95161-6.00017-5
81. Dai, H., Hafner, J.H., Rinzler, A.G., et al.: Nanotubes as nanoprobes in scanning probe microscopy. Nature **384**(6605), 147–150 (1996). https://doi.org/10.1038/384147a0
82. Medina, C., Santos-Martinez, M.J., Radomski, A., et al.: Nanoparticles: pharmacological and toxicological significance. Br. J. Pharmacol. **150**(5), 552–558 (2007). https://doi.org/10.1038/sj.bjp.0707130
83. Blecher, K., et al.: The growing role of nanotechnology in combating infectious disease. Virulence **2**(5), 395–401 (2011). https://doi.org/10.4161/viru.2.5.17035
84. Emilian Leucuta, S.: Nanotechnology for delivery of drugs and biomedical applications. Curr. Clin. Pharmacol. **5**(4), 257–280 (2010)
85. Bharathala, S., Sharma, P.: Biomedical applications of nanoparticles. In: Nanotechnology in Modern Animal Biotechnology, pp. 113–132. Elsevier (2019)
86. Labhasetwar, V., Leslie-Peleck, D.L. (eds.): Biomedical Applications of Nanotechnology (2007)
87. Labhasetwar, V., Leslie-Pelecky, D.L. (eds.): Biomedical Applications of Nanotechnology (2007)
88. Pasika, S.R., Bulusu, R., Rao, B.V.K., et al.: Nanotechnology for biomedical applications. In: Nanomaterials: Advances and Applications, pp. 297–327. Springer Nature Singapore, Singapore (2023)
89. Ramos, A.P., Cruz, M.A.E., Tovani, C.B., et al.: Biomedical applications of nanotechnology. Biophys. Rev. **9**(2), 79–89 (2017). https://doi.org/10.1007/s12551-016-0246-2
90. Thwala, L.N., Ndlovu, S.C., Mpofu, K.T., et al.: Nanotechnology-based diagnostics for diseases prevalent in developing countries: current advances in point-of-care tests. Nanomaterials (Basel, Switzerland) **13**(7), 1247 (2023). https://doi.org/10.3390/nano13071247
91. Baiguera, S., et al.: Tissue engineered scaffolds for an effective healing and regeneration: reviewing orthotopic studies. Biomed. Res. Int. 398069 (2014). https://doi.org/10.1155/2014/398069

92. Han, X., Xu, K., Taratula, O., et al.: Applications of nanoparticles in biomedical imaging. Nanoscale **11**(3), 799–819 (2019). https://doi.org/10.1039/c8nr07769j
93. Sabit, H., Abdel-Hakeem, M., Shoala, T., et al.: Nanocarriers: a reliable tool for the delivery of anticancer drugs. Pharmaceutics **14**(8), 1566 (2022). https://doi.org/10.3390/pharmaceutics14081566
94. Wang, L., et al.: The antimicrobial activity of nanoparticles: present situation and prospects for the future. Int. J. Nanomed. **12**, 1227–1249 (2017). https://doi.org/10.2147/IJN.S121956
95. Vaghasiya, J.V., et al.: Wearable sensors for telehealth based on emerging materials and nanoarchitectonics. NPJ Flex. Electron. **7**, 26 (2023). https://doi.org/10.1038/s41528-023-00261-4
96. Debbage, P.: Targeted drugs and nanomedicine: present and future. Curr. Pharm. Des. **15**(2), 153–172 (2009)
97. Kim, D.K., Dobson, J.: Nanomedicine for targeted drug delivery. J. Mater. Chem. **19**(35), 6294–6307 (2009)
98. Tewabe, A., Abate, A., Tamrie, M., et al.: Targeted drug delivery—from magic bullet to nanomedicine: principles, challenges, and future perspectives. J. Multidiscip. Healthc. 1711–1724 (2021)
99. Ray, S.S., Bandyopadhyay, J.: Nanotechnology-enabled biomedical engineering: current trends, future scopes, and perspectives. Nanotechnol. Rev. **10**(1), 728–743 (2021). https://doi.org/10.1515/ntrev-2021-0052
100. Ibrahim, K.E., Bakhiet, A.O., Khan, A., et al.: Recent trends in biomedical applications of nanomaterials. Biosci. Biotech. Res. Asia **15**(2), 235–243 (2018)
101. Barman, B., Banjare, M.K.: Study viscometric, surface tension behaviors of 5 surfactant (CTAB), anionic surfactant (SDS) and non-ionic surfactant (Tx-100) at different concentration. In: Conference: Role of Applied Sciences in Social Implications, International Conference IC-RASSI-2023, ISBN 978-81-956770-8-5 (2023)
102. Kim, D.K., Dobson, J.: Nanomedicine for targeted drug delivery. J. Mater. Chem. **19**(35), 6294–6307 (2009). https://doi.org/10.1039/B902711B
103. Xie, X., Zhang, Y., Li, F., et al.: Challenges and opportunities from basic cancer biology for nanomedicine for targeted drug delivery. Curr. Cancer Drug Targets **19**(4), 257–276 (2019). https://doi.org/10.2174/1568009618666180628160211
104. Javaid, M., Haleem, A., Singh, R.P., et al.: Exploring the potential of nanosensors: a brief overview. Sens. Int. **2**, 100130 (2021). https://doi.org/10.1016/j.sintl.2021.100130
105. Banjare, M.K., Tandon, D.: Study the solute-solvent molecular interactions of L-Glutamine and conventional surfactants. J. Mol. Liq. **395**, 123893 (2024). https://doi.org/10.1016/j.molliq.2023.123893
106. Rajana, V.K., et al.: Environmental nano-toxicology measuring device. GB Patent 6408599 (2024)
107. Chaturwedi, A.K., Kashyap, N.K., Gupta, P., et al.: Industrial effluents and their impact on water pollution—an overview. ES General **5**(1203), 1–8 (2024)
108. Kim, E.S., Ahn, E.H., Dvir, T., et al.: Emerging nanotechnology approaches in tissue engineering and regenerative medicine. Int. J. Nanomed. **9**(sup1), 1–5 (2014)
109. Mclaughlin, S., Podrebarac, J., Ruel, M., et al.: Nano-engineered biomaterials for tissue regeneration: what has been achieved so far? Front. Mater. Sci. **3**, 27 (2016)
110. Chaturwedi, A.K., Hait, M.: Hydrochemical investigation of water bodies from the effluents of sirgitti industrial area of Bilaspur (C.G.). Bull. Env. Pharmacol. Life Sci. **12**(10), 177–183 (2023)
111. Gupta, P., Biswas, S., Chaturwedi, A.K., et al.: Health impact of ground water sample and its effect on flora of Kanker District, Chhattisgarh, India. ES Food Agrofor. **13**, 940 (2023)
112. Hasan, A., Morshed, M., Memic, A., et al.: Nanoparticles in tissue engineering: applications, challenges and prospects. Int. J. Nanomed. **13**, 5637–5655 (2018). https://doi.org/10.2147/IJN.S153758
113. Fabbro, A., et al.: Carbon nanotubes in neuro regeneration and repair. Adv. Drug Deliv. Rev. **65**(15), 2034–2044 (2013). https://doi.org/10.1016/j.addr.2013.07.002

114. Gunasekaran, T., et al.: Silver nanoparticles as real topical bullets for wound healing. J. Am. Coll. Clin. Wound Spec. **3**(4), 82–96 (2012). https://doi.org/10.1016/j.jcws.2012.05.001
115. Kim, S.J., Choi, S.J., Jang, J.S., et al.: Innovative nanosensor for disease diagnosis. Acc. Chem. Res. **50**(7), 1587–1596 (2017). https://doi.org/10.1021/acs.accounts.7b00047
116. Banjare, M.K., Tandon, D.K.: Study the effect of L-arginine and L-tyrosine on the surface adsorption and micellar properties of long-chain imidazolium-based ionic liquid. Results Chem. 101452 (2024). https://doi.org/10.1016/j.rechem.2024.101452
117. Cross, L.M., Thakur, A., Jalili, N.A., et al.: Nanoengineered biomaterials for repair and regeneration of orthopedic tissue interfaces. Acta Biomater. **42**, 2–17 (2016). https://doi.org/10.1016/j.actbio.2016.06.023
118. Lau, H.C., et al.: Nanotechnology for oilfield applications: challenges and impact. J. Pet. Sci. Eng. **157**, 1160–1169 (2017). https://doi.org/10.1016/j.petrol.2017.07.06
119. Ray, P.C., et al.: Toxicity and environmental risks of nanomaterials: challenges and future needs. J. Environ. Sci. Health C **27**(1), 1–35 (2009)
120. Shafique, M., Luo, X.: Nanotechnology in transportation vehicles: an overview of its applications, environmental, health and safety concerns. Materials **12**(15), 2493 (2019)
121. Roco, M.C., Hersam, M.C., Mirkin, C.A., et al.: Nanotechnology environmental, health, and safety issues. In: Nanotechnology Research Directions for Societal Needs in 2020: Retrospective and Outlook, pp. 159–220 (2011)
122. Singh, H.: Nanotechnology applications in functional foods; opportunities and challenges. Prev. Nutr. Food Sci. **21**(1), 1 (2016)
123. Iavicoli, I., Leso, V., Ricciardi, W., et al.: Opportunities and challenges of nanotechnology in the green economy. Environ. Health **13**, 1–11 (2014)
124. Ventre, M., Coppola, V., Iannone, M., et al.: Nanotechnologies for tissue engineering and regeneration. In: Nanotechnologies in Preventive and Regenerative Medicine, pp. 93–206. Elsevier (2018). https://doi.org/10.1016/B978-0-323-48063-5.00002-2
125. Capponi, G., et al.: Breakthrough innovations and where to find them. Res. Policy **51**(1), 104376 (2022). https://doi.org/10.1016/j.respol.2021.104376
126. Nikolova, M., et al.: Nanotechnology in medicine. In: Drug Discovery and Evaluation: Methods in Clinical Pharmacology, pp. 533–546 (2020)
127. Köhler, J.M.: Challenges for Nanotechnology. Encyclopedia **1**(3), 618–631 (2021). https://doi.org/10.3390/encyclopedia1030051
128. Jaldurgam, F.F., et al.: Synthesis and performance of large-scale cost-effective environment-friendly nanostructured thermoelectric materials. Nanomater **11**(5), 1091 (2021). https://doi.org/10.3390/nano11051091
129. Morley, J., Murphy, L., Mishra, A., et al.: Governing data and artificial intelligence for health care: developing an international understanding. JMIR Form. Res. **6**(1), e31623 (2022). https://doi.org/10.2196/31623
130. Anonymous.: Institute of Medicine (US). Assessing the Effects of the Gulf of Mexico Oil Spill on Human Health: A Summary of the June 2010 Workshop. National Academies Press (US), Washington (DC); Short- and Long-Term Effects on Human Health (2010). https://www.ncbi.nlm.nih.gov/books/NBK209917/
131. Li, W.: Advanced manufacturing and precision machining. Appl. Sci. **14**(24), 11642 (2024). https://doi.org/10.3390/app142411642
132. Allhoff, F., Lin, P., Moor, J.H., et al.: Nanoethics: The Ethical and Social Implications of Nanotechnology. Wiley & Sons (2007)
133. Sandler, R.: Nanotechnology and social context. Bull. Sci. Technol. Soc. **27**(6), 446–454 (2007). https://doi.org/10.1177/0270467607308288
134. Barman, B., Banjare, M.K.: Review of green designer deep eutectic solvents (DESs) production and prospective material science applications. In: Futuristic Trends in Chemical Material Science & Nano Technology, e-ISBN: 978-93-5747-824-3, IIP series, vol. 3, Book 6, Part 2, Chapter 8 (2024). https://doi.org/10.58532/V3BJCS6P2CH8
135. Allhoff, F., Lin, P., Moor, J.H., et al.: Nano Ethics: The Ethical and Social Implications of Nanotechnology. Wiley & Sons (2007)

136. Khan, A.S. (ed.): Nanotechnology: Ethical and Social Implications. CRC Press (2012)
137. Babatunde, D.E., Denwigwe, I.H., Babatunde, O.M., et al.: Environmental and societal impact of nanotechnology. IEEE Access **8**, 4640–4667 (2019)
138. Pushparaj, K., Liu, W.C., Meyyazhagan, A., et al.: Nano-from nature to nurture: a comprehensive review on facets, trends, perspectives and sustainability of nanotechnology in the food sector. Energy **240**, 122732 (2022). https://doi.org/10.1016/j.energy.2021.122732
139. Wiek, A., et al.: Nanotechnology for sustainability: what does nanotechnology offer to address complex sustainability problems? In: Nanotechnology for Sustainable Development, pp. 371–390. Springer International Publishing (2014). https://doi.org/10.1007/978-3-319-05041-6_30
140. Palit, S., Hussain, C.M.: Green sustainability, nanotechnology and advanced materials–a critical overview and a vision for the future. Green Sustain. Adv. Mat. Appl. **2**, 1–18 (2018)
141. Weiss, P.S., Lewis, P.A. (eds.): Sustainability through nanotechnology. ACS Nano **4**(3), 1249–1250 (2010)
142. Steinfeldt, M., Petschow, U., Haum, R., et al.: Nanotechnology and sustainability. Schriftenreihe des IÖW **167**, 03 (2004)
143. Barman, B., Banjare, M.K.: Review study the physicochemical properties of ionic liquids and its tremendous significant. In: Futuristic Trends in Chemical Material Science & Nano Technology, e-ISBN: 978-93-5747-824-3, IIP series, vol. 3, Book 6, Part 2, Chapter 9 (2024). https://doi.org/10.58532/V3BJCS6P2CH9
144. Huston, M., DeBella, M., DiBella, M., Gupta, A.: Green synthesis of nanomaterials. Nanomater **11**(8), 2130 (2021). https://doi.org/10.3390/nano11082130
145. El Zein, B., Elrashidi, A., Dahlan, M., et al.: Nano and Society 5.0: Advancing the Human-Centric Revolution (2024). https://doi.org/10.5772/intechopen.1004221
146. Tawiah, B., et al.: Nanotechnology in societal development. In: Nanotechnology in Societal Development, pp. 1–64. Springer Nature Singapore, Singapore (2024)
147. Barman, B., Banjare, M.K.: A review on the study of physicochemical properties and application of surfactants. In: Futuristic Trends in Chemical, Material Sciences & Nano Technology, e-ISBN: 978-93-5747-824-3, IIP Series, vol. 3, Book 4, Part 2, Chapter 5 (2024). https://doi.org/10.58532/V3BJCS6P2CH5
148. Ramírez-Márquez, C., Posadas-Paredes, T., Raya-Tapia, A.Y., et al.: Natural resource optimization and sustainability in society 5.0: a comprehensive review. Resources **13**(2), 19 (2024). https://doi.org/10.3390/resources13020019
149. Ray, P.C.: Toxicity and environmental risks of nanomaterials: challenges and future needs. J. Environ. Sci. Health C **27**(1), 1–35 (2009). https://doi.org/10.1080/10590500802708267
150. Bahrulolum, H., Nooraei, S., Javanshir, N., et al.: Green synthesis of metal nanoparticles using microorganisms and their application in the agrifood sector. J. Nanobiotechnol. **19**, 86 (2021). https://doi.org/10.1186/s12951-021-00834-3
151. Yurtay, Y.: Carbon footprint management with industry 4.0 technologies and ERP systems in sustainable manufacturing. Appl. Sci. **15**(1), 480 (2025). https://doi.org/10.3390/app15010480
152. Lingaitiene, O., Burinskiene, A.: Development of trade in recyclable raw materials: transition to a circular economy. Economies **12**(2), 48 (2024). https://doi.org/10.3390/economies12020048
153. Schwab, F., Rothen-Rutishauser, B., Scherz, A., et al.: The need for awareness and action in managing nanowaste. Nat. Nanotechnol. **18**(4), 317–321 (2023). https://doi.org/10.1038/s41565-023-01331-4
154. Khatoon, U.T., Velidandi, A.: An overview on the role of government initiatives in nanotechnology innovation for sustainable economic development and research progress. Sustainability **17**(3), 1250 (2025). https://doi.org/10.3390/su17031250
155. Elmer, W., White, J.C.: The future of nanotechnology in plant pathology. Annu. Rev. Phytopathol. **56**(1), 111–133 (2018). https://doi.org/10.1146/annurev-phyto-080417-050108
156. Satyanarayana, T.S.V., Rai, R.: Nanotechnology: the future. J. Interdiscip. Dent. **1**(2), 93–100 (2011). https://doi.org/10.4103/2229-5194.85026

157. Jones, R.: The future of nanotechnology. Phys. World **17**(8), 25 (2004). https://doi.org/10.1088/2058-7058/17/8/34
158. Patra, G.K., Hait, M., Vaishnav, M.M., et al.: Biosensing nano materials-based composition for smart drug delivery in parkinson's disease. Patent GR, 2024082219020000DE (2024)
159. Banjare, M.K., Banjare, B.S.: Study of the molecular interaction of a phosphonium-based ionic liquid within myo-inositol and non-steroidal anti-inflammatory drugs. RSC Adv. **14**(5), 2961–2974 (2024). https://doi.org/10.1039/D3RA07721G
160. Mulder, W.J., Strijkers, G.J., Van Tilborg, G.A., et al.: Nanoparticulate assemblies of amphiphiles and diagnostically active materials for multimodality imaging. Acc. Chem. Res. **42**(7), 904–914 (2009). https://pubs.acs.org/doi/abs/10.1021/ar800223c#
161. Banjare, B.S., Banjare, M.K.: Impact of carbocyclic sugar-based myo-inositol on conventional surfactants. J. Mol. Liq. **384**, 122278 (2023). https://doi.org/10.1016/j.molliq.2023.122278
162. Acharya, S., Sahoo, S.K.: PLGA nanoparticles containing various anticancer agents and tumor delivery by EPR effect. Adv. Drug Deliv. Rev. **63**(3), 170–183 (2011). https://doi.org/10.1016/j.addr.2010.10.008
163. Zhang, F., Nangreave, J., Liu, Y., et al.: Structural DNA nanotechnology: state of the art and future perspective. J. Am. Chem. Soc. **136**(32), 11198–11211 (2014)
164. Zahid, M., Kim, B., Hussain, R., et al.: DNA nanotechnology: a future perspective. Nanoscale Res. Lett. **8**, 1–13 (2013)
165. Wang, D.K., et al.: Nanotechnology applications for cardiovascular disease treatment: current and future perspectives. Nanomed. Nanotechnol. Biol. Med. **34**, 102387 (2021)

Multifunctional NM: Integrating Optical, Thermal, and Electrical Properties for Biomedical Devices

S. Rahila, M. Nishanthi, A. Priyadharshini, P. Manikandan, H. Mohammed Ameen, G. Ganesh, and M. Syed Muzammil

Abstract The development of multifunctional NanoMaterials (NM) and devices for biomedical applications has garnered significant attention due to their potential to revolutionize healthcare. This study synthesizes and characterizes novel NM with specific properties for targeted drug delivery, imaging, and biosensing. Using a bottom-up approach, we designed and fabricated nanocomposites (NC) made of biocompatible polymers, metallic nanoparticles (NPs), and quantum dots, exhibiting unique optical, magnetic, and electronic characteristics. These NCs were functionalized with specific ligands to facilitate active targeting of cancer cells and pathogens. Furthermore, we developed microfluidic devices to efficiently capture and analyze circulating tumor cells (CTCs) using the synthesized NM. The performance of these NM and devices was evaluated both in vitro and in vivo, demonstrating enhanced drug delivery efficiency, high-resolution imaging, and sensitive biosensing capabilities. This chapter addresses the biocompatibility and long-term stability of the NM under physiological conditions. It highlights that these multifunctional NM and devices are promising for advancing personalized medicine, early diagnosis, and targeted therapy. Additionally, it offers a comprehensive understanding of the design principles and potential applications of multifunctional NM in the biomedical field, paving the way for future research and clinical translation.

Keywords Nanotechnology · Nanomedicine · Biomedical applications

1 Introduction

Nanotechnology has ushered in a new era of possibilities in biomedical applications. The unique properties of NM, such as their small size, high surface area-to-volume ratio, and the ability to be engineered with specific functionalities, make

S. Rahila · M. Nishanthi · A. Priyadharshini · P. Manikandan · H. M. Ameen · G. Ganesh · M. S. Muzammil (✉)
Department of Chemistry (S&H), Bharathidasan Engineering College, Natrampalli, Tirupattur District (Affiliated to Anna University, Chennai, TamilNadu, India) 635 854, India
e-mail: syedmuzammil@bec.ac.in

S. N. Saud Al-Humairi et al. (eds.), *Smart Nanomaterials in Biomedical Applications*, Engineering Materials, https://doi.org/10.1007/978-981-96-8023-8_14

them ideal candidates for various medical applications. In particular, the development of multifunctional NM and devices can revolutionize how we diagnose and treat diseases, enabling more precise, targeted, and effective interventions. One of the most promising applications of multifunctional NM is in the field of targeted drug delivery. Traditional drug delivery methods often suffer from low specificity, leading to off-target effects and systemic toxicity [1]. NM can be engineered with specific ligands that selectively target cancer cells or pathogens, delivering the drug payload directly to the site of interest. This increases the efficacy of the treatment and reduces the risk of side effects. Moreover, the ability to control the release of the drug over time can further enhance the therapeutic outcome. Bioactive materials, capable of eliciting biological responses upon interacting with proteins, cells, or tissues within a living organism, have become a focus of increasing interest in recent years. Unlike traditional biomaterials, bioactive materials are characterized by their unique bioactive properties. These properties include the ability to bond with both hard and soft tissues, stimulate cell adhesion, differentiation, and proliferation, mimic biological matrices for tissue regeneration, target specific proteins or cells for biomimicry, release bioactive ions or molecules, possess catalytic functions, and support targeted drug delivery. These features enable bioactive materials to impact cellular behavior and trigger specific reactions in living tissues, making them highly useful in diagnostics, therapy, and regenerative medicine. The term "bioactive materials" was first introduced in the late 1960s when Larry Hench discovered that bioglass could bond with bone tissue, forming a strong interface [2]. Larry Hench later defined bioactive materials as distinct biomaterials that can trigger biological responses at the interface between the material and tissue by forming strong bonds or interactions with surrounding tissues. His discovery marked a significant change in understanding how materials interact with tissues. Following this, it became clear that such interactions did not necessarily pose a risk to human health as long as the reactions produced beneficial outcomes. This realization opened up new opportunities for the use of bioactive materials in biomedical applications. For example, studies have shown that modifying the composition of bioactive glasses or combining them with materials like metals, polymers, or inorganic substances can create bioactive materials with remarkable properties for tissue repair and regeneration. Prior to this breakthrough, biomaterials were thought to be inert and not react with cells or tissues, with the only known tissue responses being inflammation or foreign body reactions. This new understanding has expanded the potential uses of bioactive materials, particularly in regenerative medicine and beyond. By 1999, the definition of biomaterials had evolved to describe them as "materials designed to interact with biological systems to assess, treat, enhance, or replace any tissue, organ, or body function." This marked a shift from traditional biomaterials to bioactive materials. With advances in material science and nanotechnology, the concept of bioactive materials expanded well beyond Larry Hench's original idea. Nanotechnology, in particular, enabled the development of bioactive NM with specialized properties. As a distinct category of biomaterials, these bioactive NM are not simply scaled-down versions of larger materials. Their nanoscale size, large surface area, and precise nanostructure create unique bioactivities that significantly impact how these materials interact with biological systems. These advancements have greatly expanded the range of potential applications for bioactive materials. The creation of structures with nanoscale precision has demonstrated significant potential in influencing cellular behaviors, including adhesion, differentiation, proliferation, and recognition. Additionally, bioactive NM has proven effective in interacting with proteins, nucleic acids, saccharides, signaling molecules, and complex extracellular matrices (ECMs). Thanks to their unique physicochemical properties and bioactivities, bioactive NM are expected to serve as cutting-edge platforms in personalized medicine, potentially transforming the pharmaceutical industry. Beyond these uses, bioactive NM can act as carriers to enhance the accuracy and effectiveness of delivering therapeutic or diagnostic agents to targeted cells and tissues. It is essential to differentiate between bioactive NM and responsive NM. Bioactive NM represents a significant subset of NM capable of inducing biological responses when interacting with proteins, cells, or tissues, influencing cellular activities, and triggering specific reactions in living tissues. In contrast, responsive NM is engineered to react to various biological signals such as tissue-specific pH, redox potentials, enzyme concentrations, and external triggers like light and heat. Upon exposure to these stimuli, responsive NM undergoes structural changes, which modify their physicochemical properties, such as surface charge, the exposure of cell-penetrating peptides or cell-targeting ligands, and controlled drug release. Responsive NM is increasingly being used to develop advanced drug delivery systems. Bioactive NM is applied in a wide range of important medical areas. However, a comprehensive review of bioactive NM and the factors influencing their bioactivity is still missing. This chapter provides an overview of the latest developments in bioactive NM, exploring the factors that affect their bioactivities, such as their physical structure, surface properties, and nanotopography [3, 4]. It also discusses different types of bioactive NM, including inorganic, carbon-based, polymeric, and supramolecular NM. Furthermore, the chapter highlights key applications of bioactive NM, such as their roles in wound healing, cancer treatment, neurodegenerative disease therapy, and biocatalysis. Finally, it addresses the challenges and potential future directions for bioactive NM, emphasizing their promise for further advancements and impact in medicine.

2 Properties

NM stands out from other materials due to its increased surface area and quantum effects. Their nanoscale size, high surface area-to-volume ratio, and ability to alter their surfaces give rise to exceptional biological, magnetic, and optical properties. These unique characteristics allow them to influence processes at the cellular level. The electrical properties of NM are determined by their size, with characteristics that range from metallic to semiconducting depending on their diameter due to their low crystal defect density [5]. NM displays exceptionally high electrical conductivity. Their thermal conductivity is also greatly enhanced, owing to the strong vibration of covalent bonds. Thermal conductivity is up to ten times higher than that of metals, and this is further improved by the reduced defect density found in quartz. Moreover, NM is notably strong and capable of withstanding extreme strain. Their flawless crystal structures enable them to endure bending without breaking.

3 Classification

NM can be classified in various ways, such as by their chemical composition, dimensionality, application, or role within a social system. When categorized by their phases, NM can include single-phase solids, examples of crystalline or amorphous particles, and layers. Multi-phase solids: This category contains coated particles and matrix composites. Multi-phase systems: Examples include aerogels, colloids, ferrofluids, nanoparticles, nanotubes, dendrimers, quantum dots, and micelle formations. Figure 1 shows the Multifunctional applications of Nanotechnology. From a chemical standpoint, NM is classified as organic or inorganic. Organic NM involves carbon-based structures like carbon nanotubes (CNTs), fullerenes, graphene, dendrimers, and polymer nanoparticles. Inorganic NM consists of materials like metal oxide NPs, semimetal oxides, metal NPs, and semiconductor quantum dots. NM can also be classified by their dimensions [6, 7]. Figure 2 designates the Multifunctional potentials of bioactive NM.

0D (zero-dimensional): Quantum dots.

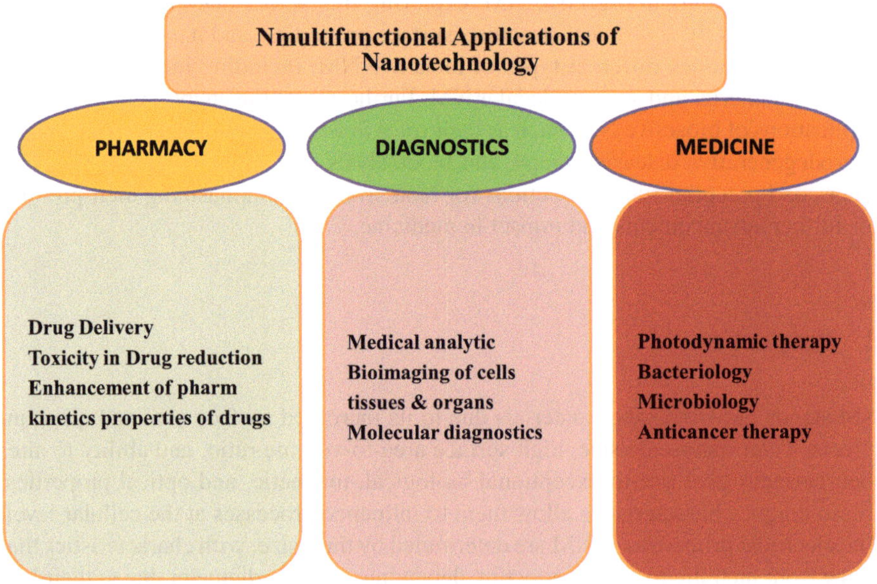

Fig. 14.1 Multifunctional applications of nanotechnology

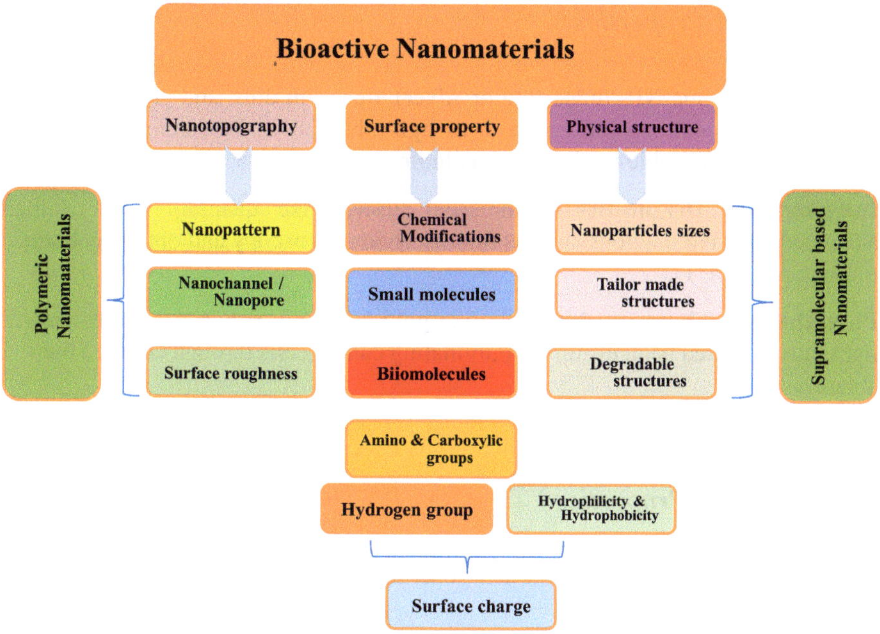

Fig. 14.2 Multifunctional potentials of bioactive NM

1D (one-dimensional): Nanorods, nanowires, and tubes.
2D (two-dimensional): Nanoplatelets and layers.
3D (three-dimensional): Nanofibers and structures.

Moreover, NM can be differentiated by their phases, such as single-phase or multi-phase solids, highlighting the variety of ways these versatile materials can be categorized.

4 Multifunctional NM

NM is drawing significant attention for their role in advancing biomedical devices due to their ability to integrate unique optical, thermal, and electrical properties. These materials promise improved performance, adaptability, and efficiency for applications in diagnostics, imaging, therapy, and tissue engineering.

4.1 Optical Properties

NM like plasmonic NPs (e.g., gold or silver), quantum dots, and graphene exhibit exceptional optical capabilities, such as light absorption, emission, or scattering. These properties enable applications including.

Imaging and Diagnostics—Fluorescent probes like quantum dots facilitate advanced imaging methods such as fluorescence microscopy and optical coherence tomography (OCT).

Photothermal Therapy—Materials like gold NPs and carbon nanotubes convert light into localized heat, useful for eliminating cancer cells through targeted thermal ablation.

4.2 Thermal Properties

NM can be designed to manipulate heat generation and transfer with precision, critical in areas such as.

Device Thermal Management—CNTs, graphene, and metallic NPs excel as thermal conductors, aiding heat dissipation in sensitive devices like implants or sensors.

Cancer Hyperthermia Treatments—NM with photothermal functionality allow targeted heating of cancer cells while minimizing damage to surrounding tissues.

4.3 Electrical Properties

NM such as graphene, conductive polymers, and CNTs are utilized for their electrical conductivity in key applications.

Biosensors—These materials enable the development of highly sensitive biosensors for detecting biomolecules or pathogens, aiding in early disease diagnosis.

Neural Interfaces—Conductive NM are pivotal for creating neural probes and brain-machine interfaces (BMIs) that monitor and stimulate neural activity, with applications in therapies like deep brain stimulation.

Wearable Sensors—Flexible and stretchable NM enhance wearable devices for monitoring signals like ECG and EMG, contributing to real-time health tracking.

4.4 *Multi-Functionality in Biomedical Applications*

Integrating optical, thermal, and electrical properties into a single NM or device enables groundbreaking multifunctionality:

Theranostic—Dual-purpose NM serve both diagnostic and therapeutic roles, combining imaging capabilities with targeted therapy, such as photothermal treatment.

Smart Implants—Multifunctional materials can create responsive implants that adapt to external stimuli for tissue regeneration and dynamic interaction with their environment.

Controlled Drug Delivery—NMs with tunable properties allow targeted drug release upon activation by light or electric fields, ensuring efficient treatment at precise locations.

4.5 *Challenges and Future Directions*

While promising, multifunctional NM faces several hurdles.

Safety—Biocompatibility and toxicity remain crucial, as some materials may cause adverse biological reactions.

Scalability—Consistent and efficient large-scale production remains challenging for these sophisticated materials.

Regulatory Barriers—Addressing long-term health impacts is necessary for regulatory approvals.

Future advancements in nanotechnology, materials science, and biomedical engineering aim to enhance these materials' biocompatibility and precise control over their properties, driving the evolution of next-generation biomedical devices that revolutionize healthcare.

4.6 *Biomedical Applications of Bioactive NM*

Various types of bioactive NM have been developed in recent decades. This section focuses on some key biomedical applications of these materials, including their use in wound healing, cancer immunotherapy, the treatment of neurodegenerative diseases, and as biocatalysts. Table 1 shows the Multifunctional advanced potentials of bioactive NM and their uses.

i. **Wound healing**

Table 1 Multifunctional advanced potentials of bioactive NM and their uses [5]

NM	Major types	Bioactive potentials	Uses
aCD47@$CaCO_3$	Inorganic (Io) NM	Release ions	Tumour immunology
Ag NP			Antibacterial agents
AmpF peptide	Supramolecular-based NM	Targeted drug delivery	Combinatorial chemo-photodynamic therapy
Ab-NPs	NM polymers	Controlled exterior properties	Tumour immunology
Au NP	Io -NM	Controlled surface properties for biomimetics	Inhibition of Aβ fibrillization
Az@MOF	Io-organic hybrid NM	Drug delivery targets	Facilitating microglia-mediated Aβ clearance
Aβ detection element-tailoredNP	NM polymers	Controlled surface properties	Attenuating Aβ induced neuronapoptosis
Bioglass NP	Io-NM	Release ions	Tumour immunology
BNNS@CuS		–	Detection of the total cholesterol in human serum
Boronate-based MINP	NM Polymers	Controlled exterior properties for biomimetics	Inhibition of tumor growth via blocking HER2-dependent signaling pathway
BSA-IrO_2 NP	Io -NM	CT imaging	Tumour theranostics
CS/nHAp/nCu-Zn scaffold		Assist cell proliferation, capability to bind hard/soft tissues	Cell penetration and bone tissue formation
Cyclodextrins and calixarenes	Supramolecular-based NM	Biorecognition	Attenuating Aβ-induced neuron damages and antidotes
Dopamine-customized MSN	Io-NM	Promote cell migration	Wound healing
Fe_3O_4 MNP		Catalytic activities	Peroxidase-Immunoassays
GOx/hemin@ZIF-8	Io-organic hybrid NM		Detection—glucose in drink, blood and urine
Graphene oxide	Carbon NM	–	Activation of autophagy of neurons andmicroglial cells

(continued)

Table 1 (continued)

NM	Major types	Bioactive potentials	Uses
Guanidinium-modified pillar [5] arene	Supramolecular-based NM	Biorecognition	Anti-wound infections
Hydroxyapatite NP	Io -NM	Facilitate cell proliferation	Promotion of osteoblast-like cell proliferation
Iron-based NP		Release ions	Ferroptotic cancer therapy
Magnetic MINP	NM polymers	Controlled facade properties for biomimetics	Prostate cancer therapy
mBiNE		Controlled surface properties	Cancer immunology
MINP		Controlled surface properties for biomimetics	Anti-toxin
MOF	Io-organic hybrid NM	Catalytic activities	Peroxidase-mimicking NM
MSN-GACs	Io-NM	Increase cell adhesion	Hemostasis
MSPMs	NM polymers	Controllable surface properties	Inhibiting amyloid protein aggregation
MWCNTs	Carbon-based NM	Induce cell differentiation	Osteogenic differentiation of mesenchymal stem cells
Nanoinhibitor	NM polymers	Controlled facade properties for altering cell behavior	Glioblastoma treatment
nChap			Cancer immunology
PC7A-NP-based nanovaccine		Controlled surface properties	
Peptide conjugated Au NP	Io-NM	Catalytic activities	
Platinum NPs/graphene oxide	Carbon-based NM		Cancer cell detection
PLGA NP	NM polymers	Controlled surface properties for antigen capture	Cancer immunology

(continued)

Table 1 (continued)

NM	Major types	Bioactive potentials	Uses
RADA16 peptide	Supramolecular NM	Identify specific proteins/cells for biomimetics	Hemostasis
RGO-PDA	Carbon NM	Induce cell differentiation	Cell adhesion, proliferation, osteogenic differentiation
SWCNTs		Stimulate cell adhesion, induce cell differentiation	Cell adhesion and neurogenic differentiation of mesenchymal stem cells
SWNTs-based nanozyme		Catalytic activities	Treatment of methamphetamine (METH) addiction
Transformable peptide	Supra molecular NM	Identify specific proteins/cells for biomimetics	Trapping the bacteria by mimicking antibacterial peptide human defensin-6
Transformable peptide			Disrupting dimerization of HER2 for cancer treatment
UCNP/ICG/RB-mal	Io-NM	Controlled surface properties for antigen capture	Cancer immunology
V-ZIF		Release ions/ molecules	Biofilm-linked infection treatment
Zn2(ppa)2(1,3-bdc)(H_2O)			Antibacterial agents
αFc-NP	NM polymers	Controlled facade properties for biomimetics	Tumour immunology

Wound healing is a natural process triggered by internal or external factors that cause tissue or organ damage. When an injury occurs, the body activates a series of self-repair mechanisms, including hemostasis, infection control, cell proliferation, tissue remodeling, and ultimately the restoration of the damaged tissue. The duration of the healing process primarily depends on the extent of the injury. In severe damage, the body may be unable to heal itself, leading to chronic or non-healing wounds. Various bioactive NMs have been developed or are under development to overcome these challenges. They are known for their biocompatibility, antimicrobial properties, and ability to deliver drugs, and these materials have shown promise in supporting and enhancing wound healing [8]. This section reviews recent advancements in bioactive

NM that aim to improve hemostasis, prevent infections, and promote cell proliferation during the healing process.

ii. **Hemostasis**

Hemostasis is the first step in wound healing, focusing on sealing injury sites or damaged blood vessels. Traditional materials such as gauze and cotton are commonly used to stop bleeding and protect the wounded area during surgeries. However, these conventional methods often provide insufficient hemostatic results. Advances in nanotechnology have introduced new possibilities for more effective bleeding control. For example, NM with larger surface areas and porous structures has been developed to absorb blood and enhance bleeding control quickly. A fast-acting hemostatic sponge based mesoporous silica nanoparticles (MSNs) combined with glycerol-modified N-alkylated chitosan (GACs). This MSNGAC sponge showed significantly better blood absorption and hemostatic efficiency than the commercial Combat Gauze (CG). In a similar effort, a gelatin-based nanofiber sponge for rapid hemostasis, featuring a macroporous structure that enables quick liquid diffusion and exceptional water absorption. Additionally, cross-linked graphene with a hierarchical porous structure has been identified as an effective hemostatic material. NM can also be directly applied to injury sites through sprays or injections, where they form nanofibers or hydrogels to stop bleeding at the site effectively [9].

Recently, bioactive NM-like self-assembling peptides (SAPs), cyanoacrylate (CA), nano-chitosan, alginate, and MSN-integrated materials have shown promising hemostatic abilities. Among these, peptides are particularly attractive for constructing nanofibers and hydrogels due to their biodegradability and biocompatibility. The structure and properties of self-assembling peptides (SAPs) can be tailored by modifying their peptide sequences. In 2006, the use of SAPs in hemostasis by discovering RADA16 (AcN-RADARADARADARADA-$CONH_2$), a peptide that rapidly assembles into nanofibers and forms a gel-like hydrogel when sprayed or injected onto an injury site, achieving hemostasis within 10–15 s. In the following decade, several SAPs were developed, including IAP (Ac KLLKLLLKLLKLLLKLLLKLLKCONH$_2$), KOD ((PLG)4(PHG)4(AHG)4), RATEA16 (CH_3CO-RATARAEARATARAEA-$CONH_2$), and SPG-178 (RLDLR-LALRLDLR), significantly improving patient outcomes. The hemostatic efficiency of SAPs has been further enhanced by incorporating functional fragments into the peptides. For example, two extracellular matrix (ECM)-related motifs, GRGDS, which binds specifically to fibronectin receptors and YIGSR (from laminin-1), into RADA16 to create RADA16-GRGDS and RADA16-YIGSR. These modified nanofibers and hydrogels showed much stronger binding to ECM, effectively stopping bleeding in seconds. However, the high cost of peptides remains a significant challenge for their widespread use in hemostatic applications. Overcoming this barrier will require developing more cost-effective peptide synthesis methods and exploring alternative NM to replace peptides for efficient hemostasis.

iii. **Anti-wound infections**

Wound infection presents a significant challenge for effective wound healing, as skin damage impairs its ability to protect against external microorganisms that can invade the injury site. Such infections can slow down the healing process and, in severe cases, lead to serious consequences like amputation or even death [10]. Therefore, preventing infections is a crucial step in healing after achieving hemostasis. Several bioactive NM, including metal-based nanoclusters (NC), graphene-based NM, and polymer-based NPs, have shown significant antimicrobial properties. These NM work through various antimicrobial mechanisms, which can be generally categorized into three types: NM containing guanidinium (GUA) and quaternary ammonium (QA) groups interact strongly with cell membranes, disrupting biofilms and causing the release of intracellular contents. The effectiveness of these materials in binding to cell membranes increases with the length of the alkyl chains in the GUA/QA groups, with the highest affinity observed at chain lengths of 16 for QA and 4 for GUA. Furthermore, the binding strength can be enhanced by incorporating GUA and QA groups into macrocyclic molecules. These developments underscore the potential of bioactive NM to combat wound infections effectively. For example, a series of pillar [5] arenes, each decorated with quaternary phosphonium (QP) and quaternary ammonium (QA) groups, all having the same alkyl chain length of three, for antimicrobial purposes. Similarly, a novel guanidinium (GUA)-modified pillar [5] arene (GP5). These pillar [5] arene derivatives showed the ability to quickly bind to anionic components in biofilms and bacterial membranes via salt bridges, disrupting membrane integrity and causing bacterial cell lysis. Another strategy involves NM that captures or immobilizes bacteria to prevent invasion [11]. For instance, a human defensin-6 mimic peptide (HDMP), which selectively recognizes bacteria and forms nanofibrous networks in situ to trap them. This approach inhibits bacterial growth and prevents them from invading host cells, significantly improving the survival rate of infected mice. Similarly, a dopamine-modified MSN (DOPA/MSN) for wound healing. In addition to its hemostatic properties, DOPA/MSN self-polymerizes on bacterial surfaces, blocking nutrient uptake and waste release, ultimately leading to bacterial lysis. A third mechanism involves NM that generates reactive oxygen species (ROS) to kill bacteria directly. ROS, particularly singlet oxygen, can induce oxidative damage inside bacterial cells, disrupt membrane potential, and release intracellular contents, leading to bacterial death. Several metal-based NC (ZnO, TiO_2, MgO, V_2O_5, AgNPs) and graphene-based NM have proven effective at generating ROS and killing bacteria without contributing to antibiotic resistance. However, metal-based NC often faces challenges such as solution instability and aggregation, reducing their antibacterial efficiency. To overcome this, these NPs are usually modified with surfactants like tea polyphenol (TP), β-cyclodextrin, chitosan, polyvinylpyrrolidone (PVP), and zwitterionic agents to improve their stability and performance.

iv. **Promote proliferation**

The proliferative phase of wound healing typically starts a few days after an injury and lasts for around 14 days. During this phase, endothelial cells and fibroblasts proliferate rapidly, forming new capillaries through angiogenesis and creating a new extracellular matrix (ECM). Additionally, myofibroblasts move from the surrounding tissues to the wound edges, assisting in the contraction of the wound area [12]. Recent studies have revealed several bioactive NM that effectively promote cell proliferation. These include metal-based NC, silver nanoparticles (AgNPs), gold nanoparticles (AuNPs), nanoceria, bioglass, carbon-based NM like nanotubes and graphene, and nitric oxide-carrier NM. These NM mainly function through biological mechanisms, particularly by enhancing the expression of genes related to growth factors such as vascular endothelial growth factor (VEGF), endothelial growth factor receptors (EGFR), and fibroblast growth factor 2 (FGF2). However, the chemical mechanisms behind these effects remain poorly understood, suggesting the need for further research. In addition, NM can act as scaffolds to support fibroblast migration. For instance, graphene-based nanoclusters (ACG NCs), with poly(vinyl alcohol) (PVA)-based nanofibers, and halloysite nanotubes (HNT), all of which have shown promise in promoting fibroblast activity and aiding in wound repair. These NM have demonstrated the ability to effectively interact with fibroblasts, facilitating their migration and accelerating the wound healing. In cancer immunotherapy, immune modulation has revolutionized treatment strategies by leveraging the body's immune system to target and eliminate tumor cells [13]. This approach offers prolonged antitumor effects with fewer metastasis and recurrence risks than traditional therapies. However, more than 70% of patients experience only temporary responses or do not respond to immunotherapy. This challenge is primarily due to two factors: (1) insufficient activation of the immune system and (2) the immunosuppressive environment within the tumor microenvironment (TME).In recent years, NM has been increasingly used as a delivery system to improve the bioavailability of immune modulators in cancer immunotherapy. Beyond serving as delivery carriers, emerging studies have shown that certain types of NM possess intrinsic properties that can modulate the immune system. This section discusses the recent advancements in developing bioactive NM for immune modulation, highlighting their potential to stimulate immune responses and alleviate the immunosuppressive effects of TME.

v. **Activate T cell-mediated immunity**

Current research in cancer immunotherapy largely centers on T cell-mediated cellular immunity. The successful activation of T cell-mediated antitumor immunity relies on a series of sequential events known as the "cancer-immunity cycle." However, this cycle is often disrupted at various stages, resulting in inadequate immune system activation. Recent studies suggest that NM can influence specific phases of the cancer-immunity cycle, such as improving antigen presentation and promoting the maturation of antigen-presenting cells (APCs), thereby initiating or reactivating a self-sustaining immune response. Regarding antigen presentation, bioactive NM can enhance the interaction between tumor-derived protein antigens (TDPAs) and APCs, improving the uptake and processing of tumor antigens [14]. The surface characteristics of bioactive NM play a critical role in this process, which can be categorized into three main types: (a) hydrophobic surfaces, (b) positively charged surfaces, and (c) surfaces with reactive groups. Bioactive NM with hydrophobic surfaces can capture TDPAs via hydrophobic-hydrophobic interactions. For example,

PLGA-based biodegradable antigen-capturing NPs (AC-NPs) that efficiently capture and deliver TDPAs to APCs, significantly enhancing antigen presentation to T cells and activating T cell-mediated antitumor immune responses. Additionally, AC-NPs boosted the effectiveness of immune checkpoint blockade (ICB)-based immunotherapy in mice with B16F10 melanoma, resulting in a survival rate of up to 20% compared to mice treated with ICB monotherapy.

Likewise, bioactive NM with positively charged surfaces can capture TDPAs through ionic interactions, providing another mechanism to promote immune activation. A mannose-modified stearic acid-grafted chitosan micelle (MChSA) for personalized immunotherapy, demonstrating the ability to capture endogenous antigens and target tumor-draining lymph nodes efficiently. This resulted in significantly improved antigen presentation, strong T-cell activation, and effective tumor growth inhibition. Bioactive nanomaterials (NM) with reactive group surfaces can capture tumor-derived protein antigens (TDPAs) through chemical reactions. For example, the development of upconversion nanoparticles (UCNP)-based antigen-capturing platform (UCNP/ICG/RB-mal) for treating metastatic cancer [15]. When irradiated, UCNP/ICG/RB-mal, containing maleimide (mal), reacted with the exposed sulfhydryl (-SH) groups on TDPAs. These antigens were then delivered to antigen-presenting cells (APCs), enhancing antigen presentation and triggering tumor-specific immune responses. Bioactive NM also aids in promoting APC maturation by initiating inflammatory responses, typically through the release of inflammatory cytokines. A polymeric nanoparticle-based nanovaccine (PC7A-NP) that effectively induced interferon secretion for efficient cancer immunotherapy. After subcutaneous injection, PC7A-NP successfully traveled to lymph nodes and activated the stimulator of interferon genes (STING) pathway, facilitating APC maturation and enhancing T cell-based antitumor immune responses. When combined with an anti-PD-1 antibody, PC7A-NP led to significant tumor suppression, with nearly 100% survival observed over 60 days. Moreover, activating the complement system can further trigger cytokine release. Reddy et al. reported pluronic-stabilized polypropylene sulfide (PPS) NPs that activated the complement system in a hydroxyl (-OH)-dependent manner. They found that lower surface charges of NM led to higher activation efficiency. In addition to promoting cytokine secretion, bioactive NM can directly activate inflammatory cytokine receptors to boost APC maturation. For example, an amphiphilic poly(amino acid) NPs (γ-PGA-Phe NPs) that directly activated toll-like receptor 4 (TLR4). Similarly, the zinc oxide NPs activated the TLR6 pathway. These NPs effectively matured dendritic cells (DCs) and stimulated robust T cell-mediated antitumor immune responses by directly interacting with inflammatory cytokine receptors.

vi. **Alleviate immunosuppressive TME**

The immunosuppressive tumor microenvironment (TME) presents a significant challenge to the success of cancer immunotherapy [16]. This complex environment comprises various inflammatory cytokines, chemokines, and immunosuppressive cells. Bioactive NM can help address these challenges by targeting and reprogramming immunosuppressive cells. Tumor-associated macrophages (TAMs), a key component of immune cell infiltration within tumors, are often polarized into M2-like TAMs that support tumor growth. The calcium carbonate NPs (aCD47@$CaCO_3$) are loaded with an anti-CD47 antibody for use in post-surgical cancer treatment. These NPs neutralize hydrogen ions (H^+) in the TME, which re-polarize TAMs from M2-like to M1-like phenotypes, enhancing their antigen presentation capabilities. The anti-CD47 antibody released from the NPs effectively blocks the "don't eat me" signal on tumor cells, aiding in regulating the immunosuppressive TME. Folate-functionalized bioactive glass NPs (BGN(F)) specifically target M2-like TAMs and release Si_4^- and Ca^{2+} ions to re-polarize them toward the M1-like phenotype. Other nanomaterials, such as glycocalyx-mimicking NPs, mannosylated liposomes, ferumoxytol, and polymeric NPs, have also shown potential in promoting TAM re-polarization.

Additionally, eliminating immunosuppressive factors helps counteract the TME's suppressive effects. For instance, a study from our team developed antibody-like polymeric NPs (APN) for the targeted removal of intra-tumoral galectin-1 (Gal-1), a key immunosuppressive factor [17–20]. These NPs are designed with a cross-linked polymer surface with two functional groups: Tufstin peptides (TKPR), which activate macrophage-mediated phagocytosis, and galactose units, which bind specifically to Gal-1. This dual-action strategy effectively addresses the challenges posed by immunosuppressive TME.APN links Galectin-1 (Gal-1) and macrophages, promoting the macrophage-mediated clearance of Gal-1. This innovative method enables APNs to effectively reduce the immunosuppressive effects of the tumor microenvironment (TME) and stimulate a potent antitumor-immune response. Notably, APN is the first known bioactive material capable of eliminating immunosuppressive factors, presenting a breakthrough that opens up new possibilities for creating bioactive materials to target and remove other harmful factors or toxins. In the field of neurodegenerative disease therapy, conditions like Parkinson's disease (PD), Alzheimer's disease (AD), and Huntington's disease (HD) are marked by the gradual loss of neuronal structure and function. These diseases are increasingly recognized to share similar underlying molecular and cellular mechanisms, primarily involving protein misfolding—misfolded proteins aggregate, triggering events that result in neuronal damage and cell death. Bioactive NM with specialized surface properties offers potential solutions by (a) regulating protein–protein interactions to prevent aggregation and (b) modulating protein-cell interactions to eliminate preformed aggregates. These approaches present promising strategies for treating neurodegenerative diseases.

vii. **Inhibit protein aggregation**

During protein aggregation, monomeric peptides or proteins, such as α-synuclein, β-amyloid, polyglutamine, and Tau, partially unfold and form intermediates that aggregate into harmful oligomers. These oligomers then act as templates for further deposition of monomeric proteins or peptides, eventually forming insoluble amyloid fibrils. To address this issue, bioactive NM can competitively bind to these monomers or oligomers, effectively preventing the aggregation of proteins [21–24]. The mixed-shell polymeric micelles (MSPMs) to treat AD. These MSPMs have a distinctive surface structure that includes hydrophilic chain segments and hydrophobic microdomains. The hydrophobic microdomains anchor hydrophobic β-amyloid (Aβ) monomers and oligomers, while the hydrophilic segments form a protective shield, preventing further Aβ deposition. As a result, MSPMs effectively capture Aβ monomers and oligomers, inhibiting the formation of amyloid fibrils. A reactive conjugated polymer (PPV-NP) capable of covalently binding to lysine (K) residues in β-amyloid (Aβ42), causing steric hindrance that prevents protein aggregation. However, these bioactive nanomaterials (NM) exhibit relatively low selectivity for Aβ monomers or oligomers, restricting their practical application in vivo. A hetero multivalent peptide recognition strategy using a system composed of calixarene (CA) and cyclodextrin (CD) amphiphiles (CACD). This system simultaneously binds to tyrosine (Y) and lysine (K) residues in Aβ42, enhancing specificity toward this peptide. Moreover, attaching peptides that specifically bind to Aβ, Tau, or α-synuclein onto bioactive NPs has increased selectivity for monomers and oligomers. For example, a recent study by our team introduced a KLVFF (Aβ-binding peptide)-integrated bioactive nanocomposite that selectively bound to Aβ42 and effectively inhibited the formation of amyloid fibrils. Likewise, the TLK ((D)-TLKIVW)-integrated polymeric micelles designed to bind specifically to the tau aggregation motif (VQIVYK), thereby preventing tau protein aggregation. These innovations represent promising strategies to address protein aggregation in neurodegenerative diseases.

In pathological brain conditions, the phagocytic function of microglia is hindered by peptide or protein oligomers and fibrils, which impede the efficient removal of protein aggregates [25–28]. Bioactive NM has demonstrated the ability to modulate interactions between microglial cells and proteins, thereby enhancing microglia-mediated clearance of these aggregates. Recently, GCA-CD is a novel Aβ inhibitor by co-assembling cyclodextrin (CD) with guanidinium-modified calixarene (GCA). This inhibitor binds to Aβ42, forming positively charged GCA-CD/Aβ co-aggregates that help microglia clear Aβ aggregates. Similarly, our team designed a neuroprotective nano-scavenger with two functional components on its surface: GLVFF (an Aβ-binding peptide) and immunoglobulin G (IgG). GLVFF binds to monomeric Aβ or its aggregates, forming Aβ/nano-scavenger co-aggregates. At the same time, IgG enhances microglia-mediated clearance via antibody-dependent cell-mediated phagocytosis (ADCP). With optimized surface properties, this nano-scavenger efficiently cleared β-amyloid aggregates in the brain, improving cognitive function in Alzheimer's disease (AD) mice. In addition to this bridging function, bioactive NM can create artificial receptors on microglial surfaces to regulate interactions with protein aggregates. For example, an engineered artificial receptor, ThS, a thioflavin dye that targets Aβ aggregates, allowed the modified microglia to bind to and clear Aβ aggregates actively.

Additionally, particular carbon-based NM, such as single-walled carbon nanotubes, graphene oxide, and C60 fullerene derivatives, have been shown to activate the autophagy pathway in microglia, further aiding in the removal of protein aggregates. However, most of these bioactive NM are administered directly into the brain due to the blood–brain barrier (BBB), which raises treatment risks and patient discomfort. Therefore, finding innovative methods to cross the BBB effectively is crucial for safely and efficiently treating neurodegenerative diseases [29, 30].

viii. **Biocatalyst**

Biocatalysis, or biological transformation, involves chemical conversion processes driven by enzymes or other biological entities like cells, organelles, and tissues. In recent years, various bioactive materials have been developed to mimic the structure and function of natural biocatalysts. One such innovation, nanozymes (NZ), refers to NM with inherent enzyme-like properties, which have attracted considerable interest. This section delves into the latest developments in NZ, highlighting their uses in biosensing, imaging, and disease treatment [31].

ix. **Biosensing**

Biosensing involves converting chemical or biological substances into measurable signals, with colorimetry being one of the most commonly used techniques. This method utilizes peroxidase as a transducer to catalyze the oxidation of colorless peroxidase substrates into colored products. In 2007, the Yan group first highlighted the peroxidase-like activity of Fe_3O_4 magnetic NPs. Since then, various NZ, including those based on vanadium, noble metals, carbon, and metal–organic frameworks (MOFs), have been discovered with similar properties. NZ offers notable advantages over natural enzymes, including higher catalytic efficiency, lower cost, and enhanced physical and chemical stability, which make them promising for biosensing applications. When combined with specific receptors, NZ can detect various chemical and biological species [32–36]. For example, NZ can efficiently identify their substrates when paired with oxidases. A ZIF-based NZ (GOx/hemin@ZIF-8) using glucose oxidase (GOx) as a receptor, effectively detecting glucose in drinks, blood, and urine. Likewise, a copper sulfide-based nanozyme (BNNS@CuS) using cholesterol oxidase as a receptor, enabling visual detection of total cholesterol in human serum. Additionally, NZ can be employed to detect specific antigens. In a notable study, the Yan group combined the peroxidase activity and magnetic properties of Fe_3O_4NZ to create a unique capture-detection immunoassay. These antibody-conjugated Fe_3O_4NZ selectively bound to antigens in mixtures, isolated them using a magnetic field, and produced a colorimetric signal. By using anti-cardiac troponin I (TnI) and anti-EBOV antibodies as receptors, this method enabled the capture, isolation, and detection of TnI and Ebola virus (EBOV) antigens. Recently, researchers developed a novel Co–Fe@heminNZ that can perform chemiluminescence immunoassays for detecting the SARS-CoV-2 antigen in serum by incorporating an anti-SARS-CoV-2 antibody. In addition to detecting proteins, NZ can identify specific cell phenotypes by targeting cell markers as antigens. For example, a NZ-based probe to assess the expression of integrin GPIIb/IIIa on cell surfaces. This probe, which consists of a peptide (H_2N-CCYKKKKQAGDV COOH) conjugated to AuNPs, selectively binds to integrin and produces a colorimetric signal, enabling the quantification of integrin expression on human erythroleukemia cells without requiring cell lysis or protein extraction. In a similar approach, Chen et al.

developed platinum NPs/graphene oxide (PtNPs/GO) for detecting cancer cells. The folic acid (FA) on PtNPs/GO binds specifically to FA receptors on cell membranes, generating a colorimetric signal in situ by catalyzing the oxidation of TMB in the presence of H_2O_2. Beyond protein receptors, other markers such as epithelial cell adhesion molecule (EpCAM) and glycans can also be used for cell detection. By combining these markers with biological recognition ligands like the anti-EpCAM aptamer (SYL3C) and lectin, nanozymes extend their ability to detect specific cell phenotypes.

x. **Imaging**

Certain metal elements possess inherent properties that make them particularly valuable in imaging applications. For example, iridium (Ir) has X-ray absorption capabilities, iron (Fe) exhibits magnetic properties, and gold (Au) offers optical features [32, 33]. By utilizing these characteristics, NZ based on Ir, Fe, and Au are commonly used in imaging techniques such as computed tomography (CT), magnetic resonance (MR), and optical imaging. A bovine serum albumin-iridium oxide NPs (BSA-IrO_2 NP) for tumor theranostics. This NPs was synthesized using a one-step biomineralization process and demonstrated a high X-ray absorption coefficient, enzyme-like activity, and excellent photothermal conversion efficiency. These attributes allowed BSA-IrO_2 NP to provide practical CT imaging and tumor therapy, as well as a magnetoferritin nanoparticle (MHFn) for targeting and visualizing tumors. The ferritin protein (HFn) in these NPs binds selectively to transferrin receptor 1 (TfR1) on tumor cells. At the same time, the iron oxide core catalyzes the oxidation of di-azo aminobenzene (DAB), producing a colorimetric signal for tumor tissue visualization. Recent developments have also led to dual-modality and multimodality imaging techniques. A folate-modified gold nanocluster (FA-AuNC) for both fluorescence and visual imaging of tumors. This NC binds to folate receptors on cell membranes, generating a colorimetric signal for visual imaging and strong fluorescence for fluorescence imaging. As a result, FA-AuNC allows for accurate molecular colocalization diagnostics for tumors. Similarly, a keratin-templated gold NC (AuNCs@Keratin) for near-infrared (NIR) fluorescence imaging and MR imaging of tumors. These advanced multimodal imaging approaches significantly improve the specificity and precision of cancer imaging, reducing the chances of false-negative and false-positive results.

xi. **Disease treatment**

Due to their intrinsic catalytic properties, NZ has proven highly effective in treating diseases [34–36]. NZ with catalase (CAT) or superoxide dismutase (SOD)-like activities can scavenge ROS, providing neuroprotection, cytoprotection, and anti-inflammatory effects. For example, a multi-functional NZ can be achieved by self-assembling vanadium pentoxide (V_2O_5) nanowires with manganese dioxide (MnO_2) NPs. While V_2O_5 nanowires exhibit glutathione peroxidase (GPx)-like activity, MnO_2 NPs mimic both SOD and CAT activities. These combined properties allow the

MnO_2 NPs to eliminate ROS, offering cytoprotection effectively. A selenium dioxide-based NZ (E-A/P-CeO_2) for stroke therapy. The angiopep-2 (TFFYGGSRGKRNN-FKTEEY) peptide attached to E-A/P-CeO_2 binds to low-density lipoprotein receptors (LRP) on endothelial cells, enabling the enzyme to cross the blood–brain barrier (BBB). Once inside the brain, cerium oxide (CeO_2) and edaravone work together to eliminate ROS, providing neuroprotection. In contrast to ROS scavenging, NZ with oxidase or peroxidase-like activities generate ROS during their catalytic process, making them effective for antitumor and antibacterial applications. A PEG-functionalized molybdenum disulfide nanoflower (PEG-MoS_2 NFs) for antibacterial wound treatment. These nanoflowers display both peroxidase-like activity and strong near-infrared (NIR) absorption, which converts hydrogen peroxide (H_2O_2) into hydroxyl radicals (·OH) and generates a photothermal effect upon NIR irradiation for combined antibacterial action. Likewise, a platinum NZ-decorated zirconium-based metal–organic framework (PCN-224-Pt) aimed at alleviating tumor hypoxia, improving the efficacy of photodynamic therapy (PDT). These innovations demonstrate NZ's wide-ranging potential in addressing various health challenges. Platinum-based nanozymes (Pt NZ) demonstrate catalase (CAT)-like activity, efficiently converting hydrogen peroxide (H_2O_2) into oxygen (O_2), which amplifies the antitumor effects of photodynamic therapy (PDT). Beyond cancer treatment, NZis is also being investigated for its potential to address drug abuse and addiction by oxidizing neurotransmitters, especially dopamine. For example, a NZ based on single-walled carbon nanotubes (SWNTs) to treat methamphetamine (METH) addiction. By adsorbing and oxidizing dopamine, the SWNTs effectively reduced the METH-induced increase in synaptic proteins and tyrosine hydroxylase levels, helping to alleviate the goal-directed behaviors often associated with drug addiction.

5 Conclusion

Bioactive NMs are a promising biomaterial with unique properties that enable them to interact effectively with biological systems, triggering specific responses in proteins, cells, and tissues. The bioactivity of these materials is influenced by factors such as their physical structure, surface characteristics, and nanotopography. Recent progress in the design of bioactive NMs has led to new solutions for treating various diseases and addressing tissue regeneration challenges. Inorganic NMs, for example, are well-suited as bone graft substitutes due to their superior mechanical strength and antibacterial properties. On the other hand, polymeric NMs can be tailored with precise structures and surfaces for specific molecular recognition, enhancing interactions between proteins, cells, and tissues while enabling controlled drug delivery. Natural polymers with bioactive properties have been investigated as drug delivery systems due to their intrinsic targeting abilities. Carbon-based NMs, such as CNTs and graphene, have significant potential in bone regeneration and can also influence cellular processes like autophagy and inflammation. Despite these advancements, challenges remain, particularly in understanding the chemical mechanisms behind NM interactions with biological systems. Exploring new biomaterials, such as DNA-based NMs and 3D-printed NMs, could open new possibilities for medical applications. Future research should focus on improving the therapeutic effectiveness of these materials in living organisms, addressing issues like selective targeting, biodistribution, biodegradation, and immune responses. A multidisciplinary approach is crucial to successful clinical applications, considering toxicity assessment, scalability, sterilization processes, and storage requirements.

References

1. Kumar, P., Kumar Jain, A., Srivastava, J.P., Kumar, R., Saxena, K.K., Prakash, C., Buddhi, D.: Multiphysics simulation of the shape prediction and material removal rate in electrochemical machining process. Adv. Mater. Process. Technol. 1–13 (2023)
2. Kumari, C.U., Murthy, A.S.D., Prasanna, B.L., Reddy, M.P.P., Panigrahy, A.K.: An automated detection of heart arrhythmias using machine learning technique: SVM. Mater. Today Proc. **45**, 1393–1398 (2021)
3. Liu, H., Yang, Y., Liu, Y., Pan, J., Wang, J., Man, F., Zhang, W., Liu, G.: Melanin-like nanomaterials for advanced biomedical applications: a versatile platform with extraordinary promise. Adv. Sci. **7**(7), 1903129 (2020)
4. Nair, A., Kumanan, S., Prakash, C., Mohan, D.G., Sxena, K.K., Kumar, S., Kumar, G.: Research developments and technological advancements in conventional and non conventional machining of superalloys—a review. J. Adhes. Sci. Technol. 1–72 (2023)
5. Noreen, S., Maqbool, A., Maqbool, I., Shafique, A., Khan, M.M., Junejo, Y., Ahmed, B., et al.: Multifunctional mesoporous silica-based nanocomposites: synthesis and biomedical applications. Mater. Chem. Phys. **285**, 126132 (2022)
6. Park, W., Shin, H., Choi, B., Rhim, W.-K., Na, K., Han, D.K.: Advanced hybrid nanomaterials for biomedical applications. Progr. Mater. Sci. **114**, 100686 (2020)
7. Pon, V.D., Wilson, K.J., Hariprasad, K., Ganesh, V., Ali, H.E., Algarni, H., Yahia, I.S.: Enhancement of optoelectronic properties of ZnO thin films by al doping for photodetector applications. Superlatt. Microstruct. **151**, 106790 (2021)
8. Raj, I.L.P., Valanarasu, S., Hariprasad, K., Ponraj, J.S., Chidhambaram, N., Ganesh, V., et al.: Enhancement of optoelectronic parameters of Nd-doped ZnO nanowires for photodetector applications. Opt. Mater. **109**, 110396 (2020)
9. Sambasivam, S., Abed, A.S., Chopde, S., Patil, P.P., Math, P., Parmar, A., et al.: Role of processing techniques related to Mg-MMCs for biomedical implantation: an overview. Mater. Today Proc. (2023)
10. Sarkhel, P., Dikshit, M.K., Pathak, V.K., Saxena, K.K., Prakash, C., Buddhi, D.: Robust deflection control and analysis of a fishing rod-type flexible robotic manipulator for collaborative robotics. Robot. Auton. Syst. **159**, 104293 (2023)
11. Singh, S., Agrawal, V., Saxena, K.K., Mohammed, K.A.: Optimization on manufacturing processes at Indian industries using TOPSIS. Ind. J. Eng. Mater. Sci. (IJEMS) **30**(1), 32–44 (2023)
12. Tang, Z., Kong, N., Ouyang, J., Feng, C., Kim, N.Y., Ji, X., Wang, C., Farokhzad, O.C., Zhang, H., Tao, W.: Phosphorus science-oriented design and synthesis of multifunctional nanomaterials for biomedical applications. Matter **2**(2), 297–322 (2020)
13. Vinoth, S., Kanimozhi, G., Hari Prasad, K., Harish, K., Srinadhu, E.S., Satyanarayana, N.: Enhanced ionic conductivity of electrospun nanocomposite (PVDF-HFP+ TiO_2 nanofibers fillers) polymer fibrous membrane electrolyte for DSSC application. Polym. Compos. **40**(4), 1585–1594 (2019)

14. Yang, L., Patel, K.D., Rathnam, C., Thangam, R., Hou, Y., Kang, H., Lee, K.-B.: Harnessing the therapeutic potential of extracellular vesicles for biomedical applications using multifunctional magnetic nanomaterials. Small **18**(13), 2104783 (2022)
15. Zhang, Y., Ding, J., Qi, B., Tao, W., Wang, J., Zhao, C., Peng, H., Shi, J.: Multifunctional fibers to shape future biomedical devices. Adv. Func. Mater. **29**(34), 1902834 (2019)
16. Prateek, S., Garg, R., Saxena,, K.K., Srivastav, V.K., Vasudev, H., Kumar, N.: Data-driven materials science: application of ML for predicting band gap. Adv. Mater. Process. Technol. 1–10 (2023)
17. Singh, B., Kumar, I., Saxena, K.K., Mohammed, K.A., Khan, M.I., Moussa, S.B., Abdullaev, S.S.: A future prospects and current scenario of aluminium metal matrix composites characteristics. Alex. Eng. J. **76**, 1–17 (2023)
18. Vashist, A., Atluri, V., Raymond, A., Kaushik, A., Parira, T., Huang, Z., Durygin, A., et al.: Development of multifunctional biopolymeric auto-fluorescent micro-and nanogels as a platform for biomedical applications. Front. Bioeng. Biotechnol. **8**, 315 (2020)
19. Wahid, F., Zhao, X.-J., Jia, S.-R., Bai, H., Zhong, C.: Nanocomposite hydrogels as multifunctional systems for biomedical applications: current state and perspectives. Compos. B Eng. **200**, 108208 (2020)
20. Awasthi, A., Gupta, A., Saxena, K.K., Dwivedi, R.K., Kundalkar, D., Abdul-Zahra, D.S., et al.: Design and analysis of equal-channel angular pressing of Al6061: a comparative study. Adv. Mater. Process. Technol. 1–10 (2022)
21. Awasthi, A., Saxena, K.K., Arun, V.: Sustainability and survivability in manufacturing sector. In: Modern Manufacturing Processes, pp. 205–219. Woodhead Publishing (2020)
22. Awasthi, A., Saxena, K.K., Arun, V.: Sustainable and smart metal forming manufacturing process. Mater. Today Proc. **44**, 2069–2079 (2021)
23. Barrett-Catton, E., Ross, M.L., Asuri, P.: Multifunctional hydrogel nanocomposites for biomedical applications. Polymers **13**(6), 856 (2021)
24. Basavapoornima, C., Kesavulu, C.R., Maheswari, T., Pecharapa, W., Depuru, S.R., Jayasankar, C.K.: Spectral characteristics of Pr3+-doped lead based phosphate glasses for optical display device applications. J. Lumin. **228**, 117585 (2020)
25. Bhardwaj, V., Kaushik, A.: Biomedical applications of nanotechnology and nanomaterials. Micromachines **8**(10), 298 (2017)
26. Bhukya, M.N., Kota, V.R.: A quick and effective MPPT scheme for solar power generation during dynamic weather and partial shaded conditions. Eng. Sci. Technol. Int. J. **22**(3), 869–884 (2019)
27. Chakraborty, D., Rathi, A., Singh, R., Pathak, V.K., Srivastava, A.K., Sharma, A., et al.: External six-bar mechanism rehabilitation device for index finger: development and shape synthesis. Robot. Auton. Syst. **161**, 104336 (2023)
28. Chauhan, R.B., Shah, T.V., Shah, D.H., Gohil, T.J., Oza, A.D., Jajal, B., Saxena, K.K.: An overview of image processing for dental diagnosis. Innov. Emerg. Technol. **10**, 2330001 (2023)
29. De Crozals, G., Bonnet, R., Farre, C., Chaix, C.: Nanoparticles with multiple properties for biomedical applications: a strategic guide. Nano Today **11**(4), 435–463 (2016)
30. Dikshit, M.K., Singh, S., Pathak, V.K., Saxena, K.K., Agrawal, M.K., Malik, V., et al.: Surface characteristics optimization of biocompatible Ti_6Al_4V with RCCD and NSGA II using die sinking EDM. J. Mater. Res. Technol. **24**, 223–235 (2023)
31. Gupta, P., Gupta, N., Saxena, K.K.: Predicting compressive strength of geopolymer concrete using machine learning. Innov. Emerg. Technol. **10**, 2350003 (2023)
32. Kalpana, G., Kumar, P.V., Aljawarneh, S., Krishnaiah, R.V.: Shifted adaption homomorphism encryption for mobile and cloud learning. Comput. Electr. Eng. **65**, 178–195 (2018)
33. Kandhare, A.D., Thakurdesai, P.A., Wangikar, P., Bodhankar, S.L.: A systematic literature review of fenugreek seed toxicity by using ToxRTool: evidence from preclinical and clinical studies. Heliyon **5**(4) (2019)
34. Kim, D., Shin, K., Kwon, S.G., Hyeon, T.: Synthesis and biomedical applications of multifunctional nanoparticles. Adv. Mater. **30**(49), 1802309 (2018)

35. Korpi, A.G., Ţălu, Ş, Bramowicz, M., Arman, A., Kulesza, S., Pszczolkowski, B., et al.: Minkowski functional characterization and fractal analysis of surfaces of titanium nitride films. Mater. Res. Exp. **6**(8), 086463 (2019)
36. Kumar, P.S.S., Allamraju, K.V.: A review of natural fiber composites [Jute, Sisal, Kenaf]. Mater. Today Proc. **18**, 2556–2562 (2019)

Nanometarial-Based Sensors: Transforming Real-Time Monitoring in Biomedical Applications

Elif Esra Altuner and Yasemin Torlak

Abstract Nanomaterial-based sensors are radically changing real-time monitoring technologies in biomedical applications. Offering higher sensitivity, selectivity, and fast response time compared to traditional sensors, these sensors have great potential for detecting various biomarkers and continuous health monitoring. Nanomaterials such as graphene, fullerenes, and metal oxide nanoparticles play an important role in disease diagnosis, monitoring of treatment processes, and personalized medicine applications by improving the performance of biosensors. These sensors, which are used primarily in detecting blood sugar, cancer markers, infection indicators, and biomolecules associated with neurodegenerative diseases, increase patient comfort by offering non-invasive and minimally invasive methods. Nanomaterial-based sensors, which can be integrated into wearable and portable devices, provide revolutionary innovations in the healthcare sector by expanding the possibilities of remote health monitoring and early diagnosis.

Keywords Nanomaterials · Biosensor · Real-time monitoring · Disease diagnosis · Personalized medicine

1 Introduction

Biomedical application is based on the principle of creating functional systems for the benefit of humanity by using biology and engineering in scientific applications [1, 2]. Therefore, this field appeals to a comprehensive perspective. Medical imaging devices such as tomography, MRI devices [3, 4], ECG [5], ultrasound devices used in hospitals to provide more detailed treatment processes for patients, robotic systems

E. E. Altuner (✉)
Program of Medical Laboratory Techniques, Department of Medical Services and Techniques, Vocational School of European, University of Kocaeli Health and Technology, Kocaeli, Turkey
e-mail: elifesraaltuner@gmail.com

Y. Torlak
Program of Organic Farming, Department of Agricultural and Livestock Production, Vocational School of Cal, University of Pamukkale, Denizli, Turkey

S. N. Saud Al-Humairi et al. (eds.), *Smart Nanomaterials in Biomedical Applications*, Engineering Materials, https://doi.org/10.1007/978-981-96-8023-8_15

for artificial intelligence, biosensors such as blood sugar measurement, hormone measurement, or smart drug design are used in various applications such as biomedical engineering applications. Therefore, this field is very popular, and studies are ongoing. As can be understood from here, sensors are essential in biomedical applications. Existing living systems and metabolomics databases can provide information on the chemical properties of metabolites as well as their functioning as complex molecular machines modulated by the dynamic biological process. Still, they do not provide detailed information on the biological processes by which these metabolites accumulate. Consequently, metabolite identification requires a combination of sensitive analyses and biological insights, with important implications for clinical applications. With referring to these properties, sensors have been developed to detect and/or monitor analytes or functions coming into the biological system. Today, the developments obtained from studies on biosensor technologies are remarkable. When the notable developments are examined, it is seen that activities such as agricultural production of biological and chemical agents, food processing, and environmental monitoring, as well as clinical diagnoses, drug testing, biological processing, and biological warfare, are carried out. The perception of changes in the ecological system by living beings and their response to them has constituted the main field of study in the development of biosensors. Wearable devices that monitor vital signs or electrolyte levels in real time serve as a tool to produce a signal from a biological response. In technological applications, sensors usually consist of two components. As the first component, the bioactive component selectively interacts with the element to be analyzed, thus transforming the recognition into a measurable state. The second component receives the signal, sends it to the transmission element, and performs the sensing (from the sensor) function [3–5].

Nanosensors are biological methods that can be used to transmit remarkable quantities such as mass, size, and temperature of nanoparticles designed for the target, selective, fast, and highly sensitive compared to traditional chemicals, and can detect, respond to, and use biological signals (biosensors) with their nanoscale properties. The presented formats exploit the unique ability of nanoparticles to resolve the active components of nanosensors by physical methods, such as quantum dots, of chemical phenomena that are often size-dependent. Traditionally developed nanosensors are molecular or in vitro and in vivo application studies, where genetically encoded indicators predominate. Thanks to production technologies and applications in quantum physics, nanomaterials that affect our daily lives have been developed. Nanosensors are a technology that emerged as a result of these developments. At least one dimension at the nanoscale is an electronic device that can detect and signal nanoscale materials or events. Nanosensors have a structural feature that can stand alone and be used as a functional part of another macroscale device. Nanosensors are electronic devices with little history. However, it has quickly gained a foothold in national security, medicine and imaging, the food industry, environmental protection, and many other fields. How far nanosensors can impact our daily lives lies in the nanoproducts we use. According to the IUPAC definition, sensors must have an analyte to be analyzed, a permanent signal or data, a transducer that converts the data into an understandable format, and a detector that records the data [6]. As can be understood from here, it is mandatory to receive the signal continuously.

Sensors are also divided into various branches. The most well-known are chemical sensors, biosensors, aptasensors, nanosensors, microsensors [7, 8] and various derivatives. Biosensors include biological agents such as DNA and bacteria as analytes; chemical sensors include chemicals as analytes; nanosensors and microsensors refer to nano- and micro-scale studies. Aptasensors are a sub-branch of biosensors and are sensor systems performed using aptamers. These sensors can be developed as electrochemical, fluorimetric, or optical sensors. There are various studies on this subject in the literature. For example, a sensor system activated by phosphate release for aptamer for the analyte 3,3′ 5,5′ tetramethylbenzidine (TMB) has been developed. A sensor catalyzed by silver nanoparticles has been designed for paracetamol used as an analyte. Fluorimetric sensors have also been developed for aluminum ions in water [9]. The role of nanoparticles and nanocomposites as sensors and catalysts cannot be denied.

Nanotechnology is a branch of science that deals with sizes in the range of 1–100 nm [10–13]. One nanometer (nm) corresponds to 10^{-9} m [14]. Therefore, talking about such small particles means scientific studies are gaining momentum rapidly. Metallic nanoparticles, especially nanoparticles (NPs), play a significant and active role. Platinum (Pt) [15], palladium (Pd) [16], gold (Au) [17], silver (Ag) [18], Ruthenium (Ru) [19], cobalt (Co) [20, 21], iron (Fe) [22] and zinc (Zn) [23] are the most commonly used metals. For example, hydrogen production and photocatalytic studies have been carried out in the literature by synthesizing trimetallic PdPtCo NPs [21]. Zn NPs have been synthesized from biological methods and used in photocatalytic studies [3]. Nanomaterials, rather than nanoparticles, are also vital since they are in the nanotechnology category. MWCNTs [24], fullerenes [25], graphenes [26], and various nanomaterials are examples of this category. Nanomaterials can be composited with metallic NPs to form a nanocomposite. Thus, catalytic activity and synergy are further increased. In the literature, a sensor was developed to support oxidized iron nanoparticles with MWCNT for tuberculosis drug detection. ZnO@AC NPs were formed by supporting ZnO NPs with AC and used in hydrogen energy production [27]. This chapter discusses nanomaterials in detail and explains them by referring to the relevant places. Then, it mentions their use and importance in sensors and biomedical applications.

2 Nanomaterials and General Properties

Nanomaterials are materials with at least one dimension between 1 and 100 nm (nm) [28, 29]. At the same time, thanks to their nanoscale, they play an essential role in optical platforms by changing the interaction of materials with light and affecting properties such as color, transparency, and absorption. Nanomaterials are used in electronics, magnetic and optoelectronics, biomedicine, pharmaceuticals, cosmetics, energy, catalytic and materials applications, as well as in defense, transportation, food safety and ecology [30, 31]. By changing the molecular structure of materials to create smart objects, nanotechnology enables enormous innovations that are revolutionizing science and industry. Targeted particles to fight cancer cells, energy-efficient and fast microchips, long-lasting batteries, to name just a few [32].

Nanomaterials have electric and magnetic properties [33, 34]. Nanomaterials, which have variable conductivity and magnetic properties due to quantum effects, are nanomaterials with electrical and magnetic properties [33, 34]. These are generally nanoparticles for transition metals [35]. Some nanomaterials can be stronger and lighter than traditional materials [27]. Nanomaterials are of great importance in the fields of [36], and cancer treatment [37]. These are also indispensable in environmental and energy platforms in water purification [38–41], solar panels [42–44], and hydrogen production [45, 46]. Nanomaterials are also preferred in electrical and electronic systems in nano-transistors, memory chips, and flexible displays. Nanomaterials are also widely used in engineering by being used in lightweight and durable materials, coatings, paint, and textile industry [47]. A general systematic illustration of the use and properties of nanomaterials in the above-mentioned platforms is as in Fig. 1.

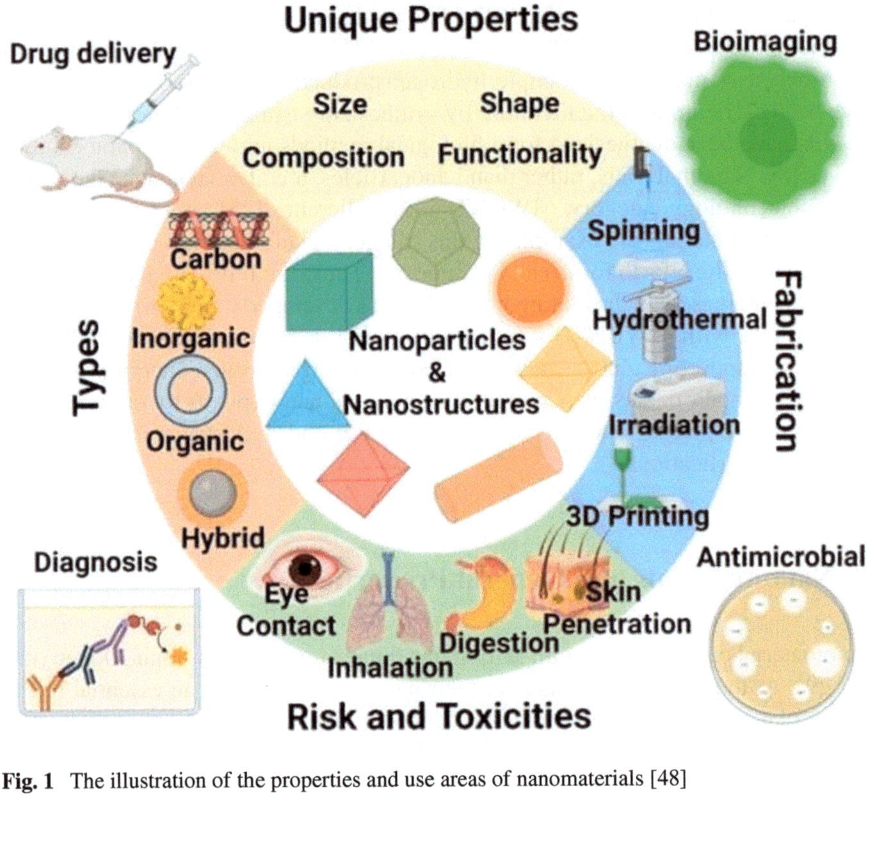

Fig. 1 The illustration of the properties and use areas of nanomaterials [48]

3 Types of Nanomaterials

Nanomaterials are one of the basic building blocks of nanotechnology and have a wide range of applications. These materials are generally between 1 and 100 nm in size and provide revolutionary innovations in many sectors. They are divided into various categories according to their structures and properties [49, 50]. This sub-title discusses the main types of nanomaterials. One of the main factors that makes nanomaterials different from materials of larger size is the increase in relative surface area. As a particle gets smaller, the ratio of surface to volume gradually increases, the numerical proportion of atoms on the surface of the particle increases relative to the number of atoms inside it. Therefore, the surface of a nanomaterial of a given mass has a greater number of favorable atoms than a material made of larger particles of the same mass. When chemical reactions occur on these surfaces, there is a more effective interaction with the environment compared to samples with larger particles. Because of all these properties, nanomaterials are more chemically reactive, potentially changing their strength and electrical properties.

3.1 Carbon-Based Materials

Different structural arrangements of carbon atoms create carbon-based nanomaterials [49–52] (Fig. 2). Carbon nanotubes (CNTs), fullerenes and graphene are examples of carbon-based nanomaterials. Fullerenes consist of 60 or more carbon atoms. Due to their chemical, biological and different electrical properties, they are widely preferred and used in both technological and biomedical applications. Carbon nanotubes are formed by the graphite layer forming a cylindrical structure. They exhibit different mechanical and electronic properties depending on the direction in which the graphite plate is bent. Carbon nanotubes are generally used in the field of materials science. The most common types are as follow.

Fullerenes: Fullerenes (C60) are carbon-based molecules with a morphologically spherical structure formed by sp^2 hybridization of carbon atoms [53]. Due to its

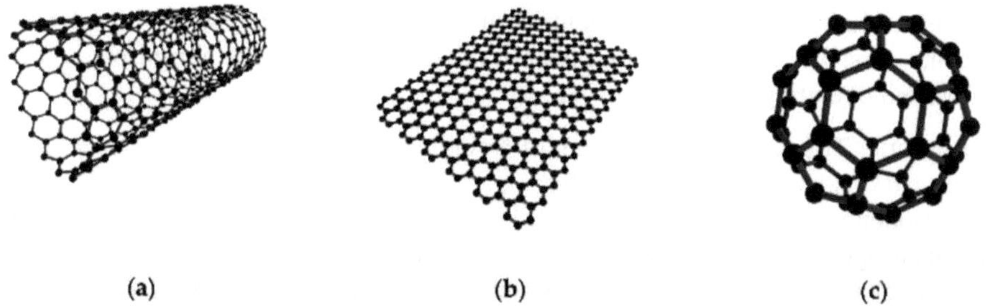

Fig. 2 Types of carbon-based nanomaterials [56]

unique electronic structure, three types of chemical bonds can be formed; sp^1, sp^2 and sp^3 hybridization. Such a feature makes it easy for C to form various polyatomic structures. Carbon has a wide variety of allotropes such as BC8 (lonsdaleite), diamond, graphite, graphene, fullerenes, carbon nanotubes with structures with different molecular configurations. Polycyclic aromatic hydrocarbon (PAH) molecules have a planar structure like graphene. Fullerenes are nano cages with different numbers of carbon allotropes. The smallest fullerene structure has 20 carbons. The largest known fullerene is C540. Endohedral fullerenes are fullerenes that carry atoms or molecules inside the cage structure, while in exohedral fullerenes, the ion is bonded to the cage structure from the outside of the cage (Fig. 2). Heterofullerenes carry atoms or molecules bonded to the cage from both inside and outside [53]. As a result, fullerenes have a wide range of applications from the medical sector to the electronics industry. In medicine, their excellent antioxidant and antibacterial properties encourage their use. In particular, they play an important role in the controlled release of drugs. In the electronics industry, fullerenes are used in areas such as solar panels, diodes and semiconductors. In addition to having great potential in the pharmaceutical sector, fullerenes also have an important place in the field of nanomedicine. In pharmacy, they can be used in the development of new pharmaceutical products. Researchers continue to work to discover more application areas of fullerenes, and the demand for various types of fullerenes is increasing over time.

Carbon Nanotubes (CNTs): These nanomaterials with a cylindrical structure exhibit extraordinary mechanical strength and thermal conductivity [54, 55]. One of the most important and current mediators used in nanobiosensor and fuel cell studies in recent years is carbon nanotubes (CNTs) with high quality mechanical and electronic properties. In conduction with CNTs, 10–100 times more current is transmitted than polyphenylene-based chains. Thermal conductivity of CNTs is 2 times that of diamond, the best thermal conductor, and can carry 1000 times more electricity than copper wire. With these properties, it is thought that CNTs will be the basis of nanoelectronics in the future. In some studies, the biosensor showed high sensitivity, low detection limit, wide linear range, fast response time and reliable measurement stability against the tested phenols. It was observed that the carbon nanotube increased the signal strength of the system by accelerating the electron flow that occurs as a result of phenolic reactions in the systems. The ability of CNTs to increase the speed of electron transfer reactions at low potentials has led researchers to conduct intensive studies on the development of enzyme electrodes prepared with CNTs in recent years. The most interesting feature of CNTenzyme-based biosensors is their improved operating stability.

Graphene: This material, consisting of a single layer of carbon atoms, offers high electrical conductivity and flexibility.

Graphene is a nanomaterial that has been the subject of many application areas, and its usage areas are expanding. The concentration of studies on this subject after the Nobel Prize in 2010 has caused graphene to be included in many commercial products today. There are many production methods for graphene nanomaterials. In some studies, it has been determined that the production method of graphene directly affects the application area. For example, since a high amount of graphene will be required for a graphene-doped polymer composite structure, if the presence of relatively minor defects in the graphene to be used will not cause serious problems, graphene nanomaterial produced by reduction methods of graphene oxide can be used.

On the other hand, graphene, which is planned to be used as an electrode in organic photovoltaic solar cells, should be produced by CVD methods, ensuring that it does not contain defects and is single-layered. Otherwise, the adverse effects that will occur in the electrical properties will directly negatively affect the efficiency of the solar cell. The application area for graphene production should be determined, and then it should be produced using the most appropriate method. Laser-generated graphene (LIG) technology has become very popular in recent years. This technique uses laser beams to form graphene layers on polymer surfaces directly. These sensors enable users to monitor biomolecules in their bodies and various physiological parameters in real time for health monitoring devices. In addition, health monitoring applications with this sensor serve many functions, from monitoring biological data such as biomolecules in sweat to tracking physiological parameters such as heart rate and respiration [54, 56].

3.2 Metallic Nanomaterials

Metal nanoparticles, which are composed of inorganic matter, metal oxides, or organic matter, have superior electronic and optical properties. These advantageous properties are synthesized by chemical and physical techniques and used intensively and successfully in catalysis, optoelectronics, chemical or biochemical sensors, and medicine. Metal nanoparticles have become popular in the health sector and sensor applications in recent years due to their simplicity, formability, cheap and environmentally friendly properties and high surface areas [57, 58] (Fig. 3). When a block of nano-form materials is broken down into smaller and smaller pieces and dispersed as nanoparticles, their surface area becomes very large. This is why a seemingly insignificant immune response can become a danger. Secondly, nanoparticles are generally much more chemically reactive due to their curvature when aggregated. Therefore, metallic nanoparticles ionize much more easily in the bulk metal state. The toxicity of lead or silver, for example, is due not only to the metal in its elemental form, but also to the ionized metals of the atom that have broken away from the blocky substance. Third, because nanoparticles are small, they can pass through structures that larger particles could not pass through. Among these various organic and inorganic nanoparticles, gold nanoparticles have become the first choice of researchers, especially in the biological and pharmaceutical fields, due to their unique optical and Surface Plasmon Resonance (SPR) properties. Inorganic nanomaterials (e.g., quantum dots, nanowires, and nanorods) can be used in optoelectronics due to their interesting optical and electrical properties. Moreover, depending on the size and shape of the nanomaterials, their optical and electronic properties are tuned by various methods, such as inducing an electric current in a closed circuit by placing it in a magnetic field, the intensity of which is constantly changing. The performance of the devices depends on the efficiency of the photoinduced process that is responsible for their function. Nanomaterials are used in different electronic application areas due to their stability and reactivity properties and their interactions with biological systems, as well as their interesting chemical behaviors and optical properties. In addition to these applications, their biocompatibility and toxicology are very important for developing safe biomedical applications, ensuring their reliable performance in applications such as catalysis and drug delivery [57, 58]. Examples containing metal nanoparticles are as follows:

(a) **Gold Nanoparticles**: Gold nanoparticles are widely used in biomedical applications and catalysis processes [59]. In the last few decades, their use in biomedical applications has attracted interest due to their unique properties, which make them suitable for the diagnosis and treatment of cancer.
(b) **Silver Nanoparticles**: These are important in the healthcare and textile industries due to their antibacterial properties [60]. Silver nanoparticles, which are evaluated within the scope of nanoparticles, can be used in food, health, and medicine due to their unique physical and chemical properties. They can also be used for consumer and industrial purposes. Among the properties that enable nanoparticles to reach a wide range of uses are stable interactions with ligands, size and shape variability, high carrying capacity, and ease of binding of both hydrophilic and hydrophobic substances, optical properties, surface plasmon

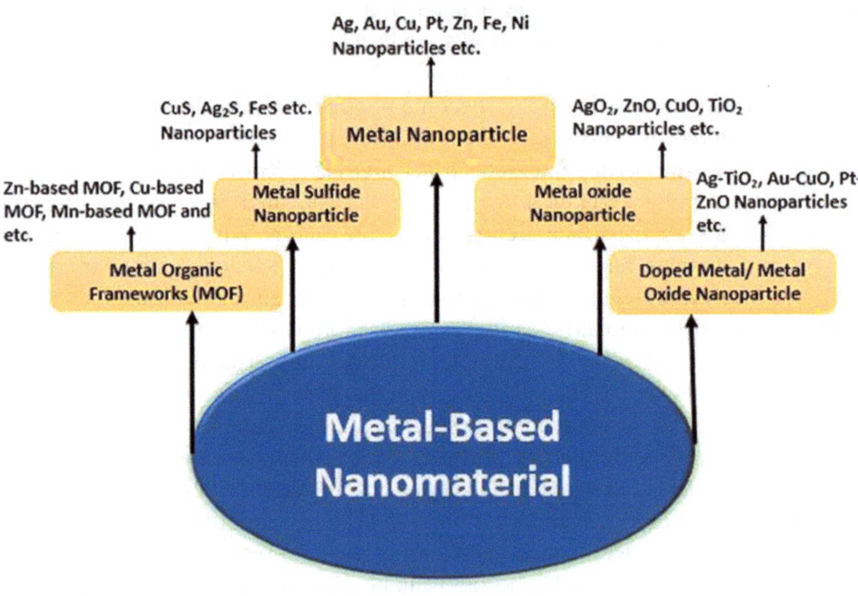

Fig. 3 Different types of metal-based nanomaterials [64]

resonance (SPR). Silver nanoparticles can penetrate bacterial cell walls, alter the structure of cell membranes, and even result in cell death. Due to these properties, nano silver particles have been widely used as antibacterial agents in the healthcare industry, food storage, textile coatings, and various environmental applications. Silver nanoparticles have been used as antibacterial agents, from disinfecting medical devices and household appliances to water purification. The use of silver in dentistry, particularly in the prevention and treatment of dental caries, has been reported since 1840. They have an essential place in nanotechnology, find applications in many fields such as physics, chemistry, and biology, and are the subject of substantial research. Studies in this field may bring further developments. Therefore, it is also possible that the usage areas of silver nanoparticles will increase.

(c) **TiO_2 Nanoparticles**: TiO_2 nanoparticles have been widely preferred as fabric coating materials in photocatalytic and antibacterial applications due to their high oxidizing power and unique photosensitive semiconductor behavior. For example, nanoparticle-modified fabrics with photocatalytic effect, when exposed to light, break down organic pollutants into carbon dioxide, water and mineral acids, a process known as "photocatalysis". Nano-sized TiO_2 particles have different semiconductor properties. In addition to this feature, they form charge carriers under appropriate irradiation due to their large surface area. These charge carriers and the high absorption properties of these nanoparticles show high photocatalytic activity. These are used in solar panels, photocatalytic cleaning, and cosmetic products [61].

(d) **Platinum Nanomaterials**: Platinum nanomaterials are used as catalysts in chemical reactions, fuel cells and medical applications. They are frequently used in research. Recently, environmental, agricultural, biomedical and physicochemical research has been conducted with metal nanoparticles. Nanoparticles, characteristic chemical sensors, antimicrobial activity, medical imaging, cancer diagnosis and treatment, cosmetic applications, specific drug release and wound healing are used for various applications. They increase the stability and solubility of the encapsulated components; drug delivery systems thanks to its advantages such as enhancement, cross-membrane transport ability and prolonged drug delivery time. Drug targeting strategies in cancer treatment are generally categorized as passive targeting and active targeting. In passive targeting, the nanoparticle is deposited in the targeted region by utilizing the physicochemical properties of the environment. Active targeting is achieved by binding an active molecule defined to target specific cells to the nanoparticle surface. For example, chemotherapeutic drugs are loaded into nanoparticles and used for active or passive targeting. Due to its increased permeability and retention effect, tumor tissue has a vascular structure that allows nanoparticles to accumulate and this pathophysiological feature makes passive targeting possible. In active targeting, a tumor-specific biomarker such as monoclonal antibody or peptide is conjugated to the nanoparticle surface. Then, the interaction of the ligand with the receptor on the tumor cell, followed by endocytosis and chemotherapeutic drug release takes place. In addition, triggered release allows nanoparticles to aggregate and affect cancer cells when exposed to an external stimulus [62].

(e) **Palladium Nanomaterials**: These plays an important role in hydrogen storage, catalytic converters and environmental applications. In recent years, Palladium nanomaterials are among the most preferred anode electrocatalysts. Fuel cells are formed with these materials. However, their limited resources and costs prevent them from being at industrial scales. Compared to Pt-based electrocatalysts, Pd-based anode catalysts have higher electrocatalytic activity and higher CO tolerance. Therefore, in some studies, when Pd nanoparticles are coated on n-doped 3D graphene architecture, Pd/N-3DG shows superior electrochemical activity by showing homogeneous distribution of small Pd nanoparticles and good electrical conductivity thanks to Pd nanoparticles. This obtained structure is more durable and more stable against methanol oxidation and showed superior performance. In addition to the fact that the hybrid structure produced can be used in fuel cells, such nanomaterials are potential candidates for other applications such as sensors, photocatalysis and energy storage systems [63].

3.3 Ceramic Nanomaterials

Ceramic-based nanomaterials are inorganic materials of oxide, carbide, or nitride compounds. These materials are generally known for their high temperature resistance and hardness. Ceramic-based nanomaterials are most commonly used in the field of renewable energy. Still, they can also be used better in the combustion, emissions, and oil refinery industries regarding environmental friendliness and ease of processing. In nanosensor application areas, they prefer to produce devices with higher efficiency. Nanosensors used in chemical, surgical, and biological fields have been developed using today's advanced technologies to collect nanoparticle data that traditional sensors cannot obtain. These are much smaller, much lighter, and portable sensors. Nanosensors work like standard sensors, but the main difference is the ability of nanosensors to detect smaller particles [65].

(a) **Zirconia Nanoparticles**: Zirconia nanoparticles are a widely used metal oxide due to their high mechanical strength, favorable surface properties, and good biocompatibility. These properties attract attention in dental materials such as denture bases and reinforcement of repair materials. At the same time, the white color of zirconia powder does not adversely affect the aesthetic appearance, unlike metal inorganic fillers such as aluminum, copper, or silver. These nanomaterials are used in biomaterials and dental applications [66].

(b) **Silica Nanoparticles**: These nanomaterials play a role in surface coatings, drug delivery systems, and optical devices. In food storage and packaging, silicate nanoparticles reduce food spoilage and rancidity. Protein luciferase nanosensor expressed on silica chip detects foodborne pathogens. In dietary supplement intake, silica mineral hydride complex increases bioavailability. Silica nanomaterials loaded with active ingredients can be incorporated into biodegradable packaging materials to help preserve seafood by preventing microbial growth and chemical degradation. Magnetic silica nanomaterials loaded with chitosan bionanocomposites based on turmeric essential oil have been reported to be effective in maintaining and extending the shelf life of surimi. The use of antimicrobial NPs in the active packaging of seafood is a subject of extensive research. The antimicrobial and antioxidant activity of nanocomposite films formed with silica was evaluated and showed antimicrobial effect by reducing lipid autoxidation, especially inhibiting Pseudomonas [67].

3.4 Composite Nanomaterials

Composite nanomaterials combine multiple nanomaterials and offer superior properties by combining the advantages of different materials [68, 69]. The design of metal nanoparticles, biofunctionalized nanoparticles, and nanocomposites has attracted research on nanosensors. Numerous analytical methods have been developed for environmental monitoring and food safety applications. It needs to be connected to a transducer to detect a visible signal. There should be no detection problems during this application. Transducers that can be incorporated into sensors today have nanomaterial properties. The size and structure of nanomaterials provide excellent opportunities to design new sensing systems and improve the performance of biosensors. Diagnosis is the first step in treating the disease. Accurate clinical diagnosis always helps to reduce the complications of early or late-stage diseases. Enzymes are the most widely used biomaterials among these biological materials due to their high sensitivity and specificity.

(a) **Carbon Nanotube Reinforced Composites**: These nanomaterials are used in aerospace, automotive, and biomedical engineering. Monolayer and multilayer CNTs are used as reinforcement materials for thermoset (epoxy, polyimide, phenolic, etc.) and thermoplastic polymers (polypropylene, polystyrene, poly methyl methacrylate (PMMA), nylon 12 (polyamide 12), poly ether ether ketone (PEEK), etc.). [58, 70]. One purpose of the material used as a matrix is to keep the reinforcement materials together under load and to distribute the load homogeneously between the reinforcements. Carbon black, graphite, and carbon nanotubes are primarily used as carbon-based reinforcement materials, and graphene has started to be preferred in recent years. The strong carbon bonds in its structure have made graphene one of the most durable materials known on Earth. Graphene in a metal matrix provides low friction and high wear resistance to the composite material due to its solid lubricant properties. Graphene is the source of all graphite-based forms. The two-dimensional material is a nanocarbon layer formed by combining six carbon rings. Graphene is quite different from carbon nanotubes and fullerene and has extraordinary properties. Graphene is a flexible, durable, super-thin, lightweight material important in today's scientific research. It has been determined that mechanical properties are improved when graphene is reinforced in matrix materials in nanolayers. Studies on the mechanical, thermal, and electrical properties of carbon nanotubes (CNTs) show that these rope-like structures have exceptional properties. This indicates that CNTs can be ideal reinforcement materials in polymer composites. The main problem encountered in producing CNT-reinforced composites is the high specific surface areas of CNTs, which cause the formation of high Van der Waals forces. This situation leads to the agglomeration of nanotubes into a small volume, in other words, to their non-homogeneous distribution and spreading within the composite. Therefore, the main problem in producing CNT-reinforced composites is ensuring good distribution and spreading of CNTs [70].

(b) **Ceramic-Polymer Composites**: These nanomaterials are used in electronic and biomedical applications [71–73]. Ceramic-polymer nanocomposites are among the inorganic materials with many applications in the health and biomedical fields. These nanocomposites are generally fabricated for thermometers, filling materials in dentistry, eyeglass lenses, and fiber optics used in endoscopy. Bioceramics are preferred materials in the clinic due to their bone coating, bone bonding, and bone tissue penetration properties. Dental bioceramics are frequently used as restorative materials in the repair or renewal of hard connective tissue used as filling materials in dentistry, as hard tissue implants.

3.5 Magnetic Nanomaterials

Nanomaterials that exhibit magnetic properties are essential in biomedical imaging, data storage, and catalysis (Fig. 4) [74–76]. Magnetic nanoparticles have been extensively used, especially in recent years, and are among the most popular topics. These materials, which were first discovered and called "magic particles", have been obtained in various geometric shapes and sizes over time. This material is symbolized with the formula $F_{e3}O_{4,}$ and its magnetic, physical, and chemical properties can be adjusted by combining it with nickel, cobalt, and copper. The preparation and application of these materials, which have attracted great interest from engineering and medical sciences as well as basic sciences, sometimes lead to the conclusion that they require complicated processes. Magnetic nanoparticles have been applied not only for drug delivery purposes but also in various fields such as additives in composite materials, heavy metal removal from wastewater containing heavy metals, enzyme support materials, tissue engineering applications, catalyst support materials, and nanoflower production, and positive results have been obtained [74].

(a) **Iron-Oxide nanoparticles**: These nanomaterials are used in MRI contrast agents and targeted drug delivery systems. The developing technology and nanotechnology field has brought about quite different innovations, such as producing implantable materials, discovering anticancer drugs with low cytotoxicity, and producing new medications at the nanoscale level. Thanks to these innovations, iron-oxide nanoparticles have started to be used mainly in biomedical fields. The physicochemical properties, sizes, and magnetic properties of nanoparticles can be directed and positioned to the areas to be treated with

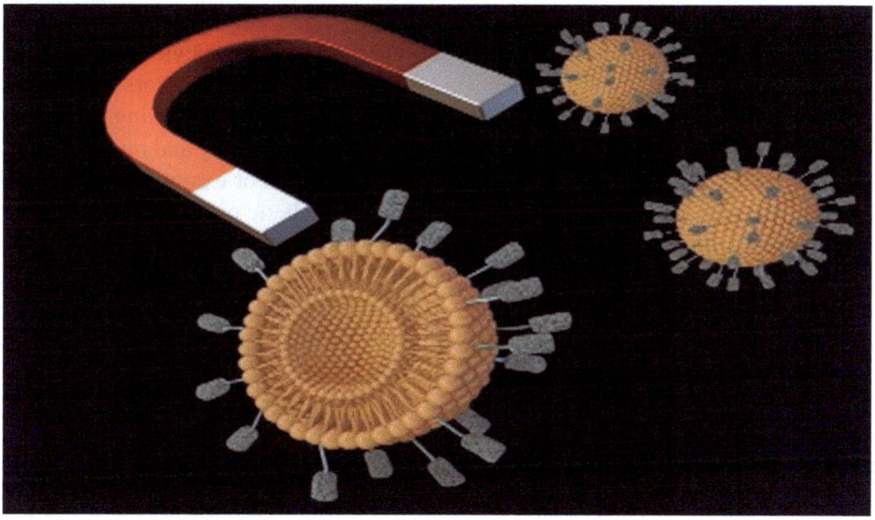

Fig. 4 The illustration of the magnetic nanoparticles [81]

external magnetic field gradients. These properties make it possible to evaluate iron-oxide nanoparticles as promising therapeutic tools for drug delivery and treating diseases. For example, Fe_3O_4 MNPs are used as hyperthermia reagents, in the development of contrast agents for magnetic field-assisted radiation [77].

(b) **Cobalt and Nickel nanomaterials**: Cobalt and nickel nanoparticles have recently been used in the production of artificial joints and dentistry. Under high loads, these nanoparticles are used in joint prostheses, such as knees and hips. Such metal compounds comprise 65% cobalt/nickel by weight, and the rest is chromium. Molybdenum is added to the structure to obtain a better grain structure. Cobalt-containing nanomaterials have better resistance to deformation than stainless steel. These nanomaterials are used in magnetic storage and magneto-optical devices [78–80].

4 Sensors and Nanomaterials

Sensors detect a specific physical or chemical parameter and convert this information into an electrical signal. Nanomaterials play an essential role in increasing the sensitivity and accuracy of sensors [82, 83].

(a) **Biosensors**: These are used in medical diagnosis and biomedical analysis [84]. All living beings are in contact with their environment through physical and chemical biological sensors used in their bodies. Biosensors are electronic devices that play a vital role in atmospheric environmental testing. They have significant advantages over traditional methods. They meet the need for portable analytical tools with selectivity and stability. The structural properties of the sensing materials used in their production critically affect device performance. Today, thanks to the developments in material science, nanosensors have gained much development and sensitivity. They generally perform using traditional techniques in the areas where they are used. High efficiency and multifunctionality are the distinguishing features of nanosensors. For example, nanosensors that measure glucose levels are convenient for diabetic patients [85].

(b) **Gas Sensors**: These are used in environmental analysis and industrial safety. For example, nanomaterials detect harmful gases such as carbon monoxide and methane [86].

(c) **Optical Sensors**: These sensors, which detect light and radiation, are widely used in security and telecommunications [87].

5 Use of Nanosensors in Biomedical Applications

Nanomaterials have great potential in biomedical applications and are used in various fields. Biosensors have been in use for many years. The most straightforward biosensor application is the stethoscope device doctors use to listen to the heart, chest, and back. Biosensors have also been used in blood pressure (blood pressure) measuring devices or glucose meters that measure the amount of sugar in the blood. They are also extremely sensitive measuring devices that can quickly provide dozens of information in micro and nano sizes with a drop of body fluids such as blood, sweat, saliva, urine, and tears [88]. In endodontic microbiology, PCR has been used to detect infection-related viruses. In this way, patients with urgent or infectious diseases can receive test results when they arrive at the hospital.

Although a simple PCR is usually used for bacterial screening, in some cases, methods such as nested PCR, multiplex PCR, and rt-PCR (COVID-19) are used. The development of biosensors thanks to nanotechnology will provide revolutionary innovations in the health sector regarding diagnosis and treatment. With biomedical applications and nanodiagnostics, patients will be given prescriptions quickly, and with developments in the sensor field, they will not have to wait too long to check their values. In this way, the patient will feel better, and since the time is shortened, the cost and intensity will be reduced. For example, thanks to nanodiagnostics, these advantageous situations are applied for the comfort of the patient diagnosed with cancer. Biosensors will provide sensitivity in detecting biological markers that can be used for early diagnosis and treatment of cancer in the future. Nano devices developed for this application are in the feasibility stage. Remote monitoring of nanotherapy will be conducted for patients without visible cancer symptoms, ensuring effective cancer surveillance. Advances in biomedical technologies facilitate earlier disease detection, including cancer, by allowing real-time measurements and prompt responses to treatment. This system will offer tailored cancer prevention strategies. As a result, early diagnosis increases the chance of treatment. Some of their critical biological applications. Functionality of nanoparticles in biomedical processes is critical in preparing biosensors with anticancer and antimicrobial activities. Due to the high efficiency of biological metal nanoparticles and the biocompatible characteristic of biological nanoparticles, biological nanoparticles reduce the required drug dose and the side effects of excessive drug amounts. It has been predicted that any anticancer drug can produce its effect by increasing the speed of drug delivery against specific cells, inhibiting them. However, particles produced by green technology may take over the tasks of gold and iron nanoparticles synthesized by physicochemical methods in photoimaging and thermal therapies. Nanoparticles are frequently used in photoimaging and photothermal fields. They are used for gene transfer and cell labeling in plants and medicine. At the same time, technological developments are still ongoing in applications such as phototherapy and magnetically sensitive drug delivery. As a result, biological systems work like nanotechnological factories in every process. Understanding the behavior of nanoparticles in biological systems opens new avenues for developing therapies, especially in the medical field, which is essential for the development of safe nanotechnology. In light of this information, advances in nanotechnology will continue to provide more advantages than promised. Iron oxide nanoparticles are contrast agents [33]. For the fluorescent labeling, the quantum dots and gold nanoparticles are used for cell and tissue imaging [89].

(a) **Targeted drug delivery**: Nanoparticles (liposomes, polymeric nanoparticles) are directed to specific cells, reducing side effects [90, 91].
(b) **Controlled drug delivery**: Nanomaterials are designed to be sensitive to certain pH, temperature, or enzymes, providing controlled drug release.
(c) **Photothermal therapy**: Gold nanoparticles absorb light of specific wavelengths, heating cancer cells and destroying them [92, 93].
(d) **Nanocarrier-based chemotherapy**: Nanoparticles that carry drugs specifically to cancer cells can reduce side effects [94].
(e) **DNA and protein detection**: Nanoscale sensors detect biomarkers quickly and sensitively [95].
(f) **Diagnosis of viral diseases such as COVID-19**: Nano-biosensors offer rapid and low-cost diagnostic methods [96].
(g) **Nanofiber scaffolds**: Biocompatible scaffolds that promote cell growth are used in tissue engineering [97].
(h) **Stem cell therapies**: Nanomaterials can direct the differentiation of stem cells [98].

The silver nanomaterials are used in wound healing and medical device coatings due to their ability to kill bacteria and viruses. For example, silver nanoparticles have been studied extensively due to their antibacterial and disinfectant properties and successful results in biosensor applications, and have been produced by various methods to date. Nanosilver and colloidal silver have long been of interest to scientists due to their antibacterial properties and the fact that bacteria cannot develop resistance to silver as they can to antimicrobial agents over time. Silver, gold, and copper are examples of nanoparticles with antibacterial activity. Nano silver particles (AgNPs) are used as alternative drugs to antibiotics, sensors, spectroscopy, and catalysis in many areas. These areas can be used in industries such as food and textiles, especially in health. AgNPs are used in wound creams to prevent possible wound infection in burned tissue or open wounds, on catheter surfaces to prevent urinary tract infection in long-term catheter use, in soaps, wall paints, and washing machines [99].

The graphene oxide and zinc oxide nanoparticles are used in antibacterial surface coatings. Many researchers have observed its inhibitory effect on bacteria. In these studies, silver affects the DNA molecules of microorganisms. This effect prevents DNA replication and interacts with the -SH groups in bacterial proteins, thus inactivating them [100, 101]. Bioapplications of nanomaterials are providing significant advances in medicine and biotechnology. However, issues such as toxicity and biocompatibility are still active research areas.

6 Bio-Nanosensors for Real-Time Conversion in Biomedical Applications

Bio-nanosensors provide real-time conversion in biomedical applications by making fast, sensitive, and continuous measurements. This process occurs through several basic mechanisms: biological recognition and conversion (via transducers), sensitivity, transduction, applications, and diagnostic diagnosis (Fig. 5) Biosensors are one of the essential electronic devices that provide advantages to the application areas of biomolecules and allow for more accurate interpretation of biological processes by monitoring the dynamic changes of biofunctions. As technology develops, biosensors offer speed and accuracy and the ability to detect diseases or drug interactions quickly, thus overcoming the difficulties in real-time monitoring. This analytical technique provides a unique insight into the real-time changes of physiological parameters, and nanomaterials are promising for health and biomedical application areas in this field, both at a lower cost and with a short response time. With the current developments in nanotechnology and advanced production technologies used in electronics, nanobiosensors are being designed, and a new era is beginning in bio/bio-nanotechnological analysis in disease diagnosis. Thanks to nanotechnology, these technologies' performance and biosensors' functionality are increasing. Biosensors are widely used in biomedical applications (diagnosis) and environmental monitoring, with their studies based on different transduction mechanisms for detecting and quantifying analyte molecules. Today, thanks to Lab-on-a-chip (LOC) technologies, new generation sensors' have increased. In Lab-on-a-chip (LOC) technology, it is possible to prepare large samples on a chip at lower costs. Today, these applications are defined as biochips [102].

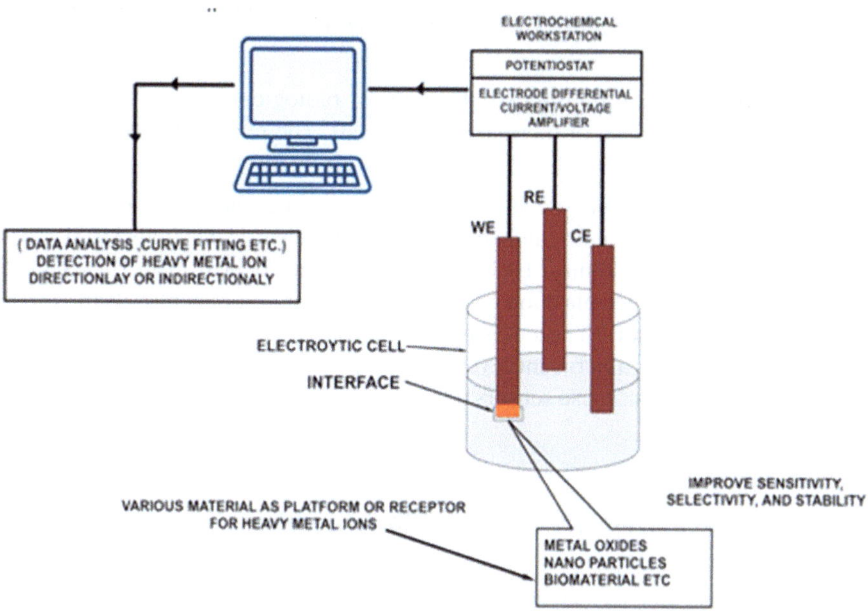

Fig. 5 The outline of an electrochemical cell system [102]

6.1 Biological Recognition and Conversion

Bio-nanosensors contain receptors that can detect a specific biomolecule or biological event. These receptors can usually be enzymes, antibodies, DNA sequences, or cells. Bioreceptors that interact with the target molecule convert these biochemical events into physical or chemical signals.

6.2 Nanoscale Sensitivity and Precision

The biggest advantage of nanosensors is their ability to perform sensitive detection at the nanoscale. Thanks to nanommost significanls (e.g., gold nanoparticles, carbon nanotubes, or quantum dots), biomolecular interactions can be detected even at lower concentrations [103]. This is of great importance for early diagnosis of diseases.

6.3 Transducer Mechanisms (Transduction)

Bio-nanosensors contain transducers that convert biological interactions into electrical, optical, mechanical, or thermal signals [104]. The main transduction mechanisms are as follows:

(a) **Electrochemical Sensors**: These sensors measure the oxidation–reduction reactions of biomolecules and convert them into electrical signals [105].
(b) **Optical Sensors**: Monitor biological events using optical properties such as fluorescence, surface plasmon resonance (SPR), or color change [106].
(c) **Mechanical Sensors**: These sensors convert biomolecule binding into mechanical changes using microcantilever or piezoelectric materials [107].
(d) **Thermal Sensors**: These sensors detect heat changes produced by biochemical reactions [108].

6.4 Real-Time Measurement and Data Analysis

Bio-nanosensors offer the advantage of providing instant data because they have the ability to produce continuous signals. In particular, wearable health sensors and implant sensors allow for instant health assessment by continuously collecting data on the patient's body [109–112]. This data is processed with artificial intelligence and big data analytics, contributing to personalized medical applications in disease diagnosis and treatment processes [113–116].

7 Conclusion

This chapter examines the role and importance of nanomaterial-based sensors in biomedical applications. The main findings of our chapter are that nanomaterial-based sensors offer significant advantages over traditional sensors by offering high sensitivity, selectivity, and fast response times. Nanomaterial-based sensors have great potential for the detection of biomarkers and continuous health monitoring. These sensors are used in the detection of various biomolecules such as blood sugar, cancer markers, and infection indicators. Nanomaterials are materials with at least one dimension between 1 and 100 nm, and this dimension range changes their physical, chemical, and mechanical properties. Their high surface area and small size increase their reactivity and surface interactions. They are of great importance in medical applications, drug delivery systems, biosensors, and cancer treatment. Sensors are devices that detect specific physical or chemical parameters and convert this information into electrical signals. They are used in medical diagnosis and biomedical analysis; for example, nanosensors that measure glucose levels provide great convenience for diabetic patients. Nanomaterial-based sensors offer revolutionary innovations in biomedical applications, expanding the possibilities of remote health monitoring and early diagnosis. Nanomaterials based nanobiosensors offer better sensitivity, selectivity and response time. Also, they have attracted great attention in biological sciences in recent years to better understand the dynamics of fundamental biological processes. With the developments in the field of tools in the entire field of nanotechnology, such as nanoscale research, nanobiosensors, etc., promising results are also being obtained in energy sources and materials. In this way, it offers the opportunity to improve and commercialize both the cost and performance of existing nanobiosensors. This chapter emphasizes the potential of nanomaterials in the healthcare sector and their importance in various biomedical applications. Research findings show that nanomaterial-based sensors contribute to developments in biomedical engineering and require further research and development in the future. As a result, the development of these sensors increases the comfort of patients in healthcare and improves treatment processes. This chapter provides theoretical information on the importance of nanotechnology and biotechnology studies in biomedical applications, their areas of use, innovative fields of study, and future biomedical applications for bio-nano sensors.

References

1. Akbari, S., Zhang, Y.-F., Wang, D., Ge, Q.: 4D printing and its biomedical applications. In: 3D and 4D Printing in Biomedical Applications, pp. 343–372. Wiley-VCH Verlag GmbH & Co. KGaA (2018). https://doi.org/10.1002/9783527813704.ch14
2. Modélisation, classification et fusion de données biomédicales—Archive ouverte HAL. (n.d.). Retrieved from https://hal.archives-ouvertes.fr/tel-00005998/
3. Dheyab, M.A., Aziz, A.A., Khaniabadi, P.M., Jameel, M.S.: Potential of a sonochemical approach to generate MRI-PPT theranostic agents for breast cancer. Photodiagn. Photodyn. Ther. **33** (2021). https://doi.org/10.1016/j.pdpdt.2021.102177
4. Mohajer, F., Mohammadi, Z.G., Badiei, A.: New advances on Au–magnetic organic hybrid core–shells in MRI, CT imaging, and drug delivery. RSC Adv. **11**(12), 6517–6525 (2021). https://doi.org/10.1039/D1RA00415H
5. Kim, N.J., Lee, S.J., Atala, A.: Biomedical nanomaterials in tissue engineering. In: Nanomaterials in Tissue Engineering: Fabrication and Applications, pp. 1–25e. Elsevier Ltd. (2013). https://doi.org/10.1533/9780857097231.1
6. Hsissou, R., Dagdag, O., Berradi, M., El Bouchti, M., Assouag, M., Elharfi, A.: Development rheological and anti-corrosion property of epoxy polymer and its composite. Heliyon **5**(11), e02789 (2019). https://doi.org/10.1016/j.heliyon.2019.e02789
7. Ahovan, Z.A., Hashemi, A., De Plano, L.M., Gholipourmalekabadi, M., Seifalian, A.: Bacteriophage based biosensors: trends, outcomes and challenges. Nanomaterials **10**(3), 501 (2020). https://doi.org/10.3390/NANO10030501
8. Kavita, V.: DNA biosensors-a review. J. Bioeng. Biomed. Sci. **7**(2) (2017). https://doi.org/10.4172/2155-9538.1000222
9. Tatli, H.N., Gunduz, S.B., Sahin, M., Altuner, E.E., Dar, U.A.: Fluorimetric methods for determination of aluminum in water resources utilizing newly synthesized N, N'-bis(2,5-dihydroxybenzylidene)-4,4′-diamino diphenyl ether. Spectrochim. Acta Part A Mol. Biomol. Spectrosc. **321**, 124726 (2024). https://doi.org/10.1016/J.SAA.2024.124726
10. Altuner, E.E., Bekmezci, M., Sen, F.: Diffusion and transport studies. In: Handbook of Magnetic Hybrid Nanoalloys and Their Nanocomposites, pp. 1–18 (2022). https://doi.org/10.1007/978-3-030-34007-0_27-1

11. Altuner, E.E., Erduran, V., Sen, F.: Green synthesized nanomaterials for bioimaging. In: Synthesis of Bionanomaterials for Biomedical Applications Micro and Nano Technologies, pp. 265–286. Elsevier (2023). https://doi.org/10.1016/B978-0-323-91195-5.00023-4
12. Altuner, E.E., Gur, T., Şen, F.: Ternary/quaternary nanomaterials for direct alcohol fuel cells. In: Nanomaterials for Direct Alcohol Fuel Cells, pp. 157–172 (2021). https://doi.org/10.1016/B978-0-12-821713-9.00001-9
13. Erduran, V., Altuner, E.E., Şen, F.: Stability and validation of bionanomaterials. In: Synthesis of Bionanomaterials for Biomedical Applications, pp. 251–263 (2023). https://doi.org/10.1016/B978-0-323-91195-5.00021-0
14. Bekmezci, M., Altuner, E.E., Sen, F.: Biological characterization of magnetic hybrid nanoalloys. In: Handbook of Magnetic Hybrid Nanoalloys and Their Nanocomposites, pp. 1–18 (2022). https://doi.org/10.1007/978-3-030-34007-0_28-1
15. Darabi, R., Alown, F.E.D., Aygun, A., Gu, Q., Gulbagca, F., Altuner, E.E., Seckin, H., Meydan, I., Kaymak, G., Sen, F., Karimi-Maleh, H.: Biogenic platinum-based bimetallic nanoparticles: synthesis, characterization, antimicrobial activity and hydrogen evolution. Int. J. Hydrogen Energy **48**(55), 21270–21284 (2023). https://doi.org/10.1016/J.IJHYDENE.2022.12.072
16. Altuner, E.E., Gulbagca, F., TiriR, N.E., Aygun, A., Sen, F.: Highly efficient palladium-zinc oxide nanoparticles synthesized by biogenic methods: characterization, hydrogen production and photocatalytic activities. Chem. Eng. J. Adv. **14**, 100465 (2023). https://doi.org/10.1016/J.CEJA.2023.100465
17. Xian, Y., Hu, Y., Liu, F., Xian, Y., Wang, H., Jin, L.: Glucose biosensor based on Au nanoparticles–conductive polyaniline nanocomposite. Biosens. Bioelectron. **21**(10), 1996–2000 (2006). https://doi.org/10.1016/J.BIOS.2005.09.014
18. Karimi, F., Altuner, E.E., Aygun, A., Bayat, R., Rajendran, S., Sen, F.: Synthesis of silver nanoparticles by biogenic methods: characterization and development of a sensor sensible to pharmaceutical medicine paracetamol. Top. Catal. **67**(9–12), 585–593 (2024). https://doi.org/10.1007/S11244-023-01887-4/TABLES/1
19. Shao, A.F., Wang, Z.B., Chu, Y.Y., Jiang, Z.Z., Yin, G.P., Liu, Y.: Evaluation of the performance of carbon supported Pt-Ru-Ni-P as anode catalyst formethanol electrooxidation. Fuel Cells **10**(3), 472–477 (2010). https://doi.org/10.1002/fuce.200900170
20. Altuner, E.E., El Houda Tiri, R.N., Aygun, A., Gulbagca, F., Sen, F., Iranbakhsh, A., Karimi, F., Vasseghian, Y., Dragoi, E.-N.: Hydrogen production and photocatalytic activities from $NaBH_4$ using trimetallic biogenic PdPtCo nanoparticles: development of machine learning model. Chem. Eng. Res. Des. **184**, 180–190 (2022). https://doi.org/10.1016/J.CHERD.2022.05.021
21. Darabi, R., Karimi-Maleh, H., Altuner, E.E., Gulbagca, F., Tiri, R.N.E., Chinnathambi, A., Brindhadevi, K., Halvaci, E., Sen, F.: Cobalt nanoparticles supported active carbon from chitosan biopolymer using thermal method: synthesis, characterization, and hydrogen production. Top. Catal. 1–10 (2024)
22. Altuner, E.E., Al Janabi, M.Y.A., Sen, F., Dar, U.A.: Synthesis and characterization of iron–cobalt@activated carbon nanoparticles for electrochemical sensing of dopamine neurotransmitters. Chem. Afr. 1–9 (2025). https://doi.org/10.1007/S42250-025-01197-0
23. Altuner, E.E.: Synthesis, characterisation and photocatalytic activation of zinc nanoparticles via. biogenic methods. Int. J. Environ. Anal. Chem. 1–11 (2025). https://doi.org/10.1080/03067319.2025.2451356
24. Alizadeh, M., Asrami, P.N., Altuner, E.E., Gulbagca, F., Tiri, R.N.E., Aygun, A., Kaynak, İ, Sen, F., Cheraghi, S.: An ultra-sensitive rifampicin electrochemical sensor based on Fe_3O_4 nanoparticles anchored multiwalled carbon nanotube modified glassy carbon electrode. Chemosphere **309**, 136566 (2022). https://doi.org/10.1016/J.CHEMOSPHERE.2022.136566
25. Goldshleger, N.F.: Fullerenes and Fullerene-based materials in catalysis. Fullerene Sci. Technol. **9**(3), 255–280 (2007). https://doi.org/10.1081/FST-100104493
26. Ozalp, V.C., Karabiyik, G., Bayrac, A.T., Guvenc Tuna, B.: Graphene-based electrochemical aptasensors. In: Palys, B. (ed.) Handbook of Graphene, Biosensors and Advanced Sensor, vol. 6, p. 465. University of Warsaw, Scrivener Publishing, Warsaw (2019)

27. Drummer, S., Madzimbamuto, T., Chowdhury, M.: Green synthesis of transition-metal nanoparticles and their oxides: a review. Materials **14**(11), 2700 (2021). https://doi.org/10.3390/MA14112700/S1
28. Altuner, E.E., Akin, M., Bayat, R., Bekmezci, M., Burhan, H., Sen, F.: Challenges in commercialization of carbon nanomaterial-based sensors. In: Carbon Nanomaterials-Based Sensors: Emerging Research Trends in Devices and Applications, pp. 381–392 (2022). https://doi.org/10.1016/B978-0-323-91174-0.00020-2
29. Altuner, E.E., Arıkan, K., Burhan, H., Ozdemir, S., Şen, F.: Commercial aspects of direct alcohol fuel cells. In: Nanomaterials for Direct Alcohol Fuel Cells, pp. 511–524 (2021). https://doi.org/10.1016/B978-0-12-821713-9.00012-3
30. Coleman, J.N., Khan, U., Blau, W.J., Gun'ko, Y.K.: Small but strong: a review of the mechanical properties of carbon nanotube-polymer composites. Carbon **44**(9), 1624–1652. Pergamon (2006). https://doi.org/10.1016/j.carbon.2006.02.038
31. Nawani, P., Burger, C., Rong, L., Chu, B., Hsiao, B.S., Tsou, A.H., Weng, W.: Characterization of nanoclay orientation in polymer nanocomposite film by small-angle X-ray scattering. Polymer **51**(22), 5255–5266 (2010). https://doi.org/10.1016/j.polymer.2010.08.048
32. Nawani, P., Burger, C., Rong, L., Chu, B., Hsiao, B.S., Tsou, A.H., Weng, W.: Characterization of nanoclay orientation in polymer nanocomposite film by small-angle X-ray scattering. Polymer **51**(22), 5255–5266 (2010b). https://doi.org/10.1016/j.polymer.2010.08.048
33. Yadav, R.B., Srivastava, S., Srivastava, R.: Identification and removal of different categories of noises from magnetic resonance image using hybrid partial differential equation-based filter. Int. J. Digit. Sig. Smart Syst. **1**(2), 87 (2017). https://doi.org/10.1504/ijdsss.2017.088050
34. Yadav, R.S., Kuřitka, I., Vilcakova, J., Urbánek, P., Machovsky, M., Masař, M., Holek, M.: Structural, magnetic, optical, dielectric, electrical and modulus spectroscopic characteristics of $ZnFe_2O_4$ spinel ferrite nanoparticles synthesized via honey-mediated sol-gel combustion method. J. Phys. Chem. Solids **110**, 87–99 (2017). https://doi.org/10.1016/j.jpcs.2017.05.029
35. Copur, F., Bekar, N., Zor, E., Alpaydin, S., Bingol, H.: Nanopaper-based photoluminescent enantioselective sensing of L-Lysine by L-Cysteine modified carbon quantum dots. Sens. Actuators B Chem. **279**, 305–312 (2019). https://doi.org/10.1016/J.SNB.2018.10.026
36. Zeng, Q., Cheng, J.S., Liu, X.F., Bai, H.T., Jiang, J.H.: Palladium nanoparticle/chitosan-grafted graphene nanocomposites for construction of a glucose biosensor. Biosens. Bioelectron. **26**(8), 3456–3463 (2011). https://doi.org/10.1016/J.BIOS.2011.01.024
37. Chatterjee, D.K., Diagaradjane, P., Krishnan, S.: Nanoparticle-mediated hyperthermia in cancer therapy. Ther. Deliv. **2**(8), 1001–1014 (2011). https://doi.org/10.4155/TDE.11.72
38. Han, M., Zhu, S., Lu, S., Song, Y., Feng, T., Tao, S., Liu, J., Yang, B.: Recent progress on the photocatalysis of carbon dots: classification, mechanism and applications. Nano Today **19**, 201–218 (2018). https://doi.org/10.1016/J.NANTOD.2018.02.008
39. Li, Q., Zhao, N., Bai, F.: Size- and shape-dependent photocatalysis of porphyrin nanocrystals. MRS Bull. **44**(3), 172–177 (2019). https://doi.org/10.1557/MRS.2019.41
40. Qumar, U., Hassan, J.Z., Bhatti, R.A., Raza, A., Nazir, G., Nabgan, W., Ikram, M.: Photocatalysis vs adsorption by metal oxide nanoparticles. J. Mater. Sci. Technol. **131**, 122–166 (2022). https://doi.org/10.1016/J.JMST.2022.05.020
41. WangXiu, H.M., Zhang, F., Zhuang, Z., Yu, Y.: Recyclable nanoscale zero valent iron doped g-C_3N_4/MoS_2 for efficient photocatalysis of RhB and Cr(VI) driven by visible light. ACS Sustain. Chem. Eng. **4**(7), 4055–4063 (2016). https://doi.org/10.1021/acssuschemeng.6b01024
42. Hames, Y., Alpaslan, Z., Kösemen, A., San, S.E., Yerli, Y.: Electrochemically grown ZnO nanorods for hybrid solar cell applications. Sol. Energy **84**(3), 426–431 (2010). https://doi.org/10.1016/J.SOLENER.2009.12.013
43. Kamat, P.V., Tvrdy, K., Baker, D.R., Radich, J.G.: Beyond photovoltaics: semiconductor nanoarchitectures for liquid-junction solar cells. Chem. Rev. **110**(11), 6664–6688 (2010). https://doi.org/10.1021/cr100243p
44. Kumar, U., Sikarwar, S., Sonker, R.K., Yadav, B.C.: Carbon nanotube: synthesis and application in solar cell. J. Inorg. Organomet. Polym. Mater. **26**(6), 1231–1242 (2016). https://doi.org/10.1007/S10904-016-0401-Z/FIGURES/6

45. Karimi, F., Altuner, E.E., Gulbagca, F., Tiri, R.N.E., Sen, F., Javadi, A., Dragoi, E.N.: Facile bio-fabrication of ZnO@AC nanoparticles from chitosan: characterization, hydrogen generation, and photocatalytic properties. Environ. Res. **216**, 114668 (2023a). https://doi.org/10.1016/J.ENVRES.2022.114668
46. Karimi, F., Altuner, E.E., Gulbagca, F., Tiri, R.N.E., Sen, F., Javadi, A., Dragoi, E.N.: Facile bio-fabrication of ZnO@AC nanoparticles from chitosan: characterization, hydrogen generation, and photocatalytic properties. Environ. Res. **216**, 114668 (2023b). https://doi.org/10.1016/J.ENVRES.2022.114668
47. Aydogdu, A., Burmaoglu, S., Saritas, O., Cakir, S.: A nanotechnology roadmapping study for the Turkish defense industry. Foresight **19**(4), 354–375 (2017). https://doi.org/10.1108/FS-06-2017-0020/FULL/XML
48. Harish, V., Tewari, D., Gaur, M., Yadav, A.B., Swaroop, S., Bechelany, M., Barhoum, A.: Review on nanoparticles and nanostructured materials: bioimaging, biosensing, drug delivery, tissue engineering, antimicrobial, and agro-food applications. Nanomaterials **12**(3), 457 (2022). https://doi.org/10.3390/NANO12030457
49. Li, F., Wang, X., Sun, X., Guo, Y.: Multiplex electrochemical aptasensor for detecting multiple antibiotics residues based on carbon fiber and mesoporous carbon-gold nanoparticles. Sens. Actuators B Chem. **265**, 217–226 (2018). https://doi.org/10.1016/j.snb.2018.03.042
50. Wang, J.: Carbon-nanotube based electrochemical biosensors: a review. Electroanalysis **17**(1), 7–14 (2005). https://doi.org/10.1002/elan.200403113
51. Genc, R., Alas, M.O., Harputlu, E., Repp, S., Kremer, N., Castellano, M., Colak, G., Ocakoglu, K., Erdem, E.: High-capacitance hybrid supercapacitor based on multi-colored fluorescent carbon-dots. Sci. Rep. **7**(1), 1–13 (2017). https://doi.org/10.1038/s41598-017-11347-1
52. Meirinho, S.G., Ferraria, A.M., do Rego, A.M.B., Fernandes, A.J.S., Viana, A.S., Fernandes, J.C.S., Oliveira, M.C.: Electrochemical properties of oxygen-enriched carbon-based nanomaterials. J. Electroanal. Chem. **873**, 114420 (2020). https://doi.org/10.1016/j.jelechem.2020.114420
53. Deza, M., Dutour, M., Fowler, P.W.: Zigzags, railroads, and knots in fullerenes. J. Chem. Inf. Comput. Sci. **44**(4), 1282–1293 (2004). https://doi.org/10.1002/chin.200439199
54. Hussain, S., Abbas, S.M., Rahman, M.: Yttria-Modified Titania-CNTs Composite as an Electrochemical Biosensor for Lysine (2013). Retrieved from https://www.researchgate.net/publication/259745013
55. Mackiewicz, N., Surendran, G., Remita, H., Keita, B., Zhang, G., Nadjo, L., Hagège, A., Doris, E., Mioskowski, C.: Supramolecular self-assembly of amphiphiles on carbon nanotubes: a versatile strategy for the construction of CNT/metal nanohybrids, application to electrocatalysis. J. Am. Chem. Soc. **130**(26), 8110–8111 (2008). https://doi.org/10.1021/ja8026373
56. Wang, X., Tang, F., Cao, Q., Qi, X., Pearson, M., Li, M., Pan, H., Zhang, Z., Lin, Z.: Comparative study of three carbon additives: carbon nanotubes, graphene, and fullerene-C60, for synthesizing enhanced polymer nanocomposites. Nanomaterials **10**(5), 838 (2020). https://doi.org/10.3390/NANO10050838
57. Narayanan, R., El-Sayed, M.A.: Effect of colloidal nanocatalysis on the metallic nanoparticle shape: the Suzuki reaction. Langmuir **21**(5), 2027–2033 (2005). https://doi.org/10.1021/la047600m
58. Wang, F., Zhang, Y., Wang, Y., Luo, Y., Chen, Y., Zhu, H.: Co-P nanoparticles supported on dandelion-like CNTs-Ni foam composite carrier as a novel catalyst for hydrogen generation from $NaBH_4$ methanolysis. Int. J. Hydrogen Energy **43**(18), 8805–8814 (2018). https://doi.org/10.1016/J.IJHYDENE.2018.03.140
59. Menon, S., Rajeshkumar, S., Venkat Kumar, S.: A review on biogenic synthesis of gold nanoparticles, characterization, and its applications. Resour. Efficient Technol. **3**(4), 516–527 (2017). https://doi.org/10.1016/j.reffit.2017.08.002
60. Klaus, T., Joerger, R., Olsson, E., Granqvist, C.G.: Silver-based crystalline nanoparticles, microbially fabricated. Proc. Natl. Acad. Sci. U.S.A. **96**(24), 13611–13614 (1999). https://doi.org/10.1073/pnas.96.24.13611

61. Altuner, E.E.: Electrochemical sensitivity, selectivity, and sensory properties of a natural polymer-supported titanium nanocomposite electrode towards lysine amino acid. Muş Alparslan Üniversitesi Fen Bilimleri Dergisi **11**(1), 28–32 (2023). https://doi.org/10.18586/MSUFBD.1268440
62. Barczuk, P.J., Lewera, A., Miecznikowski, K., Zurowski, A., Kulesza, P.J.: Enhancement of catalytic activity of platinum-based nanoparticles towards electrooxidation of ethanol through interfacial modification with heteropolymolybdates. J. Power. Sour. **195**(9), 2507–2513 (2010). https://doi.org/10.1016/j.jpowsour.2009.11.066
63. Kim, S.W., Park, J., Jang, Y., Chung, Y., Hwang, S., Hyeon, T., Kim, Y.W.: Synthesis of monodisperse palladium nanoparticles. Nano Lett. **3**(9), 1289–1291 (2003). https://doi.org/10.1021/NL0343405/SUPPL_FILE/NL0343405SI20030627_040056.PDF
64. Yaqoob, A.A., Ahmad, H., Parveen, T., Ahmad, A., Oves, M., Ismail, I.M.I., Qari, H.A., Umar, K., Mohamad Ibrahim, M.N.: Recent advances in metal decorated nanomaterials and their various biological applications: a review. Front. Chem. **8**, 528583 (2020). https://doi.org/10.3389/FCHEM.2020.00341/PDF
65. Yang, J.X., Dai, B.H., Chiang, C.Y., Chiu, I.C., Pao, C.W., Lu, S.Y., Tsao, I.Y., Lin, S.T., Chiu, C.T., Yeh, J.W., Chang, P.C., Hung, W.H.: Rapid fabrication of high-entropy ceramic nanomaterials for catalytic reactions. ACS Nano **15**(7), 12324–12333 (2021). https://doi.org/10.1021/ACSNANO.1C04259/SUPPL_FILE/NN1C04259_SI_001.PDF
66. Tao, L., Chen, Z., Li, Z., Wang, J., Xu, X., Xu, J.B.: Enhancing light-matter interaction in 2D materials by optical micro/nano architectures for high-performance optoelectronic devices. InfoMat **3**(1), 36–60 (2021). https://doi.org/10.1002/INF2.12148
67. Tan, W., Wang, K., He, X., Zhao, X.J., Drake, T., Wang, L., Bagwe, R.P.: Bionanotechnology based on silica nanoparticles. Med. Res. Rev. **24**(5), 621–638 (2004). https://doi.org/10.1002/MED.20003
68. Kanoun, O., Lazarević-Pašti, T., Pašti, I., Nasraoui, S., Talbi, M., Brahem, A., Adiraju, A., Sheremet, E., Rodriguez, R.D., Ali, M.B., Al-Hamry, A.: A review of nanocomposite-modified electrochemical sensors for water quality monitoring. Sensors **21**(12), 4131 (2021). https://doi.org/10.3390/S21124131
69. Lagashetty, A., Resonance, A.V.: Polymer Nanocomposites. In: Springer (2005). Retrieved from https://idp.springer.com/authorize/casa?redirect_uri=https://link.springer.com/content/pdf/https://doi.org/10.1007/BF02867106.pdf&casa_token=13r2BeRM12EAAAAA:BSXjJdCa1UvaB2TjulzhoNUlymdt-VeVqvIbmUlhivYvnJ-uwXIeO8rv49WWn-p2dBN_7Fs76bjFlQ
70. Zhou, H.W., Mishnaevsky, L., Yi, H.Y., Liu, Y.Q., Hu, X., Warrier, A., Dai, G.M.: Carbon fiber/carbon nanotube reinforced hierarchical composites: effect of CNT distribution on shearing strength. Compos. Part B Eng. **88**, 201–211 (2016). https://doi.org/10.1016/j.compositesb.2015.10.035
71. Colombo, P., Mera, G., Riedel, R., Sorarù, G.D.: Polymer-derived ceramics: 40 years of research and innovation in advanced ceramics. J. Am. Ceram. Soc. **93**(7), 1805–1837 (2010). https://doi.org/10.1111/J.1551-2916.2010.03876.X
72. Eckel, Z.C., Zhou, C., Martin, J.H., Jacobsen, A.J., Carter, W.B., Schaedler, T.A.: Additive manufacturing of polymer-derived ceramics. Science **351**(6268), 58–62 (2016). https://doi.org/10.1126/SCIENCE.AAD2688
73. Feng, Y., Lai, S., Yang, L., Riedel, R., Yu, Z.: Polymer-derived porous Bi_2WO_6/SiC(O) ceramic nanocomposites with high photodegradation efficiency towards Rhodamine B. Ceram. Int. **44**(7), 8562–8569 (2018). https://doi.org/10.1016/J.CERAMINT.2018.02.061
74. Baker, C., Shah, S.I., Hasanain, S.K.: Magnetic behavior of iron and iron-oxide nanoparticle/polymer composites. J. Magn. Magn. Mater. **280**(2–3), 412–418 (2004). https://doi.org/10.1016/j.jmmm.2004.03.037
75. Raghava Reddy, K., Reddy, P.A., Reddy, C.V., Shetti, N.P., Babu, B., Ravindranadh, K., Shankar, M.V., Reddy, M.C., Soni, S., Naveen, S.: Functionalized magnetic nanoparticles/biopolymer hybrids: synthesis methods, properties and biomedical applications. In: Methods in Microbiology, vol. 46, pp. 227–254. Academic Press Inc. (2019). https://doi.org/10.1016/bs.mim.2019.04.005

76. Zheng, M., Wu, X.C., Zou, B.S., Wang, Y.J.: Magnetic properties of nanosized $MnFe_2O_4$ particles. J. Magn. Magn. Mater. **183**(1–2), 152–156 (1998). https://doi.org/10.1016/S0304-8853(97)01057-3
77. Liu, X.G.Q.A.Y.Z.W.X.L.: Hydrothermal synthesis and characterization of α-FeOOH and α-Fe_2O_3 uniform nanocrystallines. J. Alloys Compd. **433**, 216–220 (2007)
78. Fararh, K.M., Atoji, Y., Shimizu, Y., Takewaki, T.: Isulinotropic properties of Nigella sativa oil in Streptozotocin plus nicotinamide diabetic hamster. Res. Vet. Sci. **73**(3), 279–282 (2002). https://doi.org/10.1016/S0034-5288(02)00108-X
79. Shkryl, Y.N., Veremeichik, G.N., Kamenev, D.G., Gorpenchenko, T.Y., Yugay, Y.A., Mashtalyar, D.V., NepomnyaschiyA, V., AvramenkoT, V., KarabtsovA, A., IvanovV, V., Bulgakov, V.P., Gnedenkov, S.V., Kulchin, Y.N., ZhuravlevY, N.: Green synthesis of silver nanoparticles using transgenic Nicotiana tabacum callus culture expressing silicatein gene from marine sponge Latrunculia oparinae. Artif. Cells Nanomed. Biotechnol. **46**(8), 1646–1658 (2018). https://doi.org/10.1080/21691401.2017.1388248
80. Zare, N., Karimi, F., Altuner, E.E., Mashkoor, H., Tiri, R.N.E., Gulbagca, F., Bijad, M., Cherif, A., Sen, F.: Synthesis of bimetallic nickel-cobalt nanoparticles supported by functional multi-walled carbon nanotubes by chemical methods: characterization and hydrogen energy generation. Top. Catal. **67**(9–12), 606–614 (2024). https://doi.org/10.1007/S11244-024-01907-X/TABLES/1
81. Suhail, A.: Magnetic nanoparticles: properties and potential applications. Pure Appl. Chem. Azo Nano **78**(9), 1793–1801 (2006). https://doi.org/10.1351/PAC200678091793
82. Mao, S., Lu, G., Chen, J.: Nanocarbon-based gas sensors: progress and challenges. J. Mater. Chem. A **2**(16), 5573–5579 (2014). https://doi.org/10.1039/c3ta13823b
83. Shetti, N.P., Malode, S.J., Nayak, D.S., Bukkitgar, S.D., Bagihalli, G.B., Kulkarni, R.M., ReddyK, R.: Novel nanoclay-based electrochemical sensor for highly efficient electrochemical sensing nimesulide. J. Phys. Chem. Solids **137**, 109210 (2020). https://doi.org/10.1016/j.jpcs.2019.109210
84. Maduraiveeran, G., Sasidharan, M., Ganesan, V.: Electrochemical sensor and biosensor platforms based on advanced nanomaterials for biological and biomedical applications. In: Biosensors and Bioelectronics, vol. 103, pp. 113–129. Elsevier Ltd. (2018). https://doi.org/10.1016/j.bios.2017.12.031
85. Erol, E., Yildirim, E., Cete, S.: Construction of biosensor for hypoxanthine determination by immobilization of xanthine oxidase and uricase in polypyrrole-paratoluenesulfonate film. J. Solid State Electrochem. **24**(7), 1695–1707 (2020). https://doi.org/10.1007/S10008-020-04715-X/FIGURES/13
86. GottamS, R., Tsai, C.T., Wang, L.W., Wang, C.T., Lin, C.C., Chu, S.Y.: Highly sensitive hydrogen gas sensor based on a MoS_2-Pt nanoparticle composite. Appl. Surf. Sci. **506**, 144981 (2020). https://doi.org/10.1016/j.apsusc.2019.144981
87. della Scuola, D., Gaetano Granozzi Supervisore, C., Andrea Rizzi, G.: Synthesis and Engineering of Easy & Cheap Silver Based Optical Sensors (2012). Retrieved from https://www.research.unipd.it/handle/11577/3422543
88. Rao, C.N.R., Müller, A., Cheetham, A.K.: Nanomaterials—an introduction. In: The Chemistry of Nanomaterials, pp. 1–11. Wiley-VCH Verlag GmbH & Co. KGaA (2005). https://doi.org/10.1002/352760247x.ch1
89. Lichtman, J.W., Conchello, J.A.: Fluorescence microscopy. Nat. Methods **2**(12), 910–919 (2005). https://doi.org/10.1038/nmeth817
90. Gong, T., Liu, T., Zhang, L., Ye, W., Guo, X., Wang, L., Quan, L., Pan, C.: Design redox-sensitive drug-loaded nanofibers for bone reconstruction. ACS Biomater. Sci. Eng. **4**(1), 240–247 (2018). https://doi.org/10.1021/acsbiomaterials.7b00827
91. Singh, R.J.L.: Nanoparticle-based targeted drug delivery. Exp. Mol. Pathol. **86**, 215–223 (2009)
92. Yuwen, L., Zhou, J., Zhang, Y., Zhang, Q., Shan, J., Luo, Z., Weng, L., Teng, Z., Wang, L.: Aqueous phase preparation of ultrasmall MoSe2 nanodots for efficient photothermal therapy of cancer cells. Nanoscale **8**, 2720–2726 (2016)

93. Zhang, L., Zhang, X., Li, L., Liu, Y., Wang, D., Xu, L., Bao, J., Zhang, A.: Fabrication of photothermally responsive nanocomposite hydrogel through 3D printing. Macromol. Mater. Eng. **305**(2) (2020). https://doi.org/10.1002/mame.201900718
94. Wisselink, D.D., Braakhuis, L.L.F., Gallo, G., van Grevenstein, W.M.U., van Dieren, S., Kok, N.F.M., de Reuver, P.R., Tanis, P.J., de Hingh, I.H.J.T.: Systematic review of published literature on oxaliplatin and mitomycin C as chemotherapeutic agents for hyperthermic intraperitoneal chemotherapy in patients with peritoneal metastases from colorectal cancer. Crit. Rev. Oncol. Hematol. **142**, 119–129 (2019). https://doi.org/10.1016/J.CRITREVONC.2019.06.014
95. Wang, J., Zhang, S., Zhang, Y.: Fabrication of chronocoulometric DNA sensor based on gold nanoparticles/poly(l-lysine) modified glassy carbon electrode. Anal. Biochem. **396**(2), 304–309 (2010). https://doi.org/10.1016/J.AB.2009.10.004
96. Zamaro, A., Kulchitsky, V.A.: Is COVID-19 a Systemic or Local Pathological Process? Combined application of nanoparticles or heterocyclic compounds with chemotherapeutic agents for reducing side effects and increasing antitumor activity in experimental models View project Book Reviews View project (n.d.). https://doi.org/10.2139/ssrn.3704738
97. Bhattacharyya, S., Kumbar, S.G., Khan, Y.M., Nair, L.S., Singh, A., Krogman, N.R., Brown, P.W., Allcock, H.R., Laurencin, C.T.: Biodegradable polyphosphazene-nanohydroxyapatite composite nanofibers: scaffolds for bone tissue engineering. J. Biomed. Nanotechnol. **5**(1), 69–75 (2009). https://doi.org/10.1166/jbn.2009.032
98. Yim, E.K.F., Darling, E.M., Kulangara, K., Guilak, F., Leong, K.W.: Nanotopography-induced changes in focal adhesions, cytoskeletal organization, and mechanical properties of human mesenchymal stem cells. Biomaterials **31**(6), 1299–1306 (2010). https://doi.org/10.1016/J.BIOMATERIALS.2009.10.037
99. Melaiye, A., Sun, Z., Hindi, K., Milsted, A., Ely, D., Reneker, D.H., Tessier, C.A., Youngs, W.J.: Silver(I)-imidazole cyclophane gem-diol complexes encapsulated by electrospun tecophilic nanofibers: formation of nanosilver particles and antimicrobial activity. J. Am. Chem. Soc. **127**(7), 2285–2291 (2005). https://doi.org/10.1021/ja040226s
100. Duan, X., Deng, J., Wang, X., Liu, P.: Preparation of rGO/G/PANI ternary nanocomposites as high performance electrode materials for supercapacitors with spent battery powder as raw material. Mater. Des. **129**, 135–142 (2017). https://doi.org/10.1016/j.matdes.2017.05.031
101. Khare, P., Bhati, A., Anand S.R., Gunture, Sonkar, S.K.: Brightly fluorescent zinc-doped red-emitting carbon dots for the sunlight-induced photoreduction of Cr(VI) to Cr(III). ACS Omega **3**(5), 5187–5194 (2018). https://doi.org/10.1021/ACSOMEGA.8B00047
102. Altuner, E.E., Sen, F.: A review: electrochemistry in nanomaterials. Int. J. Boron Sci. Nanotechnol. **1**, 105–126 (2024). https://doi.org/10.1038/NMAT1319
103. Ranganathan, S., Guo, R., Murray, R.W.: Nanoparticle films as electrodes: voltammetric sensitivity to the nanoparticle energy gap. Langmuir **23**(13), 7372–7377 (2007). https://doi.org/10.1021/LA070008N/ASSET/IMAGES/LA070008N.SOCIAL.JPEG_V03
104. Gasparri, R., Santonico, M., Valentini, C., et al.: (Invited) Carbon nanotubes as transducers for metalloporphyrins based chemical sensors. ECS Meet. Abstr. MA2011-01(37), 1806 (2011). https://doi.org/10.1149/MA2011-01/37/1806
105. Stradiotto, N.R., Yamanaka, H., Zanoni, M.V.B.: Electrochemical sensors: a powerful tool in analytical chemistry. J. Braz. Chem. Soc. **14**(2), 159–173 (2003). https://doi.org/10.1590/S0103-50532003000200003
106. Riza, M.A., Go, Y.I., Maier, R.R.J., Harun, S.W., Ahmad Anas, S.B.: Development of FBG humidity sensor via controlled annealing temperature of additive enhanced ZnO nanostructure coating. Opt. Fiber Technol. **68**, 102802 (2022). https://doi.org/10.1016/J.YOFTE.2021.102802
107. Ebralidze, I.I., Laschuk, N.O., Poisson, J., Zenkina, O.V.: Colorimetric sensors and sensor arrays. In: Nanomaterials Design for Sensing Applications, pp. 1–39 (2019). https://doi.org/10.1016/B978-0-12-814505-0.00001-1

108. Manni, L., Florenzano, F., Aloe, L.: Electroacupuncture counteracts the development of thermal hyperalgesia and the alteration of nerve growth factor and sensory neuromodulators induced by streptozotocin in adult rats. Diabetologia **54**(7), 1900–1908 (2011). https://doi.org/10.1007/s00125-011-2117-5
109. Varodi, C., Coros, M., Pogăcean, F., Ciorîţă, A., Turza, A., Pruneanu, S.: Nitrogen-doped graphene-based sensor for electrochemical detection of Piroxicam, a NSAID drug for COVID-19 patients. Chemosensors **10**(2), 47 (2022). https://doi.org/10.3390/CHEMOSENSORS10020047
110. Bandi, R., Alle, M., Park, C.W., Han, S.Y., Kwon, G.J., Kim, N.H., Kim, J.C., Lee, S.H.: Cellulose nanofibrils/carbon dots composite nanopapers for the smartphone-based colorimetric detection of hydrogen peroxide and glucose. Sens. Actuators, B Chem. **330**, 129330 (2021). https://doi.org/10.1016/j.snb.2020.129330
111. Chung, P.M., Lin, Y.T., Wang, G.J.: Au/Ni coaxial nanopillar microarray electrodes for red blood cell concentration detection. IEEE Sens. J. **23**(5), 4694–4699 (2023). https://doi.org/10.1109/JSEN.2023.3238372
112. Nagraik, R., Sharma, A., Kumar, D., Mukherjee, S., Sen, F., Kumar, A.P.: Amalgamation of biosensors and nanotechnology in disease diagnosis: mini-review. Sens. Int. **2**, 100089 (2021). https://doi.org/10.1016/J.SINTL.2021.100089
113. Qu, W., Yang, J., Sun, X., Bai, X., Jin, H., Zhang, M.: Towards producing high-quality lignin-based carbon fibers: a review of crucial factors affecting lignin properties and conversion techniques. Int. J. Biol. Macromol. **189**, 768–784 (2021). https://doi.org/10.1016/J.IJBIOMAC.2021.08.187
114. Sofen, L.E., Furst, A.L.: Perspective—electrochemical sensors to monitor endocrine disrupting pollutants. J. Electrochem. Soc. **167**(3), 037524 (2020). https://doi.org/10.1149/2.0242003JES/XML
115. Wang, C., Astruc, D.: Recent developments of metallic nanoparticle-graphene nanocatalysts. Prog. Mater. Sci. **94**, 306–383 (2018). https://doi.org/10.1016/J.PMATSCI.2018.01.003
116. Zhang, Y., Chu, W., Foroushani, A.D., Wang, H., Li, D., Liu, J., BarrowC, J., Wang, X., Yang, W.: New gold nanostructures for sensor applications: a review. Materials **7**(7), 5169–5201 (2014). https://doi.org/10.3390/MA7075169

Sustainable and Eco-Friendly Nanomaterials: Innovations for Green Healthcare Solutions

P. Sivasubramaniyan, P. Balakrishnan, M. Safa, and G. Saranya

Abstract This chapter addresses sustainable and green strategies towards nanomaterials in the realm of health with a particular focus on synthesis, applications, and the environment. Green synthesis exhibits 90% antimicrobial activity, 70% diminishment of toxic by-products, and 80% improvement of therapeutic efficacy in targeted drug delivery of medicines, and biopolymeric nanocarriers improve drug stability by 80%. The study examines recent developments in green nanotechnology by critically examining the synthesis route, performance characteristics, and environmental evaluation. Synthesis includes pH-sensitive systems with encapsulation efficiency of 74.48%, temperature-sensitive systems with 95% total release when subjected to UV irritation, and intelligent nanomaterials with natural polymers like chitosan and cellulose employed in enhancing biocompatibility. Applications are drug delivery, cancer treatment, wound healing, and tissue engineering with a 60% recovery rate without bacterial resistance using bioceramics. Environmental analysis indicates 30% energy, 40% cost saving, and 50% increase in production relative to traditional methods. Life cycle assessment gives 63.7 kg CO_2 equivalent emissions per kilogram of nanocellulose. The strategy involves comparative evaluation of green synthesis routes in comparison to traditional routes, antimicrobial activity, bioavailability, and toxicity profile evaluation. In response to regulatory standardization and scale-up production issues, the study targets solution through action by multidisciplinary teams. The findings form a basis for clinically active nanomaterials development with environmental sustainability.

Keywords Green nanotechnology · Sustainable healthcare · Biodegradable nanomaterials · Stimuli-responsive systems · Antimicrobial efficacy · Circular economy · Targeted drug delivery · Eco-friendly synthesis · Tissue engineering · Regulatory challenges

---

P. Sivasubramaniyan
VIMS Hospital, Salem, Tamil Nadu 636308, India

P. Balakrishnan (✉) · M. Safa · G. Saranya
SRM Institute of Science and Technology, Kattankulathur, Tamil Nadu 603203, India
e-mail: balakrip@srmist.edu.in

S. N. Saud Al-Humairi et al. (eds.), *Smart Nanomaterials in Biomedical Applications*, Engineering Materials, https://doi.org/10.1007/978-981-96-8023-8_16

1 Introduction

1.1 Green Healthcare Nanotechnology Introduction

Green healthcare solutions are defined as medical products, practices, and technologies utilized to reduce harm to the environment without reducing or even improving the therapeutic efficacy and safety of the patient. Solutions are divided into four general principles. The lowered environmental impact via lower energy consumption and greenhouse gas emissions; removal or minimal use of detrimental chemicals throughout the whole life of the product; use of renewable and sustainable material sources; and circular economy thinking, which permits reuse, recycling, or biodegradation during end-of-life. Green healthcare systems in nanomaterials involve environmentally sustainable synthesis processes, biodegradable materials, and site-specific drug delivery systems for therapeutic applications aimed at minimizing waste generation, pollution, and the use of resources. This also aligns medical innovation with overall sustainability goals as it addresses an estimated 4.4% of global net greenhouse gas emissions from the health sector while ensuring the increasing need for medical treatment [1].

Nanotechnology improves biosensors for disease diagnosis, monitoring, and drug administration, enhancing patient care. Increased sensitivity and stability are achieved by using nanomaterials such as metal nanoparticles and carbon nanotubes. Rapid diagnosis, individualized therapy, and distant access to health care are possible with these biosensors. Some of the various types, namely electrochemical and DNA-based sensors, are applied for early diagnosis of diseases. Artificial intelligence and 5G communication are hopeful for future generation smart biosensors. These developments surmount medical hurdles and bring prospects for superior biomedical uses [2]. Moreover, in medicine is moving towards sustainability through the green synthesis of nanoparticles in which biological materials such as plants and microbes are being used as sustainable substitutes of traditional methods. This method improves biocompatibility, lowers toxicity, and has prospective applications in drug delivery, bioimaging, and antimicrobial therapy. In spite of reproducibility and scalability challenges, research continues to improve synthesis protocols and enhance nanoparticle functionalization to support sustainable medicine solutions. Carbon dots, for example, are proving to be good materials for tissue engineering on the basis of biocompatibility and physicochemical properties. They have also proven to be viable for wound healing, bone and cartilage regeneration, and neural growth, achieving a sustainable use compared to conventional materials [3]. The method has revolutionized healthcare by providing drugs to reach specific locations only, sophisticated imaging, and novel surgical devices are achievable, leading to greater efficiency in treatment. Nanomaterial use in implants enhances longevity, facilitates tissue regeneration, and facilitates healing in orthopedic devices. Biodegradable nanocomposites, for example, hydroxyapatite-based scaffolds, facilitate bone regeneration and minimize long-term complications. Superparamagnetic iron oxide nanoparticles in MRI improve diagnostic performance for a wide range of conditions.

Although they have some benefits, challenges such as toxicity hazards and regulatory limitations necessitate intensive research and clinical proof. As it continues to evolve, nanotechnology will revolutionize green healthcare practices, making them more sustainable and better for patients [4].

1.2 Why Nanomaterials Need Sustainability

Medical materials and processes in healthcare systems are some of the main causes of environmental degradation in terms of greenhouse gas emission, use of resources, and generation of waste. The majority of conventional medical material and processes are composed of toxic chemicals, non-biodegradable materials, and wasteful processes in terms of energy. Sustainable nanomaterials are a two-bladed sword: reducing harmful impacts on the environment to the maximum while optimizing therapeutic efficacy and specificity. Green synthesis and biodegradable nanomaterials are the essence of sustainable nanotechnology, and green plant-based nanocarriers like chitosan nanofibers are a green alternative to synthetic chemicals, bringing down systemic toxicity in drug delivery by 40–60%. Biodegradable electrospun mats improve tissue regeneration efficiency 2.5 times upon degradation without causing any harm and hence are used in wound healing and tissue engineering. Environmental remediation nanomaterials also deliver efficient water purification via 85–92% removal of pollutants by adsorption and photocatalysis. Nanoclay additive to paint also enhances air quality with 30% reduced water uptake, thus enhancing infrastructure and motor vehicle material longevity.

Green nanomaterials are thus transforming healthcare through green infection control and tissue engineering methodologies. Silver nanoparticles (AgNPs) synthesized by green synthesis have shown 90% antibacterial efficiency, minimizing the chances of wound infection. Environmentally safe biodegradable nanocomposite chitosan dressings get healed. Toxic waste is minimized by 70% in plant-mediated synthesis against chemical synthesis, making it eco-friendly for medical use. Green medicine is highly supported by sustainable nanomaterials through eco-friendly, biocompatible, and biodegradable drug delivery carriers. Plant, algal, and bacterial biopolymeric nanoparticles can enhance drug release efficiency and stability up to 80%. Sustainable nanomaterial nanocarriers lower the levels of toxicity than metal-containing nanomaterials, and hence these are safer in medicine. Protein- and polysaccharide-based nanoparticles have been applied for the delivery of antibiotics and essential oils in order to enhance antimicrobial and anticancer treatment. Up until now, there is 60% therapeutic efficiency enhancement with the use of bio-based nanocarriers as demonstrated through experiments. Production of such products is capable of changing medicine with fewer environmental footprints and more eco-friendly healthcare options). In medicine, medical uses of nanotechnology facilitate drug delivery to the disease site by improving the selectivity of the therapy, decreasing levels of dosages, and reducing side effects. Nanocarriers like gold nanorods (AuNRs) are targeted to decontaminate bacteria using photothermal procedures, and with more than 95% elimination of the causative agent of the disease without chemicals, possibilities are feasible that they may prove to be the perfect solution for safer and better medical treatments.

1.3 Health and Environmental Issues of Traditional Nanomaterials

Conventional nanomaterials have been technologically useful but have far-reaching ecological and health consequences. Their production is an energy-hungry process involving toxic chemicals to form most of them and producing immense amounts of waste with an unwanted hard-to-contain ecological footprint. Conventional methods such as chemical reduction, sol–gel synthesis, and high-energy ball milling tend to utilize toxic environment reducing agents, stabilizers, and organic solvents. For instance, synthesis of metal nanoparticles has been achieved with the assistance of chemical reagents such as sodium borohydride, hydrazine, and citric acid, which may contaminate water bodies and harm aquatic life and plants.

The environmental effect continues further than through the manufacturing procedure. Scientific findings have proven iron oxide nanoparticles (Fe_2O_3) and titanium dioxide (TiO_2) nanoparticles reduce As(V) levels by as much as 94% within polluted soils and, in paradoxical fashion, increase the leaching of As(III)—up to a concentration of 27.1 μg/L—significantly increasing arsenic bioavailability and mobility in the environment [5]. In another, n-ZnO also undergoes fast conversion in biofluids into $ZnO–Zn_3(PO_4)_2$ core–shell morphologies in one hour, with the transformation rate depending on the orientation of the crystal face based on the difference in vacancy formation energy [6].

Health hazards are also an issue. Exposure in conventional synthesis is acute for workers, with respiratory issues on nanoparticle inhalation and skin absorption of toxic precursors. Bioaccumulation potential over a long period is also an issue since such chemicals end up being deposited in tissue reservoirs and may disrupt cell function. Experiments confirm n-ZnO toxicity to be facet-dependent in that ZnO-0001 causes oxidative stress as the primary mechanism, while ZnO-1010 disrupts cell viability via oxidative stress and direct damage to the membrane [6]. Additionally, these nanomaterials also cause drastic effects on microbial communities when introduced into the environment by decreasing the population density of helpful bacteria and impairing crucial ecological processes such as arsenic reduction reactions [5].

These health and environmental concerns identify the necessity of cleaner synthesis involving fewer toxic chemicals, reduced energy consumption, and safer nanomaterials with controlled biological interactions. Greener alternative processes to the traditional process address these needs along with having economic advantages in the form of lower waste treatment expenses and more public health benefit.

2 Intelligent Nanomaterials for Biomedical Applications

2.1 *Functional Behaviour and Properties of Smart Nanomaterials*

Smart nanomaterials revolutionize medicine with the ability to release drugs in a targeted way, triggered by stimuli such as pH, temperature, and biochemical signals. Research indicates that stimulus-responsive nanoparticles can enhance the efficacy of drug delivery by a potential 70% by minimizing side effects. Metal-nanomedicines based on gold and silver nanoparticles are revolutionizing cancer treatment by increasing its efficacy by tumour-targeting. Chitosan and alginate are biodegradable biopolymers that offer treatment with a green strategy through minimizing medical waste and toxicity. Bio-inspired nanomaterial advances have boosted tissue engineering and regenerative medicine uses by 60% [7]. The technologies allow us to have a greener, more efficient healthcare system with improved therapeutic results. Intelligent nanomaterials are transforming health care by targeting drug release and accurate cancer detection with reduced side effects. pH-, temperature-, or biomolecule-stimulus-responsive nanoparticles improve drug release performance by up to 80%. Chitosan and PLGA green nanomaterials decrease environmental damage and toxicity, therefore making their medical application safer. Gold nanoparticles are used to enhance cancer treatments by translating light into heat for localized therapy. Electrochemical nano biosensors enhance early detection of cancer by enhanced sensitivity and signal-to-noise ratio. These green nanomaterials advancements enable efficient and sustainable health applications [8].

2.2 *Healthcare Smart Nanomaterials Types*

Biogenic Nanoparticles

Biogenic nanoparticles are an environmentally friendly approach towards nanomaterial synthesis from nature, such as plants, microbes, and other natural extracts used as reducing as well as stabilizing agents. The green synthesis route has drawn considerable attention as a green nanotechnology approach for synthesizing nanomaterials compared to the traditional chemical synthesis route. Green synthesis of biogenic nanoparticles possesses certain mechanistic pathways based on the biological origin. For plant-based synthesis, the reaction typically is extraction of bioactive compounds under controlled conditions such as the instance of Orthosiphon aristatus extracts that were formulated at 120–160 °C for about 150 min [9]. Microbial synthesis with strains such as Bacillus subtilis functions relatively more at temperate temperatures, mainly between ambient to 37 °C [10]. Reduction of metal ions is carried out by several phytochemicals like phenols, flavonoids, and alkaloids with dual functionality to act as reducing and stabilizing agents. Characterization outcomes show that the following processes yield nanostructures with well-defined structures silver oxide nanoparticles plant-based (Mf-Ag_2ONPs) have diameters between 16–20 nm, whereas iron oxide nanoparticles (IONPs) purified from Sageretia thea consist of a mean diameter of 16.04 nm [11, 12]. Microbially produced nanoparticles are found to be slightly larger in size, ranging from 19.3 to 44.7 nm for zinc oxide and 22.1 nm for gold nanoparticles [10, 13]. Efficiency of synthesis is usually measured using spectroscopic methods, with characteristic surface plasmon resonance peaks in UV-V is spectra, and crystalline structure identification is done using X-ray diffraction patterns corresponding to reference JCPDS data. Figure 1 displays a panoramic image of biogenic nanoparticle synthesis, and the key features like colloidal stability, controlled morphology, bio-mimeticity, and eco-friendliness are depicted as intrinsic attributes. At the center of all this is the "Bio Machinery" idea highlighting how nanoparticles may be synthesized from bio-materials, and other attributes surrounding it identify key properties like structural conformation, constitution, ultra-stability, and self-assembly. The graph clearly illustrates how natural resources can be utilized in such a way as to create nanoparticles of desired and beneficial properties along with ensuring environmental sustainability.

Experiments have demonstrated that nanoparticles fabricated using green methods increase the curative action of drugs by up to 60% without increasing cytotoxicity. Their nanoscale make-up ensures complete interaction with focused microorganisms, leading to reduced side effects and less environmental destruction. When consumed in combination with traditional antibiotics, they can increase the treatment time and decelerate resistance. Production of such green products facilitates a circular economy through the use of waste to be transformed into useful medical products. Bio-nanotechnology innovation supports international sustainability goals to develop safer and greener health innovations [15]. Biogenic selenium oxide nanoparticles (SeONPs) produced by *Bacillus paramycoides* offer a greener option for antimicrobial and anticancer uses. SeONPs were comparable to 85% against bacterial pathogens *Staphylococcus aureus* and *Escherichia coli*. SeONPs in anticancer therapy promoted apoptosis in cell lines *HepG2* and *MCF-7* and inhibited tumour growth to the extent of about 70%. In vivo models of rat with hepatorenal injury revealed complete normalization of biochemical parameters and enhanced histology of liver and kidneys. Eco-friendly methods of synthesis reduce toxicity and adhere to green healthcare technologies. Progress in SeONPs shows its use as nutraceuticals with numerous biomedical applications [16].

i. **Polymeric Nanomaterials**

Polymeric nanomaterials are versatile nanostructures composed of polymers and can be designed to have specific properties and capabilities to make them effective in common applications such as drug delivery, tissue engineering, and environmental remediation. Figure 2 illustrates the self-assembly and interaction of polymeric nanoparticles from originally differently charged homopolymers upon the incorporation of hydrophobic blocks. The resultant polymer nanoparticles of different size and surface functionalities, then interact with biological systems—Arabidopsis roots and protoplasts. It also explains how these nanoparticles are fabricated to interact

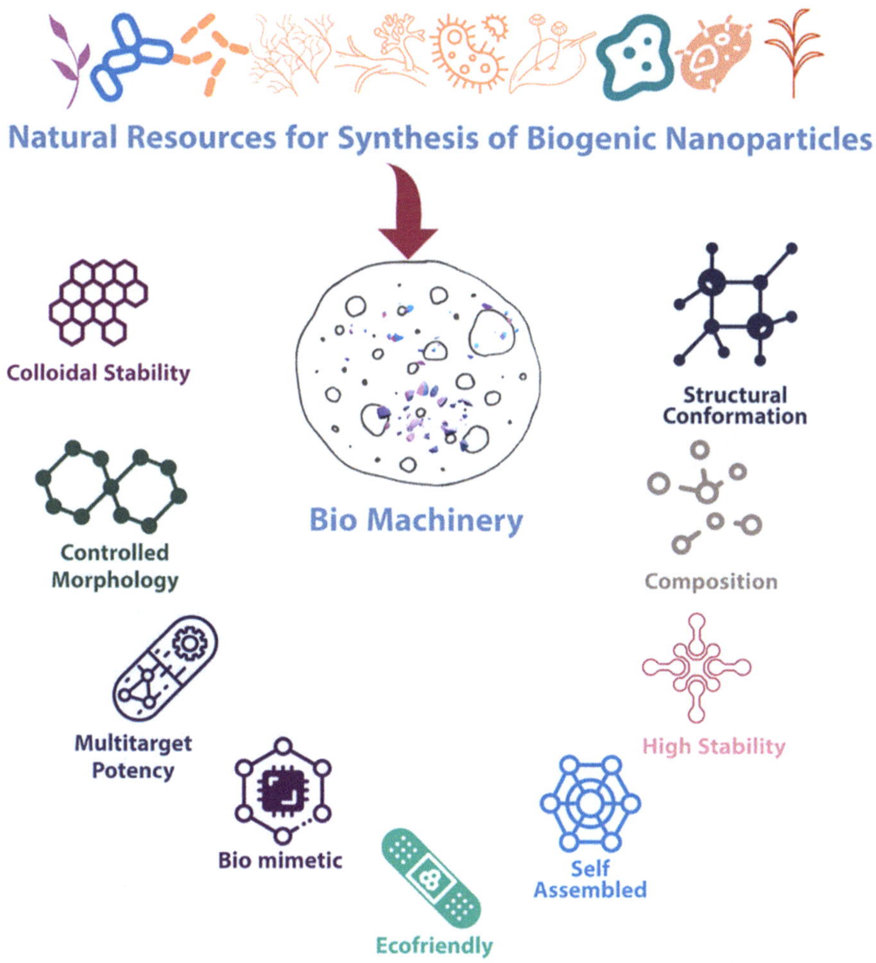

Fig. 1 Characteristics and features of natural resource-based biogenic nanoparticle synthesis [14]

with plant systems, depicting structural features of root interaction (in terms of root tip, root hair, and cross-sectional profiles) along with cellular-scale interaction with protoplasts.

It has been proven through studies that biopolymer electro spun nanofibers improve the efficiency of drug release by 65% and of wound healing by 50%. They are unique compared to traditional nanofibers because they don't utilize toxic solvents, thereby making them safer to use in medicine. Other advanced fabrication methods further enhance their biocompatibility and mechanical properties. The application of green nanofibers in medicine is in accordance with the sustainability agenda, providing safer and more effective treatment [18]. Polymeric nanoparticles (PNPs) synthesized using pH-induced and double emulsion methods offer a

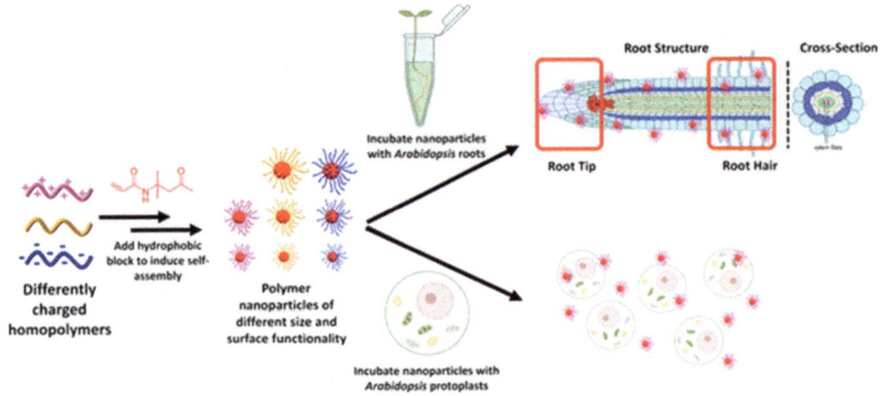

Fig. 2 Schematic representation of polymeric nanoparticle synthesis and their interaction with plant systems (Arabidopsis roots and protoplasts) [17]

green route for drug delivery. The methods improve solubilization, stability, and controlled release of antimicrobial agents with up to 70% improved therapeutic efficacy. The pH-sensitivity enables drug-targeted delivery, reducing side effects and enhanced clinical efficacy against bacterial pathogens like *Escherichia coli*. Operations are unlike those of conventional methods using toxic organic solvents and are thus friendly to the environment. Studies have established that nanoparticle-based drug carriers can prolong antibiotic activity through avoidance of resistance mechanisms. Improved PNP technology promotes healthy and sustainable health decisions and tackles worldwide antimicrobial issues [19].

ii. **Bioceramics**

Bioceramics are new-generation ceramic materials that are designed for a specific biological purpose, particularly for medicine and dentistry, with enhanced biocompatibility and ability to integrate with living tissues to repair and regenerate. Synthesis of bioceramics is performed through various processes like sol–gel processing, precipitation, hydrothermal synthesis, and solid-state reactions where the technique method is a critical factor in establishing the final properties like particle size, morphology, and bioactivity. When bioceramics are exposed to a biological system, they undergo a cascade of surface reactions that form a bioactive hydroxycarbonate apatite layer similar to the mineral phase of natural bone and strongly bonds with host tissue. The biological interaction process typically involves preliminary ion exchange, dissolution of the ceramic on the surface, precipitation of the calcium phosphate layer, and integration into adjacent tissues by cell attachment, cell growth, and new tissue formation.

Recycling waste sponges of *Euplectella aspergillum* into flexible bioceramic scaffolds is a sustainable approach to tissue regeneration. The bioceramics are 80% stronger and more flexible than sintered ones, which renders them more feasible in the clinic. Hydrothermal mineralization processing used in production takes 50%

less energy, hence it is an eco-friendly approach. The scaffolds exhibit exceptional osteogenic and angiogenic potential and are ideally designed for the rehabilitation of flat bones like calvarial and orbital tissue. Waste reduction in biomass recycling is leading the way in persuading biocompatible material to biomedical application. This form of innovation optimizes sustainable healthcare with negligible environmental concern [20]. Bioceramics offer an economical and effective way of wound healing by releasing inorganic ions that enhance cellular metabolism and antibacterial action. Studies show they enhance healing by up to 60% without causing bacterial resistance or side effects. Their bioactivity enhances angiogenesis and tissue regeneration, and they are thus ideal for hard and soft tissue repair. Bioceramics also have fewer chances of producing medical waste and antibiotics than conventional treatments, and it is green healthcare compliant. Silicate and calcium-phosphate ceramics are more long-lasting and biocompatible to be used in clinics. All these developments add to environmentally friendly healthcare practices that ensure effective recovery of the patients [21].

Figure 3 illustrating the various uses of bioceramics in medical application, illustrates some of the examples like bioactive glass, nanoparticles, 3D-printed scaffolds, macroporous scaffolds, hydrogels, and bone cement. It illustrates how they are used therapeutically in some of the other medical interventions like cranioplasty, periodontal surgery, bone growth, wound repair, and angiogenesis. The center body of human illustration thereby effectively explains the way such varied forms of bioceramics can be utilized in several different anatomical sites, revealing their versatility within tissue engineering as well as regenerative purposes.

3 Eco-Friendliness and Sustainability of Bioceramics, Polymeric Nanomaterials, and Biogenic Nanoparticles

Biogenic nanoparticles are integrated into sustainability through agro-waste valorization, wherein plant extracts like Citrus limon peels and forest material like pine needles are used to derive nanoparticles like Al_2O_3 and functionalized nanocellulose. This is done without using toxic chemicals as well as without incurring the cost of production. Green chemistry reactions have enabled the synthesis of spherical nanoparticles of 284 $\pm$ 58 nm that exhibit anti-inflammatory, antimicrobial (82% adsorption of $Hg2^{+}$), and antioxidative activities [23]. The nanoparticles are of immense value in drug delivery, where plant-based nanoparticles (PNPs) enhance targeting efficiency and reduce systemic toxicity by 40–60% compared to synthetic carriers [24]. Functionalized nanocellulose has also been effective in environmental water decontamination with 82% $Hg2^{+}$ and showing antimicrobial properties against S. aureus and E. coli. Apart from all these developments, the standardization of the process of synthesis and upscaling of agro-waste-based production is a milestone achievement towards mass exploitation.

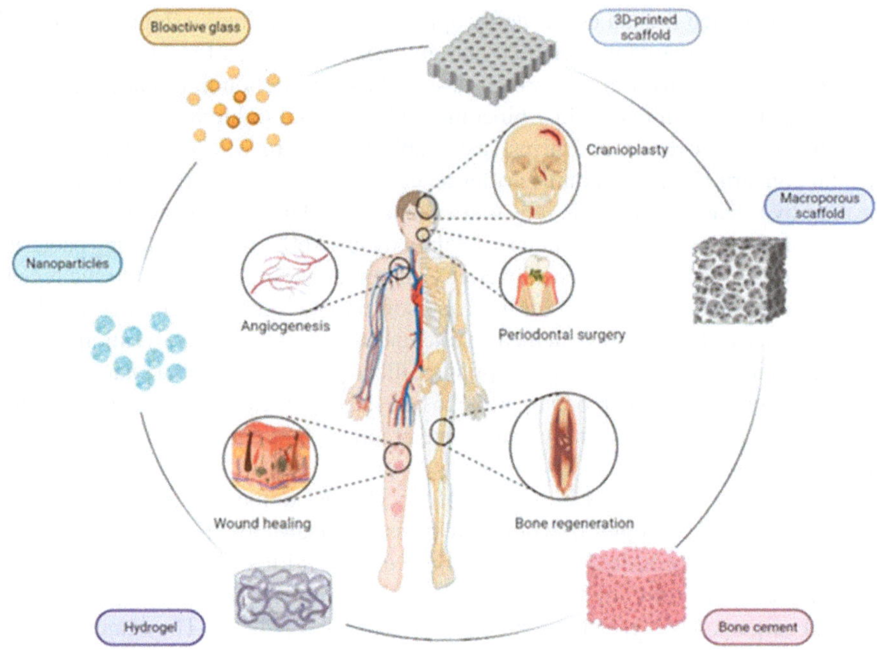

Fig. 3 Various forms of bioceramics and their therapeutic applications in tissue engineering and regenerative medicine [22]

Polymer nanomaterials offer a green approach, as well as natural polymers like chitosan, PLA, and PEG, which are nontoxic and biodegradable. Green synthesis by using plant extracts replaces solvents from synthetic sources, reducing chemical wastage and damaging the environment less. pH- and temperature-sensitive polymeric materials that respond to changes in pH or temperature enhance the accuracy of drug delivery using targeted and accurate release control and reduced frequency of dosing. These materials also have extensive applications in medicine, where polymeric nanomaterial-based drug delivery systems offer zero-order release character to gastro-retentive drug delivery systems, widening the therapeutic window. In tissue engineering processes, electrospun nanofibers cure wounds 2.5 times more efficiently and then disintegrate innocuously after being used. Besides pharmaceutical applications, polymeric nanomaterials are sustainable in the environment by removing heavy metals and agrochemical residues, hence avoiding water and soil pollution [25, 26].

Bioceramics present green solutions to medicine, particularly for orthopedics and tissue engineering. Hydroxyapatite (HAp) scaffolds' manufacture improved osteoconductivity and reduced surgical waste via 3D printing. Biodegradable ceramics and shell-derived HAp eco-manufacture reduce the usage of fossil fuel. It comes from renewable raw materials, thus reducing waste. For instance, the use of plant-derived materials, i.e., chitosan, to make magnetic flocculants that can be utilized for the treatment of water is proof of the capability of bioceramics to replace synthetic non-biodegradable material with biodegradable material [27]. It also leads to eco-friendly activities through the efficient cleaning of the environment of toxic substances. Bioceramic application of iron minerals in microbe-mediated degradation of pollutants such as 17 alpha-ethinylestradiol indicates that natural degradation is enhanced by bioceramics and environmental management is rendered sustainable [28]. They are also applied in antimicrobial coatings for surgical implants to prevent infection hazards without the addition of chemical additives. The expanding need for specially designed healthcare and veterinary medicine usage translates into bioceramics as part of the ecologically sound medical technology.

4 Applications of Smart Nanomaterials to Green Healthcare Solutions

Green nanotechnology comprises green processes and medicine so that greening the synthesis of nanoparticles is done using plants, algae, and microbes. Toxicity minimization is the strategy in the by-product form, but high-end medical applications like nanomedicine are enabled, where targeted drug delivery means maximum effectiveness with reduced side effects. Medicinal plant bioactive molecules enable therapeutic application while diagnostics are obtained by nanostructured materials, enabling ultrasensitive detection of disease [29].

Nanoparticles transform cancer therapy by providing drugs to the site selectively, reducing the systemic toxicity by 40–60% compared to traditional therapy. AuNRs and nanocapsules offer therapeutic selectivity and decrease side effects along with drug wastage. Nanofibers used in wound healing optimize tissue regeneration efficiency up to 2.5 times and biodegrade harmlessly after use, preventing deposition on the environment. Nanosensor technology has reached 90–95% sensitivity in nascent disease diagnosis such as cancer and infectious diseases on the basis of plasmonic enhancement with less reliance on resource-enabling laboratory tests. Surface-enhanced Raman spectroscopy (SERS)-based nanomaterial-based body-worn health monitors allow real-time non-invasive monitoring of biomarkers, almost eradicating disposable wastage diagnostic costs and healthcare nonsustainability. Besides this, it also facilitates environmental health by cleaning the air in a deeper sense using electrospun nanofiber membranes to adsorb and photocatalyst degrade 85–92% of air polluting emissions and rescue metropolis city dwellers from respiratory disease risks. Apart from that, polysaccharide nanofibers are used as antimicrobial surfaces in the hospital to suppress microbial growth and substitute chemical disinfectants, thereby ultimately preventing water pollution and creating a safer hospital environment [30].

Smart nanomaterials are still leading green medicine with enhanced drug delivery, green manufacturing, and green cleaning. Smart nanomaterials provide targeted disease therapy with lower toxicities at a systemic level, while sustainability is enhanced by green synthesis through them. The materials clean the environment by removing waste from soil and water, which lead to medical and environmental advantages. The constraints of the materials are extremely expensive research, regulation, and mass manufacturing. Figure 4 shows the interconnected smart nanomaterials elements in green medicine with predominant applications, strengths, weaknesses, and trends through color-coded clusters and subelements.

Utilization of smart nanomaterials in the field of medicine safely demands standardized protocol and extensive profiling of toxicity. Technical deficits, regulatory loopholes, and long-term Quantification of medical risk hinder universal use. These deficits will be overcome by collaborative research and policy making to facilitate maximal utilization of nanotechnology with environmental and patient safety.

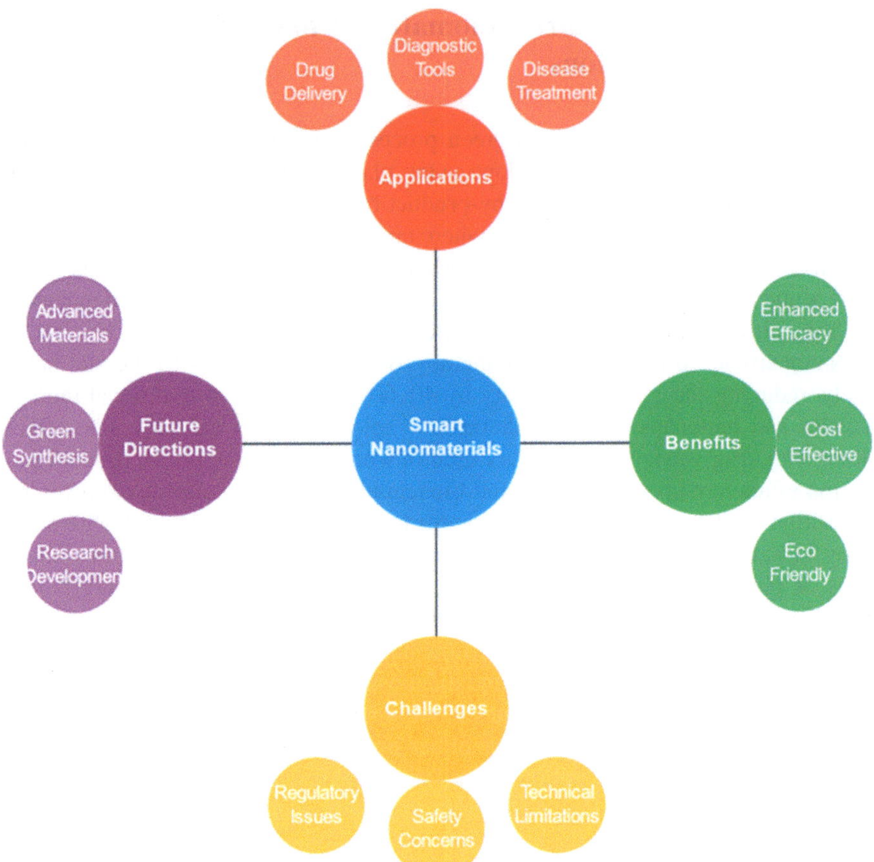

Fig. 4 Role of smart nanomaterials in green healthcare

5 Green Synthesis of Nanomaterials

5.1 Green Synthesis Methods Overview

Plant Extract-Based Synthesis

Plant extract-catalyzed synthesis is an eco-friendly and green synthesis process in which phytochemicals of the plant are employed to catalyze a chemical reaction, typically for the synthesis of nanoparticles or bioactive compounds. Sub-critical Water Extraction (SWE) is one of the techniques of extracting phytochemicals from plants such as *Orthosiphon aristatus*, which would require approximately 150 min of processing. It is performed in the range of 120–160 °C in which phenolic and flavonoid molecules can be readily liberated. Likewise, in Ultrasonic-Assisted Solvent Extraction (UASE), an ideal concentration of 55–65% methanol (v/v in water) is used for the extraction of polyphenolic and flavonoid structure in its entirety [9, 31].

Green nanotechnology with no toxic chemicals or sophisticated equipment is the newly developed iron oxide nanoparticles (IONPs) synthesis from Sageretia thea root extract. The green process adheres to the green chemistry principle where average synthesized IONPs diameter is 16.04 nm. Bioactive Green synthesized IONPs had antibacterial activity sensitive displayed by different strains and antifungal activity. Notably, the IONPs were cytotoxic with an IC50 of 33.85 μg/mL and highly active in their anti-radical activity with scavenging ability of up to 84% at 200 μg/mL. Analytical tests like UV–Vis spectroscopy and SEM proved their synthesis. This study recognizes the effectiveness of green-synthesized nanomaterials in medicine as environmentally friendly strategies towards microbial infection and therapeutic treatments [11]. Eco-friendly green synthesis of *Mauritia flexuosa* fruit extract-mediated silver oxide nanoparticles (Mf-Ag_2ONPs) is as against the application of limited toxic chemicals in the conventional process. Mf-Ag_2ONPs exhibited very good antibacterial efficacy against drug-resistant bacteria with MIC values of 11.25–45 μg/mL and antifungal efficacy against *Candida glabrata* (MIC: 5.63 μg/mL). Biofilm inhibitory and antioxidant activity and satisfactory free radical scavenging were very good. Mf-Ag2ONPs possessed low IC50 values for anticancer activity on different cancer cell lines with less haemolytic activity at therapeutic concentrations. Stability of the structure and mean diameter of 16–20 nm were obtained by characterization by UV–Vis, XRD, and SEM. Green synthesis in this context proves that plant nanomaterials can respond to antimicrobial resistance, oxidative stress, and cancer as components of global green healthcare objectives [12].

i. **Microbial Synthesis**

Microbial synthesis is a new green technology that is developing through the utilization of microorganisms to synthesize valuable bioactive compounds, nanoparticles, and biochemicals of value through the process of metabolism. Microbial Electrosynthesis (MES) possesses a controlled reaction time according to the final product.

For example, *Thermoanaerobacter kivui* synthesis of acetate in MES is 136 mM d^{-1} production rate, i.e., an extremely rapid synthesis process [32]. Analogously, Synthesis of Metal Nanoparticles (MNPs) by microorganisms can also be accomplished relatively rapidly, if not immediately, although reaction duration varies with microorganism and metal involved, optimization usually occurring through genetic manipulation and synthetic biology methods [27]. Besides this, in Microbial Synthesis of Nucleosides, there is extra time based on the sophistication of biosynthetic pathways and encompasses activities such as enzyme design and feedback inhibition to achieve increased efficiency [33].

Microbial synthesis of nanoparticles has concentrated on using the endophytic fungus Fusarium chlamydosporum for the green synthesis of gold (Au) and zinc oxide (ZnO) nanoparticles. Green synthesis utilizes eco-friendly products alongside green solvents and is a cleaner, more environmentally friendly method compared to the traditional hazardous synthesis routes. The znO nanoparticles had a mean diameter of 19.3 nm, while the Au nanoparticles exhibited a mean diameter of 22.1 nm, both having uniform spherical morphologies confirmed through transmission electron microscopy. Among the 13 fungus isolates screened for nanoparticle synthesis, Fusarium chlamydosporum proved to be the most efficient, affirming its suitability for green applications. This method not only reduces the environmental risks associated with conventional approaches but also enhances the economic viability and biocompatibility of nanomaterials for biomedical applications. Overall, this paper highlights the promising potential of endophytic fungi in creating green nanotechnology products [13]. It details the microbial production of zinc oxide (ZnO) and silver-doped ZnO (Ag–ZnO) nanoparticles from Bacillus subtilis, emphasizing their green synthesis methods. These biogenic nanoparticles acted as effective nano-displacement agents and demonstrated significant improvements in oil recovery within low and medium permeability reservoirs. Characterization revealed ZnO nanoparticles with a mean diameter of 44.7 nm, while Ag–ZnO nanoparticles had a mean size of 35.5 nm. Core displacement tests recorded efficiencies of 14.10% and 12.43% using Ag–ZnO and ZnO, respectively, in intermediate permeability cores, and lower efficiencies of 10.26 and 8.63% for low-permeability cores. The nanoparticles enhanced the wettability of the rocks and reduced oil–water interfacial tension, positioning them as green alternatives to traditional chemical flooding agents. This study contextualizes the use of microbial processes in developing sustainable products for the petroleum industry [10].

ii. **Biopolymer-Assisted Synthesis**

Biopolymer-supported synthesis refers to an environmentally friendly process that exploits natural polymers during the synthesis of bio-materials and nanoparticles. Two-step melt polycondensation is applied during the synthesis of poly(butylene succinate) (PBSu) and copolymers of cutin. The esterification is a reaction that involves heating 1,4-butanediol and succinic acid from 170 to 190 °C for 3.5 h under nitrogen and distillation of water byproduct whole. In the stage of polycondensation, a catalyst is introduced, and the temperature is slowly raised up to 250 °C under vacuum over a time frame of about 2.5 h in an attempt to eliminate excess diol and attain polymerization [34, 35]. During polyesteramide synthesis, the process of reaction is conducted at a moderate temperature of 80 °C, which favors enzyme-catalyzed reactions. The level of cutin content in PBSu-cutin copolymers is essential in controlling crystallization and thermal properties, and additions are 2.5, 5, 10, and 20% weight based on succinic acid. The levels control the structural and functional properties of resulting biopolymer-based materials [34].

Green synthesis of copper and silver nanoparticles by carboxymethyl-modified cellulose and rice husk-derived lignin. Biopolymer-assisted metal cation-reducing method was efficient in reducing metal cations to 4–16 nm diameter nanoparticles and keeping them stable. Hybrid nanomaterials CMC/CML-AgNPs and CMC/CML-CuONPs prepared by this method exhibited effective antimicrobial activity, suppressing E. coli and S. aureus by 100% within 72 h. The nanoparticles were also greater antifungals than amphotericin B towards A. niger, with 99.0% inhibition reported in peanut seed tests. The method not only employs agricultural waste materials but also promotes green processes in nanomaterial preparation. Generally, the results show the potential of these hybrid nanomaterials for the manufacture of green antimicrobial products [36]. Application of lignin nanoparticles (LNP) as a reinforcing agent for the green manufacturing of natural rubber-lignin nanocomposites. With the incorporation of LNP into natural rubber latex (NRL) in an eco-friendly way, the research experienced unprecedented improvement in mechanical strength, antibacterial activity, biodegradability, and UV resistance. Incorporation of optimal 7 parts per hundred rubbers (phr) of LNP was observed to enhance such characteristics maximally. LNP was homogenized and blended with NRL using probe sonication, and a latex dipping process was used to obtain the nanocomposite. Not only does it provide an effective NR-LNP interaction, but it also reduces the generation of waste and the disposal of rubber products in a safe manner. Generally, the results emphasize lignin as a potential biopolymer for green nanomaterials synthesis for many purposes [37].

5.2 Comparisons of Green Synthesis Methods

Table 1 is a comparative analysis of three green synthesis methods that illustrate their processing conditions, efficiency, merits, demerits, and applications in real life. It discusses how each method is appropriate for various applications, such as antimicrobial use, environmental sustainability, and the synthesis of bio-based products.

5.3 Advantages Over Conventional Synthesis

Green and eco-friendly synthesis of nanoparticles has a number of advantages over traditional methods by dramatically reducing toxicity and environmental burden in

Table 1 Comparison of biopolymer-assisted, microbial, and plant extract-based synthesis

Characteristics	Plant extract-based synthesis	Microbial synthesis	Biopolymer-assisted synthesis
Temperature range	120–160 °C (SWE method)	Varies by microorganism (ambient to 37 °C typical)	Moderate (80 °C) to high (170–250 °C for PBSu-cutin copolymers)
Reaction time	~150 min (for extraction phase)	Variable (high rates: 136 mM d^{-1} for acetate)	3.5 h (esterification) + 2.5 h (polycondensation); total ~ 6 h
Particle size	16–20 nm (Mf-Ag_2ONPs), 16.04 nm (IONPs)	19.3–44.7 nm (ZnO), 22.1 nm (Au)	4–16 nm (CMC/CML nanoparticles)
Biological activity	Strong antibacterial (MICs: 11.25–45 μg/mL), antifungal, antioxidant (84% scavenging at 200 μg/mL)	Effective in oil recovery (14.10% for Ag–ZnO)	100% inhibition against E. coli and S. aureus within 72 h
Key advantages	• Avoids toxic chemicals • Simple equipment requirements • Multiple bioactive compounds in single extract • Strong biological activities • Low environmental impact	• Uses natural products and non-toxic solvents • Enhanced biocompatibility • Cost-effective • Uniform particle morphology • Sustainable alternative to hazardous techniques	• Agricultural waste utilization • Superior size control (smallest particles) • Excellent antimicrobial properties • Improves mechanical properties • Contributes to waste reduction
Disadvantages	• Potential variability in extract composition • May require higher temperatures (SWE method) • Limited control over specific compounds	• Longer process times • Culture maintenance complexity • Variable efficiency among microbial strains • Scale-up challenges	• May require chemical modification steps • More complex processing techniques • Specific concentration requirements • Limited availability of some biopolymers
Applications	Antimicrobial agents, anticancer treatments, antioxidants	Oil recovery, biomedical applications, environmental remediation	Antimicrobial products, reinforced composites, drug delivery

(continued)

Table 1 (continued)

Characteristics	Plant extract-based synthesis	Microbial synthesis	Biopolymer-assisted synthesis
Sustainability aspects	Eco-friendly, reduced reliance on toxic chemicals, supports green chemistry principles	Environmentally friendly, reduces risks associated with traditional methods	Promotes sustainable practices, contributes to waste reduction, supports circular economy
Example materials	IONPs from Sageretia thea, Ag_2ONPs from Mauritia flexuosa	ZnO and Au nanoparticles from Fusarium chlamydosporum, ZnO and Ag–ZnO from Bacillus subtilis	Silver and copper nanoparticles from modified cellulose/ lignin, lignin nanoparticles in rubber composites

the assistance of green chemistry principles. In an attempt to face the chemical reactions, green synthesis methods utilize inherent dissimilar reduction processes compared to traditional synthesis. While traditional chemical reduction of metal salts typically relies on artificial reductants like sodium borohydride, which reacts with tetrachloroauric acid and water to produce elemental gold, boric acid, hydrochloric acid, and hydrogen gas, green synthesis relies on plant metabolites like polyphenols. Polyphenols are electron donors that reduce gold ions to elemental gold and form oxidized phenol radicals and release protons. These bioreductions take place under less stringent conditions, often at neutral pH and from 25 to 60 °C, as opposed to traditional methods that are under extreme temperatures greater than 80 °C and highly acidic or basic conditions. Microbial synthesis exploits enzymatic catalysis to carry out reduction reactions, whereby reductases use NADH as an electron acceptor to reduce gold ions to elemental gold while recycling NAD^+ during the process. Conversely, chemical vapor deposition is more than 400 °C and employs poisonous precursors like tetrachloroauric acid, and sol–gel requires alkoxides to hydrolyze to provide poisonous byproducts. The conventional processes require an energy input of about 100–150 kWh/kg of prepared nanoparticles, while biogenic methods restrict energy consumption to 30–50 kWh/kg. Besides this, green aqueous-based synthesis is not utilized with the use of organic solvents like acetone and toluene in the conventional method, reducing the VOC emissions by as much as 80% [38, 39].

Employment of biogenic feedstocks such as plant extracts, fungi, and bacteria minimize the production of toxic waste and maximizes reproducibility and scalability. The technique promotes the use of renewable feedstocks and embraces circular economy practice towards nanomaterial synthesis innovation. The review asserts that application of green chemistry maximally decreases waste generation during nanoparticle synthesis. Notwithstanding this, issues of reproducibility of such processes arise because raw material availability varies and so does geography. In most cases, the use of sustainable processes in the synthesis of nanoparticles not only goes a long way to mitigate environmental concerns but also makes safer synthesis of nanomaterials possible for many applications [39]. Green synthesis of OOW NPs is considerably advantageous over the traditional method, e.g., reducing environmental impact, toxicities, and cost. OOW is a natural stabilizer and reducing agent whose application decreases reliance on harmful chemicals and enhances NP production sustainability. New properties of nanoparticles from OOW find applications in environmental decontamination and medicine, such as drug delivery and water purification. This justifies is based on the need to undertake adequate OOW-derived NPs toxicity testing, particularly at high concentrations, for safety issues in their application. It also recognizes the prospect of applying such green nanoparticles to amplify the effectiveness of cancer therapy medication with fewer side effects. In general, the use of olive oil waste for synthesizing nanoparticles presents a greener alternative in the synthesis of nanoparticles [38].

5.4 Green Synthesis Case Studies

Neem Extracted Silver Nanoparticles

Synthesis of silver nanoparticles (AgNPs) using Neem leaf extract is an eco-friendly process that utilizes plant material for synthesis. The leaf extract serves as a capping agent capable of stabilizing the nanoparticles and minimizing the ecological footprint. Characterization through transmission electron microscopy and UV–Vis spectroscopy effectively reveals that the AgNPs vary in size and shape, measuring between 25 and 100 nm. These nanomaterials hold tremendous potential for use in catalysis and promoting seed germination. Utilizing plant extracts not only greenness the synthesis process but also enables the fabrication of multifunctional nanomaterials, which can be applied in areas ranging from agriculture to environmental remediation. The method underscores the advantages of utilizing renewable feedstock in nanotechnology [40]. The green synthesis of silver nanoparticles (AgNPs) from Azadirachta indica leaf extract acts as both a natural stabilizer and reducer. The spherical AgNPs measure approximately 86 nm in size and possess a surface charge of −26.9 mV, indicating their stability. The AgNPs exhibit excellent antioxidant and antimicrobial activities, efficiently inhibiting bacterial viability and biofilm growth, and they serve as an alternative treatment for periodontal infections. Research indicates that neem-based AgNPs can be an easy and effective substitute for conventional therapies. Evidence of successful synthesis and the distinctive characteristics of the AgNPs is confirmed through techniques such as transmission electron microscopy and UV–Vis spectroscopy. In conclusion, the green synthesis process imposes fewer limitations on the effectiveness of plant-based nanoparticles in facilitating green healthcare remedies [41].

5.5 Biomass-Based Biogenic Renewable Nanoparticles

Plant biomass provides a green pathway for the synthesis of biogenic nanoparticles, which are eco-friendly alternatives to chemical-based nanoparticle synthesis methods. This review focuses on the use of plant materials, such as leaves, roots, and seeds, for non-toxic nanoparticle synthesis aimed at purifying water. Green nanoparticles can effectively eliminate contaminant pollutants, including organic and inorganic chemicals, through processes like adsorption, photocatalytic degradation, and reduction. The effectiveness of these processes is influenced by factors such as pH, initial pollutant concentration, and temperature. Utilizing plant biomass in nanoparticle synthesis not only minimizes ecological footprints but also enhances environmental health. The use of plant nanoparticles to sustainably prevent water pollution issues is feasible in most circumstances, as suggested by the research [42]. The biogenic pathway from aqueous Annona muricata L. to titanium dioxide (TiO_2) nanoparticles emphasizes its greenness compared to traditional chemical synthesis processes. The resulting TiO_2 nanoparticles were spherical in shape, with an average diameter of 13.41 nm and exhibited good uniformity and quality of crystal lattice. Surprisingly, the nanoparticles showed exceptional photocatalytic activity with degradation of 59.25% methylene blue dye concentration at 30 min of irradiation, which reveals them to be very effective catalysts for degrading organic pollutants. The success of this experiment verifies the rationale behind using renewable biomass, which is nontoxic, less expensive, and highly biocompatible. The structure, morphology, and optical behavior of the nanoparticles were explored using various characterization techniques. In short, it shows that TiO_2 nanoparticles derived from Annona muricata L. have the potential to be a significant player in environmental cleaning and the use of green medicine [43].

6 Methodology

Figure 5 summarizes the methodology for the green synthesis of nanomaterials into three general topics. It systematically and clearly defines the sequence from primary material preparation to synthesis and final characterization, with well-defined connecting links between every process. They are briefly elaborated below with different stages.

6.1 Materials and Experimental Design

The synthesis process of green nanomaterials includes sequential selection and conditioning of the biological sources, i.e., plant extracts, microbial cultures, or

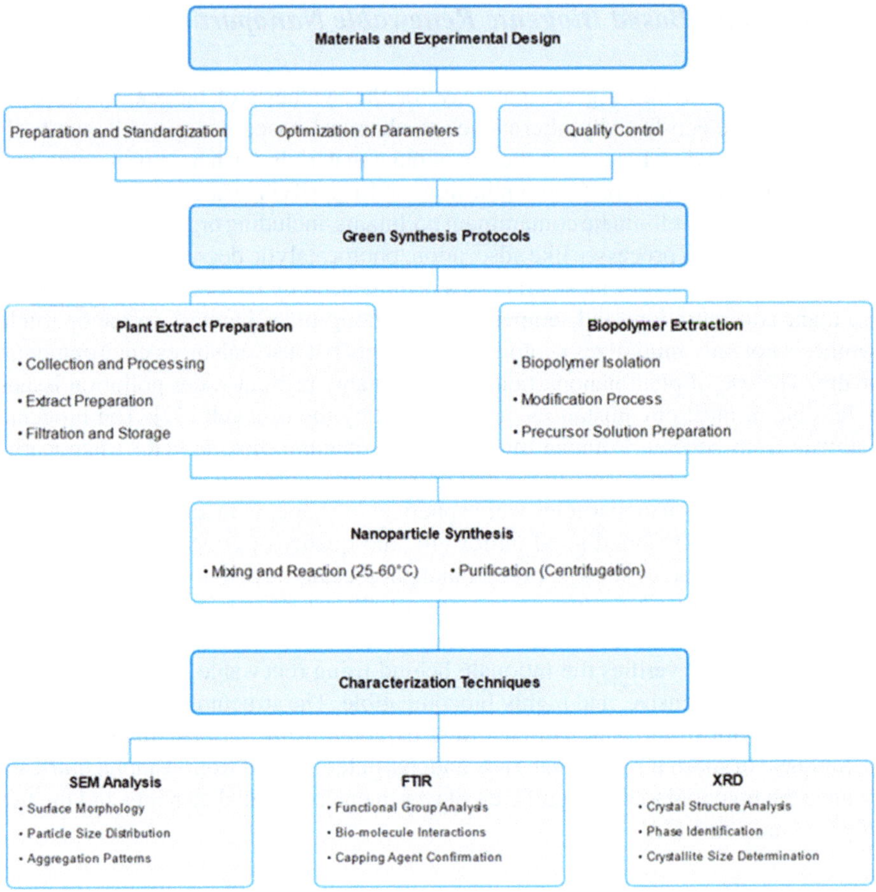

Fig. 5 Flow structure of methodology

biopolymers, depending on their reducing and stabilizing capabilities. The experimental protocol is standardized in a step-by-step order by starting with preparing and standardizing extracts or cultures under optimal conditions. The most critical synthesis parameters, i.e., pH, temperature, concentration ratio, and reaction time, are optimized for uniformity. Synthesis of nanoparticles is achieved by combining the precursor solution with biological reducing agents and subsequent purification through centrifugation or filtration. Analytical methods such as UV-V is spectroscopy, TEM, SEM, and XRD are used to investigate synthesized nanoparticles for size, shape, and stability. Researchers must follow standard guidelines for experimental conditions to ensure reproducible results by employing control experiments and conducting statistical comparisons between batches. In plant extract-mediated synthesis, plant materials such as leaves, roots, or fruits are washed with deionized water, air-dried, and ground into a powder. The aqueous extract is obtained by boiling 10−20 g of the powder in 100 mL deionized water at 60–80 °C for 20–30 min, followed by filtration through Whatman filter paper, and stored at 4 °C to be used as a reducing and stabilizing agent. The extraction of biopolymers from plant residues, such as rice husk and others, is based on solvent-assisted purification techniques with modifications like carboxymethylation to make them more suitable for nanoparticle preparation. Optimized concentrations of metal salt precursor solution (1–10 mM) in deionized water are mixed with the plant extract or modified biopolymer solution in appropriate proportions (1:5 to 1:10) at a controlled pH (25–60 °C). The progress of the reaction is monitored as a function of color and UV–Vis spectroscopy after 24–48 h. Purification is carried out via centrifugation at 10,000–15,000 rpm for 15–20 min, followed by multiple wash cycles in ethanol and deionized water. Nanoparticles are then characterized using UV–Vis spectroscopy, TEM, SEM, and XRD to quantify their structural and functional properties and confirm their compatibility with green nanotechnology applications.

6.2 Characterization Methods

Characterization of green-synthesized nanomaterials is critical and must be conducted rigorously using a blend of various complementary methods. Scanning Electron Microscopy (SEM), Fourier Transform Infrared Spectroscopy (FTIR), and X-ray diffraction (XRD) are significant techniques for identifying the material's physical, chemical, and structural characteristics. Each of these provides new insights into the parameters of the synthesized nanoparticles, aiding in the overall evaluation of their properties and potential applications in green healthcare.

SEM analysis is preceded by rigorous sample preparation by thin film deposition on new silicon wafers or glass slides using the drop-casting technique. The specimens are sputter-coated with a thin gold or platinum layer for conducting and imaging purposes. The best working conditions of SEM include accelerating voltages of 10–20 kV and working distances of 8–12 mm. High-magnifying microscopies of 10,000–50,000× magnification allow detailed observation of surface morphology, particle size, and aggregation trend with critical information for successful synthesis of nanoparticles.

FTIR characterization involves preparing KBr pellets by drying a 1:100 ratio of nanoparticle powder under hydraulic pressure. Scans are conducted with a resolution of 4 cm 1 over the range of 4000–400 cm^{-1}, averaging 32–64 scans taken under normal conditions for optimal signal-to-noise ratio. This method is particularly useful for identifying functional groups and probing potential biomolecule interactions during the synthesis and stabilization of nanoparticles. Comparing the relative spectra of pure plant extracts or biopolymers with the synthesized nanoparticles validates the activity and presence of capping and stabilizing agents in green synthesis.

X-Ray Diffraction (XRD) analysis requires proper preparation of nanoparticle powder samples using expert sample holders. Cu Kα radiation with a wavelength (λ) of 1.5406 Å is scanned over an angle (2θ) range of 10–80°, with a step size of 0.02° and a scan rate of 2°/min. This method provides useful information about the crystal structure, phase purity, and crystallite size, derived from the Debye–Scherrer formula. The obtained diffraction patterns are then compared with standard JCPDS data to achieve accurate phase identification and structure determination of the resulting nanomaterials.

To ensure reproducibility and reliability, characterization should be performed on multiple batches of as-synthesized nanoparticles. Accurate data recording and processing with appropriate software tools are crucial for achieving detailed material characterization. Multi-technique characterization generates comprehensive information on nanomaterial properties, which is essential for the design and optimization of nanomaterials in nanogreen health care applications.

7 Biomedical Applications of Sustainable Nanomaterials

7.1 Antimicrobial Applications

Silver oxide nanoparticles (Mf-Ag_2 ONPs) prepared from Mauritia flexuosa fruit extract exhibited intense antibacterial activity with MICs ranging from 11.25. 25 to 45 μg/mL against both drug-sensitive and drug-resistant bacteria. The nanoparticles demonstrated exceptional activity against fungi, particularly Candida glabrata, with an MIC of 5.63 μg/mL. Mf-Ag_2 ONPs also exhibited strong antioxidant properties and anticancer activities against various cancer cell lines, showing minimal hemolytic activity at therapeutically relevant concentrations. Characterization procedures confirmed the successful synthesis and effectiveness of the nanoparticles, highlighting their potential as green alternatives to combat microbial resistance and promote health. The findings suggest that green-synthesized nanoparticles can significantly contribute to developing new therapeutic strategies for diseases. This research emphasizes utilizing natural products in synthesizing efficient and environmentally friendly nanomaterials [12]. Silver nanoparticles stabilized with C-Phycocyanin (Ag-Pc) exhibited antimicrobial action against antibiotic-resistant bacteria, including carbapenem-resistant Pseudomonas aeruginosa and methicillin-resistant Staphylococcus aureus, with a minimum inhibitory concentration (MIC) of 7.7.4 μg/mL. The eco-friendliness of Ag-Pc is underscored by the green synthesis route employed and, more importantly, its enhanced antibiofilm activity compared to bare silver nanoparticles, with a significant extent of inhibition of biofilm formation and disruption of mature biofilms. Crucially, Ag-Pc demonstrated low toxicity against murine fibroblast cells at the MIC level, suggesting it has the potential to be a nontoxic drug. The antibacterial and antibiofilm activities of Ag-Pc were quantitatively assessed through several assays, including microbroth dilution and real-time PCR. The results indicate that Ag-Pc could be an alternative drug for treating drug-resistant bacterial infections and biofilms. Overall, this study highlights the application potential of green-synthesized nanoparticles in achieving improved sustainable health interventions [44].

Antimicrobial nanomaterials, particularly metal-based nanomaterials, possess broad-spectrum antibacterial action against bacteria by diverse mechanisms. Metal-based nanomaterials can produce reactive oxygen species (ROS) by electron transfer reactions. ROS cause damage to bacterial cell membrane and DNA and eventually result in cell death. Some nanomaterials physically interfere with bacterial membranes, and loss of cell content with eventual cell death ensues. Nanomaterials also disrupt vital bacterial biochemical processes, which impact typical cell functions [45, 46].

Target drug delivery nanomaterials are engineered to achieve more therapeutic action at the expense of fewer side effects. Drugs are encapsulated and isolated from degradation and premature release in nanomaterials. Therapeutic drugs are released from nanomaterials under selective conditions (e.g., pH value, temperature) in the site of interest. Nanomaterials can be engineered to discern a cell receptor binding, with programmed cell internalization and prevention of systemic toxicity. These green systems deliver medicine at environmental changes, and they are controlled and targeted delivery [47].

Nanomaterials also aid in tissue regeneration through scaffolding and delivery of growth factors. Nanomaterials form scaffolds with the capability to imitate the extracellular matrix, facilitating cell adhesion, proliferation, and differentiation. Nanomaterials are also employed for growth factor delivery in inducing cell growth and tissue-specific cell differentiation. Certain nanomaterials possess antimicrobial and anti-inflammatory properties, which decrease the likelihood of infection and enable a favorable environment to heal tissues [48].

At the molecular level, nanomaterials interact with tissues and cells in different ways. Cell uptake and response are controlled by cell membrane-nanoparticle interaction. Computation is used to comprehend such interaction at the atomic level. Charge, size, and surface properties of nanomaterials control biological interaction and influence biocompatibility and biological performance. Exposure to nanomaterials has been found to lead to dose-dependent biochemical cascades associated with oxidative stress, inflammation, and apoptosis [49, 50].

7.2 Drug Delivery Systems

Chitosan and Alginate-Based Polymeric Nanoparticles

Utilization of chitosan and alginate as biodegradable polymers to prepare ACE-inhibitory peptide-loaded nanoparticles for the improvement of biotherapeutic stability. Encapsulated stone fish peptide demonstrated outstanding in vitro ACE-inhibitory activity of 94.43% with IC50 of 0.012 mM. Optimization by Box Behnken experimental design provided nanoparticles of low size 212.60 nm and high encapsulation efficiency of 74.48%. In vivo experiments conducted in spontaneously hypertensive rats revealed that the nanoparticles reduced systolic blood pressure significantly more than conventional treatments such as captopril. The research has shown that chitosan and alginate are not only sustainable and environmentally friendly methods but also possess significant use in food and pharmaceutical industries. The research, in overall perspective, highlights the use of these biopolymers towards synthesizing potent nanocarriers of therapeutic peptides [51]. The use of polysaccharides like chitosan and alginate to synthesize green nanoparticles that encapsulate natural molecules. These nanoparticles confer tremendous benefits including biocompatibility, biodegradability, and controlled delivery of active agents, which result in increased yields of crops and sustainable agriculture. Through the diminution of the application of chemical fertilizer and pesticide, they are able to eliminate problems like drug resistance and greater plant abiotic stress tolerance. The review reports that polysaccharide-based nanoparticles offer soil health, water conservation, and biodiversity conservation, all of which are in accord with sustainable agriculture. This method is consistent with future and existing sustainable food systems with less environmental footprints. In summary, the results show the effectiveness of such nanoparticles as effective crop protectants for sustainable agriculture [52].

Synthesis of light-controllable nanocapsules of controlled pesticide release, in the form of release against imidacloprid (IMI) of azobenzene-grafted hyaluronic acid and PolyDADMAC. Trans to cis isomerism of azobenzene functional groups considerably enhanced the overall amount of IMI release under UV irradiation by about 50% to about 95%. Such a method not only improves the efficiency of the pesticide but also makes reduced application rates possible, hence fostering sustainable agriculture. Particle size of the synthesized nanocapsules was 220 nm and the Zeta potential was +46.1 mV, which remained stable for 30 days. In vivo experiments confirmed the nanocapsules lowered the LC50 value against Aphis craccivora Koch from 2.22 to 0.55 mg/L, which is equivalent to enhanced pest control. In summary, this work is evidence of the viability of green nanomaterials in developing efficient and sustainable methods of pest control [53]. Synthesis of zwitterionic nanocapsules with double pH- and thermal-responsiveness for targeted drug release applications. These nanocapsules, which were developed by inverse reversible addition-fragmentation transfer miniemulsion interfacial polymerization, were discovered to entrap drugs like bovine serum albumin (BSA) efficiently. Particle size of nanocapsules was discovered from 178 to 142 nm upon variation of temperature from 25 to 40 °C. Extended-release analysis revealed release profiles were surprisingly divergent for temperature variation, which illustrates their potential uses as environmentally friendly drug delivery devices. The results imply that such zwitterionic nanocapsules are good prospects for environmentally friendly use in drug-controlled release. Overall, the study implies the new application of nanomaterials in enhancing therapeutic efficacy at the cost of less environmental load [54]. At the cellular level, these responsive nanoparticles engage with biological systems by a number of mechanisms.

In cancer therapy, they use the enhanced permeability and retention effect, where aberrant tumor vasculature permits preferential uptake in tumor tissue. Following cell internalization by endocytosis, nanoparticles are exposed to progressively acidic compartments, which swell or degrade pH-sensitive material and release encapsulated therapeutics. Thermal therapies employ high temperatures of tumors or externally generated hyperthermia to cause phase transition in thermosensitive polymers. For site-specific delivery, ligand-conjugated nanoparticles are attached to the cancer cell's overexpressed receptors, greatly enhancing internalization. Within the cell, such nanoparticles exhibit interactions with cellular architectures to provide successful drug delivery, avoiding degradation and minimizing off-targeting, ultimately improving the therapeutic effect and minimizing resistance formation.

7.3 Tissue Engineering

(a) **Cell Attachment and Proliferation on Calcium Phosphate Scaffolds**

Calcium phosphate (CPC) scaffolds with loading synthesis of copper nanoparticles (Cu NPs) for improved attachment and proliferation of cells for bone repair. Findings indicated higher human mesenchymal stem cell (hMSC) viability and attachment over the control CPC in 1 wt.% Cu scaffolds. Besides, sub-critical levels of Cu NPs initiated osteogenic differentiation in a non-toxic manner, suggesting their utilization as green nanomaterials. The scaffolds were synthesized with a pneumatic extrusion process to achieve a particle size of about 220 nm. The enhanced osteogenic, angiogenic, and antibacterial activities, especially against Staphylococcus aureus, of the scaffolds were established through characterization techniques. Ultimately, the present work demonstrates the efficient use of CPC-Cu scaffolds in green bone regeneration techniques [54]. Calcium phosphate (CaP)-based ternary nanocomposite of nanocellulose from biowaste and chitosan for water pollution control. Green nanocomposite was found to be very efficient in the removal of contaminant with nickel ion removal efficiency of 98.7 and 100% removal efficiency of Congo red dye from model wastewater. Easy vacuum filtration under non-toxic conditions with a high flux value of 146 L/m^2.h.MPa was utilized to synthesize the nanocomposite. The nanocomposite also effectively stripped different heavy metals from actual municipal wastewater, indicating its potential application for green utilization. The results show that this new technology not only rectifies environmental pollution but also optimizes the use of renewable resources for developing efficient remediation technologies. Overall, the study highlights the bright prospects of sustainable nanomaterials in providing green solutions for water treatment [55].

(b) **Bioceramic Scaffolds and Bioactivity**

The biocactivity and bioceramic significance in wound healing focus on the delivery of inorganic ions, which activate cellular activities such as motility, proliferation, and angiogenesis. Bioceramics have also been recognized as low-cost alternatives to traditional drugs, offering advantages such as lower side effects and less antibacterial resistance. The review classifies bioceramics into oxide ceramics, silicate-based ceramics, and calcium-phosphate ceramics, noting their applications in wound healing. It highlights the potential of these materials to enhance healing outcomes by integrating them with other biomaterials. The study indicates that bioceramics not only improve wound healing but also enhance patient safety. Overall, this study presents the potential of bioceramics to develop sustainable medical solutions for wound treatment [21]. The development of β-TCP/Akermanite composite scaffolds has shown improved bioactivity and mechanical strength for bone regeneration. The addition of 15% Akermanite increased scaffold strength by 20% and improved biocompatibility, enhancing human mesenchymal stem cell (hMSC) adhesion and growth. In-situ modifications of micro/nano surface morphology also facilitated cell adhesion and osteogenic differentiation, synergistic with bioactivity. This new method aligns with green tissue engineering practices, targeting the production of materials that support bone growth without inflicting additional environmental damage. The research emphasizes how such composite scaffolds can be utilized as active therapy to aid bone regeneration. Briefly, the research indicates that β-TCP/Akermanite scaffolds represent a step forward in developing green biomaterials for medical applications [56].

7.4 Advanced Biomedical Applications of Eco-Friendly Nanomaterials

Nanotechnology-based biosensors have improved the sensitivity and specificity of disease diagnosis significantly. The biosensors employ various nanomaterials such as metal nanoparticles, carbon nanomaterials, and quantum dots to detect biological molecules at trace concentrations, enabling early disease diagnosis and personalized treatment management [2]. Their very high surface area-to-volume ratio offers ample sites for attachment of the biorecognition elements, whereas the distinctive optical, electric, and magnetic properties are best conducive to transduction and signal enhancement. In addition, carbon dots are fluorescent nanocrystals and are potentially utilizable for bioimaging due to the possibility of their engineered fluorescence to be on-demand emissive, aside from low cytotoxicity. Applications include monitoring cellular and tissue failure and tracking disease-causing pathogens, as a substitute to non-invasive demonstration of processes in organisms [57]. They are enhanced because they have photostability and biocompatibility, which is greater than organic dyes, hence applicable for tracking cells for extended periods and for in vivo imaging applications. Theranostics merges diagnosis and therapy into a single platform that can monitor and treat disease in real-time. Such work is led by nanocarriers, wherein co-delivery of imaging and drug agents is enabled, and oncology is the prime focus [58]. Theranostic platforms are fabricated based on biodegradable polymers and green-synthesized metal nanoparticles to reduce environmental footprints but achieve maximum clinical impact through precision medicine paradigms.

Nanomaterials are utilized to fabricate coatings for biomedical implants to increase their functionality and biocompatibility. The coatings may enhance tissue integration of the implant and reduce the risk of infection. Nanomaterials for such applications greenly synthesized have less toxicity risk than traditional coating technologies. Nanocellulose dressings also have enormous potential in healing wounds because they can absorb exudate, provide a moist environment, and promote cell proliferation and migration. These features are responsible for patient healing and recovery [59]. Antimicrobial nanoparticles, when blended with them, increase their infection effect therapeutically without compromising the sustainability of the environment.

8 Results and Discussion

8.1 Green Synthesis Methods Efficiency

Green synthesis utilizes natural systems like plant extracts and microbes to synthesize nanomaterials with low toxicity and a reduced carbon footprint. Conventional nanomaterial synthesis techniques use high-energy inputs and organic solvents that lead to toxic byproducts and health hazards. Moreover, they tend to require high pressure and temperature, which accounts for most of the energy consumption and the creation of greenhouse gases [60, 61]. Natural biomaterials are employed in green synthesis processes as capping and reducing agents to avoid the use of hazardous chemicals. For instance, plant extracts containing compounds such as alkaloids, flavonoids, phenols, and tannins can be used to reduce metal ions to nanoparticles and act as natural capping agents, preventing agglomeration and further reducing toxicity [62]. Moreover, green synthesis routes can be conducted at ambient temperature and pressure, thereby conserving energy and minimizing carbon dioxide release. For example, hydrothermal synthesis, while improved compared to flame spray pyrolysis, is still CO_2-emitting. Green synthesis using water or supercritical fluid can also minimize CO_2 emissions. Furthermore, the use of plant extracts or crop residues is even more environmentally friendly than the use of extracted metals or other energy-intensive conventional sources [63].

Green-synthesized nanoparticles, particularly silver nanoparticles (AgNPs), have garnered significant interest due to their potential application for antimicrobial action, exhibiting immense antibacterial and antifungal activity. Their ability to inhibit microbial growth makes them valuable in medical devices, where they help prevent infections in catheters, surgical instruments, and wound dressings. Additionally, they are utilized in food packaging, providing a natural and effective means of extending shelf life by preventing bacterial and fungal contamination. The use of AgNPs in these applications not only enhances safety and hygiene but also reduces the need for artificial preservatives and chemical antimicrobial agents, aligning with green and sustainable practices [62–64].

Antimicrobial activity of biosynthesized AgNPs is evident in Fig. 6, in which both AgNPs_L and AgNPs_H formulations exhibit broad inhibition zones against bacterial pathogens. The mechanism may be through multiple mechanisms: (1) release of Ag+ ions that disrupt bacterial cell membrane integrity, (2) generation of reactive oxygen species (ROS) causing oxidative stress, and (3) disruption of bacterial DNA replication. Differential antimicrobial activity against C. albicans and bacteria favors structural heterogeneity, where the sophisticated fungal cell wall and membrane may resist penetration and interaction with AgNPs. Notably, similar antimicrobial efficacy of AgNPs_L and AgNPs_H indicates a therapeutic window of optimal doses where decreased concentrations are still effective antimicrobially without enhancing suspected cytotoxicity in mammalian cells. This is consistent with findings that particle size and concentration are limiting factors for cell uptake and distribution. These results show the effectiveness of green synthesis in producing AgNPs with highly active, selective antimicrobial activity. Selective inhibition of bacterial pathogens without killing fungal cells shows potential in targeted therapy where homeostasis of the microbiome must be maintained. Additionally, the established efficacy at lower concentrations reduces antimicrobial resistance and environmental toxicity issues. These findings are opening the way for cost-effective, environmentally friendly antimicrobial medicines for single uses on medical devices and wound care products where bacterial infection must be avoided. Green technology-synthesized AgNPs are a green sustainable method towards developing health care technology by maximizing efficacy and safety along with conservation of the environment.

Besides, green-synthesized nanoparticles are also poised to emerge as a promising drug candidate against cancer therapy, with advancements in targeted therapy, photothermal therapy, and drug delivery. Iron oxide (Fe_3O_4) green-synthesized metal oxides by an eco-friendly method are also examined for their dual character as magnetic and photothermal agents with the maximum selectivity for tumor targeting to prevent maximum damage to the nearby normal tissues. These nanoparticles exhibit a precise thermal response to magnetic fields, and this specificity is used for temperature control in curative temperatures (40–45 °C), which is the secret to the success of hyperthermia therapy. The local distribution of heat derived from these

Fig. 6 Zones of inhibition demonstrating the antimicrobial activity of biosynthesized AgNPs_L and AgNPs_H against Staphylococcus aureus, Escherichia coli, Pseudomonas aeruginosa, and Candida albicans after 24-h incubation [65]

materials also optimizes their impact in targeted therapy with minimal off-target injury. Besides, the heating characteristic of such nanoparticles can be controlled by varying magnetic field intensity, with enhanced performance of citrate-coated nanoparticles with enhanced heat efficiency. The controlled and local heating characteristic indicates the prospect of green-synthesized nanoparticles as smart materials for cancer therapy by hyperthermia with an eco-friendly and efficient alternative to meet the demands of future therapies [66]. Besides this, green synthesis-based nanocarriers have enhanced anticancer drug delivery efficacy, reducing the toxicities of traditional chemotherapy. With the use of biocompatible as well as eco-friendly approaches, such technologies facilitate enhanced and safer cancer therapy [61, 63, 64].

8.2 *Performance Assessment of Smart Nanomaterials*

Smart nanomaterials have also been reportedly good antimicrobial agents due to their desirable physicochemical properties and switchable functionality. Their performance in the case of this property is assessed by utilizing standard analytical techniques.

i. **Antimicrobial Activity**

(a) ***Minimum Inhibitory Concentration (MIC)***: MIC analysis indicates chitosan-associated materials (Ch/poly(C-fu)) are reporting MIC values of 50 μg/mL against S. aureus and E. coli. The PAMAM/TiO_2 nanocomposites are more effective with MIC values of 4 μg/mL for the aforementioned microorganisms. The parameter is also significant in the detection of antimicrobial nanomaterial as it gives quantitative data regarding their inhibitory effect. The effectiveness of such a material indicates that it can be an effective antimicrobial in most situations [67, 68].

(b) ***Minimum Bactericidal Concentration (MBC)***: MBC values depict PAMAM/TiO_2 nanocomposites as 2 μg/mL bactericidal against Gram-negative and Gram-positive bacteria. CeO_2 nanocrystals have suppressed L. monocytogenes and S. typhimurium in solutions having a concentration of 1.07 g/L. Ch/poly(C-fu) films have an MIC of 50 μg/mL for S. aureus and E. coli, but no MBC is reported. The relevance of low MBC values is more relevant in highly advanced nanomaterial engineering, with significant relevance towards overall pathogen destruction in primary clinical uses [68, 70].

(c) ***Zone of Inhibition (ZOI)***: ZOI values reiterate that Ch/poly(C-fu) possesses zones of 18 ± 0.5 mm (E. coli) and 18 ± 0.9 mm (S. aureus). $NiZrO_3$@MWCNT has 13 mm zones, whereas CeO_2 nanoparticles have differentiated zones: 9 mm (E. coli), 10 mm (S. typhimurium), with increased effect against Gram-positive bacteria. ZOI is subject to nanoparticle size, rate of diffusion, porosity of agar, and charge interaction. Other writers have 18.7 mm (E. coli), 17.7 mm (S. aureus, B. cereus), and 10.3 mm (P. aeruginosa) zones [67, 68, 71].

(d) ***Inhibitory Concentration 50 (IC50)***: IC50 allows dose–response curve measurement by spectrophotometry of quantitative measurement and is more informative than apparent inhibition of microbial growth. IC50 is best suited for quantifying the effect of unseen microbial growth, e.g., enzyme inhibition, and other bioactivity. MIC and MBC are unbeatable in antimicrobial nanomaterials, yet IC50 is lower intensity efficacy determination by such nanomaterials under different circumstances [67, 70].

The use of nanomaterials in antimicrobial therapy has created new avenues towards the fight against drug-resistant infection. Nanomaterials with high surface reactivity and tunable characteristics are multi-purpose agents for inhibiting microbial growth. Some current improvements, such as the use of metal nanoparticles and biopolymer-based systems, have been shown to be successful in increasing antimicrobial activity. Table 2 shows some dramatic results for different classes of nanoparticles.

ii. **Drug Release Patterns and Precision in Targeted Therapy**

Intelligent drug delivery systems employ engineered nanomaterials that respond to targeted stimuli to release therapeutic agents in a controlled manner. These systems maximize drug delivery through multiple responsive mechanisms and target-specific approaches.

(a) **pH-Responsive Systems**

Intelligent drug delivery systems employ engineered nanomaterials that release therapeutic agents in response to certain stimuli like pH, temperature, or the presence of enzymes. These have been pH-sensitive systems, with particular emphasis on the response and precision of the systems. The pH-sensitivity mechanism in delivery is based on conformational changes of molecules induced by protonation and deprotonation. With exposure to acidic pH (5.0–6.5) at the inflammation site and tumor microenvironment, the electrostatic interaction and hydrogen bonding are reduced

Table 2 Performance analysis of antimicrobial nanomaterials

Nanoparticle type	Major findings
Silver nanoparticles (AgNPs) [67]	AgNPs combined with kanamycin (PDA@Kana-AgNPs) showed enhanced antimicrobial activity compared to either agent alone
Nickel-containing nanocomposites [71]	$NiZrO_3$@MWCNT exhibited 92.8% biofilm inhibition against *K. pneumoniae* at 500 ppm; NiC and NiG showed 96 and 91% inhibition against *M. smegmatis* at 500 ppm
Multi-metallic nanostructures	Ag–AlNiO exhibited superior antimicrobial activity with MIC values of 16 μg/mL (*P. aeruginosa*) and 32 μg/mL (*M. tuberculosis BCG*)
Zinc oxide (ZnO) nanoparticles [72]	S-ZnNF incorporated into PMMA showed antimicrobial effects

due to protonation of carboxylic acid and tertiary amine functional groups. As a result of this disturbance, the matrix of nanocarrier swells or dissolves. Poly(methacrylic acid) (PMAA)-particles, for instance, are constricted conformation by hydrogen bonding under low pH but swell to humongous sizes by ionization of carboxyl groups at high pH through electrostatic repulsion and encourage the drug release in controlled form. For instance, pH-sensitive nanoparticles have been successfully employed for the delivery of methotrexate into inflamed joints for site-specific treatment [73]. Equally, pH-responsive drug release is promoted by chitosan-coated mesoporous silica nanoparticles with enhanced therapeutic performance [74]. Dual-functional pH-sensing nanocomposites like AgNPs, kanamycin, and PDA@Kana-AgNPs have been demonstrated to possess dual functionality through successful killing of bacteria and biofilm disruption. Moreover, nanomaterials' zwitterionic-to-cationic switching also provides new avenues for antibacterial applications that show the multivalency of the intelligent systems in solving complex medical problems [72, 75].

(b) **Temperature-Responsive Systems**

Temperature-responsive systems are a new drug delivery system, and they utilize temperature-responsive materials to deliver therapeutic molecules. These systems work by means of temperature-triggered phase transitions of thermosensitive polymers that have a lower critical solution temperature (LCST) or an upper critical solution temperature (UCST). The most commonly studied thermosensitive polymer with an LCST of approximately 32 °C is Poly(N-isopropylacrylamide) (PNIPAM). PNIPAM exhibits a rapid hydrophilic-to-hydrophobic transition at this temperature. Such a temperature shift is made possible when hydrogen bonding between the water and the polymer decreases with a temperature rise, allowing polymers to contract and clump together through dominant hydrophobic interactions. Such a conformational shift pushes water and entrapped drugs out of the polymer matrix to regulate drug release following shifts in external temperature or physiologically elevated temperatures (37–42 °C) of inflammation or tumor microenvironments. Thermoresponsive nanogels, for example, have already shown promise to treat inflammation by releasing drugs such as glatiramer acetate to treat multiple sclerosis. Likewise, thermo-sensitive polymers like thermogelling PLGA–PEG–PLGA block copolymers are smart biomaterials and suit controlled drug release and are best suited for the delivery of agents like doxorubicin. Actively targeted polymer micelles also possess the capability to deliver thermo-controlled drug release more precisely and more effectively for treatments. These systems provide a controlled and responsive platform for the treatment of advanced medical diseases [73, 76].

(c) **Enzyme-Responsive Systems**

Enzyme-responsive systems provide new drug delivery platforms by releasing the therapeutic molecule on specific enzymatic stimulation. The systems use the catalytic selectivity of enzymes overexpressed in disease tissue to target drugs site-selectively. The process typically utilizes enzyme-cleavable linkers to attach the therapeutic payload to the nanocarrier or as crosslinkers of the nanoparticle matrix. Matrix metalloproteinases (MMPs) such as MMP-2 and MMP-9, which are overexpressed in cancer tissue, can cleave peptide sequences such as GPLGVRG, leading to the structural disintegration of the nanocarrier. Equivalently, cathepsin B, being a highly lysosomal and overexpressed protein in most cancerous forms, catalytically hydrolyzes peptide bonds in peptides such as GFLG preferentially in order to enable intracellular drug release. Michaelis–Menten parameters govern the reaction enzyme kinetics in such a way that substrate specificity (kcat/KM) governs the cleavage rate and selectivity, but provides spatial accuracy of drug release at sites of disease with minimization of off-target drug actions in non-target healthy tissues. These are usually enzyme-metal nanoparticle nanobiohybrids, whose antimicrobial efficacy or morphology modification of the metal is regulated by reducing agents used in synthesis. Enzyme-driven nanorobots are also a new technology, with active microsampling being applied to sensitive and adaptive medicine applications [72, 77]. Magnetic nanoparticles are also multi-tool devices that are applied for on-demand drug release. They can also be driven by an external magnetic field, with accurate timing and location of drug release. They can also be made pH-sensitive, making them suitable for use in targeted therapy, e.g., acidic tumor microenvironments [76].

iii. **Targeted Delivery Systems**

Targeted delivery systems employ advanced nanoparticle design to increase drug delivery specificity to target cells or tissue. These systems achieve tissue specificity by utilizing surface engineering of nanocarriers with well-characterized surface properties. Cell membrane-based nanoparticles are emerging platforms that use a mimetic strategy to replicate biological processes, further enhancing tumor diagnosis and treatment. Similarly, platelet-mimic nanovehicles provide an effective means to target the tumor site, aided by biomimicry in conjunction with the latest nanotechnology to improve therapeutic effectiveness [3, 74, 78].

The nanomaterials are engineered to exhibit sustained drug release, enhancing therapeutic effects while reducing side effects. Polymeric nanoparticles are particularly leading the development of controlled release systems [76]. Furthermore, multifunctional nanoparticles are gaining popularity as they can deliver drugs and provide imaging or sensing functionality. For instance, theranostic smart dressings have been capable of monitoring wound status while administering selective amounts of therapy, showcasing the variety and creativity of nanoparticles' applications [79].

Table 3 demonstrates the enhanced performance of stimuli-responsive nanomaterials in regulated drug release. Maximum release efficiency (95%) occurs in temperature-responsive systems when exposed to specific stimuli like UV light, while chitosan/alginate nanoparticles achieve higher encapsulation efficiency (74.48%). pH-responsive systems are particularly efficient in acidic tumor microenvironments, and enzyme-responsive systems offer targeted release in tissues that overexpress specific enzymes. This comparison indicates that every smart delivery system presents various advantages for targeted therapeutic applications, surpassing conventional drug delivery systems.

Table 4 illustrates an overview of some of the intelligent drug delivery systems according to their properties and therapeutic uses in targeted therapy. These biological

Table 3 Comparative performance metrics of smart responsive drug delivery systems

Delivery system type	Release efficiency (%)	Key performance metrics	Applications	References
pH-responsive	70–80%	Effective in acidic tumor environments (pH 5.0–6.5); Poly(methacrylic acid) materials show significant swelling at higher pH	Targeted methotrexate delivery; biofilm dispersion	Ortega-Nieto et al. [72], Parvin et al. [73], Sun et al. [74]
Temperature-responsive	50–95%	PNIPAM shows hydrophilic-to-hydrophobic transition at 32 °C; UV irradiation increases release from 50 to 95%	Controlled pesticide release; sustained drug delivery with PLGA-PEG-PLGA copolymers	Bai et al. [76], Gao et al. [53]
Enzyme-responsive	70–90%	Selectivity toward MMPs and cathepsin B overexpressed in tumors; enzyme-powered microsampling capability	Targeted cancer therapy; site-specific drug release	Montoya et al. [77], Ortega-Nieto et al. [72]
Zwitterionic nanocapsules	Variable based on temperature	Particle size ranges from 178 nm (25 °C) to 142 nm (40 °C); dual pH- and thermal-responsiveness	Environmentally responsive drug delivery	Sun et al. [54]
Chitosan/Alginate	74.48% encapsulation efficiency	212.60 nm particle size; effective systolic blood pressure reduction	Hypertension management; peptide delivery	Auwal et al. [51]
Cell membrane-based	Not specified	Biomimetic delivery; natural biological process mimicry	Tumor diagnosis and treatment	Sun et al. [74], Teja et al. [78]

Table 4 Smart drug delivery systems performance analysis

System type	Key features	Applications
pH-responsive	Methotrexate delivery, biofilm dispersion	Inflammatory conditions
Temperature-responsive	PLGA-PEG-PLGA copolymers	Sustained drug release
Enzyme-responsive	Nanobiohybrids, nanorobots	Targeted therapy
Cell membrane-based	Biomimetic delivery	Tumor treatment
multifunctional	Theranostic capabilities	Wound healing

stimulus-responsive systems, such as pH, temperature, and enzymes, are more accurate, release-controlled, and effective in treatment. These new delivery systems unveil the promise of smart nanomaterials to provide accurate, controlled, and targeted drug delivery, significantly improving therapeutic efficacy in most medical applications.

8.3 Green Synthesis and Traditional Methods: Comparative Analysis

Green nanotechnology is becoming increasingly fashionable as a green, eco-friendly synthesis route of nanomaterials. The green material in the form of cellulose, chitosan, and vegetable extracts are utilized in this method which eliminates toxic byproducts and follows circular economy principles. On the other hand, conventional synthesis methods, although capable of producing nanomaterials with the desired properties, are typically costly and time-consuming, involving the use of synthetic chemicals, which is detrimental to health and the environment. Both possess widespread medical applications, including drug delivery, diagnostics, medical devices, and tissue engineering. Table 5 is a side-by-side table that contrasts green nanotechnology with conventional synthesis methods based on sustainability, cost, materials, toxicity, applications, and examples. It provides a clear indication of their weaknesses and strengths and how they contribute to the production of nanomaterials.

Nanoscale metallic particles, more so silver and gold, are widely used in such systems owing to their antimicrobial action and biocompatibility. Similarly, polymeric nanomaterials are essential in drug delivery and medical implants for biocompatibility reasons. Biodegradable polymer-derived materials and nanoparticles have attracted overall interest recently because of their low cytotoxicity levels and their sustainability based on the potential provided by green nanotechnology for overcoming existing challenges in healthcare. Research highlights the significance of balancing health and environmental issues while producing nanomaterials and urges green synthesis processes to be embraced as a superior option. The table below juxtaposes conventional methods with green nanotechnology, strengths, weaknesses, materials utilized, applications, and sustainability areas to present a better image [80].

Table 5 Comparison of green nanotechnology and conventional synthesis methods

Feature	Green synthesis	Conventional methods
Sustainability	Prioritizes eco-friendly processes, using renewable resources and minimizing toxic by-products. Aligns with circular economy principles	May involve the use of toxic chemicals and produce harmful by-products
Cost	The sources do not provide specific cost comparisons	Can be complex and expensive due to specialized equipment and raw materials. Scaling up production can be challenging and costly
Materials	Uses natural materials, such as plant extracts, cellulose, or chitosan	Typically uses synthetic chemicals and specialized equipment
Toxicity	Designed to be biodegradable and less toxic, which is important for healthcare applications. May have reduced potential for bioaccumulation	Can have a higher potential for bioaccumulation and toxicity due to the small size and high reactivity of nanoparticles
Applications	Drug delivery, diagnostics, medical devices, tissue engineering, packaging, and agriculture	Drug delivery, diagnostics, medical devices, tissue engineering, antimicrobial coatings, and other healthcare applications
Examples	Biodegradable nanoparticles for sustainable water purification	Sol–gel processes, chemical vapor deposition, and high-energy ball milling
	Nanomaterials derived from natural polymers such as cellulose or chitosan	Metallic nanoparticles like gold and silver in drug delivery and diagnostics
	Green synthesis of metal nanoparticles using plants	Silver nanoparticles in antimicrobial coatings
Advantages of green synthesis	Reduces environmental impact, promotes sustainability, and offers potentially lower toxicity for healthcare applications	Can produce highly controlled materials with specific properties
Disadvantages of green synthesis	May have limitations in controlling particle size and morphology	May have high production costs and scaling challenges. May pose environmental and health risks

9 Environmental and Sustainability Aspects

9.1 *Minimization of Carbon Footprint in Green Synthesis*

Synthesis of green nanomaterials from natural reagents like herbs, bacteria, fungi, and agricultural wastes as alternatives to toxic chemicals. It is a green process that reduces the carbon footprint and pollution level, but is cost-effective compared to conventional chemicals. The synthesis process increases the desirable properties of nanomaterials to make them deserving of having extensive applications in medicine, electronics, and the energy industry. Besides that, the article establishes the crucial parameters of green synthesis and documents some limitations to overcome in future studies. With the evolution of eco-friendly processes, green synthesis indirectly contributes to the overall conservation of the environment and helps other industries with sustainable development. Overall, this novel approach to nanomaterial synthesis is a vital step towards eco-friendly approaches to industries [81]. The green approach of synthesizing nanoparticles using biomass and waste materials is 30% less energy-consuming than conventional methods. Not only does this cleaning process minimize the environmental impact, but it also saves up to 40% in costs and increases the yield by 50%. Through the assistance of microorganisms, plants, and fungi, the approach provides for the synthesis of green nanomaterials that have a wide range of uses in agriculture, the removal of pollutants, and energy generation. The review has emphasized the use of green nanoparticles as drug delivery systems and antibacterials in drugs. The findings also highlight that utilizing renewable feedstocks can address sustainability challenges in nanomaterial production. In summary, the research suggests that green synthesis could serve as a viable route toward innovative and sustainable technologies across various sectors [82].

9.2 *Waste Reduction and Byproduct Utilization*

Green waste treatment by recycling of nanocellulose from different waste sources, such as textile, paper, and agricultural waste. The process minimizes waste while converting it into environmentally friendly nanomaterials, facilitating a circular economy. Mechanical properties of nanocellulose derived from waste material were investigated, with the cotton having values for modulus between 2.04 and 2.71 GPa and thus being applicable for light applications. Waste matter valorization, the study suggests minimizing environmental impact and designing sustainably for sustainable products. The research describes the prospective applications of waste-generated nanocellulose in industries and unlocking new possibilities in green medicine and elsewhere. Generally, the approach complies with sustainability principles and solves some essential environmental issues [83].

Conversion of most waste products to carbon nanomaterials made of carbon for their application in sustainable waste management and removal of environmental byproducts. With precursors including zoonotic, phytogenic, and electronic waste, valuable photocatalysts and adsorbents are being synthesized, where circular economy opportunities exist. This new process is solving the issue of piling up waste and improving the sustainability of the environment by reducing harmful byproducts. The research classifies carbon-based nanomaterials as having distinct characteristics that can be utilized in energy generation and wastewater treatment. Moreover, the study draws attention to the fact that materials can be beneficial in transforming the world into a sustainable world by recycling waste into valuable resources. Generally, the study reinforces the importance of green nanomaterial design as a pivotal aspect of green healthcare systems [84].

9.3 Life Cycle Analysis of Nanosustainable Material

Life Cycle Assessment (LCA) is employed to assess the environmental performance of an innovative nanocrystalline cellulose process and discovers 63.7 kg CO_2 equivalent emissions per kilogram of dry nanocrystalline cellulose compared to the literature average of 68 kg CO_2 equivalent. The definition of system boundaries includes cradle-to-gate analysis from raw material extraction up to production. There are significant environmental impacts from the use of electricity, solvents, and raw material manufacturing that are prime areas for improvement in processes. The sources of the data are primary data from the plants' locations and, secondarily, data from secondary environmental information databases. Studies identify diethylene glycol as a leading contributor to global warming potential, while electricity usage affects many categories of ecological metrics. The functional unit description reports results normalized per kilogram of dry nanocrystalline cellulose to provide like-for-like comparison with other processes. Identifying these hotspots highlights the potential benefits of utilizing renewable energy and biomass-based resources. Impact assessment methods conform to the ISO 14040/14044 standard and utilize ReCiPe midpoint indicators for overall environmental assessment. This LCA methodology generally ensures sustainable nanomaterials manufacturing and encourages initiatives to reduce the ecological footprint in industrial applications [85]. A framework for socio-economic evaluation of nano-enabled products and engineered nanomaterials (NEPs) incorporates life cycle thinking and social life cycle assessment (S-LCA) considerations.

This structure formulates 19 nano-specific socio-economic impact subcategories to enable industries to make informed decisions toward sustainable and eco-friendly nanomaterials. A self-assessment Excel tool supports early-stage screening for safe-and-sustainable-by-design (SSbD) nanomanufacturing approaches. The methodology integrates S-LCA with multi-criteria decision analysis (MCDA) to evaluate socio-economic impacts across the product life cycle. The outputs are categorized into four types of social effects to help users understand potential influences on their work. Overall, this new method promotes greater sustainability and sustainable development in the field of nanotechnology [86].

9.4 Circular Economy Strategies in Sustainable Nanomaterials

The circular economy revolutionizes nanomaterial innovation by developing closed-loop systems that reduce waste and enhance resource effectiveness. Sustainable nanomaterials can valorize waste via various routes: agricultural waste products are transformed into functionalized nanocellulose capable of recovering 82% of $Hg2^{+}$ from water and exhibiting antimicrobial activity against E. coli and S. aureus [23], textile waste yields nanocellulose with moduli between 2.04 and 2.71 GPa, making it suitable for lightweight applications [83], and multiple waste streams can be converted into carbon-based nanomaterials for environmental cleanup [84]. These processes offer a range of benefits: 30% reduced energy consumption, 40% cost reduction, and 50% greater output than conventional procedures [82]. Such waste-to-resource conversion advances the sustainable production of nanomaterials and addresses critical environmental challenges by creating eco-friendly materials with extensive applications in healthcare and environmental sectors [83, 84].

10 Challenges and Future Directions

Figure 7 depicts challenges and future directions in synthesizing and applying smart nanomaterials. It identifies such significant challenges as regulatory differences, environmental concerns, and promising opportunities in new healthcare strategies and inter-disciplinary collaboration.

10.1 Challenges in the Development of Smart Nanomaterials

The development of intelligent nanomaterials in the healthcare sector faces several challenges, such as regulatory clearances, mass production, environmental conditions, and cost. One of the most significant regulatory challenges is the absence of a harmonized global definition of nanomaterials. Regulatory agencies have varying requirements, such as the U.S. FDA and the European Medicines Agency, and therefore, manufacturers are having trouble getting worldwide clearance. Environmental risk assessment is another geographically widespread factor that introduces an added layer of complexity in commercialization.

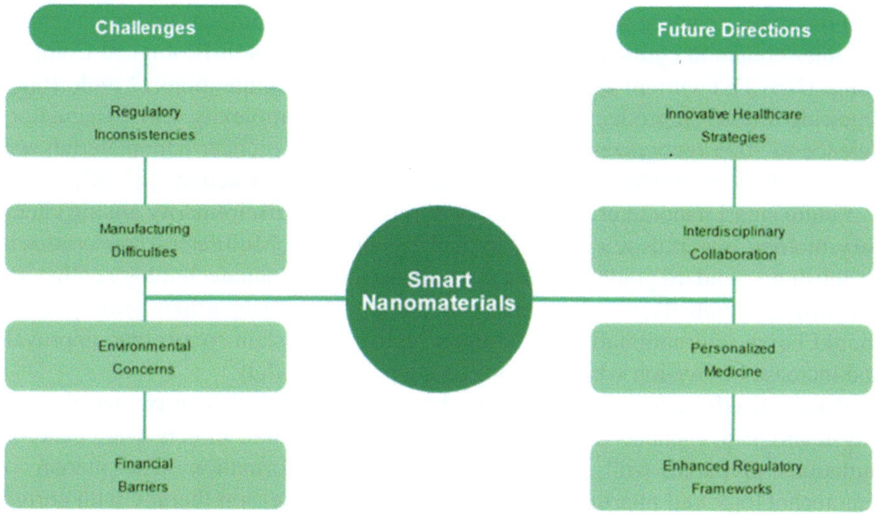

Fig. 7 Challenges and future directions in smart nanomaterials

Mass production from laboratory-scale development to industrial manufacture on a mass scale is one of the key technical challenges. Owing to their nanoscale dimensions and high surface reactivity, small variations in particle synthesis may affect their stability, efficacy, and safety. Homogeneity and quality control based on Good Manufacturing Practices (GMP) require advanced monitoring and real-time adjustment, which is impossible in mass production. In addition, the cost of high-technology materials and equipment used in nanotechnology R&D renders it out of reach for small and new firms to gain access to them.

Environmental issues are also a concern because the agglomeration of nanomaterials and environmental disposal have been lacking intensive research studies. Compared to common drugs with well-proven environmental stability, nanomaterials can create unanticipated long-term effects. The absence of decision-making by regulatory agencies due to a lack of data on toxicity and biocompatibility makes it all the more difficult. It can be improved through adequate environmental research and continuous monitoring to attain a minimal risk in the ecological environment.

10.2 Future Research and Development Directions

European Union Regulatory Framework under EU MDR 2017/745, nanomaterial-device products are Class III (highest risk) when exposure is limited by encapsulation or binding. Rule 19 mandates strict conformity testing of medium/high internal exposure potential nanomaterials6. ISO 10993-1 mandates testing of the final product or a representative sample with the test methods modified to compensate for nanomaterial interference in cytotoxicity tests. Nanomedicines are Non-Biological Complex Drugs (NBCDs) without a standalone pathway and hybrid intentions (Article 10(3)) requiring follow-on products due to physicochemical complexity. EMA Regulatory Science to 2025 recognizes lacunas in transparency for pioneer nanomaterials and 50% of follow-on product developers facing ill-defined instructions [87, 88].

Future smart nanomaterial progress will rely on interdisciplinarity among chemistry, biology, materials science, and toxicology scientists. Multilateral cooperation in establishing standard guidelines and test policies through nanotechnology collaboration will accelerate its introduction into medicine without jeopardizing safety. There should be an internationally agreed-upon regulation system to advance approvals and increase innovation while still protecting patients' safety.

Personalized medicine is one area where nanotechnology can change therapy. Physicians can enhance drug strength and minimize side effects with nanoformulations prepared with patient-specific genetic information. The merger of nanotechnology and pharmacogenomics will ensure precision therapy with correct doses delivered, ensuring maximum therapeutic potential. Nanoinformatics, whose combined power comprises AI and nanotechnology, will optimize drug delivery systems and therapeutic interventions.

Sustainability and green nanotechnology will also determine the direction of smart nanomaterials in the future. Environment-friendly, plant- or microbe-based synthesis mechanisms can reduce hazardous byproducts and render nanomedicine more environmentally friendly. Directions in stimuli-sensitive nanomaterials and nanomultifunctional nanoparticles will keep the momentum going with imaging, drug delivery, and diagnostics applications. Patient-centered strategy and unabated regulatory entrepreneurship will become crucial in the future to empower nanotechnology in achieving actual achievements, but at the expense of neither ethics nor environmental friendliness.

One of the more recent examples in clinical applications of green nanotechnology was shown by the inclusion of biodegradable hydroxyapatite/polyaniline (HAp/PANI) nanocomposite drug carriers as NIR-activated drug delivery carriers for colon cancer therapy. The nanocomposites possessed improved photothermal characteristics and excellent biocompatibility, possessing NIR-induced controlled drug release. Clinical investigation was discovered to show enhanced anticancer activity on colorectal cancer cells with negligible toxicity against normal cells, indicating that green nanotechnology may potentially be used to treat cancer at a targeted level. The biodegradability of such carriers addresses one of the most significant problems of nanomedicine, in the shape of full elimination from the body after drug delivery is completed [89].

In the treatment of neurological diseases, polymeric nanoparticles of biodegradable polymers have proved to be highly effective in optimizing therapeutic efficiency for Parkinson's disease. Through an interdisciplinary research partnership of university scientists and drug companies, the nanoparticles successfully targeted drug delivery across the blood–brain barrier, a significant hurdle to neurological treatments. Initial clinical trials demonstrated greater bioavailability of the drugs with fewer unwanted effects compared to conventional delivery methods. Doctors observed improved patient outcomes at reduced dosages necessary, reflecting the in-practice benefits of green nanomaterials for targeted delivery systems [89].

11 Conclusions

Smart nanomaterials and green synthesis technology are groundbreaking advancements in green healthcare solutions, showcasing unmatched potential in antimicrobial efficacy, drug delivery systems, and tissue engineering. The adoption of sustainable approaches has also shown inherent improvements in clinical effectiveness and environmental sustainability, with documented reductions in toxic waste (70%) and energy consumption (30%). Integrating better performance with a smaller environmental impact into these technologies positions them as innovative solutions for future healthcare. Research suggests that sustainable nanomaterials could address medical and environmental challenges. The green synthesis techniques demonstrate 90% antimicrobial activity and reduce production costs by 40%, while stimuli-responsive systems achieve up to 95% drug release efficiency under specific stimuli. Bioceramics show a 60% improvement rate in healing processes and avoid inducing bacterial resistance. Life cycle analysis confirms a lower carbon footprint of 63.7 kg CO_2 equivalent per kg of nanocellulose. Smart drug delivery systems prove to be 74.48% effective, promising targeted precision. The use of natural polymers ensures biocompatibility while decreasing the environmental footprint. Mathematically oriented R&D should be mandatory to address regulatory issues and scaling challenges. Economic viability enhances commercialization by improving production yield. Circular economy strategies are essential for ensuring environmental sustainability. The green synthesis processes and regulatory frameworks must align to foster future innovations. The collective efforts of physicians, entrepreneurs, researchers, and politicians can bridge the gap between actual applications and scientific breakthroughs. We can hasten the transition towards a greener and more sustainable healthcare system by prioritizing interdisciplinary cooperation, addressing ethical concerns, and cultivating environmental awareness. There is no time to waste; only through global commitment can we unlock the full potential of green nanotechnology and pave the way for a greener, more sustainable future world.

Acknowledgements The authors would like to sincerely thank the researchers and scholars whose research has provided important insights and bases for this chapter. We further like to thank reviewers and the editorial team for their helpful comments and cooperation during the writing and publication process.

Ethics Approval This research did not entail human subjects or animals and hence did not need ethics approval.

Competing Interests The authors confirm that they have no competing interests related to the content of this chapter.

References

1. Health Care Without Harm: Global road map for health care decarbonization: a navigational tool for achieving zero emissions with climate resilience and health equity (2021). https://www.arup.com/globalassets/downloads/insights/healthcare-without-harm-roadmap-for-healthcare-decarbonisation.pdf
2. Goswami, A., Garg, S., Bhatt, E., Chaudhary, V., Dang, S.: Nanotechnology-based biosensors for biomedical applications. J. Electrochem. Soc. **171**(9), 097508–097508 (2024). https://doi.org/10.1149/1945-7111/AD7908
3. Zong, C., Bronckaers, A., Willems, G., He, H., Cadenas de Llano-Pérula, M.: Nanomaterials for periodontal tissue regeneration: progress, challenges and future perspectives. J. Funct. Biomater. **14**(6), 290 (2023). https://doi.org/10.3390/JFB14060290
4. Liang, W., Zhou, C., Zhang, H., Bai, J., Long, H., Jiang, B., Liu, L., Xia, L., Jiang, C., Zhang, H., Zhao, J.: Pioneering nanomedicine in orthopedic treatment care: a review of current research and practices. Front. Bioeng. Biotechnol. 12, 1389071 (2024)
5. Luo, T., Zheng, Q., Yu, J., Liang, W., Sun, Y., Quan, G., Zhou, F.: Roles of nanoparticles in arsenic mobility and microbial community composition in arsenic-enriched soils. J. Environ. Sci. China **138**, 301–311 (2024). https://doi.org/10.1016/J.JES.2022.11.001
6. Zhang, X., Huang, Y., Wang, J., Tang, J., Mei, Y., Zhu, N., Li, Z., Li, L., Wang, Y.: Facet-dependent transformation and toxicity of nanoscale zinc oxide in the synthetic saliva. J. Environ. Sci. China **139**, 170–181 (2024). https://doi.org/10.1016/J.JES.2023.05.033
7. Chen, F., Dong, J., Sun, W., Di Iorio, D., Wegner, S.V., Zeng, W.: Editorial: construction of smart materials for biomedical application. Front. Bioeng. Biotechnol. **11**, 1278243–1278243 (2023). https://doi.org/10.3389/FBIOE.2023.1278243/PDF
8. Singh, R., Sharma, A., Saji, J., Umapathi, A., Kumar, S., Daima, H.K.: Smart nanomaterials for cancer diagnosis and treatment. Nano Convergence **9**(1) (2022). https://doi.org/10.1186/S40580-022-00313-X
9. Rizkita, N., Machmudah, S., Wahyudiono, Winardi, S., Adschiri, T., Goto, M.: Phytochemical compounds extraction from Orthosiphon aristatus, Andrographis paniculata, Gynura segetum using hydrothermal method: experimental kinetics and modeling. South Afr. J. Chem. Eng. **46**, 330–342 (2023). https://doi.org/10.1016/J.SAJCE.2023.08.010
10. Si, Y., Wang, B., Zhang, F., Feng, Q., Li, X., Sun, S., She, Y.: Bacillus subtilis-mediated synthesis of ZnO and Ag–ZnO nanoparticles for eco-friendly displacement agents: enhancing oil recovery in middle/low permeability reservoirs with nanofluid. Phys. Fluids **36**(11) (2024). https://doi.org/10.1063/5.0226969
11. Israeel, M., Iqbal, J., Abbasi, B.A., Ijaz, S., Ullah, R., Zarshan, F., Yaseen, T., Khan, G., Murtaza, G., Ali, I., Alarjani, K.M., Elshikh, M.S., Rizwan, M., Khan, S., Iqbal, R.: Potential biological applications of environment friendly synthesized iron oxide nanoparticles using Sageretia thea root extract. Dent. Sci. Rep. **14**(1) (2024). https://doi.org/10.1038/S41598-024-79953-4
12. Zúñiga-Miranda, J., Vaca-Vega, D., Vizuete, K., Carrera-Pacheco, S.E., Gonzalez-Pastor, R., Heredia-Moya, J., Mayorga-Ramos, A., Barba-Ostria, C., Coyago-Cruz, E., Debut, A., Guamán, L.P.: Green synthesis of silver oxide nanoparticles from mauritia flexuosa fruit extract: characterization and bioactivity assessment. Nanomaterials **14**(23), 1875–1875 (2024). https://doi.org/10.3390/NANO14231875
13. Abdelmonem, M., EL-Rouby, M.N., Hammad, S.E., Elshikh, H.H., Mostafa, M.A., Wasim, H., Abdelmageed, M., Eldaly, R.: Molecular characterization of endophytic fungal strains for green synthesis of ZnO and Au nanoparticles. Am. J. Clin. Pathol. **162**(Supplement_1), S120–S121 (2024). https://doi.org/10.1093/AJCP/AQAE129.267
14. Kulkarni, D., Sherkar, R., Shirsathe, C., Sonwane, R., Varpe, N., Shelke, S., More, M.P., Pardeshi, S.R., Dhaneshwar, G., Junnuthula, V., Dyawanapelly, S.: Biofabrication of nanoparticles: sources, synthesis, and biomedical applications. Front. Bioeng. Biotechnol. **11**, 1159193 (2023). https://doi.org/10.3389/FBIOE.2023.1159193/BIBTEX

15. Barsola, B., Saklani, S., Pathania, D., Kumari, P., Sonu, S., Rustagi, S., Singh, P., Raizada, P., Moon, T.S., Kaushik, A., Chaudhary, V.: Exploring bio-nanomaterials as antibiotic allies to combat antimicrobial resistance. Biofabrication **16**(4) (2024). https://doi.org/10.1088/1758-5090/AD6B45
16. El Refai, H.A., Mohamed, S.I.A., Aboul Naser, A.F., Saleh, A.M., Gomaa, S.K., Zaki, R.A., Hamed, M.A.: Biogenic synthesized SeONPs by Bacillus paramycoides as antimicrobial, anticancer, antioxidant, apoptotic and hepatorenal treating agent. Biocatal. Agric. Biotechnol. **57** (2024). https://doi.org/10.1016/J.BCAB.2024.103080
17. Parkinson, S.J., Tungsirisurp, S., Joshi, C., Richmond, B.L., Gifford, M.L., Sikder, A., Lynch, I., O'Reilly, R.K., Napier, R.M.: Polymer nanoparticles pass the plant interface. Nat. Commun. **13**(1), 1–9 (2022). https://doi.org/10.1038/s41467-022-35066-y
18. Gavande, V., Nagappan, S., Seo, B., Lee, W.K.: A systematic review on green and natural polymeric nanofibers for biomedical applications. Int. J. Biol. Macromol. **262**, 130135–130135 (2024). https://doi.org/10.1016/J.IJBIOMAC.2024.130135
19. de Almeida Campos, L.A., de Souza, J.B., de Queiroz Macêdo, H.L.R., Borges, J.C., de Oliveira, D.N., Cavalcanti, I.M.F.: Synthesis of polymeric nanoparticles by double emulsion and pH-driven: encapsulation of antibiotics and natural products for combating Escherichia coli infections. Appl. Microbiol. Biotechnol. **108**(1) (2024). https://doi.org/10.1007/S00253-024-13114-5
20. Yang, Z., Xue, J., Shi, Z., Zhang, H., Yu, X., Du, L., Zhu, Y., Huan, Z., Wu, C.: Naturally derived flexible bioceramics: biomass recycling approach and advanced function. Matter **7**(3), 1275–1291 (2024). https://doi.org/10.1016/J.MATT.2024.01.016
21. Al-Naymi, H.A.S., Al-Musawi, M.H., Mirhaj, M., Valizadeh, H., Momeni, A., Danesh Pajooh, A.M., Shahriari-Khalaji, M., Sharifianjazi, F., Tavamaishvili, K., Kazemi, N., Salehi, S., Arefpour, A., Tavakoli, M.: Exploring nanobioceramics in wound healing as effective and economical alternatives. Heliyon **10**(19), e38497–e38497 (2024). https://doi.org/10.1016/J.HELIYON.2024.E38497
22. Sprio, S., Antoniac, I., Chevalier, J., Iafisco, M., Sandri, M., Tampieri, A.: Editorial: recent advances in bioceramics for health. Front. Bioeng. Biotechnol. **11**, 1264799 (2023). https://doi.org/10.3389/FBIOE.2023.1264799/BIBTEX
23. Jamwal, P., Chauhan, S., Chauhan, G.S., Kumar, K., Ranote, S., Kumar, R., Gupta, N.: Functionalized nanocellulose as a bifunctional material for efficient adsorption of mercuric ions with antimicrobial properties. Polym. Eng. Sci. **64**(5), 2049–2063 (2024). https://doi.org/10.1002/PEN.26671
24. Karnwal, A., Jassim, A.Y., Mohammed, A.A., Sharma, V., Al-Tawaha, A.R.M.S., Sivanesan, I.: Nanotechnology for healthcare: plant-derived nanoparticles in disease treatment and regenerative medicine. Pharmaceuticals **17**(12), 1711 (2024). https://doi.org/10.3390/PH17121711
25. Idris, D.S., Roy, A., Pandit, S., Alghamdi, S., Almehmadi, M., Alsaiari, A.A., Abdulaziz, O., Alsharif, A., Khandaker, M.U., Faruque, M.R.I.: Polymer-based nanocarriers for biomedical and environmental applications. E-Polymers **23**(1) (2023). https://doi.org/10.1515/EPOLY-2023-0049/ASSET/GRAPHIC/J_EPOLY-2023-0049_FIG_002.JPG
26. Weerarathna, I.N., Kumar, P., Luharia, A., Mishra, G.: Sustainable biomaterials for pharmaceutical and medical applications. Multidiscip. Rev. **8**(5), 2025141–2025141 (2025). https://doi.org/10.31893/MULTIREV.2025141
27. Liu, C., Wang, X., Du, S., Liang, W.: Synthesis of chitosan-based grafting magnetic flocculants for flocculation of kaolin suspensions. J. Environ. Sci. China **139**, 193–205 (2024). https://doi.org/10.1016/J.JES.2023.05.031
28. He, H., Shi, M., Yang, X., Zhan, J., Lin, Y., Guo, Z., Liao, Z., Lai, C., Ren, X., Huang, B., Pan, X.: Dissolved organic matter accelerates microbial degradation of 17 alpha-ethinylestradiol in the presence of iron mineral. J. Environ. Sci. China **139**, 364–376 (2024). https://doi.org/10.1016/J.JES.2023.05.042
29. Ma, X., Tian, Y., Yang, R., Wang, H., Allahou, L.W., Chang, J., Williams, G., Knowles, J.C., Poma, A.: Nanotechnology in healthcare, and its safety and environmental risks. J. Nanobiotechnol. **22**(1), 1–81 (2024). https://doi.org/10.1186/S12951-024-02901-X

30. Shi, S., Bai, W., Chen, X., Si, Y., Zhi, C., Wu, H., Su, Y., Cai, W., Fei, B., Kan, C.W., Hu, J., Wang, X.: Advances in nanofiber filtration membranes: from principles to intelligent applications. Adv. Func. Mater. (2025). https://doi.org/10.1002/ADFM.202423284
31. Petchimuthu, P., Sumanth, G.B., Kunjiappan, S., Kannan, S., Pandian, S.R.K., Sundar, K.: Green extraction and optimization of bioactive compounds from Solanum torvum Swartz. using ultrasound-aided solvent extraction method through RSM, ANFIS and machine learning algorithm. Sustain. Chem. Pharm. **36**, 101323–101323 (2023). https://doi.org/10.1016/J.SCP.2023.101323
32. Deutzmann, J.S., Spormann, A.M.: High acetate titer obtained from CO_2 by thermophilic microbial electrosynthesis with Thermoanaerobacter kivui. Bioresour. Technol. Rep. **25** (2023). https://doi.org/10.1016/J.BITEB.2023.101740
33. Dai, J., Geng, M., Du, Y., Iqbal, M.W., Yang, H., Shen, X., Wang, J., Sun, X., Yuan, Q.: Microbial synthesis of nucleosides: advances and prospects. ACS Synth. Biol. https://doi.org/10.1021/ACSSYNBIO.4C00530
34. Balla, E.D., Klonos, P.A., Kyritsis, A., Bertoldo, M., Guigo, N., Bikiaris, D.N.: Novel biobased copolymers based on Poly(butylene succinate) and cutin: in situ synthesis and structure properties investigations. Polymers **16**(16), 2270 (2024a). https://doi.org/10.3390/POLYM16162270
35. Papadopoulou, K., Klonos, P.A., Kyritsis, A., Mašek, O., Wurzer, C., Tsachouridis, K., Anastasiou, A.D., Bikiaris, D.N.: Synthesis and study of fully biodegradable composites based on Poly(butylene succinate) and biochar. Polymers **15**(4), 1049 (2023). https://doi.org/10.3390/POLYM15041049
36. Tran, N.T., Vo, T.V., Nguyen, V.P., Nguyen, M.T., Le-Phuoc, M.T., Nguyen, P.L.M., Nguyen, T.T.T., Nguyen, N.T., Pham, L.H., Le, T.H.P., Nguyen, T.T.T., Van Tran, T.T., Nguyen, D.M., Hoang, D.Q.: Advanced hybrid nanomaterials based on carboxymethyl-modified biopolymer: green synthesis and application in sustainable antimicrobial products. Int. J. Biol. Macromol. **281**, 136633–136633 (2024). https://doi.org/10.1016/J.IJBIOMAC.2024.136633
37. Sethulekshmi, A.S., Joseph, K., Santhosh Aprem, A., Sisupal, S.B., Saritha, A.: Green synthesis of multifunctional natural rubber-lignin nanocomposites: a sustainable approach for waste reduction. Int. J. Biol. Macromol. **280**, 135887–135887 (2024). https://doi.org/10.1016/J.IJBIOMAC.2024.135887
38. Afonso, I.S., Cardoso, B., Nobrega, G., Minas, G., Ribeiro, J.E., Lima, R.A.: Green synthesis of nanoparticles from olive oil waste for environmental and health applications: a review. J. Environ. Chem. Eng. **12**(5), 114022–114022 (2024). https://doi.org/10.1016/J.JECE.2024.114022
39. Jain, K., Takuli, A., Gupta, T.K., Gupta, D.: Rethinking nanoparticle synthesis: a sustainable approach vs. traditional methods. Chem. Asian J. (2024). https://doi.org/10.1002/ASIA.202400701
40. Saini, A., Verma, R., Tiwari, R., Jain, A., Dandia, A., Gour, V.S., Lamba, N.P., Srivastava, S.C., Chauhan, M.S.: Green synthesis of silver nanoparticle for catalytic applications and priming study by seed germination. Dent. Sci. Rep. **14**(1) (2024). https://doi.org/10.1038/S41598-024-69120-0
41. Barik, B., Satapathy, B.S., Pattnaik, G., Bhavrao, D.V., Shetty, K.P.: Sustainable synthesis of silver nanoparticles from Azadirachta indica: antimicrobial, antioxidant and in silico analysis for periodontal treatment. Front. Chem. **12** (2024). https://doi.org/10.3389/FCHEM.2024.1489253
42. Bhushan, D., Shoran, S., Kumar, R., Gupta, R.: Plant biomass-based nanoparticles for remediation of contaminants from water ecosystems: recent trends, challenges, and future perspectives. Chemosphere **365**, 143340–143340 (2024). https://doi.org/10.1016/J.CHEMOSPHERE.2024.143340
43. Asmat-Campos, D., Lindsay Rojas, M., Carreño-Ortega, A.: Toward sustainable nanomaterials: an innovative ecological approach for biogenic synthesis of TiO2 nanoparticles with potential photocatalytic activity. Clean. Eng. Technol. **17** (2023). https://doi.org/10.1016/J.CLET.2023.100702

44. Chegini, Z., Shariati, A., Alikhani, M.Y., Safaiee, M., Rajaeih, S., Arabestani, M., Azizi, M.: Antibacterial and antibiofilm activity of silver nanoparticles stabilized with C-phycocyanin against drug-resistant Pseudomonas aeruginosa and Staphylococcus aureus. Front. Bioeng. Biotechnol. **12** (2024). https://doi.org/10.3389/FBIOE.2024.1455385
45. Ouyang, B., Wei, D., Wu, B., Yan, L., Gang, H., Cao, Y., Chen, P., Zhang, T., Wang, H.: In the view of electrons transfer and energy conversion: the antimicrobial activity and cytotoxicity of metal-based nanomaterials and their applications. Small (Weinheim an Der Bergstrasse, Germany) **20**(4) (2024). https://doi.org/10.1002/SMLL.202303153
46. Xie, M., Gao, M., Yun, Y., Malmsten, M., Rotello, V.M., Zboril, R., Akhavan, O., Kraskouski, A., Amalraj, J., Cai, X., Lu, J., Zheng, H., Li, R.: Antibacterial nanomaterials: mechanisms, impacts on antimicrobial resistance and design principles. Angewandte Chemie (International Ed. in English) **62**(17) (2023). https://doi.org/10.1002/ANIE.202217345
47. Nsairat, H., Lafi, Z., Al-Najjar, B.O., Al-Samydai, A., Saqallah, F.G., El-Tanani, M., Oriquat, G.A., Sa'bi, B.M., Ibrahim, A.A., Dellinger, A.L., Alshaer, W.: How advanced are self-assembled nanomaterials for targeted drug delivery? A comprehensive review of the literature. Int. J. Nanomed. **20**, 2133 (2025). https://doi.org/10.2147/IJN.S490444
48. Van Rijt, S., Habibovic, P.: Enhancing regenerative approaches with nanoparticles. J. R. Soc. Interface **14**(129) (2017). https://doi.org/10.1098/RSIF.2017.0093
49. Klaper, R., Arndt, D., Bozich, J., Dominguez, G.: Molecular interactions of nanomaterials and organisms: defining biomarkers for toxicity and high-throughput screening using traditional and next-generation sequencing approaches. Analyst **139**(5), 882–895 (2014). https://doi.org/10.1039/C3AN01644G
50. Zhang, X.Q., Xu, X., Bertrand, N., Pridgen, E., Swami, A., Farokhzad, O.C.: Interactions of nanomaterials and biological systems: implications to personalized nanomedicine. Adv. Drug Deliv. Rev. **64**(13), 1363 (2012). https://doi.org/10.1016/J.ADDR.2012.08.005
51. Auwal, S.M., Ghanisma, S.B.M., Saari, N.: Optimization, physicochemical stability and in vivo study of alginate-chitosan composites as nanocarriers for low molecular weight angiotensin I-converting enzyme (ACE)-inhibitory peptide. J. Food Drug Anal. **32**(3), 358–370 (2024). https://doi.org/10.38212/2224-6614.3522
52. Mondéjar-López, M., García-Simarro, M.P., Navarro-Simarro, P., Gómez-Gómez, L., Ahrazem, O., Niza, E.: A review on the encapsulation of "eco-friendly" compounds in natural polymer-based nanoparticles as next generation nano-agrochemicals for sustainable agriculture and crop management. Int. J. Biol. Macromol. **280**, 136030–136030 (2024). https://doi.org/10.1016/J.IJBIOMAC.2024.136030
53. Gao, Y., Xiao, W., Gu, C., Yu, S., Li, S., Zou, A.: Layer-by-layer assembled decomposable nanocapsules for light responsive release of pesticide imidacloprid on Aphis craccivora Koch. Pest Manag. Sci. **80**(7), 3207–3214 (2024). https://doi.org/10.1002/PS.8023
54. Sun, Z., Yang, L., Xu, C., Cai, C., Li, L.: Zwitterionic nanocapsules with pH- and thermal-responsiveness for drug-controlled release. Nanotechnology **34**(15), 155101–155101 (2023). https://doi.org/10.1088/1361-6528/ACB215
55. Peter, S., Lyczko, N., Thomas, S., Leruth, D., Germeau, A., Fati, D., Nzihou, A.: Fabrication of eco-friendly nanocellulose-chitosan-calcium Phosphate ternary nanocomposite for wastewater remediation. Chemosphere **363**, 142779–142779 (2024). https://doi.org/10.1016/J.CHEMOSPHERE.2024.142779
56. Li, X., Zhang, H., Zhang, H.: Fabrication of β-TCP/ Akermanite composite scaffold via DLP and in-situ modification of micro-nano surface morphology for bone repair. Ceram. Int. **50**(2), 2659–2669 (2023). https://doi.org/10.1016/J.CERAMINT.2023.10.276
57. Wang, J., Fu, Y., Gu, Z., Pan, H., Zhou, P., Gan, Q., Yuan, Y., Liu, C.: Multifunctional carbon dots for biomedical applications: diagnosis, therapy, and theranostic. Small **20**(3), e2303773–e2303773 (2023). https://doi.org/10.1002/SMLL.202303773
58. Bauri, S., Tripathi, S., Choudhury, A.M., Mandal, S.S., Raj, H., Maiti, P.: Nanomaterials as theranostic agents for cancer therapy. ACS Appl. Nano Mater. **6**(23), 21462–21495 (2023). https://doi.org/10.1021/ACSANM.3C04235

59. Khorsandi, D., Jenson, S., Zarepour, A., Khosravi, A., Rabiee, N., Iravani, S., Zarrabi, A.: Catalytic and biomedical applications of nanocelluloses: a review of recent developments. Int. J. Biol. Macromol. **268**, 131829–131829 (2024). https://doi.org/10.1016/J.IJBIOMAC.2024.131829
60. Huston, M., Debella, M., Dibella, M., Gupta, A.: Green synthesis of nanomaterials. Nanomaterials **11**(8), 2130 (2021). https://doi.org/10.3390/NANO11082130
61. Singh, H., Desimone, M.F., Pandya, S., Jasani, S., George, N., Adnan, M., Aldarhami, A., Bazaid, A.S., Alderhami, S.A.: Revisiting the green synthesis of nanoparticles: uncovering influences of plant extracts as reducing agents for enhanced synthesis efficiency and its biomedical applications. Int. J. Nanomed. **18**, 4727–4750 (2023). https://doi.org/10.2147/IJN.S419369
62. Zehra, S.H., Ramzan, K., Viskelis, J., Viskelis, P., Balciunaitiene, A.: Advancements in green synthesis of silver-based nanoparticles: antimicrobial and antifungal properties in various films. Nanomaterials **15**(4), 252 (2025). https://doi.org/10.3390/NANO15040252
63. Madani, M., Hosny, S., Alshangiti, D.M., Nady, N., Alkhursani, S.A., Alkhaldi, H., Al-Gahtany, S.A., Ghobashy, M.M., Gaber, G.A.: Green synthesis of nanoparticles for varied applications: green renewable resources and energy-efficient synthetic routes. Nanotechnol. Rev. **11**(1), 731–759 (2022). https://doi.org/10.1515/NTREV-2022-0034/ASSET/GRAPHIC/J_NTREV-2022-0034_FIG_005.JPG
64. Samuel, M.S., Ravikumar, M., John, A., Selvarajan, E., Patel, H., Chander, P.S., Soundarya, J., Vuppala, S., Balaji, R., Chandrasekar, N.: A review on green synthesis of nanoparticles and their diverse biomedical and environmental applications. Catalysts **12**(5), 459 (2022). https://doi.org/10.3390/CATAL12050459
65. Skóra, B., Krajewska, U., Nowak, A., Dziedzic, A., Barylyak, A., Kus-Liśkiewicz, M.: Noncytotoxic silver nanoparticles as a new antimicrobial strategy. Sci. Rep. **11**(1), 1–13 (2021). https://doi.org/10.1038/s41598-021-92812-w
66. Vassallo, M., Martella, D., Barrera, G., Celegato, F., Coïsson, M., Ferrero, R., Olivetti, E.S., Troia, A., Sözeri, H., Parmeggiani, C., Wiersma, D.S., Tiberto, P., Manzin, A.: Improvement of hyperthermia properties of iron oxide nanoparticles by surface coating. ACS Omega **8**(2), 2143–2154 (2023). https://doi.org/10.1021/ACSOMEGA.2C06244/ASSET/IMAGES/LARGE/AO2C06244_0007.JPEG
67. Christ, C., Waldl, A., Liu, Y., Johnson, L., Auer, V., Cardozo, O., Farias, P.M.A., Andrade, A.C.D.S., Stingl, A., Himly, M., Punz, B., Li, S., Wang, G., Li, Y.: Nano-scaled advanced materials for antimicrobial applications—mechanistic insight, functional performance measures, and potentials towards sustainability and circularity. Environ. Sci. Nano. https://doi.org/10.1039/D4EN00798K
68. Asrafali, S.P., Periyasamy, T., Lee, J.: A comprehensive review on bio-based polybenzoxazines emphasizing their antimicrobial property. Microorganisms **13**(1), 164 (2025)
69. Serov, D.A., Gritsaeva, A.V., Yanbaev, F.M., Simakin, A.V., Gudkov, S.V.: Review of antimicrobial properties of titanium dioxide nanoparticles. Int. J. Mol. Sci. **25**(19), 10519 (2024). https://doi.org/10.3390/IJMS251910519
70. Hosseini, S.M., Soleimani, A., Khaleghi, S. and Hajrasouliha, S.: Green synthesized ag nanoparticles as promising antibacterial and antitumor agents: in vitro studies. Indian J. Microbiol., 1–15 (2024)
71. Benitto, J.J., Vijaya, J.J., Saravanan, T.G., Manikkam, R., Budhi, B.H.: Microwave-synthesized $NiZrO_3$@GNP and $NiZrO_3$@MWCNT nanocomposites: enhanced antimicrobial efficacy against biofilms and Mycobacterium smegmatis. Biotech **15**(2) (2025). https://doi.org/10.1007/S13205-024-04201-5
72. Ortega-Nieto, C., Losada-Garcia, N., Prodan, D., Furtos, G., Palomo, J.M.: Recent advances on the design and applications of antimicrobial nanomaterials. Nanomaterials **13**(17), 2406 (2023). https://doi.org/10.3390/NANO13172406
73. Parvin, N., Joo, S.W., Mandal, T.K.: Biodegradable and stimuli-responsive nanomaterials for targeted drug delivery in autoimmune diseases. J. Funct. Biomater. **16**(1) (2025). https://doi.org/10.3390/JFB16010024

74. Sun, L., Liu, H., Ye, Y., Lei, Y., Islam, R., Tan, S., Tong, R., Miao, Y.B., Cai, L.: Smart nanoparticles for cancer therapy. Sig. Transduct. Target. Ther. **8**(1), 1–28 (2023). https://doi.org/10.1038/s41392-023-01642-x
75. Wei, H., Cui, J., Lin, K., Xie, J., Wang, X.: Recent advances in smart stimuli-responsive biomaterials for bone therapeutics and regeneration. Bone Res. **10**(1), 1–19 (2022). https://doi.org/10.1038/s41413-021-00180-y
76. Bai, X., Smith, Z.L., Wang, Y., Butterworth, S., Tirella, A.: Sustained drug release from smart nanoparticles in cancer therapy: a comprehensive review. Micromachines **13**(10), 1623 (2022). https://doi.org/10.3390/MI13101623
77. Montoya, C., Du, Y., Gianforcaro, A.L., Orrego, S., Yang, M., Lelkes, P.I.: On the road to smart biomaterials for bone research: definitions, concepts, advances, and outlook. Bone Res. **9**(1), 1–16 (2021). https://doi.org/10.1038/s41413-020-00131-z
78. Teja, K., Reddy, K., Reddy, A.S.: Recent breakthroughs in drug delivery systems for targeted cancer therapy: an overview. Cell. Mol. Biomed. Rep. **5**(1), 13–27 (2025). https://doi.org/10.55705/CMBR.2025.456494.1246
79. Moradifar, F., Sepahdoost, N., Tavakoli, P., Mirzapoor, A.: Multi-functional dressings for recovery and screenable treatment of wounds: a review. Heliyon **11**(1), e41465 (2025). https://doi.org/10.1016/J.HELIYON.2024.E41465
80. Eyube, M., Enuesueke, C., Alimikhena, M.: The Future of Nanomaterials in Manufacturing (2025). https://doi.org/10.26434/CHEMRXIV-2025-MLB2W
81. Gupta, D., Boora, A., Thakur, A., Gupta, T.K.: Green and sustainable synthesis of nanomaterials: recent advancements and limitations. Environ. Res. **231**, 116316–116316 (2023). https://doi.org/10.1016/J.ENVRES.2023.116316
82. Osman, A.I., Zhang, Y., Farghali, M., Rashwan, A.K., Eltaweil, A.S., Abd El-Monaem, E.M., Mohamed, I.M.A., Badr, M.M., Ihara, I., Rooney, D.W., Yap, P.S.: Synthesis of green nanoparticles for energy, biomedical, environmental, agricultural, and food applications: a review. Environ. Chem. Lett. **22**(2), 841–887 (2024). https://doi.org/10.1007/S10311-023-01682-3
83. Ghamari, M., Sun, D., Dai, Y., See, C.H., Yu, H., Edirisinghe, M., Sundaram, S.: Valorization of diverse waste-derived nanocellulose for multifaceted applications: a review. Int. J. Biol. Macromol. **280**, 136130–136130 (2024). https://doi.org/10.1016/J.IJBIOMAC.2024.136130
84. Poonia, K., Singh, P., Ahamad, T., Le, Q.V., Phan Quang, H.H., Thakur, S., Mishra, A.K., Selvasembian, R., Hussain, C.M., Nguyen, V.H., Raizada, P.: Sustainability, performance, and production perspectives of waste-derived functional carbon nanomaterials towards a sustainable environment: a review. Chemosphere **352**, 141419–141419 (2024). https://doi.org/10.1016/J.CHEMOSPHERE.2024.141419
85. Malovrh Rebec, K., Turk, J., Kunaver, M.: Quantifying the environmental implication of cotton-fiber-based nanocrystalline cellulose: a life-cycle assessment. Sustain. Prod. Consump. **45**, 192–202 (2023). https://doi.org/10.1016/J.SPC.2023.12.026
86. Stoycheva, S., Zabeo, A., Pizzol, L., Hristozov, D.: Socio-economic life cycle-based framework for safe and sustainable design of engineered nanomaterials and nano-enabled products. Sustainability **14**(9), 5734–5734 (2022). https://doi.org/10.3390/SU14095734
87. D'Avenio, G., Daniele, C., Grigioni, M.: Nanostructured medical devices: regulatory perspective and current applications. Materials **17**(8), 1787 (2024). https://doi.org/10.3390/MA17081787
88. Rodríguez-Gómez, F.D., Monferrer, D., Penon, O., Rivera-Gil, P.: Regulatory pathways and guidelines for nanotechnology-enabled health products: a comparative review of EU and US frameworks. Front. Med. **12**, 1544393 (2025). https://doi.org/10.3389/FMED.2025.1544393
89. Virameteekul, S., Lees, A.J., Bhidayasiri, R.: Small particles, big potential: polymeric nanoparticles for drug delivery in Parkinson's disease. Mov. Disord. Official J. Mov. Disord. Soc. **39**(11) (2024). https://doi.org/10.1002/MDS.29939

MIX
Papier aus verantwortungsvollen Quellen
Paper from responsible sources
FSC® C105338

If you have any concerns about our products,
you can contact us on
ProductSafety@springernature.com

In case Publisher is established outside the EU,
the EU authorized representative is:
Springer Nature Customer Service Center GmbH
Europaplatz 3, 69115 Heidelberg, Germany

Printed by Libri Plureos GmbH
in Hamburg, Germany